Linear Algebra with Applications

International Series in Pure and Applied Mathematics

Ahlfors: *Complex Analysis*
Bender and Orszag: *Advanced Mathematical Methods for Scientists and Engineers*
Bilodeau and Thie: *An Introduction to Analysis*
Boas: *Invitation to Complex Analysis*
Brown and Churchill: *Complex Variables and Applications*
Brown and Churchill: *Fourier Series and Boundary-Value Problems*
Buchanan and Turner: *Numerical Methods and Analysis*
Buck: *Advanced Calculus*
Burton: *Elementary Number Theory*
Burton: *The History of Mathematics: An Introduction*
Chartrand and Oellermann: *Applied and Algorithmic Graph Theory*
Colton: *Partial Differential Equations*
Conte and de Boor: *Elementary Numerical Analysis: An Algorithmic Approach*
Edelstein-Keshet: *Mathematical Models in Biology*
Farlow: *An Introduction to Differential Equations and Their Applications*
Goldberg: *Matrix Theory with Applications*
Gulick: *Encounters with Chaos*
Hill: *Experiments in Computational Matrix Algebra*
Keisler and Robbin: *Mathematical Logic and Computability*
Kurtz: *Foundations of Abstract Mathematics*
Lewin and Lewin: *An Introduction to Mathematical Analysis*
Malik, Mordeson, and Sen: *Fundamentals of Abstract Algebra*
Morash: *Bridge to Abstract Mathematics: Mathematical Proof and Structures*
Parzynski and Zipse: *Introduction to Mathematical Analysis*
Pinsky: *Partial Differential Equations and Boundary-Value Problems*
Pinter: *A Book of Abstract Algebra*
Ralston and Rabinowitz: *A First Course in Numerical Analysis*
Ritger and Rose: *Differential Equations with Applications*
Robertson: *Engineering Mathematics with Maple*
Robertson: *Engineering Mathematics with Mathematica*
Rudin: *Functional Analysis*
Rudin: *Principles of Mathematical Analysis*
Rudin: *Real and Complex Analysis*
Scheick: *Linear Algebra with Applications*
Simmons: *Differential Equations with Applications and Historical Notes*
Small and Hosack: *Calculus: An Integrated Approach*
Small and Hosack: *Explorations in Calculus with a Computer Algebra System*
Vanden Eynden: *Elementary Number Theory*
Walker: *Introduction to Abstract Algebra*

Also available from McGraw-Hill

Schaum's Outline Series in Mathematics and Statistics

Most outlines include basic theory, definitions, hundreds of example problems solved in step-by-step detail, and supplementary problems with answers.

Related titles on the current list include:

Advanced Calculus
Advanced Mathematics for Engineers & Scientists
Analytic Geometry
Basic Mathematics for Electricity & Electronics
Basic Mathematics with Applications to Science & Technology
Beginning Calculus
Boolean Algebra & Switching Circuits
Calculus
Calculus for Business, Economics, & the Social Sciences
College Algebra
College Mathematics
Combinatories
Complex Variables
Descriptive Geometry
Differential Equations
Differential Geometry
Discrete Mathematics
Elementary Algebra
Finite Differences & Difference Equations
Finite Element Analysis
Finite Mathematics
Fourier Analysis
General Topology
Geometry
Group Theory
Laplace Transforms
Linear Algebra
Mathematical Handbook of Formulas & Tables
Mathematical Methods for Business & Economics
Mathematics for Nurses
Matrix Operations
Modern Abstract Algebra
Numerical Analysis
Partial Differential Equations

Probability
Probability & Statistics
Real Variables
Review of Elementary Mathematics
Set Theory & Related Topics
Statistics
Technical Mathematics
Tensor Calculus
Trigonometry
Vector Analysis

Schaum's Solved Problems Series

Each title in this series is a complete and expert source of solved problems with solutions worked out in step-by-step detail.

Related titles on the current list include:

3000 Solved Problems in Calculus
2500 Solved Problems in Differential Equations
2000 Solved Problems in Discrete Mathematics
3000 Solved Problems in Linear Algebra
2000 Solved Problems in Numerical Analysis
3000 Solved Problems in Precalculus

Bob Miller's Math Helpers

Bob Miller's Calc I Helper
Bob Miller's Calc II Helper
Bob Miller's Precalc Helper

McGraw-Hill Paperbacks

Arithmetic and Algebra . . . Again
How to Solve Word Problems in Algebra
Mind Over Math

Available at most college bookstores, or, for a complete list of titles and prices, write to:

Schaum Division
The McGraw-Hill Companies, Inc.
11 West 19th Street
New York, NY 10011

Linear Algebra with Applications

John T. Scheick

Duke University

THE McGRAW-HILL COMPANIES, INC.

New York   St. Louis   San Francisco   Auckland   Bogotá   Caracas   Lisbon
London   Madrid   Mexico City   Milan   Montreal   New Delhi
San Juan   Singapore   Sydney   Tokyo   Toronto

This book was set in Times Roman by Publication Services, Inc.
The Editors were Jack Shira, Karen M. Minette, and John M. Morriss;
the production supervisor was Kathryn Porzio.
Project supervision was done by Publication Services, Inc.
R. R. Donnelley & Sons Company was printer and binder.

McGraw-Hill

A Division of The McGraw·Hill Companies

LINEAR ALGEBRA WITH APPLICATIONS

This book is printed on acid-free paper.

1 2 3 4 5 6 7 8 9 0 DOC DOC 9 0 9 8 7 6

ISBN 0-07-055184-7

Library of Congress Cataloging-in-Publication Data

Scheick, John T.
    Linear algebra with applications / John T. Scheick.
        p.   cm.—(International series in pure and applied
    mathematics)
        Includes index.
        ISBN 0-07-055184-7
        1. Algebras, Linear.   I. Title.   II. Series.
    QA184.S34   1997
    512′ .5—dc20   96-27488

http://www.mhcollege.com

ABOUT THE AUTHOR

J. T. SCHEICK received his Ph.D. in mathematics, in approximation theory, from Syracuse University in 1966. His interests turned to applied mathematics and numerical analysis in the mid 1970s. He has done consulting for General Motors for seven years and has worked on research projects with the Electroscience Laboratory at Ohio State University for several years. Professor Scheick has a strong interest in teaching engineering students at the undergraduate and graduate levels and has served on hundreds of Ph.D. examination and dissertation committees in the College of Engineering at Ohio State University. He is also a member of the Society for Industrial and Applied Mathematics (SIAM). He recently retired from Ohio State University and now teaches at Duke University.

TABLE OF CONTENTS AND OUTLINE

PREFACE

Linear algebra is a tool used extensively in a wide variety of mathematical, science, and engineering fields. Most textbooks focus on an elementary introduction to the topic, an abstract approach with emphasis on the algebraic aspects of the field, or on technical aspects of the subject. The purpose of this book is to present the core topics of linear algebra that are essential to its use in contemporary mathematics, applied mathematics, engineering, and the sciences. Also, in an effort to illustrate the applicability of this core material, many applications are woven into the text in a wide variety of settings.

The book is designed for strong upper-level undergraduates or beginning graduate students. It can be used in several ways:

- As a one-semester course for advanced undergraduates
- As a topics course for undergraduates
- As a one-semester course for beginning mathematics graduate students, possibly followed by a more algebraically oriented course
- As a one-semester course for graduate engineers with emphasis on application of the theory and computations

A wide variety of exercises allows ample selection of problems for each category of students. There are sections and chapters of the book that focus on applications. These may be used as special topics in lecture or as projects for students. Often students will read them out of curiosity. Many good engineering students will find material in the book that is a valuable resource for their work.

The book is self-contained and rigorous, but not designed to be covered one section per lecture. It is thematically arranged around the important concepts and language of abstract vector spaces and operator theory (e.g., the concepts of basis, dimension, inner product, orthogonal expansion, projection, eigenvectors, Hermitian operators, and unitary operators). Many of these ideas will carry over to subsequent courses, and references are cited for follow-up reading. An essential premise is that students need to know about function spaces as well as n-tuple spaces (matrix theory), since many applications involve function spaces.

The theory is given in an abstract setting, but is enhanced by many examples taken from vector spaces indispensable to the sciences and mathematics. This shows the value of an abstract theory, as well as illustrating the ideas behind the theory. Real and complex scalars, n-tuple spaces, and function spaces are stressed from the beginning. The use of several examples to illustrate one abstract idea will help the student to see the general principles and to learn to think abstractly. The theory for some serious contemporary applications is developed. For example, the linear algebra needed by a finite elements specialist is included.

The interplay between matrix theory and vector spaces is constantly stressed, and the dual viewpoints are frequently used to show complementary methods of problem solving and thinking. Students should be adept at the use of matrices and matrix methods as well as the use of vector space and operator methods after studying this book.

The text is flexible enough to allow the instructor to choose which chapters to cover and in which order. For example, there are sections and chapters that can be omitted, if the instructor so chooses, without interfering with the main ideas presented later in the book. For example, Chapter 7 can be used without Chapter 6, or both Chapters 6 and 7 can be omitted. The following sections could be considered optional: 3.4, 3.5, 3.6, 4.4, 4.5, 4.6, 5.5, 5.6, 6.4, 6.5, 7.4, and 8.3.

If students have had an elementary course, then Chapter 0 could be skipped or quickly reviewed, or its essential facts could be mentioned by the instructor as Chapter 1 is traversed. Perhaps some or all of Section 4.1 could be assumed to be known as well. If Chapter 0 is skipped, the main results that will be used later in the book are Theorem 3 of Section 0.2, Theorem 2 of Section 0.3, the matrix identities (2a)–(2f) and (3) in Section 0.4, and Theorem 4 and Corollary 5 of Section 0.5.

Some sample courses are given in the following table. Recommended sections are listed in each column under the student category.

Chapter	Undergraduate	Math graduate	Engineering graduate
0	1–5		review as needed
1	all	all	all
2	all	all	all
3	1–3	1–3, 5–6	1–5
4	1–3	1–3, 5	1–4
5		1–4	1–5
6		1–3, 5	1–3 (4, 5 optional)
7			1–3
8		all	1–2

The sections that may be considered primarily applications are 3.4, 4.4, 4.6, 5.6, and Chapter 7. These sections and some sections not typically found in undergraduate courses could serve as topics in a reading course for a second course or as a topics course for advanced undergraduates.

ACKNOWLEDGMENTS

I would like to thank the following reviewers for their perceptive and valuable comments:

Michael Ecker	Pennsylvania State University
Hans Engler	Georgetown University
Larry Grove	University of Arizona
Murli Gupta	George Washington University
Vyron M. Klassen	California State University
Glen Ledder	University of Nebraska
Justin Peters	Iowa State University
Stewart M. Robinson	Cleveland State University
Alexander Soifer	University of Colorado
Horace Wiser	Washington State University

I would also like to thank Jack Shira at McGraw-Hill for his encouragement and unflagging efforts to improve the book during review process.

Finally, I would like to thank the many energetic and thoughtful engineering students at Ohio State University who have read previous drafts of the book and who offered many insightful and valuable suggestions for its improvement, both while taking the course and later, during the various stages of their research. Their enthusiasm contributed immensely to the pleasure of writing the book and to developing the contents of the book.

Linear Algebra with Applications

Systems of Equations and Matrices

As a prelude to the study of vector spaces in general, we briefly consider systems of equations, procedures for solving them, the basic rules of matrix algebra, and matrix inverses. Most of our examples involve real or complex numbers, but our considerations are valid for any field. The last section of this chapter contains a brief overview of fields. In this chapter, only the essentials of systems of equations and matrices will be developed. In later chapters, especially Chapters 1 and 2, the theory of systems of equations and matrices will be extended to more sophisticated results as a byproduct of the study of vector spaces and linear operators.

0.1
SYSTEMS OF LINEAR EQUATIONS AND MATRICES

A system of linear equations is a set of m equations in n unknowns:

$$a_{11}x_1 + a_{12}x_2 + \cdots + a_{1n}x_n = b_1$$
$$a_{21}x_1 + a_{22}x_2 + \cdots + a_{2n}x_n = b_2 \tag{1}$$
$$\vdots$$
$$a_{m1}x_1 + a_{m2}x_2 + \cdots + a_{mn}x_n = b_m.$$

Using summation notation, this is more compactly written as

$$\sum_{j=1}^{n} a_{ij}x_j = b_j \qquad i = 1, 2, \ldots, m. \tag{2}$$

A simple example of a system of two equations in three unknowns is

$$x_1 + 3x_2 - 5x_3 = 1$$
$$2x_1 - x_2 + 4x_3 = 7. \tag{*}$$

1

System (1) is said to be *homogeneous* iff all $b_i = 0$ and *inhomogeneous,* or *nonhomogeneous,* otherwise.

In discussing procedures to solve a system of linear equations, it is convenient to introduce the concepts of a matrix, a column, and the multiplication of a matrix and a column. The matrix involved consists of the coefficients of the unknowns x_i, and the columns consist of the unknowns and the b_i. This results in an even more compact notation for system (1) and assists in the discussion of solving the system. In addition, procedures for solving a system of equations often are given solely in terms of the matrix of coefficients and the numbers b_j.

A *column* of length n is an $n \times 1$ array

$$x = (x_1, x_2, \ldots, x_n)^{\mathrm{T}} = \begin{bmatrix} x_1 \\ x_2 \\ \vdots \\ x_n \end{bmatrix},$$

where the superscript ^{T} means transpose, i.e., convert the row $(x_1, x_2, \ldots, x_n)$ to a column. An $m \times n$ matrix is an array of objects consisting of m rows and n columns

$$A = [a_{ij}] = \begin{bmatrix} a_{11} & a_{12} & \cdots & a_{1n} \\ a_{21} & a_{22} & \cdots & a_{2n} \\ \cdots & \cdots & \cdots & \cdots \\ a_{m1} & a_{m2} & \cdots & a_{mn} \end{bmatrix},$$

so that a_{ij} is the entry in the ith row and jth column. Thus a column of length n is simply an $n \times 1$ matrix and a row of length n is a $1 \times n$ matrix.

The product Ax is defined to be the column of length m whose entry in the ith row is

$$(Ax)_i = \sum_{j=1}^{n} a_{ij} x_j = a_{i1} x_1 + a_{i2} x_2 + \cdots + a_{in} x_n. \tag{3}$$

Observe that, if $A = [\alpha_1 \alpha_2 \cdots \alpha_n]$ is a single row ($m = 1$), (3) becomes

$$Ax = [\alpha_1 \alpha_2 \cdots \alpha_n] \begin{bmatrix} x_1 \\ x_2 \\ \vdots \\ x_n \end{bmatrix} = \sum_{j=1}^{n} \alpha_j x_j.$$

Thus, if the ith row of A is denoted by

$$a_i = [a_{i1} a_{i2} \cdots a_{in}],$$

then

$$Ax = \begin{bmatrix} (Ax)_1 \\ (Ax)_2 \\ \vdots \\ (Ax)_m \end{bmatrix} = \begin{bmatrix} a_1 x \\ a_2 x \\ \vdots \\ a_m x \end{bmatrix}. \tag{4}$$

If A is obtained from system (1), it is called the *coefficient matrix of system* (1). For system (*), the coefficient matrix is

$$A = \begin{bmatrix} 1 & 3 & -5 \\ 2 & -1 & 4 \end{bmatrix}.$$

The rows of A are $a_1 = \begin{bmatrix} 1 & 3 & -5 \end{bmatrix}$, $a_2 = \begin{bmatrix} 2 & -1 & 4 \end{bmatrix}$. Putting $x = (x_1, x_2, x_3)^T$ yields

$$\begin{bmatrix} 1 & 3 & -5 \\ 2 & -1 & 4 \end{bmatrix} \begin{bmatrix} x_1 \\ x_2 \\ x_3 \end{bmatrix} = \begin{bmatrix} \begin{bmatrix} 1 & 3 & -5 \end{bmatrix} \begin{bmatrix} x_1 \\ x_2 \\ x_3 \end{bmatrix} \\ \begin{bmatrix} 2 & -1 & 4 \end{bmatrix} \begin{bmatrix} x_1 \\ x_2 \\ x_3 \end{bmatrix} \end{bmatrix} = \begin{bmatrix} x_1 + 3x_2 - 5x_3 \\ 2x_1 - x_2 + 4x_3 \end{bmatrix}.$$

Now, system (1) may be written more compactly as

$$\begin{aligned} a_1 x &= b_1 \\ a_2 x &= b_2 \\ &\vdots \\ a_m x &= b_m. \end{aligned} \tag{5}$$

For example, system (*) can be written as

$$\begin{bmatrix} 1 & 3 & -5 \end{bmatrix} x = \begin{bmatrix} 1 & 3 & -5 \end{bmatrix} \begin{bmatrix} x_1 \\ x_2 \\ x_3 \end{bmatrix} = x_1 + 3x_2 - 5x_3 = 1$$

$$\begin{bmatrix} 2 & -1 & 4 \end{bmatrix} x = \begin{bmatrix} 2 & -1 & 4 \end{bmatrix} \begin{bmatrix} x_1 \\ x_2 \\ x_3 \end{bmatrix} = 2x_1 - x_2 + 4x_3 = 7.$$

Now define two columns to be equal just when their lengths are the same and their corresponding entries are equal. Then the system of equations (1) can be very compactly written as

$$Ax = b. \tag{6}$$

For system (*) this becomes

$$\begin{bmatrix} 1 & 3 & -5 \\ 2 & -1 & 4 \end{bmatrix} \begin{bmatrix} x_1 \\ x_2 \\ x_3 \end{bmatrix} = \begin{bmatrix} 1 \\ 7 \end{bmatrix}.$$

Some simple operations on matrices and the notion of equality are essential to this discussion.

DEFINITION. Let $A = [a_{ij}]$ and $B = [b_{ij}]$ be two matrices of the same size. Then equality, addition and subtraction, and multiplication by a real or complex number α are defined entrywise as follows:

$$A = B \Leftrightarrow a_{ij} = b_{ij} \quad \forall\, i, j$$
$$A \pm B = [a_{ij} \pm b_{ij}]$$
$$\alpha A = [\alpha a_{ij}].$$

That is, to add A and B, add their corresponding entries, and to multiply A by a number α, multiply every entry in A by α. Two matrices of different sizes cannot be equal and cannot be added or subtracted.

For example,

$$\begin{bmatrix} 1 & 0 \\ 2 & 3 \end{bmatrix} + 2 \begin{bmatrix} -2 & 3 \\ 5 & 1 \end{bmatrix} = \begin{bmatrix} 1 & 0 \\ 2 & 3 \end{bmatrix} + \begin{bmatrix} -4 & 6 \\ 10 & 2 \end{bmatrix} = \begin{bmatrix} -3 & 6 \\ 12 & 5 \end{bmatrix}.$$

$$2 \begin{bmatrix} 3 \\ 4 \end{bmatrix} - 5 \begin{bmatrix} 1 \\ -2 \end{bmatrix} = \begin{bmatrix} 6 \\ 8 \end{bmatrix} - \begin{bmatrix} 5 \\ -10 \end{bmatrix} = \begin{bmatrix} 1 \\ 18 \end{bmatrix}.$$

The following laws of matrix-column multiplication are easily established. Let α, β be real or complex numbers; let A, B be $m \times n$ matrices, and let x, y be columns of length n. Then

$$(\alpha A)x = \alpha(Ax) = A(\alpha x) \qquad \text{(Associative law)}$$
$$A(\alpha x + \beta y) = \alpha Ax + \beta Ay \qquad \text{(Distributive law)} \qquad (7)$$
$$(\alpha A + \beta B)x = \alpha Ax + \beta Bx \qquad \text{(Distributive law)}$$

The associative law allows the abbreviation $\alpha Ax = (\alpha A)x = \alpha(Ax) = A(\alpha x)$.

0.1 Exercises

1. Write each of the following systems in the form $Ax = b$.

(a) $\quad 2x_1 + x_2 = 2$
$\quad\quad x_1 - 3x_2 = 0$

(b) $\quad x_1 + 2x_3 = 1$
$\quad\quad 3x_2 + x_3 = 2$
$\quad\quad -x_1 + x_2 = 3$

(c) $\quad x_1 + 2x_2 = 0$
$\quad\quad 2x_1 + x_2 + 3x_3 = 1$
$\quad\quad 2x_2 + x_3 + 3x_4 = 2$
$\quad\quad 2x_3 + x_4 = -1$

(d) $\displaystyle\sum_{j=i}^{6} j^2 x_j = 1 \quad (i = 1, \ldots, 6)$

(e) $\displaystyle\sum_{j=1}^{i} 2^{j-i} x_j = i \quad (i = 1, \ldots, 4)$

(f) $\displaystyle\sum_{j=1}^{5} \frac{1}{i+j} x_j = 2^{1-i} \quad (i = 1, \ldots, 5)$

2. Prove the algebraic laws in (7).

0.2
SOLUTION OF HOMOGENEOUS SYSTEMS

In order to solve the homogeneous system of equations $Ax = 0$ for x, some simple operations, called *elementary operations,* are performed on the system of equations:

1. Add one equation that is multiplied by a scalar to a different equation.
2. Multiply an equation by a nonzero scalar.
3. Interchange two equations.

Suppose the system of equations has been written in the compact notation

$$a_i x = 0 \qquad i = 1, 2, \ldots, m,$$

where a_i is the ith row of the coefficient matrix. Then the elementary operations can be restated succinctly as

1. Replace the ith equation by $(a_i + \alpha a_j)x = 0$. $\qquad\qquad (i \neq j)$
2. Replace the ith equation by $(\alpha a_i)x = 0$. $\qquad\qquad\quad (\alpha \neq 0)$
3. Interchange $a_i x = 0$ and $a_j x = 0$. $\qquad\qquad\qquad\quad (i \neq j)$

Observe that these operations are reversible in the sense that if one of them has been done to obtain the system $A'x = 0$, then the original system of equations $Ax = 0$ can be recovered by performing one of the elementary operations on $A'x = 0$. For example, assume operation 1 had been done to obtain a new equation i: $(a_i + \alpha a_j)x = 0$. Then subtracting α times equation j from the new equation i recovers the old equation i. Operation 2 is undone by multiplying the same equation by the reciprocal of the scalar used. Operation 3 is easily undone by interchanging the same two equations. The next theorem is crucial and very simple. It asserts that these three elementary operations neither create nor destroy solutions, as sometimes happens in other algebraic manipulations.

THEOREM 1. If one system of equations $A'x = 0$ is obtained from another system of equations $Ax = 0$ by a finite sequence of elementary operations, then both systems have exactly the same set of solutions. That is, the column x satisfies one system of equations iff it satisfies the other system of equations.

Proof. It is sufficient to prove this for each elementary operation, for if none of these operations perturbs the solutions, then neither will a finite sequence of the operations. Let x be a solution of the system $Ax = 0$. For operation 1, the new equation is $(a_i + \alpha a_j)x = 0$, which is clearly satisfied by x since $(a_i + \alpha a_j)x = a_i x + \alpha(a_j x) = 0 + 0 = 0$. For operation 2, the new equation is satisfied by x since $(\alpha a_i)x = \alpha(a_i x) = \alpha \cdot 0 = 0$. For operation 3, $a_i x = 0$, $a_j x = 0$ are merely reindexed and so they still hold. Conversely, if x satisfies the new equations, it must satisfy the original equations, since the original equations can be recovered from the new equations by the elementary operations just described.

In doing the elementary operations on a homogeneous system of equations, observe that the unknowns x_i are never involved, i.e., that the operations affect only the coefficients. It is thus sufficient to do the elementary operations on the rows of the coefficient matrix $[a_{ij}]$. The strategy is to do the elementary operations until it is obvious how to solve the corresponding system. The next example illustrates this strategy.

EXAMPLE 1. Solve

$$x_1 + x_2 + x_3 - x_4 = 0$$
$$2x_1 + 2x_2 + x_3 - 3x_4 = 0$$
$$-x_1 - x_2 + x_3 + 3x_4 = 0.$$

Form the coefficient matrix

$$\begin{bmatrix} 1 & 1 & 1 & -1 \\ 2 & 2 & 1 & -3 \\ -1 & -1 & 1 & 3 \end{bmatrix}.$$

First subtract 2 $*$ row 1 from row 2 and add row 1 to row 3. Doing these operations on the original equations eliminates x_1 from the last two equations. The notation (commonly used in pseudocode) is as follows:

$R_i \leftarrow R_i + \alpha R_j$ means replace row i by (row i) $+\alpha$(row j).
Swap R_i, R_j means interchange row i and row j, $(i \neq j)$.
$R_i \leftarrow \alpha R_i$ means replace row i by α(row i), $(\alpha \neq 0.)$

Thus the preceding operations can be symbolized by

$$\begin{bmatrix} 1 & 1 & 1 & -1 \\ 2 & -2 & 1 & -3 \\ 1 & -1 & 1 & 3 \end{bmatrix} \rightarrow \begin{bmatrix} 1 & 1 & 1 & -1 \\ 0 & 0 & -1 & -1 \\ 0 & 0 & 2 & 2 \end{bmatrix}.$$

$$R_2 \leftarrow R_2 - 2R_1$$
$$R_3 \leftarrow R_3 + R_1$$

Next add row 2 to row 1 and twice row 2 to row 3. Tidy up the result by multiplying row 2 by -1. Schematically,

$$\begin{bmatrix} 1 & -1 & 1 & -1 \\ 0 & 0 & -1 & -1 \\ 0 & 0 & 2 & 2 \end{bmatrix} \rightarrow \begin{bmatrix} 1 & -1 & 0 & -2 \\ 0 & 0 & -1 & -1 \\ 0 & 0 & 0 & 0 \end{bmatrix} \rightarrow \begin{bmatrix} 1 & -1 & 0 & -2 \\ 0 & 0 & 1 & 1 \\ 0 & 0 & 0 & 0 \end{bmatrix} \equiv B,$$

$$R_1 \leftarrow R_1 + R_2 \qquad\qquad R_2 \leftarrow -R_2$$
$$R_3 \leftarrow R_3 + 2R_2.$$

Now write the system of equations $Bx = 0$

$$x_1 - x_2 - 2x_4 = 0$$
$$x_3 + x_4 = 0,$$

which must have the same set of solutions as the original system. Take x_2 and x_4 as arbitrary, and determine x_1 and x_3 by $x_1 = x_2 + 2x_4$ and $x_3 = -x_4$. The general solution, x, is then

$$x = \begin{bmatrix} x_2 + 2x_4 \\ x_2 \\ -x_4 \\ x_4 \end{bmatrix},$$

with x_2, x_4 arbitrary.

Example 1 illustrates the *Gauss–Jordan elimination procedure*. This procedure reduces the matrix A of coefficients to a standard form called the row-reduced echelon form of A, which will be precisely defined shortly. From the row-reduced echelon form it is possible to deduce many useful facts about systems of equations. For a nonzero matrix A, the procedure is as follows:

Step 1. Find the first column j_1 of the matrix A that has a nonzero entry. Select your favorite nonzero entry a_{ij_1} and interchange rows 1 and i. Next, divide row 1 by its leading entry, a_{ij_1}, to produce the number 1 in row 1, column j_1. Now make all the entries in column j_1, rows $2, \ldots, m$, equal to 0 by subtracting from each subsequent row the appropriate multiple of row 1. This is equivalent to relabeling the ith equation as the first equation and eliminating x_{j_1} from all the other equations. The resulting matrix now looks like

$$A' = \begin{bmatrix} 0 & \cdots & 1 & * & * & \cdots & * \\ 0 & \cdots & 0 & & & & \\ & & & & A_1 & & \\ 0 & \cdots & 0 & & & & \end{bmatrix},$$

where $*$ denotes an entry that may be nonzero and A_1 is the $m - 1$ by $n - j_1$ submatrix remaining after deleting from the reduced matrix the first row and the first j_1 columns.

Step 2. Now we restrict attention to the submatrix A_1. If it is nonzero, continue as follows: Let j_2 denote the index of the first nonzero column of this submatrix. Choose a row of A' below the first row whose entry in the j_2 column is nonzero, and swap it with row 2. Now the $2, j_2$ entry is made equal to one. Next, use the new row 2 to make the entries in *all* other rows of A' that are in column j_2 equal to 0 by subtracting the appropriate multiple of the new row 2 from these rows. The result looks like this:

$$A'' = \begin{bmatrix} 0 & \cdots & 1 & * & 0 & * & \cdots & * \\ 0 & \cdots & \cdots & & 1 & * & \cdots & * \\ & & & & 0 & & A_2 & \\ 0 & \cdots & 0 & & 0 & & & \end{bmatrix}.$$

The process is continued inductively in this way until the rows or columns of the matrix are exhausted.

When the process is done by hand, the favorite entry in column j_1 in step 1 is the entry that requires the least work. If the process were to be conducted on a computer, the favorite entry in column j_1 in step 1 should be that entry with the largest magnitude in column j_1. This choice reduces the effects of round-off error.

The resulting matrix is called the *row-reduced echelon form of A.* The row-reduced echelon form may be attained by many different sequences of elementary row operations, but the resulting form can be shown to be unique and is characterized by the following properties:

1. The first r rows are nonzero, and the last $m - r$ rows are zero.
2. The leading entry in row p, $1 \le p \le r$, is 1.
3. Leading 1s occur to the right of the preceding leading 1s: i.e., if the leading 1 in row p is in column j_p, $1 \le p \le r$, then $j_1 < j_2 \cdots < j_r$.
4. The entries in column j_p that are not in row p are all zero.

The row-reduced echelon form looks like this:

$$
\begin{array}{c}
\\
1\\
2\\
3\\
\cdots\\
r\\
\cdots\\
m
\end{array}
\begin{bmatrix}
\overset{\textstyle}{0} & \cdots & 0 & \overset{\displaystyle j_1}{1} & * & \cdots & \overset{\displaystyle j_2}{0} & * & \cdots & \overset{\displaystyle j_3}{0} & * & \cdots & \overset{\displaystyle j_r}{0} & * & * \\
0 & \cdots & & 0 & \cdots & 0 & 1 & * & \cdots & 0 & * & \cdots & 0 & * & * \\
0 & \cdots & & & & & 0 & 1 & * & \cdots & 0 & * & * \\
\cdots & \cdots & & & & & & & & & & & & & \\
0 & \cdots & & & \text{all} & 0 & & & & & & & 1 & * & \\
0 & \cdots & & & & & & & & & & & & & 0 \\
0 & \cdots & & & & & & & & & & & & & 0
\end{bmatrix}.
$$

Definition. The number r of nonzero rows in the row-reduced echelon form is called the *row rank* of the matrix A.

In Chapter 2, much more will be said about the significance of the row rank.

PROPOSITION 2. Suppose r is the row rank of an $m \times n$ matrix A. Then r cannot exceed the smaller of m and n.

Proof. Let r be the row rank of A. Clearly $r \le m$. On the other hand, the r columns with nonzero leading entries have indices j_i satisfying $1 \le j_1 < j_2 < \cdots < j_r \le n$. Hence $r \le n$, since one cannot fit more than n integers between 1 and n.

The equivalent system of equations $Rx = 0$, when written out, is (compare (1) to the preceding row-reduced echelon form)

$$x_{j_i} + \sum_{\substack{j \ne j_1, \ldots, j_r \\ j > j_i}} r_{ij} x_j = 0, \qquad i = 1, \ldots, r. \tag{1}$$

From this, it is clear that $x_{j_1}, \ldots, x_{j_r}$ (called *leading variables* or *pivot variables*) are determined from the complementary set of the other $n - r$ variables x_j (called *free variables*) which may be taken to be arbitrary. That is, the general solution of $Ax = 0$ is the vector $x = (x_1, \ldots, x_n)^\mathrm{T}$ in which the free variables $x_j (j \ne j_1, \ldots, j_r)$ are left in place as arbitrary parameters, and the leading variables $x_j (j = j_1, \ldots, j_r)$ are eliminated by using

$$x_{j_i} = \sum_{\substack{j \ne j_1, \ldots, j_r \\ j > j_i}} r_{ij} x_j \qquad i = 1, \ldots, r, \tag{2}$$

which easily follows from (1).

EXAMPLE 2. If A were such that its row-reduced echelon form is

$$R = \begin{bmatrix} 1 & -2 & 0 & 1 & 4 \\ 0 & 0 & 1 & 3 & 0 \\ 0 & 0 & 0 & 0 & 0 \end{bmatrix},$$

find the free variables and general solution.

Here $j_1 = 1$, $j_2 = 3$, and the free variables are x_2, x_4, x_5. The corresponding system of equations is

$$x_1 - 2x_2 + x_4 + 4x_5 = 0$$
$$x_3 - 3x_4 = 0,$$

so that

$$x_1 = 2x_2 + x_4 + 4x_5$$
$$x_3 = 3x_4.$$

Thus x_1 and x_3 are determined from the free variables x_2, x_4, x_5. The general solution of the original equations $Ax = 0$ is

$$x = (2x_2 - x_4 - 4x_5, \; x_2, \; -3x_4, \; x_4, \; x_5)^T.$$

In Section 2.2, the nature of the solutions of $Ax = 0$ and the significance of the row rank r will be examined thoroughly. We begin to examine some of these considerations in Chapter 1 as well.

EXAMPLE 3. Find the general solution of

$$x_1 + x_2 - x_3 = 0$$
$$-x_2 + 3x_3 = 0$$
$$x_1 - x_2 + 5x_3 = 0.$$

Form the coefficient matrix

$$A = [A_1 \quad A_2 \quad A_3] = \begin{bmatrix} 1 & 1 & -1 \\ 0 & -1 & 3 \\ 1 & -1 & 5 \end{bmatrix}.$$

Gauss–Jordan row reduction yields

$$R = \begin{bmatrix} 1 & 0 & 2 \\ 0 & 1 & -3 \\ 0 & 0 & 0 \end{bmatrix}.$$

But the solutions of $Ax = 0$ are the same as those of $Rx = 0$, i.e.,

$$x_1 + 2x_3 = 0$$
$$x_2 - 3x_3 = 0.$$

Thus $r = 2$, $n - r = 1$, $j_1 = 1$, $j_2 = 2$, and x_3 is the free variable, which can be taken as an arbitrary parameter. The general solution of $Ax = 0$ is $(x_1, x_2, x_3)^T$, where $x_1 = -2x_3$, $x_2 = 3x_3$, with x_3 arbitrary; i.e., $x = (-2x_3, 3x_3, x_3)^T$.

The next theorem draws some conclusions about the system $Ax = 0$ from the number of free variables and (2).

THEOREM 3. Consider the $m \times n$ system of equations $Ax = 0$. Let r be the row rank of A.

(a) If $r = n$, then the only solution is $x = 0$.
(b) If $r < n$, then there are infinitely many solutions.
(c) If $n > m$, then there is at least one nontrivial solution to the system.

Proof

(a) If $n = r$, there are no free variables, so that every variable is a leading variable. Thus $j_i = i$, and (2) becomes $x_i = 0$ for $i = 1, 2, \ldots, n$.
(b) If $n > r$, there are $n - r$ free variables, each of which may be given infinitely many values.

(c) If $m < n$, then since $r \leq m < n$, the number $n - r$ of free variables must be positive. These variables can be given arbitrary nonzero values so that there are infinitely many nonzero solutions.

0.2 Exercises

1. Find the general solution of $Ax = 0$ for the following matrices.

(a) $\begin{bmatrix} 1 & 1 & -1 \\ 2 & 0 & 2 \\ 1 & 2 & -3 \end{bmatrix}$ (b) $\begin{bmatrix} 1 & -1 & 1 & 1 \\ 1 & 2 & -1 & 1 \\ 1 & 5 & -3 & 1 \end{bmatrix}$ (c) $\begin{bmatrix} 1 & 1 & 2 \\ -1 & 2 & 5 \\ 1 & -1 & -3 \\ 1 & 0 & -1 \end{bmatrix}$

(d) $\begin{bmatrix} 1 & 2 & 2 & 1 & 1 & 0 & 5 \\ 1 & 2 & 0 & -1 & -1 & 0 & 1 \\ 1 & 2 & 1 & 0 & 1 & 1 & 2 \\ 2 & 4 & 1 & -1 & 2 & 3 & 1 \\ 2 & 4 & 2 & 0 & 3 & 3 & 3 \end{bmatrix}$ (e) $\begin{bmatrix} 2 & 2 & 0 \\ 1 & -2 & -i \\ i & 1 & 1 \end{bmatrix}$

(f) $\begin{bmatrix} 1 & -2 & 0 & 1 \\ 2 & -4 & 1 & 4 \\ -1 & 2 & 3 & 5 \end{bmatrix}$ (g) $\begin{bmatrix} i & -1 \\ 1 & i \end{bmatrix}$ (h) $\begin{bmatrix} \sqrt{2}i & 1 & 0 \\ -1 & \sqrt{2}i & 1 \\ 0 & -1 & \sqrt{2}i \end{bmatrix}$.

2. Let A be an $m \times n$ matrix, and let r be the row rank of A. Show that $Ax = 0$ has only the solution $x = 0$ iff $r = n$.

3. Define $A \sim B$ to mean that B can be obtained from A by elementary row operations (possibly none). Show that
(a) (i) $A \sim A$ (ii) $B \sim A$ (iii) If $A \sim B$ and $B \sim C$, then $A \sim C$.
(b) If A has been row-reduced to R_1 and also to R_2, then R_1 can be obtained from R_2 by elementary row operations.

4. Show that the row rank of A is unique.

0.3
SOLUTION OF INHOMOGENEOUS SYSTEMS

In this section, a procedure for solving $Ax = b$ is discussed along with some very simple theoretical observations. More detail about the nature of the solutions and the possibility of solving the system will be adduced in Chapters 1 and 2.

In contrast with the system $Ax = 0$, which always has $x = 0 \equiv (0, 0, \ldots, 0)^{\mathrm{T}}$ as a solution, linear systems $Ax = b$ with $b \neq 0$ may or may not have solutions. The system $Ax = b$ is said to be *consistent* iff it has as least one solution, and *inconsistent* otherwise.

In order to solve

$$Ax = b \tag{1}$$

or equivalently,

$$\sum_{j=1}^{n} a_{ij}x_j = b_i, \qquad i = 1, \ldots, m$$

for x, we again use the elementary operations on the system of equations. For inhomogeneous systems, the elementary operations become

1. Add one equation that is multiplied by a scalar to a different
   equation; i.e., replace the ith equation by $(a_i + \alpha a_j)x = b_i + \alpha b_j$. $\qquad (i \neq j)$
2. Multiply an equation by a nonzero scalar; i.e., replace the ith
   equation by $(\alpha a_i)x = \alpha b_i$. $\qquad (\alpha \neq 0)$
3. Interchange two equations; i.e., interchange $a_i x = b_i$ and $a_j x = b_j$. $\qquad (i \neq j)$

Just as for homogeneous systems, the following principle holds, and the proof is essentially the same.

THEOREM 1. If one system of equations, $A'x = b'$, is obtained from another, $Ax = b$, by a finite sequence of elementary operations, then both systems have exactly the same set of solutions. That is, the column x satisfies one system of equations iff it satisfies the other system of equations.

In performing the elementary operations on an inhomogeneous system of equations, one observes that the x_i are never involved, i.e., that the operations affect only the coefficients and the b_i. It is thus advantageous to retain only the a_{ij} and y_i during the process of performing the elementary operations. Thus, form the *augmented matrix* $\tilde{A}$ of the system (1):

$$\tilde{A} \equiv [A \mid b] \equiv [A_1 \ A_2 \ \cdots \ A_n \ b]. \tag{2}$$

If one of the elementary operations is done on the system of equations $\sum_{i=1}^{n} a_{ij}x_j = b_i$ to obtain a new system of equations $\sum_{j=1}^{n} a'_{ij}x_i = b'_i$, augmented matrix $\tilde{A}'$ of this equation is obtained by doing the corresponding elementary row operation on the original augmented matrix $\tilde{A}$. Thus it is sufficient to do row operations on the augmented system. When the A part of $[A \mid b]$ has been converted to a sufficiently simple form $[B \mid w]$, the equivalent system of equations $Bx = w$ is written and easily solved.

EXAMPLE 1. Solve the following, if possible:

$$\begin{aligned} x_1 + x_2 + 2x_3 &= 0 \\ 2x_1 + x_2 + x_3 &= 1 \\ x_1 + 2x_3 &= -2 \end{aligned} \qquad \text{or} \qquad \begin{bmatrix} 1 & 1 & 2 \\ 2 & 1 & 1 \\ 1 & 0 & 2 \end{bmatrix} \begin{bmatrix} x_1 \\ x_2 \\ x_3 \end{bmatrix} = \begin{bmatrix} 0 \\ 1 \\ -2 \end{bmatrix}.$$

Form the augmented matrix $\tilde{A}$ and do the following sequence of operations:

1. Produce zeros in the last two rows of the first column (which is the same as eliminating x_1 from the last two equations) by multiplying the first equation by -2 and adding it to the second, and then subtract the first equation from the last. Call this matrix $\tilde{A}_1$.
2. Produce a zero in the third row and second column by subtracting row 2 from row 3. Call this matrix $\tilde{A}_2$. This is the same as eliminating x_2 from the last equation. The

steps are illustrated as follows:

$$\tilde{A} = \begin{bmatrix} 1 & 1 & 2 & 0 \\ 2 & 1 & 1 & 1 \\ 1 & 0 & 2 & -2 \end{bmatrix} \to \tilde{A}_1 = \begin{bmatrix} 1 & 1 & 2 & 0 \\ 0 & -1 & -3 & 1 \\ 0 & -1 & 0 & -2 \end{bmatrix} \to \tilde{A}_2 = \begin{bmatrix} 1 & 1 & 2 & 0 \\ 0 & -1 & -3 & 1 \\ 0 & 0 & 3 & -3 \end{bmatrix}.$$

$$R_2 \leftarrow R_2 - 2R_1 \qquad\qquad R_3 \leftarrow R_3 - R_2$$
$$R_3 \leftarrow R_3 - R_1$$

Let $\tilde{A}_2 = [A_2 \mid b']$. The new system of equations $A_2 x = b'$ (which was called $Bx = w$ previously) is easy to solve and has exactly the same solutions as $Ax = b$:

$$x_1 + x_2 + 2x_3 = 0$$
$$-x_2 - 3x_3 = 1$$
$$3x_3 = -3.$$

From the last equation $x_3 = -1$, and then $x_2 = -1 - 3x_3 = 2$ from the second equation, and $x_1 = -x_2 - 2x_3 = 0$ from the first.

This process of working backward to find all the variables is called *back substitution*, or *back solving*.

For theoretical considerations, and to make the back solving easier for hand calculations, it is advantageous to carry the reduction further at each stage. In fact, Gauss–Jordan elimination is done on the coefficient submatrix A of $\tilde{A}$. This is illustrated schematically for Example 1 by the following:

$$\tilde{A} = \begin{bmatrix} 1 & 1 & 2 & 0 \\ 2 & 1 & 1 & 1 \\ 1 & 0 & 2 & -2 \end{bmatrix} \to \begin{bmatrix} 1 & 1 & 2 & 0 \\ 0 & -1 & -3 & 1 \\ 0 & -1 & 0 & -2 \end{bmatrix} \to \begin{bmatrix} 1 & 1 & 2 & 0 \\ 0 & 1 & 3 & -1 \\ 0 & -1 & 0 & -2 \end{bmatrix} \to$$

$$R_2 \leftarrow R_2 - 2R_1 \qquad\qquad R2 \leftarrow -R_2 \qquad\qquad R_3 \leftarrow R_3 + R_2$$
$$R_3 \leftarrow R_3 - R_1 \qquad\qquad\qquad\qquad\qquad\quad R_1 \leftarrow R_1 - R_2$$

$$\begin{bmatrix} 1 & 0 & -1 & 1 \\ 0 & 1 & 3 & -1 \\ 0 & 0 & 3 & -3 \end{bmatrix} \to \begin{bmatrix} 1 & 0 & -1 & 1 \\ 0 & 1 & 3 & -1 \\ 0 & 0 & 1 & -1 \end{bmatrix} \to \begin{bmatrix} 1 & 0 & 0 & 0 \\ 0 & 1 & 0 & 2 \\ 0 & 0 & 1 & -1 \end{bmatrix}.$$

$$R_3 \leftarrow \tfrac{1}{3}R_3 \qquad\qquad R_1 \leftarrow R_1 + R_3$$
$$R_2 \leftarrow R_2 - 3R_3$$

The resulting system is very simple:

$$x_1 = 0$$
$$x_2 = 2$$
$$x_3 = -1.$$

Remark. When solving two systems $Ax = b_1$ and $Ax = b_2$ with the same coefficient matrix A, it is convenient to doubly augment the matrix, $\tilde{A} = [A|b_1|b_2]$, and to row-reduce the A submatrix using the same strategy. One need only remember to convert the resulting final matrix into two systems of equations at the end. If, in Example 1, the system $Az = (1, 1, 1)^T$ were to be solved as well, the doubly augmented matrix and the same row reductions yield

$$\tilde{A} = \begin{bmatrix} 1 & 1 & 2 & 0 & 1 \\ 2 & 1 & 1 & 1 & 1 \\ 1 & 0 & 2 & -2 & 1 \end{bmatrix} \rightarrow \cdots \rightarrow = \begin{bmatrix} 1 & 0 & 0 & 0 & 1/3 \\ 0 & 1 & 0 & 1 & -2 \\ 0 & 0 & 1 & -1 & 1/3 \end{bmatrix}.$$

The resulting equivalent equations are

$$
\begin{array}{ll}
x_1 = 0 & x_1 = \frac{1}{3} \\
x_2 = 2 & x_2 = -2 \\
x_3 = -1 & x_3 = \frac{1}{3}.
\end{array}
$$

EXAMPLE 2. Attempt to solve

$$
\begin{array}{c}
x_1 + 3x_2 + 2x_3 = 1 \\
2x_1 + 8x_2 + 6x_3 = 6 \\
x_1 + 2x_2 + x_3 = -1
\end{array}
\qquad \text{or} \qquad
\begin{bmatrix} 1 & 3 & 2 \\ 2 & 8 & 6 \\ 1 & 2 & 1 \end{bmatrix} \begin{bmatrix} x_1 \\ x_2 \\ x_3 \end{bmatrix} = \begin{bmatrix} 1 \\ 6 \\ -1 \end{bmatrix}.
$$

Using the same strategy and the abbreviated notation,

$$\tilde{A} = \begin{bmatrix} 1 & 3 & 2 & 1 \\ 2 & 8 & 6 & 6 \\ 1 & 2 & 1 & -1 \end{bmatrix} \rightarrow \begin{bmatrix} 1 & 3 & 2 & 1 \\ 0 & 2 & 2 & 4 \\ 0 & -1 & -1 & -2 \end{bmatrix} \rightarrow$$

$$
\begin{array}{ll}
R_2 \leftarrow R_2 - 2R_1 & R_2 \leftarrow \frac{1}{2} R_2 \\
R_3 \leftarrow R_3 - R_1 &
\end{array}
$$

$$\begin{bmatrix} 1 & 3 & 2 & 1 \\ 0 & 1 & 1 & 2 \\ 0 & -1 & -1 & -2 \end{bmatrix} \rightarrow \begin{bmatrix} 1 & 0 & -1 & -5 \\ 0 & 1 & 1 & 2 \\ 0 & 0 & 0 & 0 \end{bmatrix}.$$

$$
\begin{array}{l}
R_3 \leftarrow R_3 + R_2 \\
R_1 \leftarrow R_1 - 3R_2
\end{array}
$$

The equivalent system is

$$x_1 - x_3 = -5 \qquad x_2 + x_3 = 2,$$

which has the solution $x_1 = -5 + x_3$, $x_2 = 2 - x_3$, with x_3 arbitrary.

EXAMPLE 3. Attempt to solve

$$
\begin{array}{c}
x_1 + 3x_2 + 2x_3 = 1 \\
2x_1 + 8x_2 + 6x_3 = 6 \\
x_1 + 2x_2 + x_3 = 1
\end{array}
\qquad \text{or} \qquad
\begin{bmatrix} 1 & 3 & 2 \\ 2 & 8 & 6 \\ 1 & 2 & 1 \end{bmatrix} \begin{bmatrix} x_1 \\ x_2 \\ x_3 \end{bmatrix} = \begin{bmatrix} 1 \\ 6 \\ 1 \end{bmatrix}.
$$

The matrix A is the same one as in Example 2. The same row reduction on the matrix A in the augmented matrix yields the equivalent system

$$
\begin{array}{c}
x_1 - x_3 = -5 \\
x_2 + x_3 = 2 \\
0 = 2,
\end{array}
$$

which is impossible. Since the final system has no solution, neither does the original system.

In general, to solve an inhomogeneous system of equations $Ax = b$, form the augmented matrix $\tilde{A}$ of the system. Apply row operations to $\tilde{A}$ and bring the A submatrix to its row-reduced echelon form R. Suppose $\tilde{A}$ is thereby brought to the form $[R \mid z]$. If the column z has a nonzero entry c in one of the rows $r+1, \ldots, m$, then there are no solutions to the system because the linear equation corresponding to that row is $0 = c$, with $c \neq 0$, which is impossible (see Example 3). Otherwise, rewrite the equivalent system $Rx = z$ for the augmented matrix $[R \mid z]$, and the general solution will be obvious. The system $Rx = z$ is

$$x_{j_i} + \sum_{\substack{j \neq j_1, \ldots, j_r \\ j > j_i}} r_{ij}x_j = z_i, \quad i = 1, \ldots, r. \tag{3}$$

From these observations, the next theorem follows.

THEOREM 2. Suppose the augmented matrix of the system $Ax = b$ is brought to the form $[R \mid z]$, where R is the row-reduced echelon form of A with r nonzero rows. Then

1. The system is consistent if and only if $z_i = 0$ for $i = r + 1, \ldots, m$.
2. When the system is consistent, one solution may be obtained by setting the leading variables $x_{j_k} = z_k, k = 1, 2, \ldots, r$ and the free variables $x_i = 0, i \neq j_1, j_2, \ldots, j_r$. If $r < n$, then there are infinitely many solutions, whereas if $r = n$, there is exactly one solution.

EXAMPLE 4. Suppose the system $Ax = b$ is brought to

$$[R \mid z] = \begin{bmatrix} 1 & 3 & 0 & -2 & 0 & 1 \\ 0 & 0 & 1 & -1 & 0 & 3 \\ 0 & 0 & 0 & 0 & 1 & 2 \\ 0 & 0 & 0 & 0 & 0 & 0 \end{bmatrix}$$

by row-reducing A to R. Determine the consistency of the original system, and find a particular solution and the general solution.

The equivalent system $Rx = z$ is

$$x_1 + 3x_2 - 2x_4 = 1$$

$$x_3 - x_4 = 3$$

$$x_5 = 2,$$

which is consistent. Here $r = 3$, $j_1 = 1$, $j_2 = 3$, and $j_3 = 5$. The free variables are x_2 and x_4. Setting $x_1 = 1$, $x_3 = 3$, $x_5 = 2$, and $x_2 = x_4 = 0$ satisfies $Rx = z$ and thus is a particular solution of the original equations. The general solution is obtained by writing the pivot variables x_1, x_3, x_5 in terms of the free variables. This gives as the general solution $x = (1 - 3x_2 + 2x_4, x_2, 3 + x_4, x_4, 2)^T$.

Clearly the Gauss–Jordan procedure makes the solutions easier to read and is useful for hand calculation and theoretical considerations. However, this procedure is not usually so useful for computer implementation because the number of operations is substantially larger than first computing a row echelon form and then back solving. However, there are occasions when the Gauss–Jordan algorithm is numerically preferable, and the corresponding algorithm is called full row and column pivoting in numerical analysis books.

For $n \times n$ systems, the ideas in Theorem 2 can be sharpened to a condition for the existence and uniqueness of solutions. For square matrices, the row-reduced echelon form can take a very special configuration in which each variable is a pivot variable ($i = j_i$). This leads to the following definition.

> **DEFINITION.** The $n \times n$ identity matrix I is the matrix whose entries are all 0, except those on the main diagonal, which are 1: $I = [\delta_{ij}]$, where
>
> $$\delta_{ij} = \begin{cases} 0 & i \neq j \\ 1 & i = j. \end{cases}$$

δ_{ij} is called the *Kronecker delta*.

The next theorem follows immediately from Theorem 2.

> **THEOREM 3.** Let A be an $n \times n$ matrix. Then $Ax = b$ has a unique solution for every b iff A row-reduces to the identity matrix. If $[A \mid b]$ row-reduces to $[I \mid z]$, then $x = z$ is the unique solution to $Ax = y$.
> The proof of this theorem is left as an exercise.

0.3 Exercises

1. Find the general solution of the following systems.

(a) $\begin{bmatrix} 1 & 1 & -1 \\ 2 & 0 & 2 \\ 1 & 2 & -3 \end{bmatrix} x = y$, (i) $y = \begin{bmatrix} 2 \\ -2 \\ 5 \end{bmatrix}$ (ii) $y = \begin{bmatrix} 1 \\ 0 \\ 0 \end{bmatrix}$

(b) $\begin{bmatrix} 1 & 1 & -1 & 1 \\ 1 & 0 & 2 & 1 \\ 1 & 2 & 1 & 2 \end{bmatrix} x = y$, $y = \begin{bmatrix} 1 \\ 1 \\ 1 \end{bmatrix}$

(c) $\begin{bmatrix} 1 & -1 & 1 & 1 \\ 1 & 2 & -1 & 1 \\ 1 & 5 & -3 & 1 \end{bmatrix} x = y$, (i) $y = \begin{bmatrix} 3 \\ 2 \\ 1 \end{bmatrix}$, (ii) $y = \begin{bmatrix} 3 \\ 2 \\ 2 \end{bmatrix}$

(d) $\begin{bmatrix} 2 & 2 & 0 \\ 0 & -1 & -1 \\ 1 & 2 & 1 \\ 1 & 0 & -1 \end{bmatrix} x = y$, (i) $y = \begin{bmatrix} 2 \\ 2 \\ -1 \\ 3 \end{bmatrix}$, (ii) $y = \begin{bmatrix} 0 \\ 1 \\ 0 \\ 0 \end{bmatrix}$

2. Prove Theorem 3: If A is an $n \times n$ matrix, then $Ax = b$ has a unique solution for every b iff A row-reduces to the identity matrix. If $[A \mid b]$ row reduces to $[I \mid z]$, then $x = z$ is the solution to $Ax = b$.

3. Let A be an $m \times n$ matrix, and let r be the row rank of A. Prove that if $r = m$, then for every b, $Ax = b$ is consistent.

4. Find examples of the following, for a small $m \times n$ matrix A:
   (a) $r = n$, but $Ax = b$ has a unique solution for some b, and no solution for others.
   (b) $r = m$, but no solution is unique.

5. Suppose that a system $Ax = b_0$ has an infinite number of solutions for a given b_0. Can there be a b such that $Ax = b$ has only one solution? Explain.

6. Suppose that a system $Ax = b_0$ has just one solution for a given b_0. Can there be a b such that $Ax = b$ has infinitely many solutions? Explain.

7. Let A be an $n \times n$ matrix such that $a_{ij} = 0$ when $i > j$.
   (a) Find conditions on A so that $Ax = b$ is consistent for all b.
   (b) Find conditions on A so that $Ax = b$ has at most one solution for each b.

8. Is it possible to find a matrix A and column y so that $Ax = b$ has exactly two solutions? Explain.

9. Let $\tilde{x}$ be a specific solution of $Ax = y$. Show that every other solution is of the form $x = \tilde{x} + z$, where z satisfies $Az = 0$.

0.4
MATRIX ALGEBRA

Matrix multiplication, some special matrices, and some simple operations on matrices are defined in this section. Some of the fundamental properties of multiplication and these operations are given.

Notation. Let A be an $m \times n$ matrix. A_j denotes the jth column of A, and we write $A = [A_1 \ A_2 \ \cdots \ A_n]$. $(A)_{ij}$ denotes the entry of A in the ith row and jth column. If x is a column, then $(x)_i$ or x_i denotes the ith entry (row) of x. Hence, if $A = [a_{ij}]$, then $(A)_{ij} = a_{ij}$, $(A_j)_i = a_{ij}$.

DEFINITION. Let $A = [a_{ij}]$ be an $m \times n$ matrix and $B = [b_{ij}] = [B_1 \ B_2 \ \cdots \ B_q]$ be a $p \times q$ matrix with columns B_j. The product AB is defined if and only if $n = p$, and in this case it is the $m \times q$ matrix

$$AB = [AB_1 \ AB_2 \cdots AB_q]. \tag{1}$$

Next, we define some special columns that play an important role in many calculations.

DEFINITION. Define δ_i as the column vector of length n that has an entry of 0 in every row but the ith, which has the entry 1:

$$(\delta_j)_i \equiv \delta_{ij} = \begin{cases} 0 & \text{if } i \neq j \\ 1 & \text{if } i = j. \end{cases}$$

For brevity, the length of the column is not incorporated in the notation and must be ascertained from the context. Note that the $n \times n$ identity matrix may now be expressed as

$$I_n \equiv [\delta_{ij}] = [\delta_1 \ \delta_2 \ \cdots \ \delta_n].$$

The subscript on I will be dropped when the size is clear from the context.

The following proposition lists some simple but important identities in one place for reference. Some are definitions or equations that have been encountered before.

PROPOSITION 1. Let $x = (x_1, \ldots, x_n)^T$, let A have columns A_j and rows a_i, and let B have columns B_j and rows b_i; i.e.,

$$A = [A_1\ A_2\ \cdots\ A_n] = \begin{bmatrix} a_1 \\ a_2 \\ \vdots \\ a_m \end{bmatrix}, \qquad B = [B_1\ B_2\ \cdots\ B_q] = \begin{bmatrix} b_1 \\ b_2 \\ \vdots \\ b_n \end{bmatrix}, \qquad x = \begin{bmatrix} x_1 \\ x_2 \\ \vdots \\ x_n \end{bmatrix}.$$

Then

$$(Ax)_i = a_i x = \sum_{j=1}^{n} a_{ij} x_j \tag{2a}$$

$$A_k = A\delta_k \tag{2b}$$

$$Ax = \sum_{j=1}^{n} x_j A_j \tag{2c}$$

$$AB = [AB_1 AB_2 \cdots AB_q] \tag{2d}$$

$$(AB)_{ij} = a_i B_j = \sum_{k=1}^{n} a_{ik} b_{kj} \tag{2e}$$

$$AB = \sum_{j=1}^{n} A_j b_j. \tag{2f}$$

Note. The product $A_j b_j$ is called a *diad* and (2f) is called the diadic expansion of AB.

Proof

(2a) is the definition of Ax.

(2b) follows from checking the ith entry of each side: $(A\delta_k)_i = \sum_{j=1}^{n} a_{ij}\delta_{kj} = a_{ik} = (A_k)_i$.

(2c) follows from $(\sum_{j=1}^{n} x_j A_j)_i = \sum_{j=1}^{n} x_j (A_j)_i = \sum_{j=1}^{n} x_j a_{ij} = (Ax)_i$.

(2d) is the definition of AB.

(2e) follows from (2a): the i, j entry of AB is $(AB)_{ij} = (AB_j)_i = \sum_{k=1}^{n} a_{ik}(B_j)_k = \sum_{k=1}^{n} a_{ik} b_{kj}$, which is clearly the same as the row-column product $a_i B_j$.

(2f) is left as an exercise.

EXAMPLE 1. If

$$A = \begin{bmatrix} 1 & 2 & 3 \\ 2 & 2 & 1 \end{bmatrix} \quad \text{and} \quad B = \begin{bmatrix} 1 & -1 & 0 \\ 2 & 0 & 3 \\ -1 & 4 & 1 \end{bmatrix},$$

then

$$AB = \begin{bmatrix} (1,2,3)(1,2,-1)^T & (1,2,3)(-1,0,4)^T & (1,2,3)(0,3,1)^T \\ (2,2,1)(1,2,-1)^T & (2,2,1)(-1,0,4)^T & (2,2,1)(0,3,1)^T \end{bmatrix}$$

$$= \begin{bmatrix} 2 & 11 & 9 \\ 5 & 2 & 7 \end{bmatrix} = \begin{bmatrix} (1,2,3)B \\ (2,2,1)B \end{bmatrix} = \begin{bmatrix} A\begin{bmatrix} 1 \\ 2 \\ -1 \end{bmatrix} & A\begin{bmatrix} -1 \\ 0 \\ 4 \end{bmatrix} & A\begin{bmatrix} 0 \\ 3 \\ 1 \end{bmatrix} \end{bmatrix}.$$

Other matrices of special interest are defined next. The $m \times n$ *zero matrix* 0_{mn} has all of its entries equal to 0. Usually the subscripts m, n are suppressed on the zero matrix.

An $n \times n$ *diagonal matrix* is a matrix of the form

$$D = \begin{bmatrix} d_1 & 0 & \cdots & \cdots & \cdots & 0 \\ 0 & d_2 & 0 & \cdots & \cdots & 0 \\ \cdots & \cdots & \cdots & \cdots & \cdots & \cdots \\ \cdots & \cdots & \cdots & \cdots & \cdots & 0 \\ 0 & \cdots & \cdots & \cdots & 0 & d_n \end{bmatrix},$$

wherein all entries not on the main diagonal are zero. It is denoted by

$$D = \operatorname{diag}(d_i)_1^n = \operatorname{diag}(d_1, d_2, \ldots, d_n)$$

and is defined rigorously by

$$D = \operatorname{diag}(d_i)_1^n, \Leftrightarrow d_{ij} = 0 \text{ if } i \neq j, \text{ and } d_{ii} = d_i \text{ if } i = j.$$

For example, $I = \operatorname{diag}(1, 1, \ldots, 1)$. Using the identities of Proposition 1, we will show that right multiplication of A by a diagonal matrix multiplies each column of A by the corresponding entry of the diagonal matrix:

$$A \operatorname{diag}(d_i)_1^n = [d_1 A_1 \; d_2 A_2 \; \cdots \; d_n A_n]. \tag{3}$$

This follows from the observation that

$$\operatorname{diag}(d_i)_1^n = [d_1 \delta_1 \; d_2 \delta_2 \; \cdots \; d_n \delta_n]$$

and from (2b) and (2d) as follows:

$$A \operatorname{diag}(d_i)_1^n = A[d_1 \delta_1 \; d_2 \delta_2 \; \cdots \; d_n \delta_n] = [A(d_1 \delta_1) \; A(d_2 \delta_2) \; \cdots \; A(d_n \delta_n)]$$
$$= [d_1 A \delta_1 \; d_2 A \delta_2 \; \cdots \; d_n A \delta_n] = [d_1 A_1 \; d_2 A_2 \; \cdots \; d_n A_n].$$

EXAMPLE 2. If $D = \operatorname{diag}(1, 2, 3)$ and

$$A = \begin{bmatrix} 1 & -1 & 0 \\ 2 & 0 & 3 \\ -1 & 4 & 1 \end{bmatrix},$$

then

$$AD = \begin{bmatrix} 1 \cdot \begin{bmatrix} 1 \\ 2 \\ -1 \end{bmatrix} & 2 \cdot \begin{bmatrix} -1 \\ 0 \\ 4 \end{bmatrix} & 3 \cdot \begin{bmatrix} 0 \\ 3 \\ 1 \end{bmatrix} \end{bmatrix} = \begin{bmatrix} 1 & -2 & 0 \\ 2 & 0 & 9 \\ -1 & 8 & 3 \end{bmatrix}.$$

The next theorem asserts some fundamental algebraic properties of these matrix operations. The proofs are straightforward from the definitions and will be omitted.

THEOREM 2. For all matrices for which the relevant operations are defined, the following identities hold:

$$A(\alpha C + \beta D) = \alpha AC + \beta AD \qquad \text{(Distributive law)}$$
$$(\alpha C + \beta D)A = \alpha CA + \beta DA \qquad \text{(Distributive law)}$$
$$A(BC) = (AB)C \qquad \text{(Associative law)}$$
$$IA = A = AI, \quad 0A = A0 = 0. \qquad \text{(Identity and zero)}$$

EXAMPLE 3. Find the jth column $(AB)_j$ of AB.

By $(2b)$, $(AB)_j = (AB)\delta_j = A(B\delta_j) = AB_j$.

Remark. Addition, equality, and multiplication of a matrix by a number are defined in Section 0.1. The relevant algebraic properties for these operations are

$$(\alpha + \beta)A = \alpha A + \beta A$$

$$\alpha(A + B) = \alpha A + \alpha B$$

$$\alpha(\beta A) = (\alpha\beta)A \equiv \alpha\beta A$$

$$\alpha(AB) = (\alpha A)B = A(\alpha B)$$

These follow easily from the definitions.

Remark. Two very important properties of the algebra of real and complex numbers fail for matrices—commutativity and cancellation. These properties, which are *not* valid for matrices are

1. Commutativity: $AB = BA$
2. Right cancellation (of C): $AC = DC$ and $C \neq 0$ implies $A = D$
3. Left cancellation (of A): $AC = AD$ and $A \neq 0$ implies $C = D$.

For example, if $Ax = 0$ had only the solution $x = 0$ when $A \neq 0$, the task of solving linear systems would be not nearly as difficult as it is. The statement $AB = BA$ may fail because one product is defined while the other is not. The next example illustrates the failure of commutativity and the cancellation rules when all quantities are defined.

EXAMPLE 4. Let

$$A = \begin{bmatrix} 1 & 1 \\ 0 & 0 \end{bmatrix}, \qquad B = \begin{bmatrix} 2 & 0 \\ 0 & 1 \end{bmatrix}, \qquad C = \begin{bmatrix} 1 \\ -1 \end{bmatrix}, \qquad D = 0.$$

Then

$$AB = \begin{bmatrix} 2 & 1 \\ 0 & 0 \end{bmatrix} \neq \begin{bmatrix} 2 & 2 \\ 0 & 0 \end{bmatrix} = BA,$$

so the commutative law fails. Further, $AC = 0 = DC$, yet $C \neq 0$, and $A \neq D$; i.e., the right cancellation law fails. And $AC = 0 = AD$, yet $A \neq 0$, and $C \neq D$; i.e., the left cancellation law fails.

At this point, it is convenient to mention some other operations on matrices and some related identities.

Definitions. Given a matrix $A = [a_{ij}]$, the *transpose* A^T of A is the matrix whose entry in the ith row and jth column is $(A^T)_{ij} = a_{ji}$. Thus the ith column of A^T is the (ith row of $A)^T$, and the jth row of A^T is the (jth column of $A)^T$. The *conjugate* of A is formed by conjugating every entry $\overline{A} = [\overline{a}_{ij}]$, and the *conjugate transpose*, or *Hermitian transpose*, of A is $A^H = \overline{A}^T$. (Some authors denote this by A^*, but we will reserve this notation for another concept.)

EXAMPLE 5. If

$$A = \begin{bmatrix} 1 & 2i \\ i & 3 \end{bmatrix}, \qquad i = \sqrt{-1},$$

then

$$A^{\mathrm{T}} = \begin{bmatrix} 1 & i \\ 2i & 3 \end{bmatrix}, \qquad \overline{A} = \begin{bmatrix} 1 & -2i \\ -i & 3 \end{bmatrix}, \qquad A^{\mathrm{H}} = \begin{bmatrix} 1 & -i \\ -2i & 3 \end{bmatrix}.$$

The following properties are listed without proof, since the proofs are easy and reasonably straightforward.

PROPOSITION 3. Whenever the indicated sums and products are defined (α, β are scalars), the following properties hold:

(a) $(\alpha A + \beta B)^{\mathrm{T}} = \alpha A^{\mathrm{T}} + \beta B^{\mathrm{T}}$ (a') $(\alpha A + \beta B)^{\mathrm{H}} = \overline{\alpha} A^{\mathrm{H}} + \overline{\beta} B^{\mathrm{H}}$
(b) $(AB)^{\mathrm{T}} = B^{\mathrm{T}} A^{\mathrm{T}}$ (b') $(AB)^{\mathrm{H}} = B^{\mathrm{H}} A^{\mathrm{H}}$
(c) $(A^{\mathrm{T}})^{\mathrm{T}} = A$ (c') $(A^{\mathrm{H}})^{\mathrm{H}} = A$

Block Multiplication

As a footnote to this section, we discuss block multiplication of matrices. This topic is not essential, but occasionally can be quite useful. Some of the previous matrix multiplication identities can be construed to fall into this category.

Suppose the matrices A and B are divided into submatrices A_{ij} and B_{ij}, respectively, by partitioning their entries using horizontal and/or vertical lines to get *block matrices*

$$A = \begin{bmatrix} A_{11} & \cdots & A_{1p} \\ & \cdots & \\ A_{r1} & \cdots & A_{rp} \end{bmatrix}, \qquad B = \begin{bmatrix} B_{11} & \cdots & B_{1q} \\ & \cdots & \\ B_{s1} & \cdots & B_{sq} \end{bmatrix}.$$

For example, a 3×3 matrix may be partitioned several ways. Here are a few:

$$A = \begin{bmatrix} 1 & 2 & 3 \\ 4 & 5 & 6 \\ 7 & 8 & 9 \end{bmatrix} = \left[\begin{array}{cc|c} 1 & 2 & 3 \\ 4 & 5 & 6 \\ \hline 7 & 8 & 9 \end{array}\right] = \begin{bmatrix} A_{11} & A_{12} \\ A_{21} & A_{22} \end{bmatrix},$$

with

$$A_{11} = \begin{bmatrix} 1 & 2 \\ 4 & 5 \end{bmatrix}, \qquad A_{12} = \begin{bmatrix} 3 \\ 6 \end{bmatrix}, \qquad A_{21} = [7 \quad 8], \qquad A_{22} = [9].$$

Also,

$$A = \left[\begin{array}{c|c|c} 1 & 2 & 3 \\ 4 & 5 & 6 \\ 7 & 8 & 9 \end{array}\right] = [A_{11} \quad A_{12} \quad A_{13}] = [A_1 \quad A_2 \quad A_3],$$

$$A_1 = \begin{bmatrix} 1 \\ 4 \\ 7 \end{bmatrix}, \qquad A_2 = \begin{bmatrix} 2 \\ 5 \\ 8 \end{bmatrix}, \qquad A_3 = \begin{bmatrix} 3 \\ 6 \\ 9 \end{bmatrix},$$

and

$$A = \begin{bmatrix} 1 & 2 & 3 \\ 4 & 5 & 6 \\ 7 & 8 & 9 \end{bmatrix} = \begin{bmatrix} A_{11} \\ A_{21} \\ A_{31} \end{bmatrix} = \begin{bmatrix} a_1 \\ a_2 \\ a_3 \end{bmatrix}, \qquad \begin{aligned} A_{11} &= a_1 = [1 \quad 2 \quad 3] \\ A_{21} &= a_2 = [4 \quad 5 \quad 6] \\ A_{31} &= a_3 = [7 \quad 8 \quad 9]. \end{aligned}$$

If $p = s$ and all the matrix multiplications $A_{ik}B_{kj}$ are defined, then product AB may be calculated by multiplying the blocks as if they were scalars, i.e., by $AB = [C_{ij}]$, where C_{ij} is the submatrix

$$C_{ij} = \sum_{k=1}^{p} A_{ik}B_{kj}, \quad i = 1, 2, \dots, r \quad \text{and} \quad j = 1, 2, \dots, q.$$

Using the matrix A as partitioned the first way and using the matrix

$$B = \begin{bmatrix} 0 & -1 \\ 2 & 1 \\ 1 & -2 \end{bmatrix} = \begin{bmatrix} B_{11} \\ B_{21} \end{bmatrix}$$

yields

$$AB = \begin{bmatrix} A_{11}B_{11} + A_{12}B_{21} \\ A_{21}B_{11} + A_{22}B_{21} \end{bmatrix} = \begin{bmatrix} 7 & -5 \\ 16 & -11 \\ 25 & -17 \end{bmatrix}.$$

Block multiplication can be very handy. Proposition 1 gives us more examples:

$$Ax = \begin{bmatrix} a_1 \\ a_2 \\ \vdots \\ a_m \end{bmatrix} x = \begin{bmatrix} a_1 x \\ a_2 x \\ \vdots \\ a_m x \end{bmatrix}, \qquad AB = [AB_1 \quad AB_2 \quad \cdots \quad AB_q]$$

$$AB = \begin{bmatrix} a_1 \\ a_2 \\ \vdots \\ a_m \end{bmatrix} [B_1 \quad B_2 \quad \cdots \quad B_q] = \begin{bmatrix} a_1 B_1 & a_1 B_2 & \cdots & a_1 B_q \\ a_2 B_1 & a_2 B_2 & \cdots & a_2 B_q \\ \cdots & \cdots & \cdots & \cdots \\ a_m B_1 & a_m B_2 & \cdots & a_m B_q \end{bmatrix}$$

$$AB = [A_1 \quad A_2 \quad \cdots \quad A_n] \begin{bmatrix} b_1 \\ b_2 \\ \vdots \\ b_n \end{bmatrix} = \sum_{j=1}^{n} A_j b_j.$$

0.4 Exercises

1. Given the following matrices, find the indicated products, if possible.

$$A = \begin{bmatrix} 1 & 0 & -1 \\ 2 & 1 & 3 \end{bmatrix}, B = \begin{bmatrix} 2 & 5 \\ -2 & 1 \end{bmatrix}, C = \begin{bmatrix} 1 & 2 & 1 \\ -1 & 3 & 2 \\ 3 & 0 & 1 \end{bmatrix}, u = \begin{bmatrix} 2 \\ 4 \\ 1 \end{bmatrix}, v = \begin{bmatrix} 2 \\ -3 \end{bmatrix}, w = \begin{bmatrix} 1 \\ 2 \\ -1 \end{bmatrix}.$$

(a) AB (e) $A^T u$ (i) $u^T v$
(b) BA (f) $A^T B$ (j) $u^T w$
(c) AC (g) Cu (k) uw^T
(d) CA (h) Cw (l) $v^T A$

2. Let A be an $m \times n$ matrix and let B be an $n \times p$ matrix. A *linear combination* of columns $A_1, A_2, \ldots, A_n$ is a sum of the form $\sum_{j=1}^{n} c_j A_j$. Show that every column of AB is a linear combination of columns of A.

3. Let A be an $m \times n$ matrix with columns A_j and rows a_i, and let B be an $n \times q$ matrix with rows b_i and columns B_j. Show that the following are true:
   (a) $a_i = \delta_i^T A$
   (b) The ith row of AB is $a_i B$; i.e., $AB = \begin{bmatrix} a_1 B \\ \vdots \\ a_m B \end{bmatrix}$.
   (c) $AB = \sum_{j=1}^{n} A_j b_j$
   (d) $A = \sum_{i=1}^{n} \delta_i a_i$
   (e) $A = \sum_{j=1}^{n} A_j \delta_j^T$

4. Let A be an $m \times n$ matrix with columns A_j and rows a_i, and let $D = \mathrm{diag}(d_1, \ldots, d_m)$. Show that
   (a) $D = \begin{bmatrix} d_1 \delta_1^T \\ \vdots \\ d_n \delta_n^T \end{bmatrix}$ (b) $DA = \begin{bmatrix} d_1 a_1 \\ \vdots \\ d_m a_m \end{bmatrix}$

5. Let A and B be $n \times n$ matrices. Show that
   (a) $A^H B = [A_i^H B_j]$ (b) $A^H BA = [A_i^H BA_j]$.

6. Let Q be an $n \times n$ matrix, and form P by multiplying the ith column of Q by the real number d_i, $i = 1, 2, \ldots, n$. Put $D = \mathrm{diag}(d_1, d_2, \ldots, d_n)$. Show that
   (a) $PP^H = QD^2 Q^H$ (b) $PP^H = \sum_{i=1}^{n} d_i^2 Q_i Q_i^H$.

7. Let A and E be $n \times n$ matrices, and let $D = \mathrm{diag}(d_1, d_2, \ldots, d_n)$. Show that $AE_i = d_i E_i \; \forall i \iff AE = ED$.

8. Prove that if $A = A^T$, then $y^T Ax = x^T Ay \; \forall x, y \in \mathbf{F}^n$.

9. Let A and B be $n \times n$ matrices. Suppose $y^T Ax = y^T Bx \; \forall x, y \in \mathbf{F}^n$. Show that $A = B$.

10. Prove the rules in Theorem 2.

11. Let P have columns P_j, Q have columns Q_j, and D be equal to $\mathrm{diag}(d_1, d_2, \ldots, d_n)$. Show that $PDQ^T = \sum_{j=1}^{n} d_j P_j Q_j^T$.

12. Find $(ABC)_{ij}$ as a double sum.

13. Let A and B be $m \times n$ matrices such that $Ax = Bx \; \forall x \in \mathbf{F}^n$. Show that $A = B$.

14. Prove Proposition 3.

0.5
SYSTEMS OF EQUATIONS AND MATRIX INVERSES

This section describes matrix inverses and systems $Ax = y$ that involve square matrices.

In this section it will always be assumed that A is a square matrix.

In Chapter 2, we will discuss the existence of matrix inverses for nonsquare matrices. See also exercises 16–23 at the end of this section.

Inverses of Square Matrices

First, matrix inverses are defined and conditions for their existence are studied.

DEFINITION. A matrix B is an *inverse* of A if and only if $AB = I$ and $BA = I$. In this case we write $B = A^{-1}$. A is said to be *invertible,* or *nonsingular,* when it has an inverse.

The requirement that $AB = BA = I$ is imposed since matrix multiplication is not commutative, in general.

EXAMPLE 1. Let

$$A = \begin{bmatrix} \cos(\theta) & -\sin(\theta) \\ \sin(\theta) & \cos(\theta) \end{bmatrix}.$$

Then

$$A^{-1} = \begin{bmatrix} \cos(\theta) & \sin(\theta) \\ -\sin(\theta) & \cos(\theta) \end{bmatrix}.$$

Verify that $AA^{-1} = I$ and $A^{-1}A = I$.

EXAMPLE 2. If $D = \operatorname{diag}(d_1, d_2, \ldots, d_n)$ and all $d_i \neq 0$, then $D^{-1} = \operatorname{diag}(d_1^{-1}, d_2^{-1}, \ldots, d_n^{-1})$. This follows from verifying that $DD^{-1} = D^{-1}D = I$.

From the definition of inverse and from Proposition 3 in Section 0.4, the next proposition immediately follows from the rules for matrix multiplication.

PROPOSITION 1. If A and B are invertible, then so are A^{-1}, A^T, $\overline{A}$, A^H, and AB. Furthermore,

(a) $(A^{-1})^{-1} = A$
(b) $(A^{-1})^T = (A^T)^{-1}$
(c) $\overline{A}^{-1} = \overline{(A^{-1})}$
(d) $(A^{-1})^H = (A^H)^{-1}$
(e) $(AB)^{-1} = B^{-1}A^{-1}$

Proof. To prove (*a*), note that from the symmetry of the requirements $AB = I$ and $BA = I$, both $A = B^{-1}$ and $B = A^{-1}$ hold. Thus $B^{-1} = (A^{-1})^{-1}$. The remainder of the assertions involve verifying the requirements of the definition. To prove (*b*), one has to show that $B \equiv (A^{-1})^{\mathrm{T}}$ satisfies $BA^{\mathrm{T}} = A^{\mathrm{T}}B = I$. But taking the transposes of $AA^{-1} = A^{-1}A = I$ and using the properties of the transpose yields $BA^{\mathrm{T}} = A^{\mathrm{T}}B = I$. Statement (*c*) is proved similarly, and (*d*) follows from (*b*) and (*c*). To prove (*e*), just verify that $(B^{-1}A^{-1})AB = I = AB(B^{-1}A^{-1})$.

The next theorem discusses the issue of uniqueness.

THEOREM 2. If $LA = I$ and $AR = I$, then $L = R = A^{-1}$. Hence the inverse of A is unique, when it exists.

Proof. One has $L = LI = L(AR) = (LA)R = RI = R$.

Matrix Inverses and $Ax = y$

Next, we turn to the connection between matrix inverses and systems of equations.

PROPOSITION 3. If A is invertible, then the system of equations $Ax = y$ has the unique solution $x = A^{-1}y$.

Proof. Suppose $Ax = y$. Then $A^{-1}Ax = Ix = x = A^{-1}y$. This shows that x is unique. On the other hand, $A(A^{-1}y) = Iy = y$, which shows that $A^{-1}y$ is a solution to $Ax = y$.

The next theorem is very important in matrix theory. It asserts the logical equivalency of a number of statements. This means that, if any one statement is true, then all are true, or equivalently, if any one statement is false, then all are false. The proof also contains a method for the construction of A^{-1}.

THEOREM 4. Let A be a square matrix, and let r denote the row rank of A. The following statements are logically equivalent:

(*a*) A is row-reducible to I.
(*b*) The only solution of $Ax = 0$ is $x = 0$.
(*c*) $r = n$.
(*d*) The system $Ax = y$ is consistent for every y.
(*e*) There is a matrix B such that $AB = I$.
(*f*) A has an inverse.

Proof. The scheme of the proof is to show that $(a) \Rightarrow (b) \Rightarrow (c) \Rightarrow (d) \Rightarrow (e) \Rightarrow (f)$.

(*a*) $\Rightarrow$ (*b*): Assume (*a*) is true. Since A row-reduces to I, $Ax = 0$ and $Ix = 0$ have the same solution, namely, $x = 0$.
(*b*) $\Rightarrow$ (*c*): Let R be the row-reduced echelon form of A. If $r < n$, then there would be free variables in the equations $Rx = 0$, and therefore a nonzero solution to $Ax = 0$. Therefore $r = n$ must hold.
(*c*) $\Rightarrow$ (*d*): By Theorem 2 of Section 0.3, $Ax = y$ is solvable for each y since $r = m = n$.
(*d*) $\Rightarrow$ (*e*): For each i, solve $Ax = \delta_i$, and call the solution $x = B_i$. Then, if $B = [B_1 \ B_2 \ \cdots \ B_n]$:

$$AB = [AB_1 \ AB_2 \ \cdots \ AB_n] = [\delta_1 \ \delta_2 \ \cdots \ \delta_n] = I. \tag{1}$$

$(e) \Rightarrow (f)$: In the preceding chain of argument it has been shown that (changing the letters), if a matrix B has the property that $Bx = 0$ has only the solution $x = 0$, then a matrix C exists such that $BC = I$. Now assume $AB = I$. If $Bx = 0$ had a solution $x \neq 0$, then $0 = A0 = ABx = Ix = x$, which is a contradiction. Therefore $Bx = 0$ has only $x = 0$ as a solution, and $BC = I$ for some C. By Theorem 2, $A = C = B^{-1}$, and by Proposition 1, $A^{-1} = B$.

The proof of Theorem 4 contains the following important corollary, which relieves us of half the work of showing that one matrix is the inverse of another.

COROLLARY 5. Let A and B be square matrices. If $AB = I$, then $A^{-1} = B$ and $B^{-1} = A$.

EXAMPLE 3. If A is an upper triangular matrix with all $a_{kk} \neq 0$, then A is nonsingular.

Proof. A is clearly row-reducible to I.

Theorem 4 has many useful applications, as we will see later in this book. We mention in passing a real life application that arises in scientific computing. In discretizations of partial differential equations, whether by finite elements or finite differences, a very large system $Ax = y$ to be solved can arise. It is important to know by theoretical considerations whether A is nonsingular, since numerical methods to solve $Ax = y$ may not be able to reliably detect whether A is singular. When A is singular, a computer may still return a solution that is meaningless. At worst, this solution may look plausible; at best it appears so outlandish that it is clearly not a solution. Plugging the solution back into $Ax = y$ and checking the error may reveal a small error, even though the solution is not good: This is the phenomenon of ill conditioning. Sometimes one can find, from theory or physical intuition, an $x_0 \neq 0$ so that $Ax_0 = 0$. Thus A is singular, and for some y, $Ax = y$ is not solvable. Forearmed, one can use special methods to deal with this situation. On the other hand, it may be possible to prove theoretically that $Ax = 0 \Rightarrow x = 0$, in which case $Ax = y$ is consistent for all y.

Computation of A^{-1}

One useful method for computing A^{-1} has been given in the proof of Theorem 4 just discussed: For $i = 1, \ldots, n$, solve $Ax = \delta_i$ and call the solution $x = R_i$. Then $A^{-1} = [R_1 \ R_2 \ \cdots \ R_n]$. If one of these systems is not solvable or has more than one solution, then the inverse cannot exist, by Theorem 4.

EXAMPLE 4. Find the inverse of $A = \begin{bmatrix} 1 & 1 \\ 2 & 3 \end{bmatrix}$.

Solving $Ax = \begin{bmatrix} 1 \\ 0 \end{bmatrix}$ yields $x = R_1 = \begin{bmatrix} 3 \\ -2 \end{bmatrix}$ and solving $Ax = \begin{bmatrix} 0 \\ 1 \end{bmatrix}$ yields $x = R_2 = \begin{bmatrix} -1 \\ 1 \end{bmatrix}$. Thus $A^{-1} = \begin{bmatrix} 3 & -1 \\ -2 & 1 \end{bmatrix}$.

As mentioned in Section 0.3, to solve $Ax = \delta_i$, it is convenient to form the multiply augmented matrix $[A \mid \delta_1 \ \delta_2 \ \cdots \ \delta_n] = [A \mid I]$ and row reduce the A part of the matrix. If $[I \mid R] = [I \mid R_1 \, R_2 \, \cdots \, R_n]$ is obtained, then the equivalent individual systems are: $Ix = R_i$. For each of these systems, $x = R_i$, and $AR_i = \delta_i$, so that by (1), $AR = I$. This yields the following computational scheme:

Form the $n \times 2n$ augmented array $[A \mid I]$. Perform the Gauss–Jordan algorithm until the augmented array becomes $[I \mid R]$, or until a zero row appears in the left half of the matrix, in which case the inverse does not exist. If I is obtained in the left half, then $R = A^{-1}$.

EXAMPLE 5. Illustrate the preceding computational scheme using the matrix A of Example 4.

$$[A \mid I] = \begin{bmatrix} 1 & 1 & 1 & 0 \\ 2 & 3 & 0 & 1 \end{bmatrix} \rightarrow \begin{bmatrix} 1 & 1 & 1 & 0 \\ 0 & 1 & -2 & 1 \end{bmatrix} \rightarrow \begin{bmatrix} 1 & 0 & 3 & -1 \\ 0 & 1 & -2 & 1 \end{bmatrix}.$$

$$R_2 \leftarrow R_2 - 2R_1 \qquad R_1 \leftarrow R_1 - R_2$$

Thus

$$A^{-1} = \begin{bmatrix} 3 & -1 \\ -2 & 1 \end{bmatrix}.$$

Another method, which is closely related to the preceding one, is described next.

DEFINITION. E is an *elementary matrix* iff E is obtained from I by one of the elementary row operations.

PROPOSITION 6. If B and E are obtained from A and I, respectively, by the same elementary row operation, then $B = EA$. If B is obtained by a finite sequence of elementary row operations on A, and E_i are the corresponding elementary matrices, then $B = E_p \, \cdots \, E_2 \, E_1 A$.

Proof. For the first part, it is easy to verify this for each of the elementary row operations. The second part follows by induction.

THEOREM 7. If A is reducible to I by a sequence of elementary row operations, then the same sequence of elementary row operations applied to the identity matrix yields the inverse of A. If A^{-1} exists, then such a sequence of elementary row operations exists, and can be found by using the Gauss–Jordan algorithm.

Proof. Suppose A is reducible to I by a sequence of elementary row operations. According to Proposition 6, $I = E_p \, \cdots \, E_2 E_1 A = (E_p \cdots E_1)A$. But then by Corollary 5, $A^{-1} = E_p \cdots E_2 E_1 = E_p \cdots E_2 E_1 I$, which says that the same sequence of elementary operations on I yields A^{-1}. The second part is a consequence of Theorem 4, since A must reduce to I by elementary row operations.

0.5 Exercises

1. Find A^{-1} if A is given by

(a) $\begin{bmatrix} 1 & 2 \\ 1 & 1 \end{bmatrix}$
(b) $\begin{bmatrix} 0 & 0 & 1 \\ 0 & 1 & 1 \\ 1 & 1 & 1 \end{bmatrix}$
(c) $\begin{bmatrix} 1 & 1 & 2 \\ 1 & 2 & 1 \\ 1 & 1 & 0 \end{bmatrix}$
(d) $\begin{bmatrix} 1 & 0 & 2 & 3 \\ 0 & 1 & 1 & 2 \\ 0 & 0 & 1 & 3 \\ 0 & 0 & 0 & 1 \end{bmatrix}$

2. Let E be the elementary matrix corresponding to multiplication of the ith row of A by $\alpha \neq 0$. Assume A and E are square.
   (a) Find E^T, E^{-1}.
   (b) Describe in words the effect of the multiplication AE.
   (c) If $B = AE, i = 3, \alpha = 2$, and

   $$B^{-1} = \begin{bmatrix} 2 & 2 & 1 \\ 1 & 1 & 1 \\ 1 & 2 & 1 \end{bmatrix},$$

   find A^{-1} two ways: (i) by computing A first and (ii) by using the results of (a) and (b) and applying them to B^{-1}.

3. Let E be the elementary matrix corresponding to the interchange of rows i and j of A. Assume A and E are square.
   (a) Find E^2, E^{-1}, and E^T.
   (b) Describe in words the effect of the multiplication AE.
   (c) If $B = AE, i = 3, j = 2$, and

   $$B^{-1} = \begin{bmatrix} 1 & 2 & 1 \\ 0 & 1 & 1 \\ 0 & 2 & 1 \end{bmatrix},$$

   find A^{-1} two ways: (i) by computing A first and (ii) using the results of (a) and (b) and applying them to B^{-1}.

4. Let A and B be $n \times n$ matrices, and let $AB = I$. Show that
   (a) $\displaystyle\sum_{j=1}^{n} a_{ij}b_{jk} = \delta_{ik}$ (b) $\displaystyle\sum_{j=1}^{n} a_{jk}b_{ij} = \delta_{ik}$ (c) $\displaystyle\sum_{j=1}^{n} a_{ji}b_{kj} = \delta_{ik}$

5. Let P be a nonsingular matrix with columns P_j, and let P^{-1} have rows Q_i^T. Show that
   (a) $P^{-1}P_j = \delta_j$
   (b) $Q_i^T P_j = \delta_{ij}$
   (c) $\displaystyle I = \sum_{i=1}^{n} P_i Q_i^T$
   (d) If $D = \text{diag}(d_1, \ldots, d_n)$, then $PDP^{-1} = \displaystyle\sum_{i=1}^{n} d_i P_i Q_i^T$.

6. Let Q be a real square matrix with nonzero columns Q_i, and let $Q_i^T Q_j = 0$ if $i \neq j$. Set $Q_i^T Q_i = d_i$. Show that $Q^{-1} = \text{diag}(d_1^{-1}, d_2^{-1}, \ldots, d_n^{-1})Q^T$.

7. Suppose $P^T G P = I$. Show that $P P^T G = I$ and $G P P^T = I$.

8. Suppose $P^T G P = D$, where D is diagonal and nonsingular. Show that $D^{-1}P^T G P = I$.

9. Let U be an upper triangular matrix, i.e., $u_{ij} = 0$ if $i > j$. Show that U^{-1} exists iff $u_{kk} \neq 0 \; \forall k$. Show that U^{-1} is upper triangular. What more can be said if all $u_{kk} = 1$?

10. Suppose that $v_1, v_2, \ldots, v_n$ are given columns of length n and that $V = [v_1 \; v_2 \cdots v_n]$ is nonsingular. Show that, if $Ax = v_i$ is consistent for $i = 1, \ldots, n$, then A is nonsingular.

11. Let A and B be $n \times n$ matrices, and let $C = AB$. Suppose that $Cx = y$ is consistent for every y. Show that A and B are nonsingular.

12. Suppose that $v_1, v_2, \ldots, v_n$ and $w_1, \ldots, w_n$ are given columns of length n and that $V = [v_1 v_2 \cdots v_n]$ has row rank n. Find a matrix A such that $Av_i = w_i \ \forall i$.

13. Let A and Q be $m \times n$ matrices, and let R be an $n \times n$ matrix. Suppose $A = QR$ and $Ax = 0$ has only the solution $x = 0$. Show that
(a) R is nonsingular.
(b) $Qx = 0$ has only the solution $x = 0$.
(c) For each column Q_j of Q, the equation $Ax = Q_j$ is consistent.
(d) If $Ax = Qz$, then $x = R^{-1}z$.

14. Show that every invertible matrix is the product of elementary matrices.

15. The inverse of complex matrices using real arithmetic.
Let $A = X + iY$, where X and Y are real. Define

$$\tilde{A} = \begin{bmatrix} X & -Y \\ Y & X \end{bmatrix}.$$

Suppose

$$\tilde{A}^{-1} = \begin{bmatrix} U & -V \\ V & U \end{bmatrix}.$$

Show that $A^{-1} = U + iV$.

The next sequence of exercises investigates the possibility of inverses for (possibly) nonsquare matrices. In some applications these are important, and examples of such situations will be encountered later. In Chapter 2 these issues will be discussed in a more general setting. Assume in Exercises 16–23 that $A = [A_1 A_2 \cdots A_n]$ is an $m \times n$ matrix.
Definitions: L is a *left inverse* for A iff $LA = I_n$. R is a *right inverse* of A iff $AR = I_m$. Note that L and R must be $n \times m$ matrices.

16. Suppose A has a left inverse L. Show that, if a solution to $Ax = y$ exists, then it is unique.

17. Suppose A has a right inverse R. Show that $Ax = y$ is consistent $\forall y$ and that $x = Ry$ is (one of) the solutions.

18. Suppose $Ax = y$ is consistent $\forall y$.
(a) Let R_i denote one of the solutions of $Ax = \delta_i$, where $i = 1, \ldots, m$. Show that $R = [R_1 \ R_2 \ \cdots \ R_m]$ is a right inverse of A.
(b) Find two right inverses for

$$A = \begin{bmatrix} 1 & 1 & 3 \\ 2 & 1 & 2 \end{bmatrix}.$$

19. Formulate a method for finding left inverses, and state conditions for their existence.

20. Suppose A has a left inverse L. Show that $m \geq n = r$.

21. Suppose A has a right inverse R. Show that $r = m \leq n$.

22. Suppose that A has a left inverse L and a right inverse R. Show that

    (a) $m = n$

    (b) $L = R = A^{-1}$

23. Let A be an $m \times n$ matrix with $m \neq n$. Show that

    (a) If A has one left inverse, then it has infinitely many left inverses and no right inverses.

    (b) If A has one right inverse, then it has infinitely many right inverses and no left inverses.

0.6
FIELDS

An algebraic *field* is a set of entities that obey laws similar to those of the real and complex numbers. Finite fields arise in coding, signal processing, and other areas of application.

> **DEFINITION.** A *field* **F** is a collection of objects together with an operation $+$, called addition, an operation $\cdot$, called multiplication, and a relation $=$, called equality, such that the following axioms hold:
>
> (a) $x + y$, $x \cdot y$ are defined and belong to **F** if $x, y \in$ **F**   (Closure)
>
> (b) $x + y = y + x$ and $x \cdot y = y \cdot x$ $\forall x, y \in$ **F**   (Commutative laws)
>
> (c) $x + (y + z) = (x + y) + z$ and $x \cdot (y \cdot z) = (x \cdot y) \cdot z$
> $\forall x, y, z \in$ **F**   (Associative laws)
>
> (d) $x \cdot (y + z) = x \cdot y + x \cdot z$ $\forall x, y, z \in$ **F**   (Distributive law)
>
> (e) There are $0, 1 \in$ **F** for which
>
> $$0 + x = x, \qquad 1 \cdot x = x \qquad \forall x \in \textbf{F}.$$
>
> (f) For each $x \in$ **F**, there is an element $-x \in$ **F** so that
>
> $$x + (-x) = 0. \qquad \text{(Additive inverse)}$$
>
> (g) For each $x \in$ **F**, $x \neq 0$, there is an element $x^{-1} \in$ **F** so that
>
> $$x \cdot x^{-1} = 1. \qquad \text{(Multiplicative inverse)}$$
>
> (h) For every $x, y, z \in$ **F**, the relation $=$ obeys the following:
>
> $$x = x \qquad\qquad \text{(Reflexive law)}$$
>
> $$x = y \Rightarrow y = x \qquad\qquad \text{(Symmetric law)}$$
>
> $$x = y, y = z \Rightarrow x = z. \qquad\qquad \text{(Transitive law)}$$

Of course it is assumed that $x = x'$, $y = y'$ imply that $x + y = x' + y'$ and $x \cdot y = x' \cdot y'$, so that the operations are compatible with the relation $=$. As usual, one writes $x + (-y) = x - y$ and $xy^{-1} = x/y$. It is easy to show the following familiar algebraic rules from the preceding axioms:

1. $x + y = x + c \Rightarrow y = c$

   $x \cdot y = x \cdot c, x \neq 0 \Rightarrow y = c$

2. $x \cdot 0 = 0$
   $(-x) \cdot y = x \cdot (-y) = -(x \cdot y)$
   $(-x) \cdot (-y) = x \cdot y$

3. $a + x = b$ has the unique solution $x = b - a$.
   $a \cdot x = b, a \neq 0$ has the unique solution $x = b \cdot a^{-1}$.

4. $0, 1$ are unique.

EXAMPLE 1. The following are fields, using the usual sense of $=, +, \cdot$.

(a) The rational numbers
(b) The real numbers
(c) The complex numbers
(d) All real numbers of the form $a + \sqrt{5}b$, with a, b real

A wide variety of fields are important in algebra, number theory, and the theory of equations. These will not be discussed here. Rather, finite fields will be considered. In particular, we will discuss only the fields $\mathbf{Z}_p$.

$\mathbf{Z}$ denotes the set of integers $\mathbf{Z} = \{\ldots, -2, -1, 0, 1, 2, \ldots\}$. Let n denote a positive integer. One writes, for $a, b \in \mathbf{Z}$,

$$a \equiv b \, (\text{mod } n) \Leftrightarrow a = n \cdot k + b \qquad \text{for some } k \in \mathbf{Z} \qquad (1)$$

and says that a is *congruent* to b modulo n. For example, $-3 \equiv 2 \,(\text{mod } 5), 7 \equiv 2 \,(\text{mod } 5), 22 \equiv 2 \,(\text{mod } 5)$.

A little calculation shows that the relation "is congruent to modulo n" is reflexive, symmetric, and transitive. It will be taken as the "sense of $=$" for the finite field under construction.

Now, as is conventional, let $\mathbf{Z}_n = \{0, 1, 2, \ldots, n - 1\}$. Let us use the symbol $\cong$ for equality in $\mathbf{Z}_n$ and define it as $a \cong b$ iff $a \equiv b \,(\text{mod } n)$. We will show that $\mathbf{Z}_n$ satisfies all the axioms for a field, except for the existence of multiplicative inverses, if addition, multiplication, and equality are suitably defined. Let us denote the field operations by $\oplus$ and $\bullet$ to distinguish them (temporarily) from the operations $+$ and $\cdot$ of the integers.

Given $a, b \in \mathbf{Z}_n$, it is possible that $a + b$ and $a \cdot b > n$. However, by the Euclidean (division) algorithm for every integer $c, c = k \cdot n + r$ for some integers k and r, with $0 \leq r < n$. Thus define $a \oplus b \cong c$ in $\mathbf{Z}_n$ iff $a + b = c \,(\text{mod } n)$. The Euclidean algorithm affirms that such a c exists. Define also $a \bullet b \cong c$ in $\mathbf{Z}_n$ iff $a \cdot b = c \,(\text{mod } n)$. It is not hard to see that $\oplus$ and $\bullet$ are compatible with $\cong$ in $\mathbf{Z}_n$.

EXAMPLE 2. AN ILLUSTRATION OF ADDITION AND MULTIPLICATION IN $\mathbf{Z}_3$.

$$1 \oplus 2 \cong 0 \text{ in } \mathbf{Z}_3 \qquad \text{since } 1 + 2 = 3 \equiv 0 \,(\text{mod } 3)$$
$$2 \bullet 2 \cong 1 \text{ in } \mathbf{Z}_3 \qquad \text{since } 2 \cdot 2 = 4 \equiv 1 \,(\text{mod } 3).$$

The verification of Axioms 1 through 4 are straightforward, if a bit tedious. Clearly 0 and 1 work for axiom 5. For $a \in \mathbf{Z}_n$, the additive inverse $-a$ for a is $n - a$ since $a + (n - a) \cong 0$; i.e., $a + (n - a) \cong 0 (\text{mod } n)$. The remaining axiom for fields demands a multiplicative inverse, and this will not exist if n is not a prime.

EXAMPLE 3. Show that 2 has no multiplicative inverse in $\mathbf{Z}_4$, but $2^{-1} \cong 2$ in $\mathbf{Z}_3$. Note that $2 \bullet x \cong 1$ in $\mathbf{Z}_4$ cannot be satisfied by any $x \in \mathbf{Z}_4$; no matter what x is taken, $2 \cdot x = 0$ or 2 (mod 4). On the other hand, if $n = 3$, then $2 \bullet x \cong 1$ in $\mathbf{Z}_3$ has the solution $x = 2$, since $2 \cdot 2 = 4 = 1$ (mod 3), and so $2^{-1} \cong 2$ in $\mathbf{Z}_3$.

In order that $\mathbf{Z}_n$ be a field, n must be taken to be prime: $n = p$. This is established in elementary number theory and is a consequence of the Euclidean algorithm for finding the greatest common divisor of two integers.

In using $\mathbf{F} = \mathbf{Z}_p$ for the scalar field in linear algebra, many arguments go through with no change. Difficulties arise in taking roots or in attempting to use various functions from calculus as if they were defined on $\mathbf{F}$. Another source of problems is illustrated in the next example.

EXAMPLE 4. Let $\mathbf{F} = \mathbf{Z}_2$, and use $+, \cdot,$ and $=$ for addition, scalar multiplication, and equality in $\mathbf{F}$, respectively. If

$$A = \begin{bmatrix} 1 & 1 \\ 0 & 1 \end{bmatrix},$$

then

$$A + A = \begin{bmatrix} 0 & 0 \\ 0 & 0 \end{bmatrix}, \qquad A + A^{\mathsf{T}} = \begin{bmatrix} 0 & 1 \\ 1 & 0 \end{bmatrix} = \begin{bmatrix} 0 & -1 \\ 1 & 0 \end{bmatrix},$$

$$2A = \begin{bmatrix} 2 & 2 \\ 0 & 2 \end{bmatrix} = \begin{bmatrix} 0 & 0 \\ 0 & 0 \end{bmatrix}.$$

CHAPTER 1

Vector Spaces

The concept of vector space pervades mathematics, engineering, and the sciences. The reader is surely familiar with vectors in the form of directed line segments or columns of numbers as studied in calculus and mechanics. The scope of a general notion of vector is much broader than these examples. For example, functions, matrices, and the operation of differentiation will turn out to be vectors. A *vector space* is a set of objects called vectors that obey certain rules. Vector spaces turn up in a wide variety of studies. For example, vector spaces arise naturally in the study of solutions of systems of equations, geometry in 3-space, solutions of differential and integral equations, finite elements, discrete and continuous Fourier transforms, quantum mechanics, approximation theory, and the study of a wide variety of other mathematical and scientific disciplines. Since so many sets of objects can be subsumed under the general notion of vector space, the pursuit of a general theory is highly useful. What is proved in general can be applied to any of the specific vector spaces.

In this book, a general theory of vector space is developed and then applied. The main areas of applications discussed are matrix theory and certain vector spaces of functions, especially those that are useful in both the sciences and mathematics. In this chapter, the notion of vector space is defined, examples are given, and some fundamental aspects of the theory are explored.

1.1
VECTOR SPACES

First the technical definition of vector space is given, and then some of our principal examples will be presented. Let **R** denote the set of real numbers and **C** the set of complex numbers. In the following definition, **F** represents **R**, **C**, or, for that matter, any field.

DEFINITION. A *vector space over* **F** is a nonempty set V of objects, called vectors, together with a relation $=$ between these objects, called equality of vectors, an operation $+$, called vector addition, and an operation $\cdot$ between complex numbers and vectors, called scalar multiplication, for which the following axioms hold for all x, y, z in V and for all α, β in **F**:

Axioms governing equality (e)

$(e1)$ Reflexive $x = x$
$(e2)$ Symmetry $x = y \Rightarrow y = x$
$(e3)$ Transitive $x = y$ and $y = z \Rightarrow x = z$

Axioms governing addition (a)

$(a1)$ Closure $x + y$ is in V
$(a2)$ Commutative $x + y = y + x$
$(a3)$ Associative $x + (y + z) = (x + y) + z$
$(a4)$ There is a vector 0, called the *zero vector,* such that $x + 0 = x$.
$(a5)$ There is a vector $-x$, called the *negative of* x, such that $x + (-x) = 0$.

Axioms governing scalar multiplication (s)

$(s1)$ Closure $\alpha \cdot x$ is in V
$(s2)$ Consistency $1 \cdot x = x$
$(s3)$ Associative $\alpha \cdot (\beta \cdot x) = (\alpha\beta) \cdot x$
$(s4)$ Distributive $\alpha \cdot (x + y) = \alpha \cdot x + \alpha \cdot y$
$(s5)$ Distributive $(\alpha + \beta) \cdot x = \alpha \cdot x + \beta \cdot x$

If $\mathbf{F} = \mathbf{C}$, the phrase "V is a vector space over $\mathbf{C}$" is often replaced by the language "V is a *complex vector space,*" with a similar modification for real vector spaces. In any case, the numbers in **F** are called *scalars,* and **F** is called the *scalar field.*

Much of the theory is the same for real or complex scalars and for more general fields, so all these cases are treated simultaneously unless they differ. When a statement is valid for all scalar fields, either the scalars are not specified, or **F** is used to denote a general field.

Notice that the symbols "$+$" and "0" have been used in two different senses in the axioms. The symbol "$+$" denotes both vector and scalar addition, which are usually different processes. For example, in $(s5)$, $\alpha + \beta$ denotes scalar addition, while $\alpha \cdot x + \beta \cdot x$ denotes addition of the vectors $\alpha \cdot x$ and $\beta \cdot x$. The symbol 0 is used to denote the number zero and the zero vector, which are usually different entities. One also writes $\alpha \cdot x = \alpha x$, suppressing the scalar multiplication operand. Usually the brackets are dropped in statements like $(a3)$, and $x + (-y)$ is written $x - y$. These notational agreements are conventional, and usually cause no confusion. Clearly $(a3)$, $(s3)$, $(s4)$, and $(s5)$ extend to any finite number of terms. Sometimes the negative of x is called the *additive inverse* of x. Equality in the preceding axioms ordinarily means "the same as" in the usual sense. Other such useful situations exist where equality has a different meaning (and is then often called an equivalence relation), and this has been permitted in the general definition. Two examples of this different sense of equality are given in this section. It is also tacitly assumed that equality and the vector space operations are compatible, i.e., that $x = x', y = y' \Rightarrow x + y = x' + y'$, and $\alpha x = \alpha x'$.

A word on notation: Various contexts of science and mathematics have their own notations, and it is not feasible to force one notation on all of these contexts. For example, a matrix, which is a rectangular array of scalars, is denoted in various disciplines by A, $\mathbf{A}$, $[A]$, $\underline{A}$ or $\underline{\underline{A}}$. A column of scalars may be denoted by x, $\mathbf{x}$, X, $\underline{x}$ or $\{x\}$. Since it is worthwhile to be able to judge from the context which entities are vectors and which are scalars, this book does not adopt a special notation for either, except in the first section of this chapter, where Greek letters are used for scalars and Roman letters for vectors.

Some immediate and useful consequences of the axioms are given in the following proposition. In most of the vector spaces encountered, these statements are obvious.

PROPOSITION 1

(a) $0 \cdot x = 0$ for all x in V.
(b) $(-1) \cdot x = -x$ for all x in V.
(c) $\alpha \cdot 0 = 0$ for all scalars α.
(d) The zero vector in V is unique.
(e) For each x in V, the negative of x is unique.
(f) For x, y in V, $z = y - x$ is the unique solution to $x + z = y$.

Proof. Statement (a) is shown and the rest are left as exercises. By properties of the scalars and by (s5),

$$0 \cdot x = (0 + 0) \cdot x = 0 \cdot x + 0 \cdot x.$$

Adding $-(0 \cdot x)$ to each side and using (a5), (a3), (a4), and (e3), one obtains

$$0 = 0 \cdot x - (0 \cdot x) = (0 \cdot x + 0 \cdot x) - (0 \cdot x)$$
$$= 0 \cdot x + [0 \cdot x - (0 \cdot x)] = 0 \cdot x + 0 = 0 \cdot x.$$

EXAMPLE 1. THE VECTOR SPACE OF "ARROWS" IN 3-SPACE. This vector space is often studied in calculus, mechanics, and physics. V is the collection of directed line segments in 3-space, and the scalars are real. Equality means that the two directed line segments have the same direction and length, i.e., that they are rigid translates of one another. Addition is performed by the usual parallelogram rule. Scalar multiplication is performed by the usual dilation and, if the scalar is negative, by a reversal. It is important that equality *not* mean "the same as," since the vectors are to be interpreted as free vectors, i.e., not attached to a fixed point. Verification of the axioms is straightforward (but tedious) plane geometry.

EXAMPLE 2. $\mathbf{C}^n$, THE SET OF ALL COMPLEX (COLUMN) n-TUPLES. The notation is as follows:

$$x = (x_1, x_2, \ldots, x_n)^{\mathrm{T}} \equiv \begin{bmatrix} x_1 \\ x_2 \\ \vdots \\ x_n \end{bmatrix} \qquad \text{with } x_i \text{ in } \mathbf{C}.$$

Here x_i is called the *ith component* of the column x. The superscript T means *transpose*, i.e., convert the row to a column. Similarly, define $x^{\mathrm{T}} = (x_1, x_2, \ldots, x_n)$ so that transposition also converts columns to rows. The relation $=$ and the operations $+$ and $\cdot$ are defined entrywise, i.e., entry by entry:

$$x = y \Leftrightarrow x_i = y_i \text{ for } i = 1, \ldots, n$$
$$x + y = (x_1 + y_1, x_2 + y_2, \ldots, x_n + y_n)^T$$
$$\alpha \cdot x = (\alpha x_1, \alpha x_2, \ldots, \alpha x_n)^T.$$

It is an easy matter to verify the axioms for a complex vector space, since closure is clear and the algebraic laws hold in each entry. Clearly, $0 = (0, 0, \ldots, 0)^T$ and $-x = (-x_1, -x_2, \ldots, -x_n)^T$.

EXAMPLE 3. F^n, THE SET OF COLUMN n-TUPLES WITH ENTRIES IN F. This is defined exactly as in Example 2, except that the entries and scalars are in **F**.

EXAMPLE 4. $\mathbf{C}^\infty$ is the set of all complex infinite sequences $x = (x_k)_{k=1}^\infty = (x_1, x_2, \ldots, x_n, \ldots)$. $\mathbf{R}^\infty$ is the set of infinite sequences with real terms. Equality, addition, and scalar multiplication are defined entrywise:

$$x = y \Leftrightarrow x_i = y_i \quad \text{for } i = 1, 2, \ldots$$
$$x + y = (x_1 + y_1, x_2 + y_2, \ldots, x_n + y_n, \ldots)$$
$$\alpha, \ldots, x = (\alpha x_1, \alpha x_2, \ldots, \alpha x_n, \ldots).$$

Clearly, $0 = (0, 0, \ldots)$, $-x = (-x_1, -x_2, \ldots)$.

EXAMPLE 5. THE SET $R^{m,n}$ OF ALL $m \times n$ MATRICES WITH REAL ENTRIES. An $m \times n$ matrix is an an array with m rows and n columns:

$$A = [a_{ij}] = \begin{bmatrix} a_{11} & a_{12} & \cdots & \cdots & a_{1n} \\ a_{21} & a_{22} & \cdots & \cdots & a_{2n} \\ \cdots & \cdots & \cdots & \cdots & \cdots \\ \cdots & \cdots & \cdots & \cdots & \cdots \\ a_{m1} & a_{m2} & \cdots & \cdots & a_{mn} \end{bmatrix},$$

where a_{ij} is the entry in the ith row and jth column. $\mathbf{F}^{m,n}$ is the set of all $m \times n$ matrices with entries from **F**. Equality, addition, and scalar multiplication are defined entrywise:

$$[a_{ij}] = [b_{ij}] \Leftrightarrow a_{ij} = b_{ij} \quad \text{for all } i, j$$
$$[a_{ij}] + [b_{ij}] = [a_{ij} + b_{ij}]$$
$$\alpha[a_{ij}] = [\alpha a_{ij}].$$

EXAMPLE 6. $\mathcal{F}_c(S)$, THE SET OF ALL COMPLEX VALUED FUNCTIONS ON A GIVEN SET S. f denotes the function or rule f, and $f(x)$ denotes the value of the function f at x. (See the remarks after Example 7.) By definition,

$$f = g \Leftrightarrow f(x) = g(x) \quad \forall x \in S$$
$$f + g \text{ is the function with values } (f + g)(x) = f(x) + g(x) \quad \forall x \in S$$
$$\alpha \cdot f \text{ is the function with values } (\alpha f)(x) = \alpha f(x) \quad \forall x \in S.$$

These are the usual pointwise operations on functions used in calculus. It is easy to verify the axioms for a complex vector space. 0 denotes the function with values $0(x) = 0$, and $-f$ has values $(-f)(x) = -f(x)$.

EXAMPLE 7. $\mathcal{F}_r(S)$, THE SET OF ALL REAL VALUED FUNCTIONS ON AN ARBITRARY SET S. The definitions are the same as in Example 4, except that the scalars are real.

Remarks. When the scalars are unimportant or the context is clear, $\mathcal{F}(S)$ is used to denote either of the preceding vector spaces of functions. A function f that is defined on a set larger than S is considered to belong to $\mathcal{F}(S)$, but the values $f(x)$ for $x \notin S$ are not considered. For example, when we consider the sine function as a vector in $\mathcal{F}([0, 1])$, only the values $\sin(x)$ with $0 \le x \le 1$ are considered, even though the sine function is defined for all complex numbers.

Remarks. Some comments concerning functional notation follow. When one is speaking, say, of t^3 or $\sin(2t)$ as functions, then t is regarded as a dummy variable or placeholder and not a specific number. In this context, the statement "the functions $t^3, \sin(2t) \ldots$" is the same as the statement "the functions $x^3, \sin(2x) \ldots$" wherein x is the placeholder. This is conventional usage and much shorter than being rigorous and saying "the functions ϕ and ζ defined by $\phi(t) = t^3, t \in \mathbf{R}$ and $\zeta(t) = \sin(2t)$, $t \in \mathbf{R}$." When the values of functions are being discussed, then t^3 and x^3 mean different things, if $x \ne t$. Since there are no conventional names for many functions (such as in the examples just given), the reader is expected to judge from the context whether functions or their values are being discussed.

EXAMPLE 8. Let V be the set of all discontinuous functions on the real line, with the usual operations. Determine if V is a vector space.

V is not a vector space for several reasons. For example, V is not closed under addition, since the sum of two discontinuous functions can be continuous: e.g., if u is a step function, then $u + (-u) = 0$, which is continuous. V is also not closed under scalar multiplication: $0 \cdot f = 0$ is continuous.

Sometimes when one works with functions, equality at every point is not crucial, and inequality at a finite number of points can be tolerated. For example, when one works with integrable functions and their expansions into Fourier series, two functions that agree at all but a finite number of points in $[0, 2\pi]$ will have the same Fourier series, and one may wish to regard them as the same function. Thus in Example 4, the definition of $f = g$ would be changed to mean "$f(x) = g(x)$ at all but a finite number of points." The definitions of vector addition and scalar multiplication remain the same. The axioms of a vector space still hold.

Exercises 1.1

1. In each of the following, V, vector addition, and scalar multiplication are defined. Equality means "the same as" if undefined. Determine whether or not a vector space results. If so, exhibit the zero vector and $-x$. If not, show which axioms are violated.

   (a) Let V be the set of all real numbers greater than 0. Take the scalars to be real. Define vector addition by $x \oplus y = xy$, and define scalar multiplication by $\alpha \otimes x = x^\alpha$.

   (b) V is the set of all pairs (x, y) of real numbers. Take the scalars to be real. Vector addition and scalar multiplication are defined by $(x, y) + (u, v) = (x + u, y + v)$ and $\alpha \cdot (x, y) = (\alpha x, 3\alpha y)$.

   (c) V is the set of all divergent sequences $(a_n)_{n=1}^\infty$ with complex entries. Use the same operations as for $\mathbf{C}^\infty$ in Example 4.

(d) V is the set of all functions of the form $p = p_0 + p_1 t + p_2 t^2$. $p = q$ means $p_i = q_i$, where $i = 0, 1, 2$. Addition and scalar multiplication are defined by $p + q = p_0 + q_0 + (p_1 + q_1)t + p_1 q_1 t^2$, $\alpha p = \alpha p_0 + (\alpha p_1)t + (\alpha p_2)t^2$, with $\alpha \in \mathbf{R}$.

2. Prove the remaining statements of Proposition 1:
   (b) $(-1) \cdot x = -x$ for all x in V.
   (c) $\alpha \cdot 0 = 0$ for all scalars α.
   (d) The zero vector in V is unique. If also $\theta + x = x$ $\forall x$, then $\theta = 0$.
   (e) For each x in V, the negative of x is unique.
   (f) For x, y in V, $z = y - x$ is the unique solution to $x + z = y$.

1.2
SUBSPACES

Often one wants to consider a more restricted set of vectors within a larger vector space. For example, we may desire to consider only continuous functions on S rather than the collection of all functions on S. Or we may wish to consider the plane in 3-space that passes through the origin and is perpendicular to a given line. In other situations, we may want to generate a certain subset of vectors in a larger vector space. For certain problems, it becomes important to know if the smaller set is again a vector space with the same sense of equality and with the same operations of addition and scalar multiplication. In this section, criteria for deciding this issue are presented.

DEFINITION. M is a *subspace* of V iff M is a nonempty subset of V and if M is a vector space using the same scalars, sense of equality, vector addition, and scalar multiplication as in V.

The following theorem says that after a large set has been shown to be a vector space, it is very easy to determine whether or not a subset of that set is also a vector space. It avoids the work of checking the long list of axioms.

THEOREM 1. Let M be a nonempty subset of a vector space V. The following statements are equivalent:

   (i) M is a subspace of V.
   (ii) M is closed under addition and scalar multiplication.
   (iii) If x, y are in M and if α, β are scalars, then $\alpha x + \beta y$ is in M.

Proof. We show that $(i) \Rightarrow (ii) \Rightarrow (iii) \Rightarrow (i)$. Suppose (i) holds. Since M is a subspace, it is also a vector space, and the axioms hold; hence closure immediately follows. Next, assume (ii) holds. If $x, y \in M$ and α, β are scalars, we know that $\alpha x, \beta y$ are in M, and then so is $\alpha x + \beta y$, and therefore (iii) is true. Next, suppose (iii) holds. Then $0 = 0x \in M$ and $-x = (-1)x \in M$ by (iii) and by Proposition 1 in Section 1.1. Closure holds, since $x + y = 1x + 1y \in M$ and since $\alpha x = \alpha x + 0y \in M$ when x and y are in M and α, β are scalars. All the axioms that assert algebraic identities hold because the vectors are in V, and these axioms hold for V. Thus (iii) implies that M is a vector space and therefore a subspace of V.

Remark. If a subset M is to be a subspace, the preceding proof shows that M must contain the zero vector of the parent space. Verifying this also shows that M is nonempty. Thus it is wise to check $0 \in M$ as a first step in testing whether or not a specific set M is a subspace.

EXAMPLE 1. Let M consist of all $x = (x_1, x_2, x_3)^T$ for which $x_1 + x_2 + x_3 = 0$. Show that M is a subspace of $\mathbf{R}^3$.

Since $0 = (0, 0, 0)^T$ clearly satisfies the given relation, M is nonempty. If x and y satisfy the given relation, and if $z = \alpha x + \beta y$, then

$$z_1 + z_2 + z_3 = (\alpha x_1 + \beta y_1) + (\alpha x_2 + \beta y_2) + (\alpha x_3 + \beta y_3)$$
$$= \alpha(x_1 + x_2 + x_3) + \beta(y_1 + y_2 + y_3) = \alpha 0 + \beta 0 = 0.$$

Hence M is a subspace.

EXAMPLE 2. Let M consist of all $(x_1, x_2, x_3)^T$ for which $x_1 + x_2 + x_3 = 1$. Show that M is not a subspace of $\mathbf{R}^3$.

In fact, $x = 0$ does not obey $0 + 0 + 0 = 1$, so $0 \notin M$, and M cannot be a subspace. Closure under addition fails as well, since $x, y \in M \Rightarrow (x_1 + y_1) + (x_2 + y_2) + (x_3 + y_3) = 2$. It is easy to see that closure under scalar multiplication also fails.

In the following examples, some commonly used vector spaces of functions are given.

EXAMPLE 3. The set $C(S)$ of all continuous functions on a set S is a vector space.

Since $C(S)$ is a nonempty subset of the vector space of all of the functions on S, and since, from calculus, $\alpha f + \beta g$ is continuous when f and g are continuous, $C(S)$ is a vector space.

EXAMPLE 4. The set $C^n(S)$ of all functions whose derivatives of order n are continuous on the set S is a vector space.

Here $C^n(S) \subset \mathcal{F}(S)$. $C^n(S)$ is a subspace, since it is nonempty and since, from calculus, the kth derivative satisfies $(\alpha f + \beta g)^{(k)} = \alpha f^{(k)} + \beta g^{(k)}$ when $f, g \in C^n(S)$ and when α, β are scalars. Moreover, $\alpha f^{(k)} + \beta g^{(k)}$ is continuous when $f^{(k)}, g^{(k)}$ are continuous. In this example, S is assumed to be an interval on the real line or a region in the plane. A similar example is valid for functions of several variables defined on a region S in $\mathbf{R}^n$. In this case, partial derivatives are involved.

EXAMPLE 5. THE SOLUTION SPACE OF A HOMOGENEOUS DIFFERENTIAL EQUATION. Let $L = \sum_{i=0}^{n} a_i D^i$, where $D = d/dx$ and each a_i is a given scalar. Let M be the *solution space* of L; that is, $f \in M \Leftrightarrow f \in C^n(\mathbf{R})$, and f satisfies the differential equation $L(f) = 0$. Show that M is a vector space.

M is a subset of $C^n(\mathbf{R})$. By definition, $L(f) = \sum_{i=0}^{n} a_i f^{(i)}$. Clearly $L(0) = 0$, so $0 \in M$. If $Lf = Lg = 0$, then

$$L(\alpha f + \beta g) = \sum_{i=0}^{n} a_i D^i(\alpha f + \beta g) = \sum_{i=0}^{n} a_i(\alpha f^{(i)} + \beta g^{(i)})$$

$$= \alpha \sum_{i=0}^{n} a_i \alpha f^{(i)} + \beta \sum_{i=0}^{n} a_i \alpha f^{(i)} = \alpha 0 + \beta 0 = 0.$$

Hence $\alpha f + \beta g \in M$ so that M is a subspace of $C^n(\mathbf{R})$.

EXAMPLE 6. THE *SOLUTION SPACE N(A)* OF A MATRIX. Let A be an $m \times n$ matrix. Let $N(A) = \{x : Ax = 0\}$; i.e., $N(A)$ consists of all solutions of the homogeneous system $Ax = 0$. Show that $N(A)$ is a subspace of $\mathbf{F}^n$.

Since $A0 = 0$, $N(A)$ is nonempty. If $Ax = Ay = 0$, then $A(\alpha x + \beta y) = \alpha Ax + \beta Ay = \alpha 0 + \beta 0 = 0$. Thus $\alpha x + \beta y \in M$, and so $N(A)$ is a subspace of $\mathbf{F}^n$.

The next example is a subspace of functions used in (very) elementary finite elements and for approximations of functions.

EXAMPLE 7. Let $x_0 < x_1 < \cdots < x_n$ be given real numbers. Define $V = CPL(\{x_0, x_1, \ldots, x_n\})$ to be the set of continuous functions on $[x_0, x_n]$ that are linear in the intervals $I_i = [x_i, x_{i+1}]$. $\{x_0, x_1, \ldots, x_n\}$ is a fixed finite set of points, and CPL stands for continuous piecewise linear.

V is a subset of the vector space $C([x_0, x_n])$. Clearly, the scalar multiple of a function that is linear on I_i is again a linear function on this interval, and the sum of two linear functions on I_i is again a linear function on I_i. Thus V is closed under addition and scalar multiplication, and so V is a subspace of $C([x_0, x_n])$ and is therefore a vector space.

In many problems, one encounters sums of the form $\sum_{k=1}^{n} \alpha_k e_k$, where the α_k are scalars and e_k are vectors. For example, polynomials are sums of this form: $p(t) = \sum_{k=0}^{n} \alpha_k t^k$, with $e_k = t^k$. The importance of such sums prompts the next definitions.

DEFINITION. A *linear combination* of vectors $e_1, \ldots, e_n$ is a sum of the form $\sum_{k=1}^{n} \alpha_k e_k = \alpha_1 e_1 + \alpha_2 e_2 + \cdots + \alpha_n e_n$ where the α_i are scalars. x is a linear combination of $e_1, \ldots, e_n$ if and only if $x = \sum_{k=1}^{n} \alpha_k e_k$ for some scalars α_i.

Linear combinations of vectors always lie in the vector space from which they were taken, by the closure axioms. Note that linear combinations are always finite sums, by definition. If infinite sums were allowed, convergence questions would arise, which could be difficult. The collection of all possible such linear combinations of a fixed set of vectors is often useful and turns out to be a method of constructing subspaces.

DEFINITION. Let $\{e_k\}_{k=1}^{n} \equiv \{e_1, \ldots, e_n\}$ be a given set of vectors in a vector space V. The *span* of $\{e_k\}_{k=1}^{n}$ is the set of all linear combinations of $e_1, \ldots, e_n$, and is denoted span$\{e_k\}_{k=1}^{n}$. According to the next proposition, span$\{e_k\}_{k=1}^{n}$ is a vector space, so span$\{e_k\}_{k=1}^{n}$ is called the *vector space spanned by the set* $\{e_k\}_{k=1}^{n}$.

PROPOSITION 2. If $e_i \in V$, for $i = 1, \ldots, n$ then span$\{e_k\}_{k=1}^{n}$ is a subspace of V.

Proof. Clearly, span$\{e_k\}_{k=1}^{n}$ is nonempty. Let $x = \sum_{i=1}^{n} \alpha_i e_i$ and $y = \sum_{i=1}^{n} \beta_i e_i$. Since $\gamma x + \delta y = \gamma \sum_{i=1}^{n} \alpha_i e_i + \delta \sum_{i=1}^{n} \beta_i e_i \sum_{i=1}^{n} (\gamma \alpha_i + \delta \beta_i) e_i$, span$\{e_k\}_{k=1}^{n}$ is a subspace of V.

The next two examples are commonly used sets of functions.

EXAMPLE 8. P_n, THE SET OF *POLYNOMIALS OF DEGREE NOT EXCEEDING n*.

$$\mathbf{P}_n = \text{span}\{1, t, \ldots, t^n\} = \left\{ \sum_{k=0}^{n} c_k t^k : c_k \in \mathbf{C}, k = 0, 1, \ldots, n \right\}.$$

EXAMPLE 9. T_n, THE SET OF *COMPLEX TRIGONOMETRIC POLYNOMIALS OF DEGREE NOT EXCEEDING n*.

$$\mathbf{T}_n = \text{span}\{e^{ikx}\}_{k=-n}^{n} = \left\{ \sum_{k=-n}^{n} c_k e^{ikx} : c_{-n}, \ldots, c_0, \ldots, c_n \in \mathbf{C} \right\}, \qquad \text{where } i = \sqrt{-1}.$$

These functions are important in the analysis of periodic functions, and they arise in the fast Fourier transform and in the study of complex Fourier series.

Expansions and Spanning Sets

A basic problem of linear algebra is as follows: Given a vector f and a set $\{e_k\}_{k=1}^n$ of vectors, is it possible to expand f as a linear combination

$$f = \sum_{i=1}^n \alpha_i e_i?$$

Or, more succinctly, is $f \in \text{span}\{e_k\}_{k=1}^n$? For example, do coefficients α_i exist so that

$$\cos^n(t) = \sum_{i=1}^n \alpha_k \cos(kt)$$

for all real t? Given columns y and $A_1, \ldots, A_n$ in $\mathbf{F}^m$, do scalars x_j exist so that

$$y = \sum_{i=1}^n x_j A_j?$$

A related question of much importance is as follows: Given a set $\{e_i\}_{i=1}^n$ in V, is *every* f in V a linear combination of the e_i? That is, is $V = \text{span}\{e_i\}_{i=1}^n$? Such a subset is called a *spanning set for V.* For example, is every vector in $\mathbf{C}^n$ a linear combination of $E_1 = (1, 0, \ldots, 0)^T, E_2 = (1, 1, 0, \ldots, 0)^T, \ldots, E_n = (1, 1, \ldots, 1)^T$?

In general, deciding whether a given vector is in the span of a set of vectors and if $V = \text{span}\{e_i\}_1^n$ are not easy tasks. Ingenuity, experience, or knowledge may be required. Some of the theory developed in the next sections will make these tasks much easier.

DEFINITION. $\delta_j \in \mathbf{F}^n$ is defined by $(\delta_j)_i = \delta_{ij} = \begin{cases} 0 & i \neq j \\ 1 & i = j \end{cases}$. δ_{ij} is called the *Kronecker delta.* Thus δ_j has zero entries except for the ith row, in which it has the entry 1.

EXAMPLE 10. $\mathbf{C}^n = \text{span}\{\delta_1, \ldots, \delta_n\}$. This follows immediately from

$$(x_1, x_2, \ldots, x_n)^T = x_1 \delta_1 + x_2 \delta_2 + \cdots + x_n \delta_n.$$

The next example is extremely valuable in connecting the theory of vector spaces with the theory of systems of linear equations, as we will see later in this chapter and in Chapter 2. It permits useful insight into many of the applications later in this book.

EXAMPLE 11. THE *RANGE* AND *COLUMN SPACE* OF A MATRIX. Let A be an $m \times n$ matrix with columns A_j: $A = [A_1 A_2 \cdots A_n]$, and let $R(A)$ be the *range* (i.e., the set of outputs) of A: $R(A) = \{y : Ax = y \text{ for at least one } x\}$. Let M be the *column space* of A: $M = \text{span}\{A_k\}_{k=1}^n$. Show that $R(A) = M$ and that $R(A)$ is a subspace of $\mathbf{F}^m$.

Let $A_j = (a_{1j}, a_{2j}, \ldots, a_{mj})^T$; i.e., let $(A_j)_i = a_{ij}$. Then (cf. Sections 0.1 and 0.4)

$$\sum_{j=1}^n x_j A_j = x_1 \begin{bmatrix} a_{11} \\ a_{21} \\ \vdots \\ a_{m1} \end{bmatrix} + x_2 \begin{bmatrix} a_{12} \\ a_{22} \\ \vdots \\ a_{m2} \end{bmatrix} + \cdots + x_n \begin{bmatrix} a_{1n} \\ a_{2n} \\ \vdots \\ a_{mn} \end{bmatrix} = \begin{bmatrix} \sum_{j=1}^n a_{1j} x_j \\ \sum_{j=1}^n a_{2j} x_j \\ \vdots \\ \sum_{j=1}^n a_{mj} x_j \end{bmatrix} = Ax. \qquad (1)$$

Thus $y \in R(A)$ iff $y \in \text{span}\{A_k\}_{k=1}^n$. By Proposition 2, M is a subspace.

Remark. Due to Example 11, $y \in \text{span}\{A_k\}_{k=1}^n$ iff $Ax = y$ is consistent.

EXAMPLE 12. Determine whether or not $y = (1, 6, -1)^T$ and $z = (1, 6, 1)^T$ are in the span of $(1, 2, 1)^T$, $(3, 8, 2)^T$, and $(2, 6, 1)^T$. Equivalently, attempt to solve $Ax = y$ and $Ax = z$, where

$$A = \begin{bmatrix} 1 & 3 & 2 \\ 2 & 8 & 6 \\ 1 & 2 & 1 \end{bmatrix}.$$

Row-reducing the augmented matrix $[A \mid y]$ yields

$$\tilde{A} = \begin{bmatrix} 1 & 3 & 2 & 1 \\ 2 & 8 & 6 & 6 \\ 1 & 2 & 1 & -1 \end{bmatrix} \rightarrow \begin{bmatrix} 1 & 3 & 2 & 1 \\ 0 & 2 & 2 & 4 \\ 0 & -1 & -1 & -2 \end{bmatrix} \rightarrow$$

$$R_2 \leftarrow R_2 - 2R_1 \qquad\qquad R2 \leftarrow \tfrac{1}{2}R_2$$

$$R_3 \leftarrow R_3 - R_1$$

$$\begin{bmatrix} 1 & 3 & 2 & 1 \\ 0 & 1 & 1 & 2 \\ 0 & -1 & -1 & -2 \end{bmatrix} \rightarrow \begin{bmatrix} 1 & 0 & -1 & -5 \\ 0 & 1 & 1 & 2 \\ 0 & 0 & 0 & 0 \end{bmatrix}.$$

$$R_3 \leftarrow R_3 + R_2$$

$$R_1 \leftarrow R_1 - 3R_2.$$

The equivalent system is therefore $x_1 - x_3 = -5$, $x_2 + x_3 = 2$, which has many solutions, so that $y \in \text{span}\{A_1, A_2, A_3\}$.

Row-reducing $[A \mid z]$ yields

$$\begin{bmatrix} 1 & 0 & -1 & -5 \\ 0 & 1 & 1 & 1 \\ 0 & 0 & 0 & 2 \end{bmatrix}.$$

Since the last column has a nonzero entry in the third row, $Ax = z$ is inconsistent and $y \notin \text{span}\{A_1, A_2, A_3\}$.

EXAMPLE 13. Show that $\mathbf{P}_2 = \text{span}\{1, 1+t, 1+t+t^2\}$. The goal is to write an arbitrary $p \in \mathbf{P}_2$ as a linear combination of the other functions. Let $p = a + bt + ct^2$. To obtain expansion, write down a linear combination and rearrange it according to powers of t:

$$a + bt + ct^2 = c_1 \cdot 1 + c_2(1 + t) + c_3(1 + t + t^2) \tag{$*$}$$
$$= (c_1 + c_2 + c_3) + (c_2 + c_3)t + c_3 t^2.$$

We see that it is sufficient to require that $c_3 = c$, $c_2 + c_3 = b$, $c_1 + c_2 + c_3 = a$. But this yields $c_2 = b - c$ and $c_1 = a - c_2 - c_3 = a - b - 2c$. Thus $p = (a - b - 2c)1 + (b - c)(1 + t) + c(1 + t + t^2)$.

EXAMPLE 14. Let $e_k(t) = \cos(kt)$, $k = 0, \ldots, n$ and $f(t) = \cos^n(t)$. Show that $f \in \text{span}\{e_i\}_{i=0}^n$.

Using the binomial theorem and Euler's identities, we have

$$\cos^n(t) = [(e^{it} + e^{-it})/2]^n = 2^{-n}\sum_{k=0}^{n}\binom{n}{k}e^{ikt}e^{-i(n-k)t} = 2^{-n}\sum_{k=0}^{n}\binom{n}{k}e^{i(2k-n)t}$$

$$= 2^{-n}\sum_{k=0}^{n}\binom{n}{k}\cos(2k-n)t = 2^{-n}\sum_{k=0}^{n}\binom{n}{k}\cos(|2k - n|t).$$

The fact that the imaginary part of the sum must be 0 was used in line 2, and the fact that cosine is an even function was used in the last line. Since $|2k - n|$ is one of the values $0, 1, \ldots, n$, if $k = 0, 1, \ldots, n$, it follows that $\cos^n(t)$ is a linear combination of $1, \cos(t), \ldots, \cos(nt)$.

EXAMPLE 15. $f(t) \equiv t$ is not in the span of $e_k(t) = \exp(-kt)$, $k = 1, \ldots, n$, where $t \in \mathbf{R}$. If it were so, then there would be some α_k so that

$$t = \sum_{k=1}^{n} \alpha_k \exp(-kt) \qquad \text{for all } t \in \mathbf{R}.$$

If $t \to \infty$, the right side tends to zero, whereas the left side tends to ∞, which is a contradiction.

Exercises 1.2

1. In each case, determine whether or not the given set is a vector space. Explain.
   (a) The x_2 axis in $\mathbf{R}^3$
   (b) All functions f such that $\lim_{x \to \infty} f(x) = \pi$
   (c) All x in $\mathbf{R}^3$ such that $x_1 - 2x_2 + x_3 = 0$
   (d) All x in $\mathbf{R}^3$ such that $x_1 - 2x_2 + x_3 = 2$
   (e) All x in $\mathbf{C}^n$ such that $\sum_{k=1}^{n} a_k x_k = 0$, where $a \in \mathbf{C}^n$ is given
   (f) All 2×2 matrices A such that $a_{11} + a_{22} = 0$
   (g) All x in $\mathbf{R}^n$ such that $\sum_{i=1}^{n} x_i^2 = 1$
   (h) All continuous functions f such that $\int_0^1 e^x f(x)\,dx = 0$
   (i) All functions defined on $\mathbf{R}$ such that $f(2) = 0$
   (j) All functions bounded on $(0, \infty)$
   (k) All functions f such that $Lg = f$ for some $g \in C^n(\mathbf{R})$, where $L = \sum_{i=1}^{n} a_i D^i$ and $D = d/dx$
   (l) All functions F of the form $F(s) = \int_0^\infty e^{-st} f(t)\,dt$, for some function f continuous and bounded on $[0, \infty)$
   (m) All $y = (y_1, y_2)^T$ for which there exists an $x = (x_1, x_2)^T$ so that $2x_1 + 3x_2 = y_1$ and $x_1 - 5x_2 = y_2$
   (n) All convergent sequences $(a_n)_{n=1}^\infty$
   (o) All functions with a finite number of discontinuities (possibly none)
   (p) All polynomials of exact degree 10
   (q) All even functions on $\mathbf{R}$ such that $f(-x) = f(x)$ $\forall x \in \mathbf{R}$

2. Let $A_1 = (2, 0, 1, 1)^T$, $A_2 = (2, -1, 2, 0)^T$, $A_3 = (0, -1, 1, -1)^T$, and $M = \text{span}\{A_1, \ldots, A_3\}$. Are $(2, 2, -1, 3)^T$ and $(2, 3, -1, 3)^T \in M$? Explain.

3. Let $M = \text{span}\{(t + 1)^2, t^2 + 3t + 2, t^2 - 1\}$. Is $t \in M$? Is $t + 1 \in M$? Explain.

4. Determine whether the following are true or false. Explain. Assume functions are defined on $\mathbf{R}$ unless otherwise stated.
   (a) $t^2 \in \text{span}\{(t + 1)^2, 1 + 2t^2, 1 + 2t + 3t^2\}$
   (b) $t + t^2 \in \text{span}\{(t + 1)^2, (t + 1)(t + 2), t^2 - 1\}$
   (c) $t^3 \in \text{span}\{1, t, t(t - 1), t(t - 1)(t - 2)\}$
   (d) $\delta_3 \in \text{span}\{(1, 1, 1)^T, (0, 1, 1)^T, (1, -1, -1)^T, (2, 1, 1)^T\}$
   (e) $\delta_1 \in \text{span}\{(1, 1, 1)^T, (0, 1, 1)^T, (0, 0, 1)^T\}$

(f) $t \in \text{span}\{\sin(t), \sin(2t), \ldots, \sin(nt)\}$

(g) $e^t \in \text{span}\{1, t, t^2, \ldots, t^n\}$

5. Prove or disprove the following:

(a) $\mathbf{R}^2 = \text{span}\left\{ \begin{bmatrix} 1 \\ 0 \end{bmatrix} \begin{bmatrix} 1 \\ 1 \end{bmatrix} \right\}$

(b) $\mathbf{R}^3 = \text{span}\{(1, 0, 0)^T, (1, 1, 1)^2, (2, 1, 1)^T\}$

(c) $\mathbf{P}_2 = \text{span}\{1, t - 1, (t - 1)^2\}$

(d) $\mathbf{P}_2 = \text{span}\{t, t(t - 1), t^2\}$

6. Prove the following:

(a) $(t - a)^n \in \text{span}\{t^k\}_{k=0}^n$

(b) $t^n \in \text{span}\{(t - a)^k\}_{k=0}^n$

(c) $\cos(nt) \in \text{span}\{\cos^k(t)\}_{k=0}^n$

(d) $\sin(nt) \in \text{span}\{\sin(t)\cos^k(t)\}_{k=0}^{n-1}$

(e) $1 \in \text{span}\{x^k(1 - x)^{n-k} : k = 0, 1, \ldots, n\}$.

7. Show that the following are spanning sets for $\mathbf{P}_n$.

(a) $\{(t - a)^k\}_{k=0}^n$

(b) $\{x^k(1 - x)^{n-k} : k = 0, 1, \ldots, n\}$.

8. Let $A = [A_1 \ A_2 \ \cdots \ A_n]$. Show directly that $R(A)$ is a subspace, without using $R(A) = \text{span}\{A_k\}_{k=1}^n$.

9. Show that the columns of A are in the column space of B iff $A = BC$ for some matrix C.

10. Let M and N be subspaces of V. Define the following:

Union	$M \cup N = \{x : x \in M \text{ or } x \in N\}$ (the "or" is inclusive)
Intersection	$M \cap N = \{x : x \in M \text{ and } x \in N\}$
Sum	$M + N = \{x + y : x \in M, y \in N\}$

In each case, show that the new set is a subspace, or give an example to show it is not.

11. (a) Suppose $x \in \text{span}\{e_1, \ldots, e_m\}$ and each $e_k \in \text{span}\{f_1, \ldots, f_n\}$. Show that $x \in \text{span}\{f_1, \ldots, f_n\}$.

(b) State conditions in order that $\text{span}\{e_1, \ldots, e_m\} = \text{span}\{f_1, \ldots, f_n\}$.

12. **Cubic Splines.** Let $x_0 < x_1 < \cdots < x_n$ be given. Let V be the set of functions f such that f, f', and f'' are continuous on $[x_0, x_n]$ and restricted to each subinterval $[x_i, x_{i+1}]$, and f is a polynomial of degree not exceeding 3. Show that V is a vector space.

1.3
LINEAR INDEPENDENCE

The concept of linear independence discussed in this section is one of the most important basic ideas in linear algebra. First it is described and defined. Then, due to its importance, many examples of it are given in various settings. Emphasis is on function spaces and the n-tuple spaces $\mathbf{F}^n$.

A fundamental problem of linear algebra centers around statements such as

$$\sum_{i=0}^{n} a_i t^i = \sum_{i=0}^{n} b_i t^i \qquad \text{for all } t \Rightarrow a_i = b_i \quad \text{for } i = 0, \ldots, n$$

and

$$\sum_{i=0}^{n} a_i \cos(i\theta) = \sum_{i=0}^{n} b_i \cos(i\theta) \qquad \text{for all } \theta \Rightarrow a_i = b_i \quad \text{for } i = 0, \ldots, n.$$

These statements assert the uniqueness of the coefficients in the preceding expansions. The problem of *uniqueness of coefficients* may be generalized: In a vector space V, for a given set of vectors $\{f_k\}_{k=1}^{n}$, when is it true that

$$\sum_{i=1}^{n} \alpha_i f_i = \sum_{i=1}^{n} \beta_i f_i \qquad \text{implies that} \quad \alpha_i = \beta_i \quad \text{for } i = 1, \ldots, n? \qquad (1)$$

Subtracting the right side from the left, and setting $\gamma_i = \alpha_i - \beta_i$ yields the equivalent question: For a given set of vectors $\{f_k\}_{k=1}^{n}$, when does

$$\sum_{i=1}^{n} \gamma_i f_i = 0 \qquad \text{imply that} \quad \gamma_i = 0 \quad \text{for } i = 1, \ldots, n?$$

These equivalent questions reflect an important property of the set of vectors $\{f_k\}_{k=1}^{n}$. Not every set of vectors has this property, as the following examples show.

DEFINITION. The set of vectors $\{f_i\}_{i=1}^{n} \equiv \{f_1, \ldots, f_n\}$ is *linearly independent* if and only if

$$\sum_{i=1}^{n} \alpha_i f_i = 0 \qquad \text{implies that} \quad \alpha_i = 0 \quad \text{for } i = 1, \ldots, n \qquad (2)$$

and is said to be *linearly dependent* otherwise.

In other words, the set is linearly independent when only the trivial linear combination of the f_i, i.e., the one with all zero coefficients, yields the zero vector. It is linearly dependent just when there is at least one nontrivial linear combination of the f_i, which is the zero vector.

The introductory observations and definitions are summarized in the next proposition.

PROPOSITION 1. The set of vectors $\{f_k\}_{k=1}^{n}$ is linearly independent if and only if it has the following property: Whenever a vector x is expressible as a linear combination of the f_i, i.e., when $x = \alpha_1 f_1 + \alpha_2 f_2 + \cdots + \alpha_n f_n$, the coefficients α_i are uniquely determined by x; that is, (1) holds.

The next proposition follows immediately from the definition and is left to the reader to prove. (See Exercise 18.)

PROPOSITION 2. A subset of a linearly independent set is linearly independent.

The next proposition gives some insight into the idea of linear dependence.

PROPOSITION 3. A set $\{f_i\}_{i=1}^n$ of vectors is linearly dependent if and only if at least one of them is a linear combination of the others.

Proof. If the vectors are linearly dependent, there exists a linear combination, $\alpha_1 f_1 + \alpha_2 f_2 + \cdots + \alpha_n f_n = 0$ with, say, $\alpha_p \neq 0$. Solving for f_p yields

$$f_p = \sum_{k \neq p} (-\alpha_k/\alpha_p) f_k. \qquad (*)$$

On the other hand, if $f_p = \sum_{k \neq p} \beta_k f_k$ holds, then rewriting this expression gives $\beta_1 f_1 + \beta_2 f_2 + \cdots + (-1) f_p + \cdots + \beta_n f_n = 0$, which is a linear combination of the f_i, with not all coefficients equal to 0, whose sum is the zero vector.

To show linear independence, one usually begins with a linear combination of vectors equal to 0 and then gives an argument to show that the coefficients must all be 0. To establish linear dependence, either Proposition 3 can be used, or one can exhibit a set of coefficients, not all of which are zero, that produces a linear combination whose sum is zero. Or finally, one can provide an argument to show that such a set of coefficients exists, without actually exhibiting them.

In general vector spaces, there is no one standard algorithm by which vectors can be shown to be linearly independent or dependent, so ingenuity must be used. Often, there are several ways to show the linear independence or dependence of a given set of vectors. However, in $\mathbf{F}^n$, there are standard arithmetic algorithms that will always decide the issue, at least in principle. One such algorithm is mentioned at the end of the section.

The examples that follow exhibit useful independent sets and show technique. We begin with an important example in $\mathbf{F}^n$.

EXAMPLE 1. $\{\delta_1, \ldots, \delta_N\}$ **IS A LINEARLY INDEPENDENT SET IN** $\mathbf{F}^n$. In fact, suppose $0 = \sum_{i=1}^n x_i \delta_i$. But then $\sum_{i=1}^n x_i \delta_i = (x_1, x_2, \ldots, x_n)^{\mathrm{T}} = 0$ so that all $x_i = 0$, by the definition of the 0 vector and the relation "$=$" in $\mathbf{F}^n$.

Examples in Function Spaces

EXAMPLE 2. $\{\sinh(x), \cosh(x), e^x\}$ is a linearly dependent set.

Since $e^x = 1/2(\cosh x + \sinh x)$ for all complex x, these functions will be linearly dependent in every function space $\mathscr{F}(S)$, where S is a nonempty subset of the complex plane, by Proposition 3.

EXAMPLE 3. Let $f_1(t) = 1 + t + t^2$, $f_2(t) = 1 + 2t$, and $f_3(t) = -t + t^2$. Then $\{f_1, f_2, f_3\}$ is a linearly dependent set in $\mathscr{F}(\mathbf{R})$.

Write $a f_1 + b f_2 + c f_3 = 0$, and rearrange this expression according to the powers of t:

$$a f_1 + b f_2 + c f_3 = (a + b) \cdot 1 + (a + 2b - c)t + (a + c)t^2 = 0. \qquad (*)$$

If a, b, and c not all zero can be found so that $a + b = 0$, $a + 2b - c = 0$, and $a + c = 0$, then the set is linearly dependent. The first and last equations imply that $a = -b = -c$, and putting this in the second equation results in nothing new. If $a = 1$ is chosen, then $b = c = -1$, and one finds that $f_1 - f_2 - f_3 = 0$, which is easily checked by substituting. Alternatively, one could have noticed immediately that $f_1 = f_2 + f_3$, which also yields the linear dependence by Proposition 3.

The next example illustrates several techniques for showing the linear independence of functions.

EXAMPLE 4. The set of functions $\{t^k\}_{k=0}^n \equiv \{1, t, t^2, \ldots, t^n\}$ is linearly independent in the vector space $\mathcal{F}(\mathbf{R})$.

Assume $\sum_{i=0}^n \alpha_i t^i = 0$; i.e., assume the polynomial $p(t) \equiv \sum_{i=0}^n \alpha_i t^i$ satisfies $p(t) = 0$ for all real t, according to the definition of "=" and the definition of the zero vector. The first method goes as follows: First, set $t = 0$, which yields $p(0) = \sum_{i=0}^n \alpha_i 0^i = \alpha_0 = 0$. Dividing $p(t)$ by t yields $p(t)/t = p_1(t) \equiv \alpha_1 + \alpha_2 t + \cdots + \alpha_n t^{n-1} = 0$ for all real t except possibly $t = 0$. By continuity, $p_1(t) \to p_1(0) = \alpha_1 = 0$ as $t \to 0$. Next, divide p by t^2 to get $p_2(t) \equiv \alpha_2 + \alpha_3 t + \cdots + \alpha_n t^{n-2} = 0$ for all real t, except possibly $t = 0$. Letting $t \to 0$ again yields $\alpha_2 = 0$. Continuing inductively in the same way, we find that that all $\alpha_i = 0$.

The second method uses derivatives. Observe that $p^{(k)}(t) = \sum_{i=0}^n i(i-1) \cdots (i-k+1)\alpha_i t^{i-k}$. Since $p(t) = 0$ $\forall t$, we have $0 = p^{(k)}(0) = k!\alpha_k$, for each k, so that all $\alpha_k = 0$.

The third method uses the observation that the powers of t grow at different rates as $t \to \infty$: $1 \ll t \ll t^2 \ll \cdots \ll t^n$ as $t \to \infty$. Divide $p(t)$ by t^n to get $p(t)t^{-n} = \sum \alpha_i t^{i-n} = 0$ for all $t > 0$. Now let $t \to \infty$, and observe that $\alpha_n = \lim_{n \to \infty} p(t)t^{-n} = 0$. Next, show that $\alpha_{n-1} = \lim_{n \to \infty} p(t)t^{-n+1} = 0$, and so forth, until all of the α_i have been shown to be 0.

Remark. Often the set of functions $\{t^k\}_{k=0}^n$ of Example 4 are considered vectors in some other vector space of functions $\mathcal{F}(S)$, where S is a subset of the complex plane, and we want to know whether it is linearly independent. Depending on the application, S may be an interval in $\mathbf{R}$, a disk in the complex plane, or even a finite set, for example. The use of finite sets occurs in curve fitting and in the discrete Fourier transform. The next proposition sharpens the result of the last example.

PROPOSITION 4. Let t^i, $i = 0, 1, \ldots, n$, be considered to belong to the vector space $\mathcal{F}(S)$, where $S \subset \mathbf{C}$ contains more than n points. Then $\{t^k\}_{k=0}^n$ is linearly independent in $\mathcal{F}(S)$.

Proof. Suppose $p(t) \equiv \sum_{k=0}^n c_k t^k = 0$ for each t in S. Then all $c_k = 0$, since no nonzero polynomial may have more than n zeros.

PROPOSITION 5. Let e^{ikt}, $k = -n, \ldots, 0, 1, \ldots, n$ be considered to belong to the vector space $\mathcal{F}(S)$, where S contains at least $2n + 1$ points that belong to an interval of the form $[a, a + 2\pi)$. Then $\{e^{ikt}\}_{k=-n}^n$ is linearly independent in $\mathcal{F}(S)$.

Proof. This follows from Proposition 4, and the outline of the proof is in the exercises. (See Exercise 6.)

Remark. If $\{f_i\}_{i=1}^n$ is linearly independent in $\mathcal{F}(S)$ and $S \subset S'$, then $\{f_i\}_{i=1}^n$ is also linearly independent in $\mathcal{F}(S')$, since, if $\sum_{i=1}^n \alpha_i f_i(t) = 0$ $\forall t \in S'$, then $\sum_{i=1}^n \alpha_i f_i(t) = 0$ $\forall t \in S$, and so all $\alpha_i = 0$. It is important to note that the converse is not true. For example, $1, t, t^2, t^3$ are linearly independent on $S' = [-1, 1]$, but not on $S = \{-1, 0, 1\}$. In fact, $t^3 - t = 0$ $\forall t \in S$, yet $t^3 - t$ is a linear combination of $\{1, t, t^2, t^3\}$ without all coefficients equal to zero. For another example, consider $f_1(x) = x$ and $f_2(x) = |x|$. If $S = [0, 1]$, these functions are linearly dependent, since $f_1(x) = f_2(x)$ if $0 \leq x \leq 1$. If $S' = [-1, 1]$ and $c_1 f_1 + c_2 f_2 = 0$ in $\mathcal{F}(S')$, then $0 = c_1 f_1(1) + c_2 f_2(1) = c_1 + c_2$, and $0 = c_1 f_1(-1) + c_2 f_2(-1) = -c_1 + c_2$, from which it follows that $c_1 = c_2 = 0$, so that $\{f_1, f_2\}$ is linearly independent on $[-1, 1]$. The idea here is that, if there are too few points in S, then $\{f_i\}_{i=1}^n$ is likely to be linearly dependent as vectors in $\mathscr{F}(S)$.

EXAMPLE 5. The set $\{\cos^k(t)\}_{k=0}^n$ is linearly independent in $f([0, \pi])$.

   Assume that $\sum_{k=0}^n a_k \cos^k(t) = 0$ for all $t \in [0, \pi]$. The task is to show that all $a_k = 0$. Make the substitution $x = \cos(t)$. Then $\sum_{i=0}^n a_k x^k = 0$ for all $x \in [-1, 1]$. But then, all $a_k = 0$, by the linear independence of $\{x^k\}_{k=0}^n$ in $f_k([-1, 1])$.

Remark. For a set of functions $\{f_k\}_{k=1}^n$ defined on a set S, the statement "$\{f_k\}_{k=1}^n$ is linearly independent in $\mathscr{F}(S)$" is often replaced by "$\{f_k\}_{k=1}^n$ is linearly independent on S."

Interpolation Functions

In many problems that arise, for example, in finite elements, numerical analysis, approximation theory, and discrete Fourier analysis (fast Fourier transform), it is desirable to *interpolate* given data, $y_1, \ldots, y_n$, at given points, $x_1, \ldots, x_n$, by a linear combination of a set of fixed functions, $f_1, \ldots, f_n$. That is, we want to find scalars α_i such that

$$\sum_{i=1}^n \alpha_j f_j(x_i) = y_i \qquad \text{for } i = 1, 2, \ldots, n. \tag{3}$$

If the y_i are values of a function g, i.e., if $g(x_i) = y_i$, then the linear combination $\sum_{j=1}^n \alpha_j f_j$ is said to *interpolate the function* g at the x_i, and (3) becomes

$$\sum_{i=1}^n \alpha_j f_j(x_i) = g(x_i) \qquad \text{for } i = 1, 2, \ldots, n.$$

Suppose, by some fortuitous stroke, the functions f_j and points x_i satisfy

$$f_j(x_i) = \delta_{ij}. \tag{4}$$

Then

$$y_i = \sum_{i=1}^n \alpha_j f_j(x_i) = \sum_{i=1}^n \alpha_j \delta_{ij} = \alpha_i.$$

Thus a linear combination of the f_j that accomplishes the task of interpolation is $\sum_{j=1}^n y_j f_j(x_i)$. We will call (4) the *interpolation property*, since it makes the task of interpolation by linear combinations of the f_j very easy. The next proposition establishes the linear independence of a set of functions satisfying (4).

PROPOSITION 6. If the set of functions $\{f_j\}_{j=1}^n$ satisfies the interpolation property (4) at the points $\{x_i\}_{i=1}^n$, then $\{f_i\}_{i=1}^n$ is linearly independent on every set S containing $\{x_i\}_{i=1}^n$.

Proof. Suppose $\sum_{j=1}^n \alpha_j f_j(x) = 0 \ \forall x \in S$. Set $x = x_i$. Then $0 = \sum_{j=1}^n \alpha_j f_j(x_i) = \sum_{j=1}^n \alpha_j \delta_{ij} = \alpha_i$, yielding the linear independence.

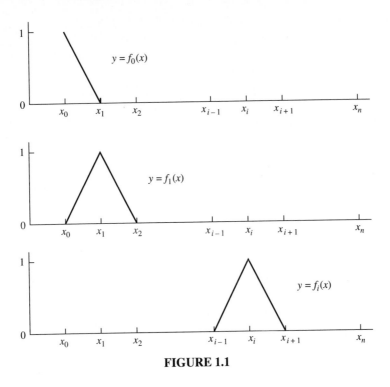

FIGURE 1.1

Next, two widely used examples are given. Many more are to be found in the subjects just mentioned.

EXAMPLE 6. THE VECTOR SPACE $CPL(\{x_i\}_{i=0}^n)$ **AND HAT FUNCTIONS.** Let $a = x_0 < x_1 < x_2 < \cdots < x_n = b$ be fixed points in $[a, b]$. Define continuous piecewise linear functions f_i $(i = 0, 1, \ldots, n)$ on $[a, b]$ by requiring that f_i is linear in each $[x_j, x_{j+1}]$ and that f_i is continuous on $[a, b]$. For example (see Fig. 1.1), $f_0(x_0) = 1$, $f_0(x_1) = 0$, and f_0 is linear on $[x_0, x_1]$ and zero on $x_1 \leq x \leq b$. Likewise, $f_1(x)$ is zero at $x = x_0$, x_2 and at all x in $[x_2, b]$, whereas $f_1(x_1) = 1$, and f_1 is linear on $[x_0, x_1]$ and $[x_1, x_2]$. These functions are called *hat functions* (or chapeau, roof, or tent functions by various imaginative authors).

In the present case, the hat functions give us a convenient way to write down a continuous piecewise linear function f, which interpolates a given set of data or a given function g, namely, $f = \sum_{i=0}^n g(x_i) f_i$. See Fig. 1.2, where $n = 6$.

It is clear from Fig. 1.2 that if g is a smooth function and if the x_i are close together, then f is a good approximation to g across the whole interval $[x_0, x_n]$. In fact, one can show that, for $x \in [x_0, x_n]$, $|f(x) - g(x)| \leq \frac{1}{8}M\Delta^2$, where $M = \max\{|g''(x)| : x \in [x_0, x_n]\}$ and $\Delta = \max\{x_{i+1} - x_i : i = 0, \ldots, n-1\}$.

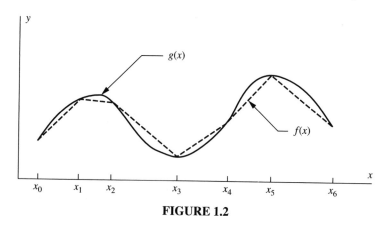

FIGURE 1.2

EXAMPLE 7. LAGRANGE INTERPOLATION POLYNOMIALS ON $\{x_i\}_{i=0}^n$. Let

$$L_i(t) = \frac{(t - x_o)\cdots(t - x_{i-1})(t - x_{i+1})\cdots(t - x_n)}{(x_i - x_0)\cdots(x_i - x_{i-1})(x_i - x_{i+1})\cdots(x_i - x_n)} \qquad i = 0, 1, \ldots, n.$$

The ith factor, in other words, is omitted in the products in the numerator and denominator. Clearly $L_i \in \mathbf{P}_n$, and $L_i(x_j) = \delta_{ij}$, so these functions have the interpolation property. Thus for a given set $\{y_i\}_0^n$, the polynomial $p(t) = \sum_{i=0}^n y_i L_i(t)$ interpolates the data at the points x_j: $p(x_j) = y_j$, $j = 0, 1, \ldots, n$. Furthermore $\{L_0, L_1, \ldots, L_n\}$ is a linearly independent set.

Linear Independence in $\mathbf{F}^m$

The study of linear independence in $\mathbf{F}^m$ is important in several ways. To begin with, it is possible to reduce the problem of testing a set of vectors for linear independence to an arithmetic algorithm. In certain circumstances, as we will see in later sections, it is also possible to reduce problems concerning linear independence in general vector spaces to a problem of linear independence in $\mathbf{F}^m$, thereby increasing the importance of the work in $\mathbf{F}^m$.

We begin with a concrete example and then consider the general case.

EXAMPLE 8. Let $A_1 = (1, 0, 2)^{\mathrm{T}}$, $A_2 = (0, -1, 3)^{\mathrm{T}}$, $A_3 = (2, 1, 1)^{\mathrm{T}}$. Show that $\{A_1, A_2, A_3\}$ is a linearly dependent set.

Begin with $x_1 A_1 + x_2 A_2 + x_3 A_3 = 0$. Writing this out gives

$$x_1 A_1 + x_2 A_2 + x_3 A_3 = x_1 \begin{bmatrix} 1 \\ 0 \\ 2 \end{bmatrix} + x_2 \begin{bmatrix} 0 \\ -1 \\ 3 \end{bmatrix} + x_3 \begin{bmatrix} 2 \\ 1 \\ 1 \end{bmatrix} = \begin{bmatrix} x_1 + 2x_3 \\ -x_2 + x_3 \\ 2x_1 + 3x_2 + x_3 \end{bmatrix} = \begin{bmatrix} 0 \\ 0 \\ 0 \end{bmatrix}.$$

Thus we must solve the following system of equations:

$$x_1 + \qquad 2x_3 = 0$$
$$-x_2 + x_3 = 0$$
$$2x_1 + 3x_2 + x_3 = 0.$$

Solving the first equation for x_1 and the second for x_2 and putting the results in the last equation yields the solution $x_1 = 2$, $x_2 = -1$, $x_3 = -1$. Thus $2A_1 - A_2 - A_3 = 0$.

This might have been noticed without forming the system. Thus the vectors are linearly dependent.

Now consider the general case. The goal here is to describe the problem in general and introduce some notation.

Let $A_1, A_2, \ldots, A_n \in \mathbf{F}^m$ be given columns. Then (cf. Section 0.4 and (1) of Section 1.2)

$$\sum_{j=1}^{n} x_j A_j = Ax. \qquad (*)$$

From this observation, the next proposition follows. Proposition 7 reduces the problem of linear independence in $\mathbf{F}^m$ to the solutions of homogeneous linear equations, for which there are arithmetic algorithms, one of which is Gauss elimination.

PROPOSITION 7. Let $A = [A_1 \ A_2 \ \cdots \ A_n]$ with $A_j \in \mathbf{F}^m$. The set of vectors $\{A_1, A_2, \ldots, A_n\}$ is linearly independent iff the system of equations $Ax = 0$ has only the trivial solution $x = 0$.

COROLLARY 8. Let A be a square matrix. Then A is nonsingular iff the columns of A are linearly independent.

Proof. By Theorem 4 of Section 0.5, A^{-1} exists iff $Ax = 0$ has only the solution $x = 0$. Combine this fact with Proposition 7.

EXAMPLE 9. Investigate the linear independence of $A_1 = (1, 0, 1, 2)^{\mathrm{T}}$, $A_2 = (1, 1, 1, 1)^{\mathrm{T}}$, $A_3 = (2, 1, 2, 1)^{\mathrm{T}}$.

Form the matrix

$$A = [A_1 \ A_2 \ A_3] = \begin{bmatrix} 1 & 1 & 2 \\ 0 & 1 & 1 \\ 1 & 1 & 2 \\ 2 & 1 & 1 \end{bmatrix}.$$

Solving $Ax = 0$ by row reduction yields

$$R = \begin{bmatrix} 1 & 0 & 0 \\ 0 & 1 & 0 \\ 0 & 0 & 1 \\ 0 & 0 & 0 \end{bmatrix},$$

so

$$Ax = 0 \Leftrightarrow x_i = 0 \qquad \text{for } i = 1, 2, 3.$$

Thus the given vectors are linearly independent.

Exercises 1.3

1. Decide whether the given sets of vectors are linearly independent. Explain. Functions are defined on $\mathbf{R}$ unless noted otherwise.
   (a) $1, \cos(t), \cos(2t), \cos(3t), \cos^3(t)$
   (b) $e^{0.1t}, e^t, e^{1.5t} \qquad t \geq 0$
   (c) $t^3, t^2(1 - t), t(1 - t), 3t + t^2 - t^3$

(d) $t^2 + 1, (1 + t)^2, 1 + t - t^2$

(e) $x^{\sqrt{2}}, x^{-1}, x^{-\pi}$ where $x > 0$

(f) $1, e^x, e^{-x}, \sinh(x)$

(g) $1, \cos(x), \sin(x), e^{ix}$

(h) $1, x, |x|, |x - 1|$

(i) $\begin{bmatrix} 1 & 0 \\ 0 & 1 \end{bmatrix}, \begin{bmatrix} 1 & 1 \\ 0 & 0 \end{bmatrix}, \begin{bmatrix} 0 & 1 \\ 1 & 1 \end{bmatrix}$

(j) $\begin{bmatrix} 1 & 0 \\ 1 & 2 \end{bmatrix}, \begin{bmatrix} 1 & 1 \\ -1 & 1 \end{bmatrix}, \begin{bmatrix} 0 & -1 \\ 2 & 1 \end{bmatrix}$

2. Investigate the linear independence of the following.

(a) $(1, i)^T, (i, -1)^T$

(b) $(1, 1, 2)^T, (2, -1, 1)^T, (1, 1, -1)^T$

(c) $(1, 1, 1)^T, (1, 1, 0)^T, (1, 1, 3)^T$

(d) $(1, 1, 1)^T, (1, 1, 0)^T, (1, 0, 1)^T$

(e) $(1, 1, 1, 2)^T, (1, -1, 1, -2)^T, (1, -3, 1, -4)^T$

(f) $(1, 1, 0, 1)^T, (1, -1, -2, 1)^T, (2, 1, -1, 2)^T$

(g) $(1, 1, 1, 1)^T, (1, 2, 2, 2)^T, (2, 1, 2, 2)^T$

(h) $(2, 3, 1)^T, (\sqrt{2}, 5, 1)^T, (8, \pi, 0)^T, (1, 1, 1)^T$

3. Show that $1, t, t^2, (t - 1)^2$ are linearly dependent. Find two distinct linear combinations for which $t^2 + 1 = a \cdot 1 + bt + ct^2 + d(t - 1)^2$.

4. Show that the following sets are linearly independent. Functions are defined on $\mathbf{R}$ unless noted otherwise.

(a) $\{(t + 1)^j : j = 0, 1, \ldots, n\}$

(b) $\{t^k(1 - t)^{n-k} : k = 0, 1, \ldots, n\}$

(c) $\{\sin^k(t) : k = 1, 2, \ldots, n\}$

(d) $\{e^{-\lambda_i t} : i = 1, \ldots, n\}$ where $0 < \lambda_1 < \lambda_2 < \cdots < \lambda_n$ and $t \geq 0$

(e) $\{\sinh(kt) : k = 1, 2, \ldots, n\}$

(f) $\{\cosh(kt) : k = 0, 1, \ldots, n\}$

(g) Let $f_k(t) = \sin(t - x_1) \sin(t - x_2) \cdots \sin(t - x_{k-1}) \sin(t - x_{k+1}) \cdots \sin(t - x_n)$. Here $\{x_1, x_2, \ldots, x_n\}$ is a fixed set of points in $[0, \pi)$.

5. Let E_{ij} be the $m \times n$ matrix with all entries equal to 0 except for a 1 in the ith row and jth column. Show that the set $\{E_{ij}, i = 1, 2, \ldots, m, j = 1, 2, \ldots, n\}$ is linearly independent in $\mathbf{F}^{m,n}$.

6. **Linear independence of $\{e^{ikx} : k = -n, \ldots, 0, 1, \ldots, n\}$ on a set S.** Assume S contains $2n + 1$ points x_j lying in an interval $[a, a + 2\pi)$. Show that these functions are linearly independent on S.

   Hint: Consider the statement

   $$\sum_{k=-n}^{n} c_k e^{ikx} = 0 \quad \forall x \in S. \tag{$*$}$$

   Substitute $t = e^{ix}$ in $(*)$ and multiply the result by t^n.

7. (a) Show that $\{1, \cos(t), \cos(2t), \ldots, \cos(nt)\}$ is linearly independent in $\mathscr{F}([0, \pi))$.

   (b) Show that $\{\sin(t), \sin(2t), \ldots, \sin(nt)\}$ is linearly independent in $\mathscr{F}([0, \pi))$.

   (c) Generalize the results of (a) and (b) to $\mathscr{F}(S)$, where S is a finite subset of $[0, \pi)$.

8. Show that if $m > n$, then m vectors in $\mathbf{F}^n$ are linearly dependent.

9. Let A be an $n \times n$ matrix with $a_{ij} = 0$ if $i < j$ and $a_{ii} \neq 0$ $\forall i$. Show that the columns of A are linearly independent.

10. Suppose A has linearly independent columns. Show that $Ax = Ay$ implies that $x = y$.

11. Let $\{A_1, \ldots, A_{n-1}\}$ be a set of linearly independent columns, and let A_n be a nonzero column such that $A_n^{\mathsf{T}} A_k = 0$ if $k = 1, 2, \ldots n-1$. Show that $\{A_k\}_{k=1}^n$ is linearly independent.

12. Let $A_1, A_2, \ldots, A_n \in \mathbf{F}^n$, and let $A = [A_1\ A_2\ \cdots\ A_n]$. Show that $\{A_k\}_{k=1}^n$ is linearly independent iff the row rank of A is n.

13. Let $\{A_k\}_{k=1}^n$ be linearly independent columns in $\mathbf{F}^m$. Show that $\{A_i A_j^{\mathsf{T}}\}_{i;j=1}^n$ is linearly independent in $\mathbf{F}^{m,m}$.

14. Let A be an $m \times n$ matrix, and let B be an $n \times p$ matrix. Suppose the columns of A and B are linearly independent. Show that the columns of AB are linearly independent.

15. Suppose $C = AB$ has linearly independent columns. Show that the columns of B are linearly independent.

16. Suppose A is an $m \times n$ matrix, B is an $n \times m$ matrix, and $AB = I$. Show that
   (a) the columns of B are linearly independent.
   (b) the columns of A span $\mathbf{F}^m$.

17. Let $C = AB$ with A, B being $n \times n$ matrices. Suppose the column space of C is $\mathbf{F}^n$. Show that A and B are nonsingular.

18. A subset of a linearly independent set is linearly independent. Suppose $p < n$ and $\{e_k\}_{k=1}^n$ is linearly independent. Show that $\{e_k\}_{k=1}^p$ is linearly independent.

19. A superset of a linearly dependent set is linearly dependent. Suppose $p > n$ and $\{e_k\}_{k=1}^n$ is linearly dependent. Show that $\{e_k\}_{k=1}^p$ is linearly dependent.

20. Show that, if one of the vectors in $\{e_k\}_{k=1}^n$ is zero, then $\{e_k\}_{k=1}^n$ is linearly dependent.

21. Let $\{f_k\}_{k=1}^n$ be a set of functions defined on a set S. Can $\{f_k\}_{k=1}^n$ be linearly independent in $\mathscr{F}(S)$ if S contains fewer than n points? Explain.

22. A set of functions $\{f_i\}_{i=1}^n$ defined on a set S having the property

$$\sum_{i=1}^n c_i f_i(x) = 0 \text{ for } n \text{ points}, \qquad x \in S \Rightarrow c_i = 0 \quad \text{for } i = 1, 2, \ldots, n, \qquad (*)$$

is called a *Chebyshev* or *unisolvent* set of functions on S. Such sets of functions generalize the notion of polynomials or trigonometric polynomials (see Propositions 4 and 5 in this section) and are important in approximation theory. It is assumed that S contains at least n points. Show that
   (a) If $\{f_i\}_{i=1}^n$ is a Chebyshev set on S, then $\{f_i\}_{i=1}^n$ is linearly independent in $\mathscr{F}(S)$.
   (b) For every set $\{x_k\}_{k=1}^n \subset S$, the matrix $[f_j(x_i)]$ is nonsingular.

(c) $\{\cos(kt): k = 0, 1, \ldots, n\}$ is a Chebyshev set on $[0, \pi)$.

(d) $\{\sin(kt): k = 1, 2, \ldots, n\}$ is a Chebyshev set on $(0, \pi)$.

23. Let $V = \mathcal{F}(\mathbf{P}_2)$. Determine the linear independence or dependence of the following sets $\{\phi_1, \phi_2, \phi_3\}$ in V:

(a) $\phi_1(p) = p(0), \phi_2(p) = p(1), \phi_3(p) = p(2)$

(b) $\phi_1(p) = p, \phi_2(p) = p', \phi_3(p) = p''$

(c) $\phi_1(p) = p(0), \phi_2(p) = \int_0^1 p(t)\,dt, \phi_3(p) = \int_{-1}^1 p(t)\,dt$.

1.4
BASIS AND DIMENSION

This section discusses the fundamental concepts of basis and dimension. The associated theory can be used to draw remarkable conclusions about various structures and problems in linear algebra. The theory is developed here and some of its applications are given.

In many problems, it is very desirable to represent every vector f in a vector space V as a unique linear combination of a fixed, finite number of given vectors e_i: $f = \sum_{i=1}^n \alpha_i e_i$. This motivates the next definition.

DEFINITION. $\{e_k\}_{k=1}^n$ is a basis for V if and only if $\{e_k\}_{k=1}^n$ is linearly independent and $V = \text{span}\{e_k\}_{k=1}^n$.

The definition requires that $1 \le n < \infty$ and that each $e_k \in V$. Therefore V is not the trivial vector space $\{0\}$. In this section, we assume $V \ne \{0\}$.

Sometimes deciding whether or not a given set $\{e_i\}_{i=1}^n$ is a basis involves a fair amount of work. Several useful theorems will assist us. Before proceeding to the theory, we give some examples. After presenting the theory, some of its applications are given. The next four examples exhibit some sets that have previously been shown to be linearly independent spanning sets for the vector spaces in question.

EXAMPLE 1. $\{\delta_1, \delta_2, \ldots, \delta_n\}$ is a basis for $\mathbf{F}^n$. $\{\delta_n\}_{i=1}^n$ called the *standard basis* for $\mathbf{F}^n$.

EXAMPLE 2. $\{1, t, t^2, \ldots, t^n\}$ is a basis for $\mathbf{P}_n$.

EXAMPLE 3. $\{e^{ijt}\}_{j=1}^n$ is a basis for $\mathbf{T}_n$.

EXAMPLE 4. The hat functions are a basis for $CPL(\{x_i\}_{i=0}^n)$.

The hat functions have been shown to be linearly independent. To see that they span, let $g \in CPL(\{x_i\}_{i=0}^n)$. Let $f = \sum_{i=1}^n g(x_i) f_i$. By the interpolation property, $f(x_j) = g(x_j)$ for $j = 0, 1, \ldots, n$. Since both functions are linear in between the x_i, they must be equal at all the points between the x_i as well. Hence $f(x) = g(x)\ \forall x \in [x_0, x_n]$, and therefore $f = g$.

EXAMPLE 5. Let M be the subset of $p \in \mathbf{P}_2$ for which $\int_0^1 p(t)\,dt = 0$. Show that M is a vector space, and find a basis for M.

Clearly $0 \in M$, so M is nonempty. If $p, q \in M$, then $\int_0^1 p(t)\,dt = \int_0^1 q(t)\,dt = 0$, and so

$$\int_0^1 (\alpha p(t) + \beta p(t))\,dt = \alpha \int_0^1 p(t)\,dt + \beta \int_0^1 q(t)\,dt = 0.$$

Thus M is a subspace of $\mathbf{P}_2$ and therefore a vector space. Let $p(t) = a + bt + ct^2 \in M$. Then $\int_0^1 p(t)dt = a + b/2 + c/3$. Eliminating $a = -b/2 - c/3$ gives $p(t) = b(t - \frac{1}{2}) + c(t^2 - \frac{1}{3})$. Thus $\{t - \frac{1}{2}, t^2 - \frac{1}{3}\}$ spans M. To show that these vectors are linearly independent, consider $b(t - \frac{1}{2}) + c(t^2 - \frac{1}{3}) = 0$. Setting $t = \frac{1}{2}$ and $t = 1/\sqrt{3}$ shows that $b = c = 0$. Thus $\{t - \frac{1}{2}, t^2 - \frac{1}{3}\}$ is linearly independent and spans M and is therefore a basis.

EXAMPLE 6. A BASIS FOR SPAN$\{A_1, A_2, \ldots, A_n\}$, $A_k \in \mathbf{F}^m$. First we establish the following fact:

If R is the row-reduced echelon form of B, then the nonzero rows of R are a basis for the vector space spanned by the rows of B.

To see this, observe that each row operation produces a new matrix whose rows are linear combinations of the rows of the original matrix. Thus the rows of R are linear combinations of the rows of B. Since the elementary row operations are reversible, the rows of B are also linear combinations of the rows of R. Thus the span of the rows of B is the same as the span of the rows of R. The nonzero rows of R are linearly independent because the leading ones are in different columns, and so the rows of R are a basis for the vector space spanned by the rows of R.

We can use this fact to find a basis for $V = \text{span}\{A_1, \ldots, A_n\}$. The procedure is as follows: Let B be the $n \times m$ matrix with rows A_k^{T}, i.e., $B = [A_1\ A_2\ \cdots\ A_n]^{\mathrm{T}}$. Find the row-reduced echelon form R of B. Then a basis for span$\{A_1, \ldots, A_n\}$ is the set of nonzero rows of R transposed.

EXAMPLE 7. Let $A_1 = (1, 1, -1)^{\mathrm{T}}$, $A_2 = (1, 2, -4)^{\mathrm{T}}$, $A_3 = (1, -1, 5)^{\mathrm{T}}$, and $A_4 = (2, 1, 1)^{\mathrm{T}}$. Find a basis for $M = \text{span}\{A_1, A_2, A_3, A_4\}$.

According to Example 6, we form the matrix B as:

$$B = \begin{bmatrix} 1 & 1 & -1 \\ 1 & 2 & -4 \\ 1 & -1 & 5 \\ 2 & 1 & 1 \end{bmatrix}.$$

B row-reduces to

$$R = \begin{bmatrix} 1 & 0 & 2 \\ 0 & 1 & -3 \\ 0 & 0 & 0 \\ 0 & 0 & 0 \end{bmatrix}.$$

Thus

$$\left\{ \begin{bmatrix} 1 \\ 0 \\ 2 \end{bmatrix}, \begin{bmatrix} 0 \\ 1 \\ -3 \end{bmatrix} \right\}$$

is a basis for M.

Now we turn to theoretical considerations. The next lemma is fundamental to the theory.

LEMMA 1. If $\{e_i\}_{i=1}^m$ spans V, then every set of more than m vectors in V is linearly dependent.

Proof. Let $f_1, \ldots, f_n$ be in V, and let $n > m$. Since the set $\{e_i\}_{i=1}^m$ spans V, there are scalars a_{ij} such that $f_k = \sum_{j=1}^m a_{jk}e_j$. Forming a linear combination of the f_k and rearranging the sum yields

$$\sum_{k=1}^{n} x_k f_k = \sum_{k=1}^{n} x_k \left[\sum_{j=1}^{m} a_{jk} e_j \right] = \sum_{j=1}^{m} \left[\sum_{k=1}^{n} a_{jk} x_k \right] e_j. \tag{*}$$

But since $m < n$, the system

$$\sum_{k=1}^{n} a_{jk} x_k = 0, \qquad j = 1, \ldots, m$$

has a nontrivial solution for $x_1, x_2, \ldots, x_n$ by Theorem 3 of Section 0.2. When these x_k are put in the sum (*), it follows that $\sum_{k=1}^{n} x_k f_k = 0$, with some $x_j \neq 0$, which implies that $\{f_1, \ldots, f_n\}$ is linearly dependent.

The next theorem and definition are central.

THEOREM 2. Let V be a given vector space that has a basis. Then every basis of V has the same number of vectors: If $\{f_i\}_{i=1}^{m}$ and $\{e_i\}_{i=1}^{n}$ are bases for V, then $n = m$.

Proof. Suppose $n > m$. Since $\{f_i\}_{i=1}^{m}$ spans V and $n > m$, $\{e_i\}_{i=1}^{n}$ must be linearly dependent according to Lemma 1. But this is false, so $n \leq m$. Similarly, $m > n$ is impossible, so $n = m$.

DEFINITION. The *dimension of a vector space* V, denoted $\dim(V)$, is defined as follows: If V has a basis, $\dim(V)$ is the number of elements in a basis. The trivial vector space $V = \{0\}$, consisting of only the zero vector, has no basis, and its dimension is defined to be zero. We write $\dim(V) = \infty$ iff for every positive integer n, there is a linearly independent set in V containing at least n vectors.

EXAMPLE 8. $C([0, 1])$ has infinite dimension.
   This follows from the fact that the linearly independent sets $\{t^k\}_{k=0}^{n}$ are a subset of $C([0, 1])$ for every $n \geq 0$.

EXAMPLE 9. $\dim(\mathbf{F}^n) = n$, $\dim(\mathbf{P}_n) = n + 1$, $\dim(\mathbf{T}_n) = 2n + 1$, $\dim(CPL\{x_i\}_{i=0}^{n}) = n + 1$.
   See Examples 1, 2, 3, and 4 of this section.

EXAMPLE 10. Let V be the set of columns $x = (x_1, x_2, x_3, x_4)^T$ satisfying

$$x_1 - 2x_2 + x_3 - 2x_4 = 0, \qquad 2x_1 - 4x_2 + x_3 - x_4 = 0.$$

Show that V is a vector space and find its dimension.
   The equations may be written as $Ax = 0$, where

$$A = \begin{bmatrix} 1 & -2 & 1 & -2 \\ 2 & -4 & 1 & -1 \end{bmatrix}.$$

If x and y satisfy $Ax = Ay = 0$, then also $A(\alpha x + \beta y) = \alpha Ax + \beta Ay = \alpha 0 + \beta 0 = 0$. Thus V is a subspace of $\mathbf{R}^3$, and therefore a vector space. Alternatively, $V = N(A)$, which is a vector space by Example 6 of Section 1.2. Reducing A to row-reduced echelon form yields the equivalent system $x_1 - 2x_2 + x_4 = 0$ and $x_3 - 3x_4 = 0$. x_2 and x_4 are the free variables, and the general solution is

$$x = \begin{bmatrix} 2x_2 - x_4 \\ x_2 \\ 3x_4 \\ x_4 \end{bmatrix} = x_2 \begin{bmatrix} 2 \\ 1 \\ 0 \\ 0 \end{bmatrix} + x_4 \begin{bmatrix} -1 \\ 0 \\ 3 \\ 1 \end{bmatrix} \equiv x_2 E_1 + x_4 E_2.$$

Thus E_1 and E_2 span V. To show that $\{E_1, E_2\}$ is linearly independent, set

$$0 = \alpha E_1 + \beta E_2 = \alpha \begin{bmatrix} 2 \\ 1 \\ 0 \\ 0 \end{bmatrix} + \beta \begin{bmatrix} -1 \\ 0 \\ 3 \\ 1 \end{bmatrix} = \begin{bmatrix} 2\alpha - \beta \\ \alpha \\ 3\beta \\ \beta \end{bmatrix} = \begin{bmatrix} 0 \\ 0 \\ 0 \\ 0 \end{bmatrix}.$$

Examining the second and fourth components (which correspond to the free variables), one finds that $\alpha = \beta = 0$. Thus $\{E_1, E_2\}$ is an independent spanning set for V and hence is a basis. Thus $\dim(V) = 2$.

In Section 2.2, the solutions of $Ax = 0$ for a general $m \times n$ matrix A will be studied in detail, and the idea exhibited in this example will be shown to hold in general.

The next corollary is an important application of Lemma 1 and of the definition of dimension.

COROLLARY 3. Let $\dim(V) = n$. Then

(a) Every set of more than n vectors in V is linearly dependent.
(b) No set of fewer than n vectors can span V.

Proof. Let $\beta = \{e_k\}_{k=1}^n$ be a basis for V and let $\gamma = \{f_k\}_{k=1}^m$. (a) If $m > n$, then since β spans V, γ must be linearly dependent by Lemma 1. (b) Let $m < n$. If γ spanned V, then β would be linearly dependent, by Lemma 1. But this is false, so γ cannot span V.

The next example gives an application of Corollary 3 to systems of equations.

EXAMPLE 11. Let A be an $m \times n$ matrix. If $m > n$, then there is a $y \in \mathbf{F}^m$ such that the system $Ax = y$ is inconsistent.

If $Ax \equiv \sum_{j=1}^n A_j x_j = y$ were solvable for every y, then the columns $A_1, A_2, \ldots, A_n$ of A would span $\mathbf{F}^m$. This cannot be by Corollary 3(b), since $\dim(\mathbf{F}^m) = m > n$.

EXAMPLE 12. Show that $\{1, t, t(t - 1), (t - 1)^2, 1 + t + t^2\}$ is linearly dependent.
In fact, five vectors in the three-dimensional space $\mathbf{P}_2$ must be linearly dependent.

The next two lemmas are useful in a variety of applications. The first one asserts that dependent spanning sets are somehow redundant; one of the vectors may be cast out, while the remaining vectors still form a spanning set.

LEMMA 4. If $m > 1$ and $\{e_k\}_{k=1}^m$ is a linearly dependent spanning set for V, then $m - 1$ of the e_k span V.

Proof. One of the vectors must be a linear combination of the others, so suppose (after a possible relabeling) that $e_1 = \sum_{k=2}^m a_k e_k$. Let f be an arbitrary vector in V. Since $\{e_k\}_{k=1}^m$ spans V, $f = \sum_{k=1}^m b_k e_k$ for some b_k. Eliminating e_1 from the expression for f yields

$$f = b_1 e_1 + \sum_{k=2}^m b_k e_k = b_1 \left(\sum_{k=2}^m a_k e_k \right) + \sum_{k=2}^m b_k e_k = \sum_{k=2}^m (b_1 a_k + b_k) e_k.$$

Thus each f in V is a linear combination of $\{e_i\}_{i=2}^m$, completing the proof.

The next lemma states that independent sets that do not span can be augmented to form a larger independent set.

LEMMA 5. If $\{e_i\}_{i=1}^m$ is linearly independent and $e_{m+1} \notin \operatorname{span}\{e_i\}_{i=1}^m$, then $\{e_1, e_2, \ldots, e_m, e_{m+1}\}$ is linearly independent.

Proof. Let $\sum_{i=1}^{m+1} a_i e_i = 0$. We will show that all $a_i = 0$. Suppose that $a_{m+1} \neq 0$. Then $e_{m+1} = -\sum_{i=1}^m (a_i/a_{m+1}) e_i$, which is contrary to the hypothesis; hence $a_{m+1} = 0$. But this implies that $\sum_{i=1}^m a_i e_i = 0$, which in turn implies that all $a_i = 0$, by the independence of $\{e_i\}_{i=1}^m$.

The next theorem is often very useful and is not intuitively obvious. It asserts that, when the dimension of the vector space is known, it is not always necessary to show that a set of vectors is both a spanning set and a linearly independent set in order to deduce that it is a basis. In fact, if the number of vectors in the given set is equal to the dimension of the vector space, the work can be cut in half.

THEOREM 6. Suppose that $\{e_i\}_{i=1}^n$ is a subset of the n-dimensional vector space V.

(a) If $\{e_i\}_{i=1}^n$ is linearly independent, then it spans V and hence is a basis.
(b) If $\{e_i\}_{i=1}^n$ spans V, then it is linearly independent and hence is a basis.

Proof. Suppose $\{f_i\}_{i=1}^n$ is a basis for V.

(a) If $\{e_i\}_{i=1}^n$ does not span V, then select an $e_{n+1} \notin \{e_i\}_{i=1}^n$ and extend the original set to a larger independent set $\{e_i\}_{i=1}^{n+1}$ by Lemma 5. But $\{f_i\}_{i=1}^n$ spans V, so by Lemma 1, the larger set $\{e_i\}_{i=1}^{n+1}$ must be linearly dependent, which is a contradiction. Hence $\{e_i\}_{i=1}^n$ must span V.
(b) Now suppose $\{e_i\}_{i=1}^n$ is not a linearly independent set. By Lemma 4, $n-1$ of the e_k must span V. But by Lemma 1, the larger set $\{f_i\}_{i=1}^n$ then cannot be linearly independent, which is a contradiction. Hence $\{e_i\}_{i=1}^n$ must be a linearly independent set.

Some applications of Theorem 6 follow.

EXAMPLE 13. $\{(t-a)^k\}_{k=0}^n$ is a basis for $\mathbf{P}_n$.
Since there are $n+1$ vectors in the given set and $n+1 = \dim(\mathbf{P}_n)$, it is sufficient to show either that the set is linearly independent or that the set spans $\mathbf{P}_n$; it is not necessary to show both statements. For each k, the binomial theorem implies that

$$t^k = [(t-a) + a]^k = \sum_{i=0}^k \binom{k}{i}(t-a)^i a^{k-i}, \qquad (**)$$

so that each $t^k \in \operatorname{span}\{1, (t-a), \ldots, (t-a)^k\}$. Thus a linear combination of $1, t, \ldots, t^k, \ldots, t^n$ can be written as a linear combination of $1, (t-a), \ldots, (t-a)^n$ by using $(**)$ to eliminate each t^k and rearranging terms. Thus $\{(t-a)^k\}_{k=0}^n$ spans $\mathbf{P}_n$ and hence is a basis. Note that a proof of linear independence is easily constructed along the lines of the proof of the linear independence of $\{t^k\}_{k=0}^n$ in Example 4 of Section 1.3, and this alone would have been sufficient to prove the conclusion.

EXAMPLE 14. $M = \{p : p \in \mathbf{P}_3, p(1) = 0\}$. Show that M is a subspace, and find the dimension of M and a basis for M.
If p and q are in M, then $(\alpha p + \beta q)(1) = \alpha p(1) + \beta q(1) = 0$; hence M is a subspace. The set $\{(t-1)^k : k = 0, 1, 2, 3\}$ is a basis for $\mathbf{P}_3$ by Example 13, so each p in M can be written $p(t) = \sum_{k=0}^3 c_k(t-1)^k$. But $p(1) = c_0 = 0$, so the set $\{(t-1)^k : k = 1, 2, 3\}$ spans M. This set is linearly independent as a subset of a linearly independent set, and so it is a basis for M. Hence $\dim(M) = 3$.

EXAMPLE 15. $\{(1, 1, 1)^T, (1, 1, 0)^T, (1, 0, 0)^T\}$ is a basis for $\mathbf{R}^3$.

It is an easy matter to prove that these vectors are linearly independent: The 3×3 matrix A, whose columns are the given vectors, row-reduces to the identity matrix so that the only solution of $Ax = 0$ is $x = 0$. Thus the given set of vectors is linearly independent by Proposition 7 of Section 1.3. Since there are $3 = \dim(\mathbf{R}^3)$ given vectors, they are a basis for $\mathbf{R}^3$.

EXAMPLE 16. The Lagrange interpolation polynomials $\{L_i\}_{i=0}^n$ for a set $\{x_j\}_{j=0}^n$ is a basis for $\mathbf{P}_n$.

The interpolation property $L_i(x_j) = \delta_{ij}$ implies that these functions are linearly independent. Since their number is the same as the dimension of $\mathbf{P}_n$, they span and are a basis.

EXAMPLE 17. A basis for $\mathbf{T}_n$ is $\beta = \{1, \cos(t), \sin(t), \dots, \cos(kt), \sin(kt), \dots, \cos(nt), \sin(nt)\}$.

Since $\dim(\mathbf{T}_n) = 2n + 1$, it is sufficient to show that β spans $\mathbf{T}_n$. By the Euler formulae, $\cos(kt) = \frac{1}{2}(e^{ikt} + e^{-ikt})$ and $\sin(kt) = 1/2i(e^{ikt} - e^{-ikt})$ so that each member of β is in $\mathbf{T}_n$. On the other hand, $e^{\pm ikt} = \cos(kt) \pm i\sin(kt)$, so that every linear combination of $\{e^{ikt}\}_{k=-n}^n$ is a linear combination of elements of β, which shows that β spans $\mathbf{T}_n$.

The next theorem may be interpreted as saying that a basis is a smallest spanning set.

THEOREM 7. From a spanning set, one may always extract a basis. More specifically, if $\text{span}\{e_i\}_{i=1}^m = V$, then $\dim(V) \le m$, and a subset of $\{e_i\}_{i=1}^m$ is a basis for V.

Proof. If $\{e_k\}_{k=1}^m$ is linearly independent, there is nothing to do. Otherwise, by Lemma 4, discard one of the e_k and a spanning set remains. If this set is linearly independent, stop. Otherwise continue this process. Eventually an independent spanning set is obtained.

EXAMPLE 18. Let A be an $m \times n$ matrix, and let W be the column space of A. Then $\dim(W) \le n$.

This follows from the fact that the column space is the span of the n columns of A and Theorem 7.

For the special case in which $V = \text{span}\{A_k\}_{k=1}^n$, where each A_k is a column in $\mathbf{F}^m$, a method of selecting a subset that is a basis for V will be presented in Section 2.2.

One may regard the following theorem as saying that a basis is a largest independent set.

THEOREM 8. A linearly independent set in a finite dimensional vector space V may be extended to a basis for V.

Proof. Suppose $\{e_i\}_{i=1}^m$ is a linearly independent set. If $\{e_i\}_{i=1}^m$ is a spanning set, there is nothing to do. Otherwise select $e_{m+1} \notin \text{span}\{e_i\}_{i=1}^m$. By Lemma 5, $\{e_i\}_{i=1}^{m+1}$ is an independent set. Continue this process if $\{e_i\}_{i=1}^{m+1}$ does not span V. The process must terminate in a basis in $\dim(V) - m$ steps, by Lemma 1.

There are many ways to make the extension of a linearly independent set to a basis, as the next simple example shows.

EXAMPLE 19. The vectors δ_1 and δ_2 are linearly independent in $\mathbf{R}^3$. Every vector $r = (x, y, z)^\mathrm{T}$, with $z \neq 0$, may be added to δ_1 and to δ_2 to make a basis $\{\delta_1, \delta_2, v\}$ for $\mathbf{R}^3$.

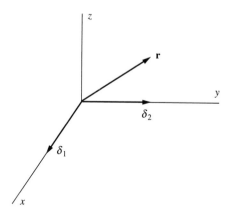

The next theorem is an extension of Theorem 8.

THEOREM 9. If W is a nontrivial subspace of the finite-dimensional vector space V, then W has a basis $\{f_k\}_{k=1}^m$ and $\dim(W) \leq \dim(V)$. Further, each basis of W may be extended to a basis for V.

Proof. The argument is similar to the one in Theorem 8. Select $f_1 \neq 0$, $f_1 \in W$. If $W = \mathrm{span}(f_1)$, there is nothing to do. Otherwise select $f_2 \in W$, $f_2 \notin \mathrm{span}(f_1)$. Then $\{f_1, f_2\}$ is a linearly independent set, and if it spans W, stop. Continue in this way until a basis for W is obtained. The process must stop in a finite number of steps, say, at $\{f_k\}_{k=1}^m$, since $\dim(V) < \infty$. By Theorem 8, this independent set in V may be extended to a basis for V, and so $\dim(W) = m \leq \dim(V)$.

EXAMPLE 20. Let A be an $m \times n$ matrix, and let W be the column space of A. Then $\dim(W)$ does not exceed the smaller of m and n.

The statement $\dim(W) \leq m$ follows from the fact that W is a subspace of $\mathbf{F}^m$ and $m = \dim(\mathbf{F}^m)$. In Example 18, it was shown that $\dim(W) \leq n$.

The next theorem is useful in many applications. Its proof is left as an exercise. (See Exercise 36.)

THEOREM 10. Suppose W is a subspace of V, and V is finite dimensional. Then $\dim(W) = \dim(V) \Rightarrow W = V$.

Theorem 10 is not valid if $\dim(V) = \infty$. In fact, if W is a subspace of V and $\dim(V) = \infty$, then $W \neq V$ is possible. For example, take $V = C([0, 1])$ and $W = \{f : f' \text{ is continuous on } [0, 1]\}$. Both contain all the independent functions t^k, and so are infinite dimensional. $W \neq V$ since there are continuous functions that are not differentiable, e.g., $f(x) = |x - 0.2|$.

Exercises 1.4

1. In each of the following, let M be the set of polynomials f in $\mathbf{P}_3$ such that the given condition(s) hold. Show that M is a vector space, and find a basis for and the dimension of M.

(a) $f'(1) = 0$ (b) $\int_0^1 f(t)\,dt = 0$
(c) $f(0) = 0,\quad \int_{-1}^1 f(t)\,dt = 0$ (d) $f(0) = 0,\quad f(1) = 0$

2. In each case, let M be the subspace of the appropriate $\mathbf{R}^n$ spanned by the given vectors. Find a basis for M and the dimension of M.
   (a) $(1, 1, 0, 1)^T, (1, -1, 1, 0)^T, (0, 2, -1, 1)^T, (2, 0, 1, 2)^T$
   (b) $(1, 1, 0, 1)^T, (0, 2, 1, 0)^T, (1, -1, -1, 1)^T$
   (c) $(1, 5)^T, (2, -3)^T, (3, -1)^T, (1, 1)^T, (5, 2)^T$
   (d) $(1, 0, 1, 2)^T, (1, 1, -1, 1)^T, (1, 2, -3, 0)^T, (2, 3, -4, 1)^T$

3. Find a basis for and the dimension of the column space of each of the following:

(a) $\begin{bmatrix} 1 & 1 & 1 \\ -1 & 2 & 5 \\ 1 & -1 & -3 \\ 1 & 0 & -1 \end{bmatrix}$
(b) $\begin{bmatrix} 1 & 1 & 1 \\ 0 & 1 & 2 \\ 2 & -1 & 1 \\ 1 & 1 & 2 \end{bmatrix}$

(c) $\begin{bmatrix} 1 & -1 & 1 & 1 \\ 1 & 2 & -1 & 1 \\ 1 & 5 & -3 & 1 \end{bmatrix}$
(d) $\begin{bmatrix} 1 & 1 & -2 & 1 \\ 2 & 0 & -1 & 1 \\ 1 & -1 & 1 & 0 \\ -1 & 3 & -4 & 1 \end{bmatrix}$.

4. For each of the matrices A in Exercise 3, let M be the set of x such that $Ax = 0$. Find a basis for M and its dimension.

5. Let M be the subspace of $\mathbf{R}^3$ consisting of vectors x satisfying $x_1 + x_2 - x_3 = 0$. Find a basis for M and dim(M).

6. Show that $\{(1, 1, 1, 1)^T, (1, 1, 1, 0)^T, (1, 1, 0, 0)^T, (1, 0, 0, 0)^T\}$ is a basis for $\mathbf{C}^4$.

7. Determine whether or not $\{(2, 2, 3, 1)^T, (1, 1, 2, 1)^T, (1, 0, 1, -1)^T, (-1, 1, 0, 2)^T\}$ spans $\mathbf{C}^4$.

8. Let

$$B = \begin{bmatrix} 1 & 0 & 0 & 3 \\ 1 & 1 & 0 & 0 \\ 1 & 1 & 1 & 2 \\ 1 & 1 & 1 & 1 \end{bmatrix}.$$

Is the system $Bx = y$ consistent $\forall y \in \mathbf{C}^4$? Explain.

9. Let $E_1 = (5, 0, 0, 1)^T, E_2 = (3, -1, 1, 2)^T, E_3 = (-1, 0, 0, 5)^T$, and $E_4 = (-1, 10, 2, 0)^T$. Show that every $x \in \mathbf{R}^4$ has a unique representation $x = \sum_{i=1}^4 c_i E_i$.

10. For each of the following sets M in $\mathbf{F}^{2,2}$, show that M is a vector space, and find a basis for M and for dim(M).
    (a) M is the subset of matrices A satisfying

$$A \begin{bmatrix} 1 \\ 1 \end{bmatrix} = 0.$$

(b) M is the subset of matrices A satisfying $A = A^T$.

(c) M is the subset of matrices A satisfying $A = -A^T$.

11. Show that $\{\delta_i \delta_j^T\}_{i;j=1}^n$ is a basis for $\mathbf{F}^{n,n}$.

12. Let $\{E_k\}_{k=1}^n$ and $\{F_k\}_{k=1}^n$ be bases for $\mathbf{F}^n$. Show that $\{E_i F_j^T\}_{i;j=1}^n$ is a basis for $\mathbf{F}^{n,n}$.

13. Find a basis for $\mathbf{C}^{m,n}$ and its dimension.

14. Let A be an $m \times n$ matrix. Show that if $m < n$, there exist $x \neq y$ so that $Ax = Ay$.

15. Let $V = CPL\{0, 1, 2, 3\}$. Define $f(t)_+ = \begin{cases} f(t) & f(t) \geq 0 \\ 0 & f(t) < 0 \end{cases}$. Show that the following are bases for V:

(a) $\{1, t, |t - 1|, |t - 2|\}$

(b) $\{1, t, (t - 1)_+, (t - 2)_+\}$

Hint: $f_+ = \frac{1}{2}\{f + |f|\}$.

16. Generalize Exercise 15 to $CPL\{x_0, x_1, \ldots, x_n\}$.

17. Show that each of the following sets is a basis for $\mathbf{P}_n$:

(a) $1, t, t(t - 1), t(t - 1)(t - 2), \ldots, t(t - 1)(t - 2)(t - 3)\ldots(t - n + 1)$

(b) $t^k(1 - t)^{n-k}, k = 0, 1, \ldots, n$

18. Let $c_0 = \{(a_k)_{k=1}^\infty : a_n \to 0 \text{ as } n \to \infty\}$. Find $\dim(c_0)$.

19. (a) Let $\mathbf{T}_n^e$ be the set of even trigonometric polynomials: $\mathbf{T}_n^e \equiv \{f \in \mathbf{T}_n : f(-x) = f(x) \; \forall x\}$. Find $\dim(\mathbf{T}_n^e)$.

(b) Let $\mathbf{T}_n^o$ be the set of odd trigonometric polynomials: $\mathbf{T}_n^o \equiv \{f \in \mathbf{T}_n : f(-x) = -f(x) \; \forall x\}$. Find $\dim(\mathbf{T}_n^o)$.

20. Let $f_k(t) = [\sin(kt)]/[k\sin(t)]$ if $t \neq j\pi$ and $f_k(j\pi) = \lim_{t \to j\pi} f_k(t)$. Show that $\{f_1, f_2, \ldots, f_n\}$ is a basis for $\mathbf{T}_{n-1}^e$.

21. (a) **Trigonometric interpolation.** Let n and $0 \leq x_0 < x_1 < \cdots < x_{2n} < 2\pi$ be given. Define $f_k(x) = \prod_{i=1; i \neq k}^n \sin(x - x_k)$. Show that $\{f_k\}_{k=0}^{2n}$ is a basis for $\mathbf{T}_n$.

(b) Show that for every set $\{y_k\}_{k=0}^{2n}$ there is a unique $p \in \mathbf{T}_n$ such that $p(x_k) = y_k$.

22. Let $\{f_k\}_{k=0}^n$ be a set of polynomials of degree not exceeding n. Suppose $f_k(1) = 0$ for each k. Show that $\{f_k\}_{k=0}^n$ is linearly dependent.

23. Let V be the set of polynomials in $\mathbf{P}_n$ for which $\int_0^1 f(t)\,dt = 0$. Show that V is a vector space and $\dim(V) \leq n$.

24. Prove that $\dim(V) = \infty$ iff there is no finite spanning set for V.

25. Show that every vector space has a dimension that is zero, a positive integer, or ∞.

26. Suppose P is nonsingular and $\beta = \{e_k\}_{k=1}^n$ is a basis for V. Define $\gamma = \{f_k\}_{k=1}^n$ by $\sum_{i=1}^n p_{ij} e_i = f_j$. Show that γ is a basis for V.

27. Let A be an $m \times n$ matrix for which the only solution to $Ax = 0$ is $x = 0$. Show that $m \geq n$.

28. Let A be an $m \times n$ matrix for which $m < n$. Show that the columns of A are linearly dependent.

29. Let A be an $n \times n$ matrix. Show that $Ax = y$ is consistent $\forall y \in \mathbf{F}^n$ iff the columns of A are linearly independent.

30. Let A be an $n \times n$ matrix. Show that $Ax = 0$ has only the solution $x = 0$ iff the columns of A span $\mathbf{F}^n$.

31. Let $A = [A_1\ A_2\ \cdots\ A_n]$ be an $n \times n$ matrix. Show that the following statements are logically equivalent. (See Theorem 4 of Section 0.5.)
   (*i*) $\{A_k\}_{k=1}^n$ is a basis for $\mathbf{F}^n$.
   (*ii*) $\{A_k\}_{k=1}^n$ is linearly independent.
   (*iii*) span$\{A_k\}_{k=1}^n = \mathbf{F}^n$.
   (*iv*) A^{-1} exists.

32. Suppose $\{v_1, v_2, \ldots, v_n\}$ is a basis for $\mathbf{F}^n$ and A is an $m \times n$ matrix such that $\{Av_1, Av_2, \ldots, Av_n\}$ is linearly independent. Show that $m \geq n$ and the columns of A are linearly independent.

33. Let A be an $n \times n$ matrix. Show that A^{-1} exists iff dim(span$\{A_1, \ldots, A_n\}) = n$.

34. Suppose $\{v_1, v_2, \ldots, v_n\}$ is a basis for $\mathbf{F}^n$, A is an $n \times n$ matrix, and $Ax = v_i$ has a solution for each i. Show that A is invertible.

35. Let A and Q be $m \times n$ matrices, and let R be an $n \times n$ matrix. Suppose $A = QR$ and the columns of A are linearly independent. Show that
   (*a*) R is nonsingular.
   (*b*) The columns of Q are a basis for the column space of A.
   (*c*) If $Ax = Qz$, then $x = R^{-1}z$.

36. Suppose W is a subspace of V, and V is finite dimensional. Prove that if dim(W) = dim(V), then $W = V$.

1.5
COORDINATES WITH RESPECT TO A BASIS

DEFINITION. Let $\beta = \{e_i\}_{i=1}^n$ be a basis for V. Then by the definitions, every x in V has a unique expansion $x = \sum_{k=1}^n c_k e_k$. The scalars c_k are called the *coordinates of x in the basis* β. The column $[x]_\beta = (c_1, c_2, \ldots, c_n)^\mathrm{T}$ is called the *coordinate column of x with respect to (in) the basis* β. When no confusion is possible, $[x]$ is written instead of $[x]_\beta$.

It is important to note that the basis and the coordinates are ordered (by the indexing of the basis); each coordinate must refer to the correct basis vector.

EXAMPLE 1. Let $V = \mathbf{P}_2$, and let $\beta = \{1, t, t^2\}$. Then $p(t) = 2 + 3t + 5t^2$ has the coordinate column

$$[p]_\beta = \begin{bmatrix} 2 \\ 3 \\ 5 \end{bmatrix}.$$

EXAMPLE 2. Let $V = \mathbf{T}_1$, and let $\beta = \{e^{-it}, 1, e^{it}\}$. Then $\sin(t) = 1/(2i)(e^{it} - e^{-it})$ has the coordinate column

$$[\sin]_\beta = \begin{bmatrix} -1/2i \\ 0 \\ 1/2i \end{bmatrix}.$$

EXAMPLE 3. Let $\beta = \{\mathbf{e}_1, \mathbf{e}_2\}$ be a basis of position vectors in the plane. Plot $\mathbf{r}$ if $[\mathbf{r}]_\beta = \begin{bmatrix} 2 \\ 1 \end{bmatrix}$.

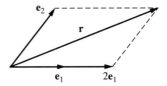

We have

$$\mathbf{r} = 2\mathbf{e}_1 + \mathbf{e}_2,$$

which may be plotted using the parallelogram rule.

The next example often provokes a great deal of confusion, but it is of fundamental importance. It may be helpful to imagine columns in $\mathbf{F}^n$ as coordinate columns of points in a (geometrical) n-space relative to a reference coordinate system. However, there are many applications in which this is not easy to do. For example, one may be studying a predator–prey problem using $x = (x_1, x_2)^T$, with x_1, x_2 being the populations of the predator and prey, respectively. In another application, x could be an n-tuple of 0s and 1s representing a digital signal. What the geometrical n-space is, is not so clear.

EXAMPLE 4. In $\mathbf{F}^n$, the coordinate column of x with respect to the standard basis $\delta = \{\delta_1, \delta_2, \ldots, \delta_n\}$ is

$$[x]_\delta = x.$$

This follows from $x = (x_1, x_2, \ldots, x_n)^T = \sum_{i=1}^n x_i \delta_i$.

EXAMPLE 5. Let $\beta = \{E_k\}_{k=1}^n$ be a basis for $\mathbf{F}^n$. Then the coordinate column of x in β is

$$[x]_\beta = E^{-1}x, \qquad \text{where } E = [E_1\ E_2\ \cdots\ E_n]. \tag{1}$$

Let $[x]_\beta = c$. Then $x = \sum_{i=1}^n c_i E_i = Ec$ by the definition of $[x]_\beta$ and by Proposition 1 of Section 0.4. According to Corollary 8 of Section 1.3, E is invertible since it has linearly independent columns.

Suppose a basis $\beta = \{e_i\}_{i=1}^n$ for V is given. The process of finding coordinates of a given vector x in V is an operation that takes vectors in V to an n-tuple in $\mathbf{F}^n : x \rightarrow [x]_\beta$. This operation simply detaches the coordinates from the sum $x = \sum_{i=1}^n c_i e_i$ to make the column $c = [x]_\beta$, and the inverse operation restores the entries of the column c back to the correct basis vectors. The order in the column and basis are crucial: The coordinates must multiply the correct basis vectors. The next proposition states some properties of this operation.

PROPOSITION 1. Suppose a basis $\beta = \{e_k\}_{k=1}^n$ for V is given. Then

(a) The operation $x \rightarrow [x]$ is a one-to-one correspondence between V and $\mathbf{F}^n$: i.e., to each x in V there is only one c in $\mathbf{F}^n$ such that $[x] = c$, and to each c in $\mathbf{F}^n$ there is just one x in V such that $[x] = c$.

Furthermore,

(b) $[x + y] = [x] + [y] \qquad \forall x, y \in V$
(c) $[\alpha x] = \alpha[x] \qquad \forall x \in V, \quad \alpha \in \mathbf{F}$
(d) $[\sum_{k=1}^n \alpha_k x_k] = \sum_{k=1}^n \alpha_k[x_k]$.

Proof. Statement (a) immediately follows from the definition and from the fact that β is a basis for V. Statements (b) and (c) follow from

$$\left[\sum c_i e_i + \sum d_i e_i\right] = \left[\sum (c_i + d_i)e_i\right] = (c_1 + d_1, \ldots, c_n + d_n)^\mathsf{T} = c + d$$

$$\left[\alpha \sum c_i e_i\right] = \left[\sum \alpha c_i e_i\right] = (\alpha c_1, \ldots, \alpha c_n) = \alpha c.$$

(d) follows from a repeated application of (b) and (c).

Remark. Let V and W be vector spaces. A function $T : V \rightarrow W$ when T satisfies $T(x + y) = T(x) + T(y)$, $T(\alpha x) = \alpha T(x)$, and when T is a one-to-one correspondence between the vectors in V and those in W is called an *isomorphism* between V and W. When two vector spaces are isomorphic, i.e., when such a T exists, then in some sense one is just a replication or copy of the other. In this sense, Proposition 1 says that all n-dimensional vector spaces over $\mathbf{F}$ are isomorphic to $\mathbf{F}^n$ by the isomorphism $T(x) = [x]_\beta$, where β is a fixed basis of V. Thus all n-dimensional vector spaces are isomorphic to each other. In some sense, this means that there is no n-dimensional vector space other than $\mathbf{F}^n$. However, this is a rather narrow view, since, for example, there are questions about $\mathbf{P}_n$ that one would never think of posing while contemplating $\mathbf{F}^{n+1}$, and vice versa.

In another sense, the isomorphism in Proposition 1 is an extremely powerful tool because it means that questions in V can be reformulated in $\mathbf{F}^n$ and possibly solved there, either theoretically or numerically. For example, if we want to test $f_0, f_1, \ldots, f_m \in \mathbf{P}_n$ for linear independence, we can select a basis $\{e_k\}_{k=0}^n$ for $\mathbf{P}_n$ and compute the coordinate columns $[f_0], [f_1], \ldots, [f_m]$. The original vectors f_i will be linearly independent iff the columns $[f_i]$ are linearly independent in $\mathbf{F}^{n+1}$. Some of our previous examples have actually been based on this technique; see Example 13 of Section 1.2 and Example 3 of Section 1.3. Another example based on this technique follows:

EXAMPLE 6. Let $f_1 = 1 - \cos(t)$, $f_2 = 1 - 3\cos(t) + \sin(t)$, and $f_3 = 1 - \cos(t) + \sin(t)$. Show that $\{f_1, f_2, f_3\}$ is a basis for $\mathbf{T}_1$.

Since $\dim(\mathbf{T}_1) = 3$, it is sufficient to show that $\{f_1, f_2, f_3\}$ is linearly independent. Choose the basis $\{1, \cos(t), \sin(t)\}$ for $\mathbf{T}_1$. Then $[f_1] = (1, -1, 0)^T$, $[f_2] = (1, -3, 1)^T$, and $[f_3] = (1, -1, 1)^T$. Using the isomorphism, $\{f_1, f_2, f_3\}$ is linearly independent iff $\{[f_1], [f_2], [f_3]\}$ is linearly independent. The independence of these columns is tested by row-reducing:

$$A \equiv [[f_1] [f_2] [f_3]] = \begin{bmatrix} 1 & -1 & 0 \\ 1 & -3 & 1 \\ 1 & -1 & 1 \end{bmatrix}.$$

Since row reduction produces the identity matrix, $Ax = 0$ has only the solution $x = 0$, and $\{[f_1], [f_2], [f_3]\}$ is linearly independent by Proposition 7 of Section 1.3.

Change of Basis

When the basis is changed, then the coordinates of a fixed vector change. Here we study how the coordinate columns change with a change of basis.

The necessity to change a basis can arise in several ways. For example, measurements of coordinates may be taken by different observers in different reference frames (i.e., using different basis vectors), and the measurements must be compared; thus the relationship between the coordinates is needed. Many problems become easy to solve when the correct basis is chosen. Many examples of this kind will appear later in this book. A special basis may also permit profound interpretations of the problem at hand. This happens in pattern recognition, statistics, and oscillation theory, for example. Examples will be given later in the book.

The next theorem gives the relationship between the coordinate columns of a fixed vector when the basis is changed.

THEOREM 2. Let $\beta = \{e_1, e_2, \ldots, e_n\}$, $\gamma = \{f_1, f_2, \ldots, f_n\}$ be two bases for the n-dimensional vector space V. Expand each f_j:

$$f_j = \sum_{i=1}^{n} p_{ij} e_i, \qquad j = 1, 2, \ldots, n. \tag{2}$$

Then $P \equiv [p_{ij}]$ is nonsingular, and the coordinate columns of $x \in V$ are related by

$$[x]_\beta = P[x]_\gamma. \tag{3}$$

Specifically,

$$x = \sum_{i=1}^{n} c_i e_i = \sum_{j=1}^{n} d_j f_j \iff Pd = c. \tag{3'}$$

Proof. Two proofs are given of (3) in order to illustrate different techniques. Let

$$x = \sum_{i=1}^{n} c_i e_i = \sum_{j=1}^{n} d_j f_j \tag{*}$$

so that $[x]_\beta = c$ and $[x]_\gamma = d$. For the first proof, eliminate f_j in (*) using (2) and rearrange:

$$x = \sum_{j=1}^{n} d_j f_j = \sum_{j=1}^{n} \left\{ \sum_{i=1}^{n} p_{ij} e_i \right\} = \sum_{i=1}^{n} \left\{ \sum_{j=1}^{n} p_{ij} d_j \right\} e_i = \sum_{i=1}^{n} c_i e_i. \tag{**}$$

The independence of the e_i implies (by uniqueness of coefficients) that

$$\sum_{j=1}^{n} p_{ij} d_j = c_i, \qquad i = 1, 2, \ldots, n.$$

This is equivalent to $Pd = c$, proving (3).

For the second proof, take the coordinates of x in (∗) and use statement 4 of Proposition 1:

$$[x]_\beta = \left[\sum_{j=1}^{n} d_j f_j\right]_\beta = \sum_{j=1}^{n} d_j [f_j]_\beta = \sum_{j=1}^{n} d_j P_j = Pd = P[x]_\gamma.$$

To show that P is nonsingular, suppose all $c_i = 0$, so that $x = 0$ in (∗∗). By the preceding equation, $Pd = c = 0$. Since γ is a basis, $\sum_{j=1}^{n} d_j f_j = 0$ implies that all of the d_i are zero. Thus the only solution of $Pd = 0$ is $d = 0$. By Theorem 4 of Section 0.5, P is nonsingular.

The terminology surrounding the matrix P is sometimes confusing. Different authors use different names, and many avoid formulating a precise name.

DEFINITION. The matrix P in (1) is called the *change of basis matrix* from the basis β to the basis γ since the vectors f_j in γ are expressed in terms of the e_i in β by (2). Some authors call P the *change of coordinate matrix* from the basis γ to the basis β since the coordinates of x in β are expressed in terms of its coordinates in γ by (3). Other authors refuse (perhaps wisely) to give P a name. Some authors call P the *modal matrix* or the *transition matrix,* and some label it with the letter S instead of the letter P. The appellation "modal matrix" will be explained in Chapter 7. A more precise notation would be $P = P^{(\beta,\gamma)}$, but we avoid this because it is unnecessarily pedantic.

The concept of change of basis causes confusion since the basis and the coordinates change in different ways and since different authors use different names, depending on whether (2) or (3) is deemed to be most natural. The best strategy is to remember the relationship between how the coordinate vectors change and how the coordinates change. The relationships are restated again, both in terms of sums and matrix multiplications. Equation (3″) is equivalent to (3) and (3′).

$$f_j = \sum_{i=1}^{n} p_{ij} e_i \Leftrightarrow (f_1, f_2, \ldots, f_n) = (e_1, e_2, \ldots, e_n)P \qquad (2')$$

$$c_i = \sum_{j=1}^{n} p_{ij} d_j \Leftrightarrow c = Pd, \qquad \text{where } x = \sum_{i=1}^{n} c_i e_i = \sum_{j=1}^{n} d_j f_j. \qquad (3'')$$

Note that in (2′) the sum index is on the row index of P, whereas in (3″) it is on the column index.

To construct P, follow this recipe: Expand each f_j in terms of the e_i. By (2′), $(p_{1j}, p_{2j}, \ldots, p_{nj})^T$ is the coordinate vector of f_j in β, so that the jth column of P is

$$P_j = (p_{1j}, p_{2j}, \ldots, p_{nj})^T = [f_j]_\beta. \qquad (4)$$

Accumulate each of these columns to get the matrix P;

$$P = [[f_1]_\beta \, [f_2]_\beta \, \cdots \, [f_n]_\beta]. \qquad (4')$$

P can be displayed diagrammatically by

$$
P = \begin{array}{c} \\ e_1 \\ e_2 \\ \\ \\ e_n \end{array}
\begin{array}{cccc} f_1 & f_2 & & f_n \\
\left[\begin{array}{cccc} p_{11} & p_{12} & \cdots & p_{1n} \\
p_{21} & p_{22} & \cdots & p_{2n} \\
\vdots & \vdots & \cdots & \vdots \\
p_{n1} & p_{n2} & \cdots & p_{nn} \end{array}\right] \end{array} = [[f_1]_\beta \ [f_2]_\beta \cdots [f_n]_\beta].
$$

This indicates that

$$f_1 = p_{11}e_1 + p_{21}e_2 + \cdots + p_{n1}e_n,$$
$$f_2 = p_{12}e_1 + p_{22}e_2 + \cdots + p_{n2}e_n,$$

and so on.

For convenience, we call the basis $\beta = \{e_i\}_{i=1}^n$ of Theorem 2 the old basis and $\gamma = \{f_i\}_{i=1}^n$ the new basis because the f_j are given in terms of the e_i.

The next example illustrates the proof of the theorem and the rule for the construction of P.

EXAMPLE 7. Let $V = \mathbf{P}_2$, let $\beta = \{1, t, t^2\} = \{e_0, e_1, e_2\}$, and let $\gamma = \{1, t - 1, (t - 1)^2\} = \{f_0, f_1, f_2\}$. If $g(t) = d_0 + d_1(t - 1) + d_2(t - 1)^2$, find the c_k in the expansion $g(t) = c_0 + c_1 t + c_2 t^2$.

To illustrate the construction of $P = [P_0 \ P_1 \ P_2]$, note that

$$f_0 = 1 + 0 \cdot t + 0 \cdot t^2, \qquad f_1 = -1 + t + 0 \cdot t^2, \qquad f_2 = 1 - 2t - t^2,$$

which yields

$$[f_0]_\beta = P_0 = \begin{bmatrix} 1 \\ 0 \\ 0 \end{bmatrix}, \qquad [f_1]_\beta = P_1 = \begin{bmatrix} -1 \\ 1 \\ 0 \end{bmatrix},$$

$$[f_2]_\beta = P_2 = \begin{bmatrix} 1 \\ -2 \\ 1 \end{bmatrix}, \qquad \text{and} \qquad P = \begin{bmatrix} 1 & -1 & 1 \\ 0 & 1 & -2 \\ 0 & 0 & 1 \end{bmatrix}.$$

To illustrate the proof, begin with an arbitrary p expressed in terms of γ, expand the members of γ in terms of those in β, and rearrange into a sum over the members of β:

$$g(t) = d_0 \cdot 1 + d_1(t - 1) + d_2(t^2 - 2t + 1)$$
$$= (d_0 - d_1 + d_2) \cdot 1 + (d_1 - 2d_2)t + d_2 t^2 = c_0 + c_1 t + c_2 t^2.$$

Thus

$$c = \begin{bmatrix} c_0 \\ c_1 \\ c_2 \end{bmatrix} = [g]_\beta = \begin{bmatrix} d_0 - d_1 + d_2 \\ d_1 - 2d_2 \\ d_2 \end{bmatrix} = \begin{bmatrix} 1 & -1 & 1 \\ 0 & 1 & -2 \\ 0 & 0 & 1 \end{bmatrix} \begin{bmatrix} d_0 \\ d_1 \\ d_2 \end{bmatrix} = P[g]_\gamma.$$

Both methods give the same P.

EXAMPLE 8. In a plane, establish an x_1-x_2 coordinate system so that $x = (x_1, x_2)^\mathsf{T}$ denotes the coordinates of a point R. Establish a new coordinate system y_1-y_2 by rotating the old one through a fixed angle θ counterclockwise. $y = (y_1, y_2)^\mathsf{T}$ denotes the coordinates of the point R in the new system. Find the relation between x and y.

Let $\beta = \{e_1, e_2\}$ and $\gamma = \{f_1, f_2\}$ be the (geometric) basis vectors in the x and y coordinate systems, respectively. From trigonometry,

$$f_1 = \cos(\theta)e_1 + \sin(\theta)e_2$$

so that the coordinate column of f_1 in the basis β is

$$P_1 = [f_1]_\beta = \begin{bmatrix} \cos(\theta) \\ \sin(\theta) \end{bmatrix}.$$

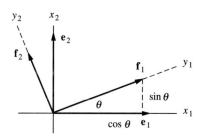

Similarly, the coordinate column of f_2 is

$$P_2 = [f_2]_\beta = \begin{bmatrix} -\sin(\theta) \\ \cos(\theta) \end{bmatrix}.$$

Thus

$$P = [[f_1]_\beta \ [f_2]_\beta] = \begin{bmatrix} \cos(\theta) & -\sin(\theta) \\ \sin(\theta) & \cos(\theta) \end{bmatrix},$$

and the relationship between x and y is

$$x = Py, \qquad \text{i.e.,} \qquad \begin{array}{l} x_1 = y_1 \cos(\theta) - y_2 \sin(\theta) \\ x_2 = y_1 \sin(\theta) + y_2 \cos(\theta) \end{array}.$$

P is called a *rotation matrix*.

PROPOSITION 3. If Q is the (reverse) change of basis matrix from the basis γ to the basis β, then $Q = P^{-1}$.

The proof is left as an exercise. (See Exercise 15.)

EXAMPLE 9. Let $V = P_2$, let $\beta = \{1, t, t^2\}$, and let $\gamma = \{1, t - 1, (t - 1)^2\}$. Find the change of basis matrix Q from γ to β.

According to Proposition 3, this is the inverse of the matrix P found in Example 7, and Q may be computed as $Q = P^{-1}$. However, Q can also be calculated directly from expanding each $e_j \in \beta$ in terms of the $f_i \in \gamma$:

$$e_0 = 1 = f_0$$
$$e_1 = t = f_1 + f_0$$
$$e_2 = [(t - 1) + 1]^2 = f_2 + 2f_1 + f_0$$

$$\Rightarrow Q = [[e_0]_\gamma \ [e_1]_\gamma \ [e_2]_\gamma] = \begin{bmatrix} 1 & 1 & 1 \\ 0 & 1 & 2 \\ 0 & 0 & 1 \end{bmatrix}.$$

It is easy to see that $QP = I = PQ$.

The next proposition is used often in this book. It involves the quirk in $\mathbf{F}^n$ that $[x]_\beta = x$ when $\beta = \delta = \{\delta_1, \delta_2, \ldots, \delta_n\}$. See Example 5, where the corresponding change of coordinates is deduced.

PROPOSITION 4. Let $V = \mathbf{F}^n$. Suppose the old basis is the standard basis $\delta = \{\delta_1, \delta_2, \ldots, \delta_n\}$ and the new basis is $\gamma = \{A_1, A_2, \ldots, A_n\}$. Then $P = [A_1 \ A_2 \ \cdots \ A_n]$ and $P[x]_\gamma = x$.

Proof. According to (3') in Theorem 2, with $e_i = \delta_i$ and $f_j = A_j$,

$$x = \sum_{i=1}^n c_i \delta_i = \sum_{j=1}^n d_j A_j \Leftrightarrow c = Pd.$$

In this context, $c_i = x_i$, so $c = x$. Since $\sum_{j=1}^n d_j A_j = [A_1 \ A_2 \ \cdots \ A_n]d$, it must be that $P = [A_1 \ A_2 \ \cdots \ A_n]$.

Alternatively, one could note that $P_j = [A_j]_\delta = A_j$ from Example 4 and by using (4').

Change of Coordinates

In some applications we will encounter, the problem is stated in a coordinate system x. This means that the discussion involves variables x_i, which are the coordinates of x in a reference basis $\{e_i\}_{i=1}^n$, which is given. In order to produce some desirable effect, we want to change to the y coordinate system by multiplying x by a nonsingular matrix Q: $y = Qx$. The next proposition shows that this is a change of basis with $P = Q^{-1}$ and identifies the basis $\{f_k\}_{k=1}^n$.

PROPOSITION 5. Let $P = [p_{ij}]$ be nonsingular, let $\beta = \{e_i\}_{i=1}^n$ be a basis for V, and define vectors f_j by

$$f_j = \sum_{i=1}^n p_{ij} e_i \qquad j = 1, 2, \ldots, n.$$

That is, $P = [[f_1]_\beta \ [f_2]_\beta \ \cdots \ [f_n]_\beta]$. Then $\gamma = \{f_i\}_{i=1}^n$ is a basis for V, and P is the change of basis matrix from β to γ.

Proof. Suppose $0 = \sum_{j=1}^n d_j f_j$. Following the method of Theorem 2 and eliminating the f_j from this equation using (1) yields

$$0 = \sum_{j=1}^n d_j f_j = \sum_{j=1}^n d_j \left\{ \sum_{i=1}^n p_{ij} e_i \right\} = \sum_{i=1}^n \left\{ \sum_{j=1}^n p_{ij} d_j \right\} e_i.$$

Hence $\sum_{i=1}^n p_{ij} d_j = 0$ for each j, by the independence of $\{e_k\}_{k=1}^n$. Thus $Pd = 0$, and since P is nonsingular, $d = 0$. Hence $\{f_i\}_{i=1}^n$ is linearly dependent. Since $n = \dim(V)$, it is a basis by Theorem 6 of Section 1.4.

EXAMPLE 10. Show that the set $f_1 = 1 + t + t^2$, $f_2 = 2t + 3t^2$, $f_3 = 4t^2$ is a basis for $\mathbf{P}_2$. Express $p(t) = 1 - t + 2t^2$ as a linear combination of the f_j.

Here we choose $\beta = \{1, t, t^2\}$. The matrix

$$P = \begin{bmatrix} 1 & 0 & 0 \\ 1 & 2 & 0 \\ 1 & 3 & 4 \end{bmatrix} = [[f_1] \ [f_2] \ [f_3]]$$

is nonsingular, and so $\{f_i\}_{i=1}^3$ is a basis. Next, find d_j so that $p(t) = d_1 f_1 + d_2 f_2 + d_3 f_3$. By (2),

$$Pd = \begin{bmatrix} 1 \\ -1 \\ 2 \end{bmatrix}.$$

Solving this yields $d_1 = 1$, $d_2 = -1$, $d_3 = 1$.

In the next example, precise notation is suspended; we will identify geometrical vectors and their coordinates in the plane, as is often done in calculus texts.

EXAMPLE 11. In $\mathbf{R}^2$, take the standard basis as a reference with coordinates x, y. Define new coordinates by the rule

$$x' = \cos(\phi)x + \sin(\phi)y$$
$$y' = -\sin(\phi)x + \cos(\phi)y.$$

Find the new basis and interpret geometrically.

Here $\beta = \{\delta_1, \delta_2\}$ and the given change of coordinates is of the form $[v]_\gamma = P^{-1}[v]_\beta$, so

$$P^{-1} = \begin{bmatrix} \cos(\phi) & \sin(\phi) \\ -\sin(\phi) & \cos(\phi) \end{bmatrix},$$

whence $\quad P = \begin{bmatrix} \cos(\phi) & -\sin(\phi) \\ \sin(\phi) & \cos(\phi) \end{bmatrix}.$

Thus the new basis is

$$f_1 = \cos(\phi)\delta_1 + \sin(\phi)\delta_2 = \begin{bmatrix} \cos(\phi) \\ \sin(\phi) \end{bmatrix}$$

$$f_2 = -\sin(\phi)\delta_1 + \cos(\phi)\delta_2 = \begin{bmatrix} -\sin(\phi) \\ \cos(\phi) \end{bmatrix}.$$

According to Example 8, f_1 and f_2 are the vectors δ_1, δ_2 rotated counterclockwise by an angle ϕ.

The next simple example illustrates how using the "correct" coordinates and basis can simplify a problem. A plethora of examples occur in subsequent chapters.

EXAMPLE 12. Let $f(x, y) = (x + y)^2 + 2(y - x)^2$. Plot $f(x, y) = 8$.
    Make the change of coordinates (i.e., substitution)

$$x' = (x + y)/\sqrt{2}, \qquad y' = (y - x)/\sqrt{2}. \tag{$*$}$$

This reduces the problem to plotting $(x')^2 + 2(y)'^2 = 4$, which is an ellipse in the x'-y' plane. This can easily be done if we recognize from Examples 8 and 11 that ($*$) rotates the basis vectors by 45° in the counterclockwise direction without changing their length. Thus we draw the positive x' axis along a ray 45° counterclockwise from the positive x axis, and we position the positive y' axis along a ray 45° counterclockwise from the positive y axis. Once these axes are sketched in the x-y plane, the ellipse is easily drawn. In Chapter 6, on quadratic forms, the ideas in this example will be generalized and extended.

Exercises 1.5

1. Show that $\{(1, 1, 1)^T, (1, 0, 1)^T, (1, -1, 2)^T\}$ is a basis for $\mathbf{C}^3$, and find the coordinates of $(2, 3, -1)^T$ relative to this basis, in the given order.

2. Let $x = (2, 0, 2, 4)^T$. Find the coordinates of x in the following ordered bases for $\mathbf{R}^4$:
   (a) $(1, 1, 0, 0)^T, (1, -1, 0, 0)^T, (0, 0, -1, 1)^T, (0, 0, 1, 1)^T$
   (b) $(0, 1, 1, 1)^T, (1, 0, 1, -1)^T, (1, -1, 0, 1)^T, (1, 1, -1, 0)^T$

3. Let $e_k = t^k$, $k = 0, 1, 2$, and let $f_0 = 1 + t^2$, $f_1 = t - t^2$, $f_2 = 1 - 2t + t^2$. Find the change of basis matrix P and the reverse change matrix Q. Write $1 + t$ as a linear combination of the f_i.

4. Let $f_1(t) = 2 - \cos(t) + \sin(t)$, $f_2(t) = 1 + \sin(t)$, and $f_3(t) = 2 - 2\cos(t) + 3\sin(t)$. Show that $\{f_i\}_{i=1}^3$ is a basis for $\mathbf{T}_1$. Express $\cos(t) = \sum_{i=1}^3 c_k f_k$, $1 = \sum_{i=1}^3 d_k f_k$.

5. Find the coordinates of $\cos^3(x)$ in each of the following ordered bases for $\mathbf{T}_3$:
   (a) $\{1, \cos(x), \sin(x), \cos(2x), \sin(2x), \cos(3x), \sin(3x)\}$
   (b) $\{e^{ikx} : k = -3, -2, \ldots, 3\}$

6. Let $p_0(t) = t^3$, $p_1(t) = t^2(t + 1)$, $p_2(t) = t(t + 1)(t - 1)$, $p_3(t) = t^3 + t^2 + t + 1$.
   (a) Show that $\gamma = \{p_0, p_1, p_2, p_3\}$ is a basis for $\mathbf{P}_3$.
   (b) Let $f(t) = \sum_{i=0}^3 a_i t^i = \sum_{i=0}^3 b_i p_i(t)$. Set $a = (a_0, a_1, a_2, a_3)^T$, $b = (b_0, b_1, b_2, b_3)^T$. Find matrices R and S so that $Ra = b$ and $Sb = a$.
   (c) Find the coordinates of x^3 and of $1 + x + x^2$ with respect to γ.

7. Let $p_k(x) = x^{3-k}(1 - x)^k$, where $k = 0, 1, 2, 3$.
   (a) Show that $\gamma = \{p_0, p_1, p_2, p_3\}$ is a basis for $\mathbf{P}_3$.
   (b) Let $f(t) = \sum_{i=0}^3 a_i x^i = \sum_{i=0}^3 b_i p_i(x)$. Set $a = (a_0, a_1, a_2, a_3)^T$, $b = (b_0, b_1, b_2, b_3)^T$. Find matrices R and S so that $Sa = b$ and $Rb = a$.
   (c) Find the coordinates of 1 and of $1 + x + x^2$ with respect to γ.

8. Let $p_0(t) = 1$, $p_1(t) = t$, $p_2(t) = t(t - 1)$, and $p_3(t) = t(t - 1)(t - 2)$.
   (a) Show that $\gamma = \{p_0, p_1, p_2, p_3\}$ is a basis for $\mathbf{P}_3$.
   (b) Let $f(t) = \sum_{i=0}^3 a_i t^i = \sum_{i=0}^3 b_i p_i(t)$. Set $a = (a_0, a_1, a_2, a_3)^T$, $b = (b_0, b_1, b_2, b_3)^T$. Find matrices M and N so that $Ma = b$ and $Nb = a$.
   (c) Find the coordinates of t^2 and of t^3 with respect to γ.

9. Show that $\{x^k(1 - x)^{n-k} : k = 0, 1, \ldots, n\}$ is a basis for $\mathbf{P}_n$.

10. Let
$$M_1 = \begin{bmatrix} 1 & 0 \\ 0 & 0 \end{bmatrix}, \quad M_2 = \begin{bmatrix} 1 & 1 \\ 0 & 0 \end{bmatrix}, \quad M_3 = \begin{bmatrix} 1 & 1 \\ 1 & 0 \end{bmatrix},$$
$$M_4 = \begin{bmatrix} 1 & 1 \\ 1 & 1 \end{bmatrix}, \quad A = \begin{bmatrix} 1 & 2 \\ 3 & 1 \end{bmatrix}.$$
   (a) Show that $\gamma = \{M_i\}_{i=1}^4$ is a basis for $\mathbf{C}^{2,2}$.
   (b) Find the coordinates of A with respect to γ.
   (c) Express A as a linear combination of the M_i.

11. Let

$$M_1 = \begin{bmatrix} 1 & 1 \\ 1 & 1 \end{bmatrix}, \quad M_2 = \begin{bmatrix} -1 & 1 \\ 1 & -1 \end{bmatrix} \quad M_3 = \begin{bmatrix} 0 & 1 \\ -1 & 0 \end{bmatrix},$$

$$M_4 = \begin{bmatrix} 1 & 0 \\ 0 & -1 \end{bmatrix}, \quad A = \begin{bmatrix} 1 & 3 \\ 1 & -1 \end{bmatrix}.$$

(a) Show that $\gamma = \{M_i\}_{i=1}^4$ is a basis for $\mathbf{C}^{2,2}$.
(b) Find the coordinates of A with respect to γ.
(c) Express A as a linear combination of the M_i.

12. Suppose $\beta = \{E_1, E_2, \ldots, E_n\}$ and $\gamma = \{F_1, F_2, \ldots, F_n\}$ are bases for $\mathbf{F}^n$, and suppose P is the change of basis matrix from β to γ. Let $E = [E_1 \ E_2 \ \cdots \ E_n]$ and $F = [F_1 \ F_2 \ \cdots \ F_n]$. Show that
(a) $F_j = EP_j$ for $j = 1, 2, \ldots, n$
(b) $P = E^{-1}F$

13. Let $V = \mathbf{P}_n$, and consider the basis $\beta = \{(t - a)^k : k = 0, 1, \ldots, n\}$. Show that the coordinates c_k of $p(t)$ with respect to β are

$$c_k = p^{(k)}(a)/k!, \quad k = 0, 1, \ldots, n.$$

14. Let $\beta = \{t^k\}_{k=0}^n$ and $\gamma = \{(t - 1)^k\}_{k=0}^n$. Find the change of basis matrix P from β to γ, and find P^{-1}.

15. Suppose P is the change of basis matrix from β to γ and Q is the change of basis matrix from γ to β. Show that $Q = P^{-1}$. *Hint:* Several proofs are possible. One begins with $e_i = \sum_{k=1}^n q_{ki} f_k$.

16. Let $\{e_k\}_{k=1}^n$ and $\{f_k\}_{k=1}^n$ be bases for V. Show that the change of basis matrix is upper triangular iff $\text{span}\{e_1, e_2, \ldots, e_k\} = \text{span}\{f_1, f_2, \ldots, f_k\}$ for $k = 1, 2, \ldots, n$.

17. Let $P = QR$ where all the matrices are $n \times n$. Suppose P is nonsingular. Show that the columns of P and of Q are bases for $\mathbf{F}^n$. Find the change of basis matrix from the Q columns to the P columns.

18. "Line of sight" basis and coordinates: A basis for three-dimensional graphing.
Let x, y, z be a reference coordinate system in $\mathbf{R}^3$. The viewer is located along a ray through the origin, which is the new x axis. The new coordinate system x', y', z' is obtained as follows:

1. Rotate the x and y axes counterclockwise about the z axis through an angle ϕ in the x-y plane; the position of the rotated y axis is the y' axis.
2. Rotate the resulting position of the x axis upwards by an angle Ψ about the y' axis; the resulting position is the x' axis. In the second rotation, the z axis is rotated as well.

Ψ is called the *tilt* or *elevation*. The resulting ray along the new x axis has spherical coordinates $(\phi, (\pi/2) - \Psi, 1)$.
(a) Let $\beta = \{e_1, e_2, e_3\}$ be the basis in the x, y, z system, and let $\gamma = \{f_1, f_2, f_3\}$ be the basis in the x', y', z' system. Express each f_i as a linear combination of the e_j.
(b) Find the change of basis matrix P from β to γ. Show that $PP^T = I$, and find the reverse change of basis matrix.
(c) If a point has coordinates $(x, y, z)^T$ in the reference system, find its coordinates $(x', y', z')^T$ in the rotated system. Conversely, if a point has coordinates $(x', y', z')^T$ in the reference system, find its coordinates $(x, y, z)^T$ in the rotated system.

CHAPTER 2

Linear Operators

Linear operators occur frequently in mathematics, engineering, and the sciences. The reader may be familiar with the Laplace and Fourier transforms and linear differential operators, which are all examples of linear operators. The behavior of linear operators on finite dimensional spaces is our central topic. This chapter introduces some of the general properties of linear operators, many examples of them, and the relationship of linear operators to matrices. It turns out that matrices can be regarded as linear operators, so that the theory of linear operators can be applied to determine properties of matrices. Another connection between linear operators and matrices allows matrix theory to be applied to linear operators to great advantage.

2.1
PRELIMINARY TOPICS

In this section, we look at some basic properties of linear operators and give some important examples. We begin with a review of arbitrary functions and some of the language used to describe them.

The Language of Functions

We say that *T acts on V and has values in W* or that *T is a function from V to W* and write $T : V \rightarrow W$ to mean that V and W are sets, and T is a rule such that for each x in V, the value $T(x)$ is defined in just one way (single valued), and that $T(x)$ belongs to W for each x in V. V is called the *domain* of T, and W is the *codomain* of T. $T(x)$ is called the *value of T at x* or the *image of x* under the function T. The words "function," "*map*," "*mapping*," "*transformation*," and "*operator*" are used synonymously, although some reserve the word "operator" for functions that act on other functions.

The operational rule T may be defined on a larger space than V, and in this case, the notation $T : V \rightarrow W$ indicates that only $x \in V$ is allowed to be acted on by T. For example, if $T(f) = f''$ (second derivative of f), and if we want to restrict this operator to act only on polynomials of degree not exceeding n, then we would write $T : \mathbf{P}_n \rightarrow \mathbf{P}_n$. Changing V or W can radically change the properties of the mapping $T : V \rightarrow W$, as we will see later, so specifying V and W is important.

The *range of T on V* is the set of all $y \in W$ for which the equation $y = T(x)$ has at least one solution for x. The range is also called the *image of T on V*, and is denoted

$$R(T) \equiv \text{Im}(T) \equiv T(V) = \{T(x) : x \in V\}.$$

T maps V onto W, or the function $T : V \rightarrow W$ is onto if and only if, for every $y \in V$, the equation $y = T(x)$ has at least one solution. That is, $T : V \rightarrow W$ is onto if and only if $R(T) = W$.

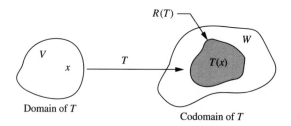

Domain of T Codomain of T

T is one to one on V, or the function $T : V \rightarrow W$ is one to one, if and only if, for every y in W, the equation $y = T(x)$ has at most one solution in V; i.e., solutions are unique, if they exist. The following statements are equivalent:

1. For each y, $T(x) = y$ has at most one solution for x.
2. T is one to one on V.
3. $T(x) = T(y)$, with $x, y \in V$ implies $x = y$.
4. $x \neq y$ and $x, y \in V$ implies $T(x) \neq T(y)$.

In words, (3) and (4) say that the images of distinct points are again distinct points.

For example, the function $\sin : [0, 2\pi] \rightarrow \mathbf{R}$ is not onto, because $\sin(x) = 2$ has no solution and is not one to one since $\sin(0) = \sin(\pi) = 0$. The image or range of this function is $[-1, 1]$, so that $\sin : [0, 2\pi] \rightarrow [-1, 1]$ is onto. However, the function $\sin : [0, \pi/2] \rightarrow [0, 1]$ is one to one and onto since, for every y in $[0, 1]$, $\sin(x) = y$ has exactly one solution.

Multiplication of operators may be defined as follows:

DEFINITIONS. The *identity operator I* is defined by $I(x) = x, \forall x \in V$, where $I : V \rightarrow W$ and $V \subset W$. The *composition $S \circ T$* is defined by $(S \circ T)(x) = S[T(x)] \, \forall x \in V$, when $T : V \rightarrow W, S : W \rightarrow X$, so that $S \circ T : V \rightarrow X$.

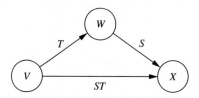

Remarks. Properly, one should write $I = I_V$ to indicate the domain of I, since technically $I_V \neq I_X$ when $X \neq V$. In contrast to calculus, in linear algebra the composition sign $\circ$ is always suppressed because of a similarity to matrix algebra that we will explore in later sections. The notation is $S \circ T = ST$ and $T^3 = T \circ T \circ T = TTT$, and so on. Similar algebraic rules for the operator combinations hold as those for matrix algebra, according to the next proposition.

PROPOSITION 1. When the operator sums and products in the following expressions are defined, then the following equalities hold:

$$S(TU) = (ST)U \qquad (S + T)U = SU + TU \qquad U(S + T) = US + UT$$

$$IS = S \qquad TI = T.$$

Remark. The commutative and cancellation rules fail; i.e., $TS \neq ST$ in general, and $ST = SU$ with $S \neq 0$ does not imply that $T = U$, nor does $TS = US$ with $S \neq 0$. Furthermore, $T \neq 0$ does not guarantee that T^{-1} exists. The matrix examples of Section 0.4 illustrate this.

Linear Operators

Now we turn to some fundamental facts about linear operators and some central examples.

DEFINITION. A *linear operator T* is a function from a vector space V to another vector space W satisfying

$$T(x + y) = T(x) + T(y) \qquad \text{for all } x, y \in V, \tag{1}$$

and

$$T(\alpha x) = \alpha T(x) \qquad \text{for all } x, y \in V \text{ and all scalars } \alpha. \tag{2}$$

Thus

$$T\left(\sum_{i=1}^{n} \alpha_i x_i\right) = \sum_{i=1}^{n} \alpha_i T(x_i); \tag{3}$$

that is, T preserves linear combinations.

It is easy to see that T is linear if and only if

$$T(\alpha x + \beta y) = \alpha T(x) + \beta T(y) \qquad \text{for all scalars } \alpha, \beta \text{ and for all } x, y \text{ in } V. \tag{4}$$

It also follows from the definitions that $T(0) = 0$ since, if $\alpha = 0$, then $T(0) = T(\alpha 0) = \alpha T(0) = 0 T(0) = 0$.

Remark. For linear operators, some authors write $T(x) = Tx$.

EXAMPLE 1. The coordinate detaching operation $[\ ] : V \to \mathbf{F}^n$ is linear, one to one, and onto.

Given a basis $\{e_k\}_{k=1}^{n}$ for V, the coordinate map is defined to be the coordinate column of a vector; that is, $[\sum_{i=1}^{n} x_i e_i] = (x_1, x_2, \ldots, x_n)^T$. In Proposition 1 in Section 1.5, these assertions were proved.

DEFINITION. A matrix operator is a map $T : \mathbf{F}^n \to \mathbf{F}^m$ defined by $T(x) = Ax$, where A is an $m \times n$ matrix with entries in $\mathbf{F}$. T is called the *matrix operator generated by A*, and is sometimes denoted by T_A.

EXAMPLE 2. Matrix operators are linear.
    This follows from $A(\alpha x + \beta y) = \alpha Ax + \beta Ay$.

EXAMPLE 3. MINIMAL DISTANCES. Given a plane M through the origin in $\mathbf{R}^3$, define $T(x) = p$ iff p is the closest point p in M to x. Is it true that $T(x + y) = T(x) + T(y)$, i.e., that the nearest point to the sum is the sum of the nearest points? This is not so clear. However, in Section 3.3, T will be shown to be linear.

DEFINITION. A *linear functional* on V is a linear operator $T : V \to \mathbf{F}$, where $\mathbf{F}$ is the scalar field for V.

EXAMPLE 4. SOME LINEAR FUNCTIONALS

$$T(x) = \sum_{i=1}^{n} x_i \qquad V = \mathbf{R}^n$$

$$T(x) = \int_0^1 x(t)e^{-t}\, dt \qquad V = C([0, 1])$$

$$T(x) = x(t_0) \qquad V = C([0, 1]), \text{ and } t_0 \text{ is a fixed number in } [0, 1].$$

Operators on function spaces

Suppose $T : V \to W$, where V and W are vector spaces of functions. $T(f)$ then denotes the new function that results when T acts on f. The value of this function at a point x is then $[T(f)](x)$. Normal mortals usually do not like this many parentheses, so we will adopt various levels of abbreviation.

$$[T(f)](x) = (Tf)(x) = Tf(x).$$

The last is read as, Tf is applied to (or evaluated at) x.
    The following example gives three common linear operators from calculus. The precise V and W of each operator is omitted for brevity.

EXAMPLE 5. Let x_0 be a fixed scalar in $\mathbf{F}$, and let a and b be given real numbers. The following operators are linear:

(a) Evaluation $T(f) = f(x_0)$
(b) Differentiation $T(f) = f'$
(c) Integration $(Tf)(x) = \int_a^x f(t)\, dt$.

    Linearity follows from the next observations: (a) $(\alpha f + \beta g)(x_0) = \alpha f(x_0) + \beta g(x_0)$, (b) $(\alpha f + \beta g)' = \alpha f' + \beta g'$, (c) $\int_a^x (\alpha f + \beta g)(t)\, dt = \alpha \int_a^x f(t)\, dt + \beta \int_a^x g(t)\, dt$.

EXAMPLE 6. LINEAR DIFFERENTIAL OPERATORS. A differential operator $T = \sum_{k=0}^{n} c_k D^k$, where $D^k = d^k/dx^k$ and where c_k is a continuous function of x, operates on a function f according to $T(f) = \sum_{k=0}^{n} c_k f^{(k)}$. The value of $T(f)$ at x is $(Tf)(x) = \sum_{k=0}^{n} c_k(x) f^{(k)}(x)$. Here the domain V of T may be taken to be $C^n(I)$, which consists of all n times continuously differentiable functions on a given interval I, and W may be taken to be $C(I)$, which is comprised of all the continuous functions on I. That T is linear follows from this familiar fact from calculus: $D^k(\alpha f + \beta g) = \alpha D^k f + \beta D^k g$.

EXAMPLE 7. LINEAR INTEGRAL TRANSFORMS. Let $K(s, t)$ be continuous, and let $V = W = C[a, b]$. $T : V \to W$ is defined by specifying the value of the function Tf at each individual point s:

$$(Tf)(s) = Tf(s) = \int_a^b K(s, t) f(t) \, dt.$$

That T is linear follows from the linearity of the integral:

$$\int_a^b K(s, t)[\alpha f(t) + \beta g(t)] \, dt = \alpha \int_a^b K(s, t) f(t) \, dt + \beta \int_a^b K(s, t) g(t) \, dt.$$

If $K(s, t) = e^{-st}$ and if $a = 0$ and $b = \infty$, then T is the Laplace transform. If $K(s, t) = e^{-ist}$ and if $a = -\infty$ and $b = +\infty$, then T is a Fourier Transform. However, in these cases V cannot consist of all continuous functions, since the integrals must converge. Some other conditions must be posited, such as $\int_a^b |f(t)| \, dt < \infty$. Many venerable operators of mathematics and the sciences are of this form.

The next two examples arise in a variety of applications.

EXAMPLE 8. FOURIER COEFFICIENTS OF CONTINUOUS FUNCTIONS. Let $V = C([0, 2\pi])$, and let $W = \{(c_k)_{k=-\infty}^{\infty} : c_k \in C\}$. That is, W is the set of all bi-infinite complex sequences. Define $T : V \to W$ by $T(f) = c$ iff $c_k = 1/2\pi \int_0^{2\pi} f(t) e^{-ikt} \, dt$. $T(f)$ is the sequence of complex Fourier coefficients of f. The linearity of T follows from

$$\frac{1}{2\pi} \int_0^{2\pi} [\alpha f(t) + \beta g(t)] e^{-ikt} \, dt = \alpha \frac{1}{2\pi} \int_0^{2\pi} f(t) e^{-ikt} \, dt + \beta \frac{1}{2\pi} \int_0^{2\pi} g(t) e^{-ikt} \, dt.$$

EXAMPLE 9. MOMENT SEQUENCES OF CONTINUOUS FUNCTIONS. Let $V = C([0, 1])$, and let $W = \{(c_k)_{k=0}^{\infty} : c_k \in C\}$. That is, W is the set of all infinite complex sequences. Define $T : V \to W$ by $T(f) = c$ iff $c_k = \int_0^1 f(t) t^k \, dt$. $T(f)$ is called the *moment sequence* of f. The linearity of T follows from

$$\int_0^1 [\alpha f(t) + \beta g(t)] t^k \, dt = \alpha \int_0^1 f(t) t^k \, dt + \beta \int_0^1 g(t) t^k \, dt.$$

EXAMPLE 10. SOME NONLINEAR OPERATORS

$T(x, y, z) = xy + z,$ where x, y, z are real

$T(f) = ff'' + 2f' + f,$ where $' = d/dx$

$Tf(x) = \int_a^b \sin[f(t)x] \, dt$

In considering an equation $T(x) = y$, where $y \in W$, $x \in V$, the questions of existence and uniqueness of solutions arise. As noted above, these questions are related to the range of T and whether T is one to one. For linear operators, it is useful to introduce more terminology.

DEFINITION. The *null space* or *kernel* of a linear operator is defined to be the set of all $x \in V$ for which $T(x) = 0$. It is denoted by

$$N(T) = \ker(T) = \{x : x \in V, T(x) = 0\}.$$

It turns out that the null space and range of a linear operator are extremely important in the study of the properties of the operator. The next proposition is a step in that direction.

PROPOSITION 2. The null space and range of a linear operator $T : V \to W$ are subspaces of V and W, respectively.

Proof. First consider the null space. If $x, y \in N(T)$, then $T(x) = T(y) = 0$ and $T(\alpha x + \beta y) = \alpha T(x) + \beta T(y) = 0$, so $\alpha x + \beta y \in N(T)$. Next consider the range. If $u, v \in R(T)$, then $u = T(x)$ and $v = T(y)$ for some x and y. Then $\alpha u + \beta v = \alpha T(x) + \beta T(y) = T(\alpha x + \beta y)$, and so $\alpha u + \beta v \in W$.

EXAMPLE 11. Consider the differential operator $T : C^n([a, b]) \to C([a, b])$ defined by $T(f) = \sum_{k=0}^{n} c_k f^{(k)}$, where each c_k is a continuous function on $[a, b]$. Then $N(T)$ is the set of solutions f to the differential equation $\sum_{k=0}^{n} c_k(x) f^{(k)}(x) = 0$, $x \in [a, b]$ that have n continuous derivatives.

EXAMPLE 12. Let A be an $m \times n$ matrix, and define the operator T by $T(x) = Ax$. Then $N(T)$ consists of the solutions of $Ax = 0$. $N(T)$ is sometimes called the solution space of A.

PROPOSITION 3. For a linear operator T, the following are equivalent:

(*a*) T is one to one.
(*b*) $T(x) = 0$ implies $x = 0$.
(*c*) $N(T) = \{0\}$.

Proof. (*a*) $\Rightarrow$ (*b*): Since $T(0) = 0$, if T is one to one, then $x = 0$ is the only solution to $T(x) = 0$. (*b*) $\Rightarrow$ (*c*): If $x \in N(T)$, then $T(x) = 0$, (*b*) implies that $x = 0$, and (*c*) holds. (*c*) $\Rightarrow$ (*a*): Suppose that $T(x) = T(y)$. Then $0 = T(x) - T(y) = T(x - y)$ implies that $x - y \in N(T)$, so $x - y = 0$, and thus $x = y$. Therefore T is one to one.

EXAMPLE 13. Let $T : \mathbf{R}^2 \to \mathbf{R}^2$ be defined by

$$T(x) = \begin{bmatrix} 2 & 1 \\ 0 & 0 \end{bmatrix} x = \begin{bmatrix} 2x_1 + x_2 \\ 0 \end{bmatrix}.$$

Determine whether T is one to one and whether T is onto, and find the range and null space of T.

   T is not onto, because $T(x) = (0, 1)^{\mathrm{T}}$ has no solution. $R(T) = \{(z, 0)^{\mathrm{T}} : z \in \mathbf{R}\}$, since for each z, $T(x) = (z, 0)^{\mathrm{T}}$ has (at least) one solution $(z/2, 0)^{\mathrm{T}}$. T is not one to one because $T((1, 0)^{\mathrm{T}}) = T((0, 2)^{\mathrm{T}}) = (2, 0)^{\mathrm{T}}$ and because $T((1, -2)^{\mathrm{T}}) = 0$. $N(T) = \{(c, -2c)^{\mathrm{T}} : c \in \mathbf{R}\} \neq \{0\}$, which again implies that T is not one to one.

EXAMPLE 14. Let $V = C(\mathbf{R})$, let $W = C^1(\mathbf{R})$, and define $J : V \to W$ by

$$J f(x) = \int_0^x f(t)\, dt.$$

Determine whether J is one to one or onto. Find the range of J.

Suppose $Jf = 0$, i.e., $Jf(x) = \int_0^x f(t)\,dt = 0$ for all x. Differentiating this yields

$$0 = \frac{d}{dx}\int_0^x f(t)\,dt = f(x) \text{ for all } x.$$

Thus $f = 0$, and J is one to one by Proposition 2. J is not onto since the equation $Jf(x) = e^x\ \forall x$ has no solution for f. To see this, note that $Jf(0) = 0$, whereas $e^0 = 1$. The range of J is the set of those functions g in W such that $g(0) = 0$ since for these functions, $(Jg')(x) = \int_0^x g'(t)\,dt = g(x) - g(0) = g(x)$.

Exercises 2.1

1. Let $V = C^1(\mathbf{R})$ and $W = C(\mathbf{R})$, and define $D : V \to W$ by $Dg(x) = g'(x)$. Determine whether D is one to one or onto.

2. Let $V = W = \mathbf{C}^\infty$. Define $T, S : V \to W$ by $T(a_1, a_2, a_3, \ldots) = (0, a_1, a_2, a_3, \ldots)$ and $S(a_1, a_2, a_3, \ldots) = (a_2, a_3, \ldots)$. Determine whether T and S are one to one or onto.

3. Define $T : \mathbf{F}^{2,2} \to \mathbf{F}^{2,2}$ by $T(A) = A + A^\mathsf{T}$. Show that T is linear. Is T one to one? onto?

4. Define $T : \mathbf{F}^{2,2} \to \mathbf{F}^{2,2}$ by $T(A) = A + 2A^\mathsf{T}$. Show that T is linear. Is T one to one? onto?

5. Define T by $Tg(x) = \int_0^x tg'(t)\,dt + f(0)$, $g \in \mathbf{P}_2$. Show that $T : \mathbf{P}_2 \to \mathbf{P}_3$ and T is linear. Is T one to one? Is T onto?

6. Define $T : \mathbf{R}^3 \to \mathbf{R}^3$ by $T(\mathbf{v}) = \mathbf{w} \times \mathbf{v}$, where $\mathbf{w}$ is a fixed nonzero vector and $\times$ is the cross product. Show that T is linear. Is T one to one? onto?

7. Let $T : V \to W$ be linear, and let $\dim(R(T)) = \dim(W) < \infty$. Show that T is onto.

8. Let $T : \mathbf{F}^n \to \mathbf{F}^m$ be the matrix operator $T(x) = Ax$. Show that
   (a) T is one to one iff the columns of A are linearly independent.
   (b) T is onto iff the columns of A span $\mathbf{F}^m$.
   For (c) and (d), assume that $n = m$.
   (c) T is one to one and onto iff the columns of A are a basis for $\mathbf{F}^n$.
   (d) T is one to one and onto iff A is nonsingular.

9. Suppose $y \in R(T)$ and $T(x_0) = y$. Show that every solution of $T(x) = y$ is of the form $x = x_0 + z$, with $z \in N(T)$.

10. Let $L(V, W)$ be the set of all linear operators $T : V \to W$. Operator addition, scalar multiplication, and equality are defined as usual: $(T + S)(x) = T(x) + S(x)$, $(\alpha T)(x) = \alpha T(x)$ and $T = S$ iff $T(x) = S(x)\ \forall x \in V$. Show that $L(V, W)$ is a vector space.

11. Consider the linear operators $D^i : \mathbf{P}_n \to \mathbf{P}_n$ defined by $D^i = (d/dx)^i$, $i = 0, 1, \ldots, m$. Let $m \le n$. Show that $\{D^i\}_{i=0}^m$ is linearly independent. Is it still linearly independent if $m > n$?

12. Let $x_0 < x_1 < \cdots < x_m$. Define linear functionals $\phi_i : \mathbf{P}_n \to \mathbf{P}_n$ by $\phi_i(f) = f(x_i)$. Let $m \le n$. Show that $\{\phi_i\}_{i=1}^m$ is linearly independent. Is it still linearly independent if $m > n$?

13. Define linear functionals $\phi_i : \mathbf{P}_n \to \mathbf{P}_n$ by $\phi_i(p) = p^{(i)}(0)$. Let $m \le n$. Show that $\{\phi_i\}_{i=1}^m$ is linearly independent. Is it still linearly independent if $m > n$?

2.2
THE RANK AND NULLITY THEOREM

Suppose $T : V \rightarrow W$ is a linear operator. Let x_p be a solution of $T(x) = y$, and let x' be another solution. Then $T(x' - x_p) = y - y = 0$. Setting $z = x' - x_p$ shows that every solution of $T(x) = y$ is of the form $x' = x_p + z$, with $z \in N(T)$. This suggests that V can be broken into two parts: $N(T)$, and special x_p with $T(x_p)$ filling up the range of T. The rank and nullity theorem and its proof make this precise. This theorem has many important applications to theory and various problems. Some of the theoretical applications are discussed first, and then applications to matrix theory are given. In this section the vector spaces in question are assumed to be nontrivial finite dimensional spaces. First, the following definition is needed:

DEFINITION. The *rank of a linear operator* $r(T)$ is the dimension of its range, and the *nullity of the operator* is the dimension of its null space:

$$r(T) = \dim(R(T)) = \dim(\mathrm{Im}(T))$$

$$n(T) = \dim(N(T)) = \dim(\ker(T)).$$

RANK AND NULLITY THEOREM. Let $T : V \rightarrow W$ be a linear operator, and let $\dim(V) = n$. Then

$$\dim(R(T)) + \dim(N(T)) = \dim(V); \text{ i.e.,}$$
$$r(T) + n(T) = \dim(\text{domain of T}). \tag{1}$$

Proof. Let $\{e_1, e_2, \dots, e_k\}$ be a basis for $N(T)$. If $n = k$, then $N(T) = V$, $R(T) = \{0\}$, and the proof is finished. Otherwise, the basis for $N(T)$ may be extended to a basis $\beta = \{e_1, e_2, \dots, e_n\}$ for V. It will be shown that $\gamma = \{T(e_i) : i = k+1, \dots, n\}$ is a basis for $R(T)$, so that $\dim(R(T)) + \dim(N(T)) = (n - k) + k = n$, and the theorem will then be proved. Let $y = T(x) \in R(T)$. Since β is a basis, $x = \sum_{i=1}^{n} \alpha_i e_i$ for certain α_i. Thus

$$y = T(x) = T\left(\sum_{i=1}^{n} \alpha_i e_i\right) = \sum_{i=1}^{n} \alpha_i T(e_i) = \sum_{i=k+1}^{n} \alpha_i T(e_i)$$

since $T(e_i) = 0$ for $i = 1, \dots, m$. Thus γ is a spanning set for $R(T)$. To show that γ is linearly independent, suppose $\sum_{i=k+1}^{n} \alpha_i T(e_i) = 0$. It must be shown that all the α_i are 0. By linearity, $0 = \sum_{i=k+1}^{n} \alpha_i T(e_i) = T(\sum_{i=k+1}^{n} \alpha_i e_i)$, which means that $\sum_{i=k+1}^{n} \alpha_i e_i$ is in the null space of T. Since $\{e_1, e_2, \dots, e_k\}$ is a basis for $N(T)$, there are scalars η_i such that $\sum_{i=k+1}^{n} \alpha_i e_i = \sum_{i=1}^{k} \eta_i e_i$. Hence $\sum_{i=1}^{k} \eta_i e_i - \sum_{i=k+1}^{n} \alpha_i e_i = 0$, and thus all the α_i and η_i are 0 since β is linearly independent. Thus γ is a basis for $R(T)$.

A diagram of the proof of the rank and nullity theorem is as follows.

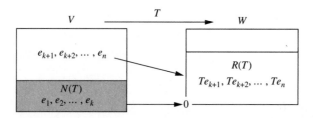

Remark. The method of constructing a basis for $R(T)$ given in the theorem is sometimes useful.

COROLLARY 1. Let $T : V \rightarrow W$ be a linear operator, and let $\dim(V) = n$ and $\dim(W) = m$. Then $r(T) \leq \min(m, n)$.

Proof. Since $R(T) \subset W$, $r(T) = \dim(R(T)) \leq \dim(W) = m$. But also $r(T) = n - n(T) \leq n$.

The next simple example shows how the rank and nullity theorem can be used to infer information about an operator.

EXAMPLE 1. Define $T : \mathbf{T}_n \rightarrow \mathbf{T}_n$ by $T(f) = f'' + f$. Find the rank of T.
    From differential equations, $T(f) = 0$ has the general solution $f = a\sin(t) + b\cos(t)$, so that $N(T) = \text{span}\{\sin(t), \cos(t)\}$. Since $\{\sin(t), \cos(t)\}$ is linearly independent, $N(T)$ is two-dimensional. Thus $r(T) = \dim(\mathbf{T}_n) - 2 = 2n - 1$.

The next theorem estimates the rank of the product ST.

THEOREM 2. Let $T : V \rightarrow W$, $S : W \rightarrow U$ be a linear operator, and let V be finite dimensional. Then $r(ST) \leq r(S), r(T)$.

Proof. Let $\{e_1, e_2, \ldots, e_n\}$ be a basis for V such that $\{T(e_1), \ldots, T(e_r)\}$ is a basis for $R(T)$, and $T(e_i) = 0$, where $i = r + 1, \ldots, n$ as constructed in the rank and nullity theorem. Thus $r(T) = r$. Then $\{ST(e_1), \ldots, ST(e_r)\}$ spans $R(ST)$ so that $r(ST) \leq r$. Furthermore $R(ST) \subset R(S)$ since $y = S(T(x))$ for some x implies that $y = S(z)$ for some z (namely, $z = T(x)$). Thus $r(ST) \leq r(S)$.

Operator Inverses

Operators that have an inverse enjoy many of the same properties as nonsingular matrices, and the rules concerning inverses are formally the same. Here we define additional inverses that are sometimes important.

DEFINITION. Let $T : V \rightarrow W$ be a linear operator. A *left inverse* L of T is a linear operator $L : W \rightarrow V$ such that $LT = I = I_V$. A *right inverse* of T is a linear operator $R : W \rightarrow V$ such that $TR = I = I_W$. An *inverse* of T is a linear operator $S : W \rightarrow V$ that is both a left and a right inverse of T : $ST = TS = I$. One writes $S = T^{-1}$. Linear operators that have an inverse are called *invertible* or *nonsingular.* They are also called *isomorphisms.*

Operator inverses obey rules similar to matrix inverses. For example, from the symmetry of the definition, $S^{-1} = T$, so that $(T^{-1})^{-1} = T$. Also $(TS)^{-1} = S^{-1}T^{-1}$, provided that both inverses exist and TS is defined. To see this, just observe that $(S^{-1}T^{-1})TS = I = TS(S^{-1}T^{-1})$.

EXAMPLE 2. Define $J : \mathbf{P}_n \rightarrow \mathbf{P}_{n+1}$ and $D : \mathbf{P}_{n+1} \rightarrow \mathbf{P}_n$ by $Df(x) = f'(x)$ and $Jf(x) = \int_0^x f(t)\,dt$. Then D is a left inverse of J, and J is a right inverse of D.
    From the fundamental theorem of calculus,

$$DJf(x) = \frac{d}{dx} \int_0^x f(t)\,dt = f(x).$$

Hence $DJ = I$.

EXAMPLE 3. Let

$$A = \begin{bmatrix} 3 & -3 & 1 \\ 0 & -1 & 1 \end{bmatrix}, \qquad B = \begin{bmatrix} 1 & 1 \\ 1 & 2 \\ 1 & 3 \end{bmatrix}.$$

Then A is a left inverse of B and B is a right inverse of A.
    In fact, $AB = I_2$.

EXAMPLE 4. Let A be an $n \times n$ nonsingular matrix, and let $T(x) = Ax$. Then T^{-1} is the matrix operator $T^{-1}(x) = A^{-1}x$.

EXAMPLE 5. Let $V = C(\mathbf{R})$, let α be a fixed real number, and let $S, T : V \to V$ be defined by $Tf(x) = f(x + \alpha)$ and $Sf(x) = f(x - \alpha)$. Then $S = T^{-1}$.
    Here $STf(x) = Sf(x + \alpha) = f(x + \alpha - \alpha) = f(x)$, so $STf = f$ and $ST = I$. Similarly, $TS = I$.

The next theorem collects some basic facts about the inverses of linear operators.

THEOREM 3. Let $T : V \to W$ be a linear operator.

(a) If T has an inverse S, then S is unique.
(b) If T has a left inverse, then T is one to one.
(c) If T has a right inverse $R : W \to V$, then T is onto. Furthermore, a solution of $T(x) = y$ is $x = R(y)$.
(d) T has an inverse iff T is one to one and onto.

Proof

(a) If S and S' are inverses of T, then $S = SI = S(TS') = (ST)S' = IS' = S'$.
(b) Suppose $LT = I$ and $T(x) = 0$. Then $L(T(x)) = (LT)(x) = I(x) = x = L(0) = 0$, so that T is one to one by Proposition 3 of Section 2.1.
(c) Suppose $TR = I$. Given $y \in W, T(R(y)) = (TR)(y) = I(y) = y$. Thus $x = R(y)$ solves $T(x) = y$.
(d) If T has an inverse, it is one to one and onto by (b) and (c). Conversely, suppose T is one to one and onto; i.e., suppose the equation $T(x) = y$ has a unique solution $x = T^{-1}(y)$ for every y in W. Define $T^{-1}(x) = y \Leftrightarrow T(y) = x$. We must show that T is linear. Suppose $T(u) = x, T(v) = y$. Then $T(\alpha u + \beta v) = \alpha T(u) + \beta T(v) = \alpha x + \beta y$, from which $T^{-1}(\alpha x + \beta y) = \alpha u + \beta v = \alpha T^{-1}(x) + \beta T^{-1}(y)$.

Remark. If $T : V \to W$ is one to one but not onto, T^{-1} does not exist. However, if the codomain W is changed to $R(T)$, i.e., $T : V \to R(T)$, then T is one to one and onto, and T^{-1} exists.
    The next theorem says that invertible linear operators cannot change dimensions; i.e., the domain and codomain must have the same dimension.

THEOREM 4. Let $T : V \to W$ be a linear operator, and let $\dim(V) = n$ and $\dim(W) = m$.
    If T is invertible, then $m = n$.

Proof. Since T is onto, $W = R(T)$ so that $m = \dim(W) = \dim(R(T))$. Since T is one to one, $N(T) = \{0\}$. By the rank and nullity theorem, $m = \dim(R(T)) = n - \dim(N(T)) = n$.

In Chapter 0, matrix inverses were defined only for square matrices. The next corollary shows why this was done.

COROLLARY 5. Let A be an $m \times n$ matrix. If A has an inverse, then $m = n$.

Proof. If there is a matrix B so that $AB = BA = I$, then the matrix operator $T(x) = Ax$ has its inverse defined by $T^{-1}(x) = Bx$. Since $T : \mathbf{F}^n \to \mathbf{F}^m$, and $\dim(\mathbf{F}^k) = k$, it follows that $m = n$ by Theorem 4.

The next theorem simplifies the work of showing that an operator inverse exists and verifying that $S = T^{-1}$. Statement (d) says that, when $\dim(V) = \dim(W)$, the left and right inverses must also be inverses.

THEOREM 6. Let $T : V \to W$ and $S : W \to V$ be linear operators, and let $n = \dim(V) = \dim(W)$. Then

(a) If T is one to one, then T is invertible.
(b) If T is onto, then T is invertible.
(c) If $r(T) = n$, then T is invertible.
(d) If $ST = I$, then $TS = I$; i.e., $S = T^{-1}$ and $S^{-1} = T$.

Proof

(a) Since $N(T) = \{0\}$, $\dim(R(T)) = n - \dim(N(T)) = n$. But $R(T) \subset W$, and both have the same dimension, hence $R(T) = W$. Thus T is one to one and onto and must have an inverse.
(b) Since $R(T) = W$, $\dim(N(T)) = n - \dim(R(T)) = 0$. Thus $N(T) = \{0\}$, is therefore one to one as well as onto, and so has an inverse.
(c) $r(T) = n$ implies that $R(T) = W$, so that T is onto and hence invertible by (c).
(d) Since T is one to one, it has an inverse T^{-1}, by (a). Now $T^{-1} = IT^{-1} = STT^{-1} = SI = S$.

Suppose f is a continuous function on $[0, 1]$ and we define $x_k = \int_0^1 t^k f(t)\, dt$. The numbers x_k are called the *moments of f* and are important in probability theory and other areas of mathematics. Questions arise such as, Which sequences $x = (x_0, x_1, \ldots, x_n, \ldots)$ are moment sequences? Does the sequence x uniquely characterize f? The next example answers these questions in a much simpler setting.

EXAMPLE 6. Let n be given, and let the scalars be real. Then every $x = (x_0, x_1, \ldots, x_n) \in \mathbf{R}^{n+1}$ is a moment sequence of an $f \in \mathbf{P}_n$, and x uniquely determines $f \in \mathbf{P}_n$.
Define $T : \mathbf{P}_n \to \mathbf{R}^{n+1}$ by $x = T(f) = (x_0, x_1, \ldots, x_n)^{\mathrm{T}}$, with $x_k = \int_0^1 t^k f(t)\, dt$. Since $\dim(\mathbf{P}_n) = \dim(\mathbf{R}^{n+1})$, if it can be shown that T is one to one, then it follows that T is onto and invertible, and the assertions follow. Suppose $T(f) = 0$. Then all $x_k = 0$. Let $f(t) = \sum_{k=0}^n c_k t^k$. Then

$$0 = \sum_{k=0}^n c_k x_k = \sum_{k=0}^n c_k \int_0^1 t^k f(t)\, dt = \int_0^1 \left\{ \sum_{k=0}^n c_k t^k \right\} f(t)\, dt = \int_0^1 f^2(t)\, dt.$$

Since $f^2 \geq 0$ and is continuous, a theorem in calculus asserts that $f^2(t) = 0$ for every t. Hence $f = 0$, and T is one to one.

Application to Matrix Theory and Systems of Equations

In this section the theory of linear operators is applied to obtain some important and interesting facts about matrices and systems of equations.

Let A be an $m \times n$ matrix with entries in $\mathbf{F}$. To tie the theory of linear operators to A, consider the linear operator generated by $A : T : \mathbf{F}^n \to \mathbf{F}^m$, with $T(x) = Ax$. We use the following notation:

$$R(A) = R(T), \qquad N(A) = N(T)$$
$$r(A) = r(T) = \text{rank of } A$$
$$n(A) = n(T) = \text{nullity of } A.$$

As usual, $A = [A_1 \, A_2 \cdots A_n] = [a_{ij}]$, where A_j denotes the jth column of A.

The next proposition restates some facts of Chapter 1 into operator language.

PROPOSITION 7. Let $T : \mathbf{F}^n \to \mathbf{F}^m$ be the matrix operator $T(x) = Ax$. Then

(a) $R(A) = \text{span}\{A_1, A_2, \ldots, A_n\}$.
(b) $N(A)$ is the set of solutions of $Ax = 0$.

Proof. Statement (a) follows from the identity $Ax = \sum_{j=1}^n x_j A_j$ and from the definitions. Statement (b) follows directly from the definitions.

Remark. Since the range of A is the span of its columns, the range is called the column space of A, and the rank of A is called the *column rank of A*. In Section 0.2 the row rank of a matrix was defined to be the number of nonzero rows in its reduced-row echelon form. In Example 6 of Section 1.4, the row rank was implicitly shown to be the number of linearly independent rows in A. Below, the relationship between these numbers is established.

In the following, methods of determining a basis for $N(A)$, $R(A)$, and the determination of the rank and nullity of an $m \times n$ matrix A are given. In all of these, we use the following matrix as an example.

$$A = \begin{bmatrix} 1 & 2 & 1 & 2 & 0 \\ 2 & 4 & 1 & 3 & -1 \\ 1 & 2 & -2 & -1 & -3 \\ 3 & 6 & -2 & 1 & -5 \end{bmatrix}.$$

Computation of a basis for $N(A)$

The process is illustrated with the example matrix A, and then we describe the general case.

To solve $Ax = 0$, compute the row-reduced echelon form R of A (cf. Section 0.2):

$$R = \begin{bmatrix} 1 & 2 & 0 & 1 & -1 \\ 0 & 0 & 1 & 1 & 1 \\ 0 & 0 & 0 & 0 & 0 \\ 0 & 0 & 0 & 0 & 0 \end{bmatrix}.$$

Thus the number of nonzero rows is $r = 2$, the leading (pivot) variable indices are $j_1 = 1$, $j_2 = 3$, and the free variable indices are $f_1 = 2$, $f_2 = 4$, and $f_3 = 5$. Solve the equations $Rx = 0$ for the leading variables in terms of the free variables to get

$$x_1 = -2x_2 - x_4 + x_5$$
$$x_3 = -x_4 - x_5.$$

Eliminating the leading variables, the general solution of $Ax = 0$ (i.e., the general element of the null space) is

$$x = \begin{bmatrix} -2x_2 - x_4 + x_5 \\ x_2 \\ -x_4 - x_5 \\ x_4 \\ x_5 \end{bmatrix}. \tag{$*$}$$

Next, expand x in a linear combination of vectors, each term of which involves only one free variable:

$$x = \begin{bmatrix} -2x_2 \\ x_2 \\ 0 \\ 0 \\ 0 \end{bmatrix} + \begin{bmatrix} -x_4 \\ 0 \\ -x_4 \\ x_4 \\ 0 \end{bmatrix} + \begin{bmatrix} x_5 \\ 0 \\ -x_5 \\ 0 \\ x_5 \end{bmatrix} = x_2 \begin{bmatrix} -2 \\ 1 \\ 0 \\ 0 \\ 0 \end{bmatrix} + x_4 \begin{bmatrix} -1 \\ 0 \\ -1 \\ 1 \\ 0 \end{bmatrix} + x_5 \begin{bmatrix} 1 \\ 0 \\ -1 \\ 0 \\ 1 \end{bmatrix}$$

$$\equiv x_2 E_1 + x_4 E_2 + x_5 E_5 = x_{f_1} E_1 + x_{f_2} E_2 + x_{f_3} E_5. \tag{$**$}$$

The vectors E_1, E_2, E_3 so defined span $N(A)$ by construction. E_1, E_2, E_3 are clearly linearly independent: If $x = 0$ in $(**)$, then clearly $x_2 = x_4 = x_5 = 0$. Note that E_1 is the vector x, which is obtained from $(*)$ by setting $x_2 = 1$, $x_4 = 0$, and $x_5 = 0$. Similarly, E_2 is obtained by setting $x_2 = 0$, $x_4 = 1$, and $x_5 = 0$ in (1), and E_3 is obtained by setting $x_2 = 0$, $x_4 = 0$, and $x_5 = 1$. Thus $\dim(N(A)) = 3$. Thus $\dim(R(A)) = 2 = r$ by the rank and nullity theorem.

Procedure to find a basis for the null space

Row-reduce A to its row-reduced echelon form R. Suppose the first r rows of R are nonzero and that the indices of the leading variables are j_k, where $k = 1, 2, \ldots, r$, and the indices of the free variables are f_i, where $i = 1, 2, \ldots, n - r$. Recall that $Ax = 0$ iff $Rx = 0$ (cf. Eq. (1) of Section 0.2). For x in $N(A)$, the equations $Rx = 0$ can be used to express each leading variable x_{j_i} as a linear combination of the $n - r$ other variables x_{f_i} called free variables. Form the general solution x of $Ax = 0$ by eliminating the leading variables, as was done in Eq. $(*)$. Note that in row f_i of the column x, only the variable x_{f_i} appears. Now expand the general solution as $x = \sum_{i=1}^{n-r} x_{f_i} E_i$, where $x_{f_i} E_i$ is the part of x containing the free variable x_{f_i} and no others: i.e., E_i is the vector obtained by setting $x_{f_i} = 1$ and setting all the other free variables equal to zero in the general solution x. The set $\{E_1, E_2, \ldots, E_{n-r}\}$ will be a basis for $N(A)$.

Proof. The $n - r$ vectors E_i span the null space by construction. Note that the entry of E_i is 1 in row f_i, and the entry of E_i is 0 in all the other free variable rows f_j. If $x = \sum_{i=1}^{n-r} x_{f_i} E_i = 0$, then all the $x_{f_i} = 0$, since in row f_i of x, only x_{f_i} appears. Thus $\{E_1, E_2, \ldots, E_{n-r}\}$ is linearly independent and so forms a basis for the null space. The fact that the dimension of the null space is $n - r$ makes good intuitive sense, since there are $n - r$ free variables, or independent parameters, that describe the null space.

Remark. From the rank and nullity theorem, the rank of the matrix is $r(A) = \dim(\mathbf{F}^n) - \dim(N(A)) = n - (n - r) = r$. This shows that the definition of row rank of a matrix in Section 0.2 coincides with the rank of the operator generated by A.

Computation of a basis for $R(A)$

Three ways to construct a basis for $R(A)$ are given.

First method. Row-reduce A to its row-reduced echelon form R. Let $j_1, j_2, \ldots,$ j_r be the columns of R with leading 1s. Then the set of columns $\{A_{j_1}, A_{j_2}, \ldots, A_{j_r}\}$ of A that corresponds to the set of leading ones is a basis for $R(A)$.

For the example matrix A, the leading variables are 1 and 3, so A_1 and A_3 form a basis for $R(A)$.

To see why this method works, suppose R is the row-reduced echelon form of A, r is the number of nonzero rows of R, and $j_1, \ldots, j_r$ are the indices of the leading variables. Now let B be the matrix A but with the columns corresponding to the free variables deleted; that is, let $B = [A_{j_1} A_{j_2} \cdots A_{j_r}]$. If the same sequence of row operations is done on B, one obtains a matrix S that is obtained by deleting from R the columns corresponding to the free variables $S = [R_{j_1} R_{j_2} \cdots R_{j_r}]$. But then

$$S = \begin{bmatrix} I_r \\ 0 \end{bmatrix},$$

since the j_i column of R is all 0, except for the ith entry, which is 1. The columns of B are linearly independent iff $Bx = 0$ has only $x = 0$ as a solution. But $Bx = 0$ iff $Sx = 0$, and it is clear from the form of S that the only solution to $Sx = 0$ is $x = 0$. Thus the columns $A_{j_1}, A_{j_2}, \ldots, A_{j_r}$ of A are linearly independent. But $r(A) = \dim(R(A)) = r$ by the remark following the construction of a basis for the null space. Thus $\{A_{j_1}, A_{j_2}, \ldots, A_{j_r}\}$ must be a basis for $R(A)$, since it consists of r linearly independent vectors in an r-dimensional space (by Theorem 6 of Section 1.4). Note that this gives another proof that $r(A) = r$.

Second method. Since $R(A) = \mathrm{span}\{A_1, A_2, \ldots, A_n\}$, a basis for $R(A)$ may be found as follows (cf. Example 6 of Section 1.4): Form A^T, row-reduce it to its row-reduced echelon form S, and suppose S has s nonzero rows. Then the s nonzero rows of S transposed are a basis for $R(A)$. Thus the rank of A is s. From the preceding work, $s = r$.

For the matrix A in the current example,

$$S = \begin{bmatrix} 1 & 0 & -5 & -7 \\ 0 & 1 & 3 & 5 \\ 0 & 0 & 0 & 0 \\ 0 & 0 & 0 & 0 \\ 0 & 0 & 0 & 0 \end{bmatrix}.$$

Hence $r(A) = 2$, $n(A) = 5 - 2 = 3$, and a basis for $R(A)$ is $(1, 0, -5, -7)^T$, $(0, 1, 3, 5)^T$.

Third method (based on the proof of the rank and nullity theorem). Find a basis for $N(A)$. Extend it in any way to a basis for the domain of A. The images of these vectors under A will be a basis for $R(A)$.

For the example matrix A being considered, the set of vectors $\{E_1, E_2, E_3\}$ with $E_1 = (-2, 1, 0, 0, 0)^T$, $E_2 = (-1, 0, -1, 1, 0)^T$, and $E_3 = (1, 0, -1, 0, 1)^T$ was found to be a basis for $N(A)$. Add any two vectors E_4 and E_5 so that a basis for $\mathbf{F}^5$ is obtained. For example, $E_4 = (0, 0, 0, 1, 0)^T$ and $E_5 = (0, 0, 0, 0, 1)^T$ work, so $AE_4 = A_4$ and $AE_5 = A_5$ form a basis for $R(A)$.

The next theorem collects some of the results just deduced.

THEOREM 8. Let A be an $m \times n$ matrix, and let r be the row rank of A. Then

(a) $r(A) = \dim(R(A)) = r$
(b) $n(A) = \dim(N(A)) = n - r$
(c) $r(A^T) = r$
(d) $m - r = \dim(N(A^T))$.

Remark. Statement (c) may be interpreted as saying that the number of linearly independent rows of A is the same as the number of linearly independent columns of A.

Finally, the theory relating to operator inverses that is developed in this chapter is applied to the matrix operator $T(x) = Ax$, where A is an $n \times n$ matrix. Some of the theory from Chapter 1 is also applied. Statements (a) through (k) follow from the vector space theory, although some of them were deduced in Chapter 0 through analysis of row reduction.

THEOREM 9. Let A be an $n \times n$ matrix. Then the following are logically equivalent:

(a) A is nonsingular.
(b) A has a left inverse.
(c) For every y, $Ax = y$ has at most one solution.
(d) $N(A) = \{0\}$.
(e) $Ax = 0$ has only the solution $x = 0$.
(f) The columns of A are linearly independent.
(g) For each y, $Ax = y$ has at least one solution.
(h) $R(A) = \mathbf{F}^n$.
(i) The columns of A span $\mathbf{F}^n$.
(j) A has a right inverse.
(k) $r(A) = n$.
(l) A row-reduces to I.

Proof. To prove statements (a) through (k), combine the relevant theorems of this chapter and Proposition 7 of Section 1.3. Statement (l) was proved in Section 0.2 using Gauss elimination.

Exercises 2.2

1. Let $A = \begin{bmatrix} 1 & 2 & 2 & 1 & 1 & 0 \\ 1 & 2 & 0 & -1 & -1 & 0 \\ 1 & 2 & 1 & 0 & 1 & 1 \\ 2 & 4 & 1 & -2 & 2 & 3 \end{bmatrix}$.

(a) Find a basis for $N(A)$.
(b) Use two methods to find a basis for $R(A)$.

2. Let T be the matrix operator $T(x) = Ax$ with A given as follows:

$$(i) \begin{bmatrix} 1 & 1 & 2 \\ 1 & -1 & 1 \\ 1 & 5 & 4 \end{bmatrix} \quad (ii) \begin{bmatrix} 1 & 0 & 2 & 3 & 5 \\ 2 & 0 & -2 & 1 & 6 \\ 3 & 0 & 2 & 1 & 3 \end{bmatrix} \quad (iii) \begin{bmatrix} 1 & 3 & 4 \\ 1 & 2 & -1 \\ 0 & -2 & -4 \\ 1 & 1 & 0 \end{bmatrix}$$

$$(iv) \begin{bmatrix} 1 & 0 & 1 & 2 & -3 \\ 1 & 1 & 3 & 1 & -1 \\ 3 & 2 & 7 & 1 & -2 \\ 2 & 1 & 4 & 1 & -2 \end{bmatrix} \quad (v) \begin{bmatrix} 1 & 1 & 1 & -1 \\ 2 & 1 & 1 & 0 \end{bmatrix}$$

(a) Find a basis for $N(T)$, and find its dimension. Find $r(T)$.
(b) Is T one to one? onto? Explain.

3. Let T be the matrix operator $T(x) = Ax$ with A given as follows:

$$(i) \begin{bmatrix} 1 & 1 \\ 1 & 0 \\ 0 & 1 \\ 1 & -1 \end{bmatrix} \quad (ii) \begin{bmatrix} 1 & 1 & 1 \\ 1 & 2 & -1 \\ 1 & -1 & 5 \end{bmatrix}$$

$$(iii) \begin{bmatrix} 1 & 1 & -1 \\ -1 & 3 & -1 \\ -2 & 2 & 6 \end{bmatrix} \quad (iv) \begin{bmatrix} 1 & 0 & 1 & 2 & -3 \\ 1 & 1 & 3 & 1 & -1 \\ 3 & 2 & 7 & 1 & -2 \\ 2 & 1 & 4 & 1 & -2 \end{bmatrix}$$

(a) Find a basis for $R(T)$, and find its dimension. Find the nullity of T.
(b) Is T one to one? onto? Explain.

4. Let $g \in \mathbf{F}^m$ and $f \in \mathbf{F}^n$ be nonzero vectors, and let $A = gf^{\mathrm{T}}$. Show that A is a rank-one matrix, i.e., that $R(A) = 1$. Find the null space and range of A. Show that every rank-one matrix is of this form.

5. Define $T : \mathbf{T}_n \to \mathbf{T}_n$ by $T(g) = g'' + a^2 g$, where a is a fixed positive number. For which a is T invertible?

6. For each $f \in \mathbf{P}_n$, define $a_k = \int_k^{k+1} f(t)\, dt$, $k = 0, 1, \ldots, n$. Show that every $a \in \mathbf{R}^{n+1}$ is of this form for some $f \in \mathbf{P}_n$ and that a determines f uniquely.
(Hint (from calculus): If f is continuous and $\int_a^b f(t)\, dt = 0$, then there is at least one $c \in (a, b)$ for which $f(c) = 0$.)

7. Define $T : \mathbf{P}_n \to \mathbf{R}^{n+1}$ by $T(f) = x$, where $x_k = \int_0^1 t^k (1 - t)^{n-k} f(t)\, dt$, $k = 0, 1, \ldots, n$. Show that T is invertible.

8. Let A and Q be $m \times n$ matrices, let R be an $n \times n$ matrix, and suppose $A = QR$. Assume that $r(A) = n$.
(a) Show that R is nonsingular.
(b) Determine, if possible, the rank of Q.
(c) What can be said about the relationship between $R(A)$ and $R(Q)$?

9. Suppose A is an $m \times n$ matrix and B is an $n \times p$ matrix. Show that
   (a) If $r(A) = r(AB)$, then the columns of A, AB span the same vector space.
       (*Hint:* Show that $R(AB) \subset R(A)$, i.e., that $x \in R(AB)$ implies that $x \in R(A)$.)
   (b) If $r(B) = r(AB)$, then $N(B) = N(AB)$.

10. Let A be an $m \times n$ matrix with rank r, and define $T(x) = Ax$. For each of the following statements, give the equivalent statement relating m, n, and r. Some statements may be impossible, so designate them. Explain.
    (a) The rows of A are linearly dependent.
    (b) The columns of A are linearly independent.
    (c) $\dim(R(T)) > m$.
    (d) $Ax = 0$ has a nonzero solution.
    (e) Not every vector in $\mathbf{C}^m$ is a linear combination of the columns of A.
    (f) For each b in $\mathbf{C}^m$, $Ax = b$ has at least one solution.
    (g) $Ax = b$ has infinitely many solutions for some vectors b, but not for others.
    (h) T is one to one.
    (i) A^{-1} exists.
    (j) There are more independent rows than columns.
    (k) For each b in $\mathbf{C}^m$, $Ax = b$ has at most one solution.
    (l) T is onto.
    (m) There is a b so that $Ax = b$ has exactly three solutions.
    (n) The span of the columns of A is $\mathbf{C}^m$.
    (o) The span of the columns of A is $\mathbf{C}^n$.

11. Let A be nonsingular, and let AB, CA be defined. Show that
    (a) $r(AB) = r(B)$.
    (b) $r(CA) = r(C)$.

12. Let A be an $m \times n$ matrix, and let $T : \mathbf{F}^n \to \mathbf{F}^m$ be defined by $T(x) = Ax$. Suppose the row-reduced echelon form of A has r nonzero rows. State conditions on r, m, and n so that
    (a) T is one to one.
    (b) T is onto.
    (c) T is one to one but not onto.
    (d) T is onto but not one to one.

13. Let $\{E_k\}_{k=1}^n$ be a basis for $\mathbf{F}^n$, let $\{F_k\}_{k=1}^n$ be a set of vectors in $\mathbf{F}^n$, and let $F = [F_1 \ F_2 \cdots F_n]$.
    (a) Define $T : \mathbf{F}^n \to \mathbf{F}^n$ by $T(x) = \sum_{i=1}^n E_i F_i^{\mathrm{T}} x$. Find the rank of T in terms of the rank of F.
    (b) Define $T : \mathbf{F}^n \to \mathbf{F}^n$ by $T(x) = \sum_{i=1}^n F_i E_i^{\mathrm{T}} x$. Find the rank of T in terms of the rank of F.

14. Let $T : V \to W$. Prove the following:
    (a) If T is one to one and has a right inverse $R : W \to V$, then $R = T^{-1}$.
    (b) If T is onto and has a left inverse $L : W \to V$, then $L = T^{-1}$.

15. Let $T : V \to W$ be a linear operator, and let $n = \dim(V)$, $m = \dim(W)$. Show that
    (a) If $n < m$, then T is not onto.
    (b) If $n > m$, then T is not one to one.

16. Suppose $T : V \to V$ is a linear operator and $\dim(V) \neq 0, \infty$. Show that $T(x_0) = 0$ for some $x_0 \neq 0$ iff $T(x) = y$ is not solvable for at least one y.

17. Let $T : V \to W$ be a linear operator and $n = \dim(V) = \dim(W)$. Suppose that $\{e_k\}_{k=1}^n$ is a basis for V and $\{f_k\}_{k=1}^n$ is a basis for W. Show that
(a) $\{T(e_k)\}_{k=1}^n$ is a basis for W iff T is invertible.
(b) $T(x) = f_k$ is solvable for $k = 1, \ldots, n$ iff T is invertible.

18. Let $R : V \to V$, $U : V \to W$ be linear operators, and let $T = UR$. Let $n = \dim(V)$ and $m = \dim(W)$. Suppose T is one to one and $\{e_k\}_{k=1}^n$ is a basis for V. Show that
(a) R is invertible.
(b) U is one to one.
(c) $R(T) = R(U)$.
(d) $\{T(e_k)\}_{k=1}^n$ is a basis for $R(T)$.
(e) $\{U(e_k)\}_{k=1}^n$ is a basis for $R(T)$.

19. Let $S : V \to W$ and $U : W \to X$ be linear operators, and suppose $T = US$. Let $n = \dim(V) = \dim(W) = \dim(X) \neq 0, \infty$. Show that
(a) If $R(T) = X$, then U and S are invertible.
(b) If $N(T) = \{0\}$, then U and S are invertible.

20. Integrals and derivatives as mutual inverses. Let $V = C(\mathbf{R})$ and $W = C^1(\mathbf{R})$, and define $J : V \to W$ and $D : W \to V$ by $Jf(x) = \int_0^x f(t)\,dt$, $Dg(x) = g'(x) = d/dx\, g(x)$. Show that
(a) D is a left inverse for J, and J is a right inverse for D.
(b) J is one to one, and D is onto.
(c) D is not one to one, and J is not onto.
(d) If W is replaced by $W_0 = \{g : g \in W, g(0) = 0\}$, then $J = D^{-1}$ and $D = J^{-1}$.

21. Shift operators. Let $V = W = \mathbf{C}^\infty$, and define T by $T(x_1, x_2, x_3, \ldots) = (x_2, x_3, \ldots)$, the shift-left operator. Define S by $S(x_1, x_2, x_3, \ldots) = (0, x_1, x_2, x_3, \ldots)$, the shift-right operator. Determine whether T and S are (left or right) inverses and whether they are one to one or onto.

22. Inverses of $Tf = f''$. (From the subject of Green's functions.) Let $V = C^2[0, 1]$, $W = C[01]$ and $T : V \to W$ be defined by $Tf = f''$.
(a) Show that T has no left inverses.
(b) Show that the operators S_1 and S_2, defined as follows, are right inverses:

$$S_1 f(x) = \int_0^1 G(x, y) f(y)\,dy, \qquad \text{where } G(x, y) = \begin{cases} x(y-1) & x < y \\ y(x-1) & y \le x \end{cases},$$

$$S_2 f(x) = \int_0^x (x - t) f(t)\,dy.$$

(c) Let V_2 be the set of functions in V satisfying $f(0) = f'(0) = 0$. Show that $S_2 = T^{-1}$ if the domain of T is restricted to V_2.
(d) Let V_1 be the set of functions in V satisfying $f(0) = f(1) = 0$. Show that $S_1 = T^{-1}$ if the domain of T is restricted to V_1.

2.3
LINEAR OPERATORS AND MATRICES

In the preceding sections, a matrix was used to create a linear operator. In this section, arbitrary linear operators on finite dimensional spaces are connected to matrices. There are two central reasons for doing this. First, computational techniques for matrices are made available for solving problems about linear operators. Second, the theory of linear operators may be applied to matrices. The first two propositions set the stage.

PROPOSITION 1. A linear operator is completely determined by its values on a basis. Specifically, let V and W be vector spaces, let $\{e_k\}_{k=1}^n$ be a basis for V, and let $\{f_k\}_{k=1}^n$ be a sequence of (not necessarily distinct) vectors in W. Then there is just one linear operator $T : V \to W$ such that $T(e_i) = f_i, i = 1, 2, \ldots, n$.

Proof. First we show that such an operator exists. Define T by $T(\sum_{i=1}^n \alpha_i e_i) = \sum_{i=1}^n \alpha_i f_i$. Then $T(e_i) = f_i$, and T is linear. Next let $T : V \to W$ be linear with $T(e_i) = f_i$. If $x = \sum_{i=1}^n \alpha_i e_i$, then $T(x) = \sum_{i=1}^n \alpha_i T(e_i) = \sum_{i=1}^n \alpha_i f_i$, so that $T(x)$ is completely known for every $x \in V$ when the f_i are known, and so T is unique.

The next proposition shows that all linear operators on $\mathbf{F}^n$ are matrix operators, i.e., they are of the form $T(x) = Ax$. This happens due to the unusual circumstance that the coordinate column of the vector x with respect to the standard basis is exactly x itself.

PROPOSITION 2. Linear maps $\mathbf{F}^n \to \mathbf{F}^m$ are always matrix operators. Specifically, let $T : \mathbf{F}^n \to \mathbf{F}^m$ be a linear operator. Then $T(x) = Ax \ \forall x \in \mathbf{F}^n$, where

$$A = [T(\delta_1), T(\delta_2), \ldots, T(\delta_n)]. \tag{1}$$

Proof. Set $A_k = T(\delta_k)$, and $A = [A_1 A_2 \cdots A_n]$. Then

$$T(x) = T\left(\sum_{i=1}^n x_i \delta_i\right) = \sum_{i=1}^n x_i T(\delta_i) = \sum_{i=1}^n x_i A_i = Ax.$$

EXAMPLE 1. Consider the operator $T : \mathbf{R}^3 \to \mathbf{R}^3$ defined by

$$T(x) = \tfrac{1}{2}(x^T E_1)E_1 + (x^T E_2)E_2, \tag{*}$$

where $E_1 = (1, 0, -1)^T$, $E_2 = (0, 1, 0)^T$. Find A such that $T(x) = Ax$ for all $x \in \mathbf{R}^3$.
  To find the matrix A by (1), compute

$$A_1 = T(\delta_1) = \tfrac{1}{2}E_1, \qquad A_2 = T(\delta_2) = E_2, \qquad A_3 = T(\delta_3) = -\tfrac{1}{2}E_1.$$

Then

$$A = [A_1, A_2, A_3] = \frac{1}{2}\begin{bmatrix} 1 & 0 & -1 \\ 0 & 2 & 0 \\ -1 & 0 & 1 \end{bmatrix}.$$

More directly,

$$T(x) = \tfrac{1}{2}E_1(E_1^T x) + E_2(E_2^T x) = (\tfrac{1}{2}E_1 E_1^T + E_2 E_2^T)x = Ax.$$

The connection between operators and matrices is addressed next. Suppose that $T : V \to W$, where $\beta = \{e_k\}_{k=1}^{n}$ is a basis for V and $\gamma = \{f_k\}_{k=1}^{m}$ is a basis for W, and suppose that two vectors x and y are related by $T(x) = y$. Let x and y have expansions $x = \sum_{j=1}^{n} c_j e_j$, and let $y = \sum_{i=1}^{m} d_i f_i$ so that the coordinate columns are $(c_1, c_2, \ldots, c_n)^{\mathrm{T}} = [x]_\beta = [x]$ and $(d_1, d_2, \ldots, d_m)^{\mathrm{T}} = [y]_\gamma = [y]$, respectively. The goal is to describe the relation $T(x) = y$ in terms of the coordinate columns. Form the expansions

$$T(e_j) = \sum_{i=1}^{n} a_{ij} f_i, \qquad j = 1, 2, \ldots, n. \tag{2}$$

Note that (2) is equivalent to

$$[T(e_j)]_\gamma = (a_{1j}, a_{2j}, \ldots, a_{mj})^T \equiv A_j, \tag{2'}$$

where $A = [a_{ij}] = [A_1 \, A_2 \cdots A_n]$.

Expanding $T(x)$, substituting (2) in the result, and rearranging terms yields

$$T(x) = T\left(\sum_{j=1}^{n} c_j e_j\right) = \sum_{j=1}^{n} c_j T(e_j) = \sum_{j=1}^{n} \left\{\sum_{i=1}^{m} a_{ij} f_i\right\} c_j$$

$$= \sum_{i=1}^{m} \left\{\sum_{j=1}^{n} a_{ij} c_j\right\} f_i = \sum_{i=1}^{m} d_i f_i = y.$$

Since γ is linearly independent,

$$\sum_{j=1}^{n} a_{ij} c_j = d_i, \qquad \text{for } j = 1, 2, \ldots, n,$$

which is the same as $Ac = d$.

Another derivation of this using the coordinate map and its linearity is

$$d = [y]_\gamma = [T(x)]_\gamma = \left[T\left(\sum_{j=1}^{n} c_j e_j\right)\right]_\gamma = \left[\sum_{j=1}^{n} c_j T(e_j)\right]_\gamma$$

$$= \sum_{j=1}^{n} c_j [T(e_j)]_\gamma$$

$$= \sum_{j=1}^{n} c_j A_j = Ac.$$

DEFINITION. The matrix A defined by (2) or (2') is called the *matrix of T with respect to the (ordered) bases β and γ*. The notation is $A = [T] = [T]_{\beta,\gamma}$. The reference to the bases β and γ will be suppressed when there is no chance of confusion. In the case where $V = W$ and $\beta = \gamma$, the notation is $[T]_{\beta,\beta} = [T]_\beta$.

Remark. *A* can be displayed diagrammatically as follows:

$$
A = \begin{array}{c}
\\
f_1 \\
f_2 \\
\\
\\
f_m
\end{array}
\overset{\displaystyle T(e_1) \quad T(e_2) \qquad\quad T(e_n)}{
\begin{bmatrix}
a_{11} & a_{12} & \cdots & a_{1n} \\
a_{21} & a_{22} & \cdots & a_{2n} \\
\vdots & \vdots & \vdots & \vdots \\
a_{m1} & a_{m2} & \cdots & a_{mn}
\end{bmatrix}}
$$

$$
= [[T(e_1)]_\gamma \ [T(e_2)]_\gamma \cdots [T(e_n)]_\gamma]
$$

to indicate that

$$
T(e_1) = a_{11} f_1 + a_{21} f_2 + \cdots + a_{m1} f_m
$$

$$
T(e_2) = a_{12} f_1 + a_{22} f_2 + \cdots + a_{m2} f_m,
$$

and so on; i.e., the *j*th column A_j is the coordinate column of $T(e_j)$ with respect to the basis $\{f_k\}_{k=1}^m$. The next theorem summarizes the discussion.

THEOREM 3. Let $T : V \to W$ be a linear operator, let $\beta = \{e_k\}_{k=1}^n$ be a basis for V, and let $\gamma = \{f_k\}_{k=1}^m$ be a basis for W. Construct the matrix $[T] = [T]_{\beta,\gamma}$ whose *j*th column is $[T(e_j)]_\gamma$. Then

$$
T(x) = y \Leftrightarrow [T][x]_\beta = [y]_\gamma. \tag{3}
$$

In other words,

$$
[T(x)]_\gamma = [T][x]_\gamma. \tag{4}
$$

The next diagram illustrates the relationships in the case $V = W$ and $A = [T]$.

The diagram has the following meaning: $y = T(x)$ may be computed two ways. Either apply the operator *T* to *x* directly (top leg), or find the coordinate column $[x]$ of *x*, compute $A[x] = [y] = (d_1, d_2, \ldots, d_n)^T$, and then obtain *y* be reattaching the coordinates d_i to the basis vectors: $y = \sum_{i=1}^n d_i e_i$ (go down to $[x]$, across to $[y]$ using *A*, and back up to *y*).

In most of the applications in this book, $V = W$ and $\beta = \gamma$. Thus, if nothing to the contrary is said, this will be assumed.

The next example shows the consistency of some of our definitions and connects Proposition 2 with Theorem 3. It says that, in the standard basis, the matrix of a matrix operator is the matrix that generates the operator. It is the analog of $[x]_\delta = x$ in $\mathbf{F}^n$.

EXAMPLE 2. Let T be the linear operator $\mathbf{F}^n \to \mathbf{F}^m$ generated by the matrix $A = [A_1 \ A_2 \ \cdots \ A_n] : T(x) = Ax$. Let both β and γ be the standard bases in $\mathbf{F}^n$ and $\mathbf{F}^m$, respectively. Then $[T] = A$.

This follows from the fact that $T(\delta_j) = A\delta_j = A_j$.

EXAMPLE 3. Let $V = \mathbf{P}_2$ and $T(x) = x'' + 2x' + x$, where $' = d/dt$. Find the matrix of T with respect to $\beta = \{1, t, t^2\}$. Verify Theorem 1. Solve $T(x) = 1 + t + t^2$ if possible. Find $N(T)$ and the rank of T.

It is easy to see that if x is a polynomial of degree ≤ 2, then so is $T(x)$; hence $T : V \to V$. Let $\{1, t, t^2\} = \{e_0, e_1, e_2\}$. $[T]$ is computed as follows. Compute $T(1) = 1$, $T(t) = 2 + t$, and $T(t^2) = 2 + 2 \cdot 2t + t^2$. Thus

$$
\begin{aligned}
Te_0 &= e_0 + 0 \cdot e_1 + 0 \cdot e_2 \\
Te_1 &= 2 \cdot e_0 + 1 \cdot e_1 + 0 \cdot e_2 \\
TE_2 &= 2 \cdot e_0 + 4 \cdot e_1 + 1 \cdot e_2
\end{aligned}
\quad \Rightarrow [Te_0] = \begin{bmatrix} 1 \\ 0 \\ 0 \end{bmatrix}, \quad [Te_1] = \begin{bmatrix} 2 \\ 1 \\ 0 \end{bmatrix}, \quad [Te_2] = \begin{bmatrix} 2 \\ 4 \\ 1 \end{bmatrix}.
$$

Therefore

$$
[T] = [[Te_0], [Te_1], [Te_2]] = \begin{bmatrix} 1 & 2 & 2 \\ 0 & 1 & 4 \\ 0 & 0 & 1 \end{bmatrix}.
$$

To verify the theorem, first compute $T(x)$ directly by putting $x = c_0 \cdot 1 + c_1 t + c_2 t^2$ into T, computing the derivatives, and regrouping terms according to powers of t:

$$
\begin{aligned}
T(x) &= T(c_0 + c_1 t + c_2 t^2) \\
&= (c_0 + c_1 t + c_2 t^2)'' + 2(c_0 + c_1 t + c_2 t^2)' + (c_0 + c_1 t + c_2 t^2) \\
&= (2c_2) + (2c_1 + 4c_2) + (c_0 + c_1 t + c_2 t^2) \\
&= (c_0 + 2c_1 + 2c_2) + (c_1 + 4c_2)t + c_2 t^2.
\end{aligned}
$$

Hence

$$
[T(x)] = \begin{bmatrix} c_0 + 2c_1 + 2c_2 \\ c_1 + 4c_2 \\ c_2 \end{bmatrix} = [T] \begin{bmatrix} c_0 \\ c_1 \\ c_2 \end{bmatrix},
$$

which verifies the theorem. To solve $T(x) = 1 + t + t^2$ using the matrix $[T]$, solve $[T]c = [1 + t + t^2] = (1, 1, 1)^{\mathrm{T}}$. This gives $c = (5, -3, 1)^{\mathrm{T}} = [x]$. Hence $x = 5 - 3t + t^2$ solves $T(x) = 1 + t + t^2$. Two methods are given to find $N(T)$. First, note that $T(x) = x'' + 2x' + x = 0$ has the general solution $x = c_1 e^t + c_2 t e^t$, which is not in $\mathbf{P}_2$ unless $c_1 = c_2 = 0$, so that $N(T) = \{0\}$. Second, observe that $[T]$ row-reduces to the identity, and is therefore nonsingular. Since $T(x) = 0 \Leftrightarrow [T][x] = 0$, and the latter linear system has only the 0 solution by the invertibility of $[T]$, it follows again that $N(T) = 0$. By the rank and nullity theorem, $r = \dim(\mathbf{P}_2) - \dim(N(T)) = 3$.

EXAMPLE 4. Let $V = \mathbf{P}_2$ and $T : V \to V$ be defined by

$$
(Tp)(x) = \frac{1}{2} \int_{-1}^{1} (3s + 6x + 6xs + 3x^2) p(s) \, ds.
$$

Determine whether T is one to one or onto. Find the null space and range of T.

The variable s in the integral is a dummy variable. It can be replaced by any symbol but x, and the value of the integral will be the same. First, compute a matrix A of T with respect to a convenient basis. By finding the null space and range of A, the null space and range of T can be reconstructed. Choose the basis to be $\{1, t, t^2\} = \{e_0(t), e_1(t), e_2(t)\}$.

Then

$$(Te_0)(x) = \frac{1}{2}\int_{-1}^{1}(3s + 6x + 6xs + 3x^2) \cdot 1\ ds = 6x + 3x^2$$

$$(Te_1)(x) = \frac{1}{2}\int_{-1}^{1}(3s + 6x + 6xs + 3x^2) \cdot s\ ds = 1 + 2x$$

$$(Te_2)(x) = \frac{1}{2}\int_{-1}^{1}(3s + 6x + 6xs + 3x^2) \cdot s^2\ ds = 2x + x^2$$

so that

$$(Te_0)(x) = 0e_1(x) + 6e_1(x) + 3e_2(x)$$
$$(Te_1)(x) = 1e_0(x) + 2e_1(x) + 0e_3(x)$$
$$(Te_2)(x) = 0e_1(x) + 2e_1(x) + e_2(x)$$

whence

$$[T] = A = \begin{bmatrix} 0 & 1 & 0 \\ 6 & 2 & 2 \\ 3 & 0 & 1 \end{bmatrix}.$$

Row reduction of A yields

$$\begin{bmatrix} 3 & 0 & 1 \\ 0 & 1 & 0 \\ 0 & 0 & 0 \end{bmatrix} = R.$$

Thus it is clear that $Ac = 0$ has nontrivial solutions, and therefore so does $T(x) = 0$. Thus T is not one to one. In particular, the solutions of $Ac = 0$ are $c_0 = -c_2/3$ and $c_1 = 0$; i.e., $c = [x] = c_2(-1, 0, 3)^{\mathrm{T}}$ is the general solution of $Ac = 0$ and $(-1, 0, 3)^{\mathrm{T}}$ is a basis for $N(A)$. The corresponding vector $p_1(t) = -1 + 3t^2$ is a basis for $N(T)$. By the rank and nullity theorem, $\dim(R(T)) = \dim(V) - \dim(N(T)) = 3 - 1 = 2$. Thus T cannot be onto, since $\dim(R(T)) < \dim(V)$. There are several ways to find a basis for $R(T)$. One method is to find a basis for the range of $[T]$. See Section 2.2, where three methods are given. For example, if A^{T} is row-reduced and the nonzero rows are transposed, then one finds that $\{F_1 = (1, 0, -1)^{\mathrm{T}}, F_2 = (0, 2, 1)^{\mathrm{T}}\}$ is a basis for $R(A)$. F_1 and F_2 are the coordinate columns of $f_1 = 1 - t^2$, $f_2 = 2t + t^2$, which therefore form a basis for $R(T)$. Alternatively, the columns of A corresponding to the pivot columns of R are $(0, 6, 3)^{\mathrm{T}}$, $(1, 2, 0)^{\mathrm{T}}$, and these form a basis for $R(A)$. Thus, normalizing, $\{2t + t^2, 1 + 2t\}$ is a basis for $R(T)$.

EXAMPLE 5. Describe the components of the stress tensor at a given point in a given coordinate frame.

In linear elasticity, there is an object T called the stress tensor. T may have different values at different points in the continuum being considered, so let us fix such a point P in our discussion. The stress tensor has the following interpretation:

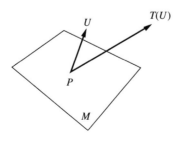

Let U be a unit vector and M be a plane through P that is perpendicular to U and oriented by U. Then $T(U)$ is the vector force per unit area (i.e., stress) across M exerted by the material on the positive (i.e., U) side of M. The stress tensor at a fixed point is a linear operator. One can find a matrix representation for T by computing its matrix with respect to a given basis. Choosing three mutually orthogonal axes through the given point is the same as choosing three mutually orthogonal vectors u_1, u_2, u_3 of unit length. Next express forces and vectors U (not necessarily a unit vector now) in terms of this basis:

$$U = x_1 u_1 + x_2 u_2 + x_3 u_3, \qquad \text{and} \qquad T(U) = y_1 u_1 + y_2 u_2 + y_3 u_3.$$

These expressions are equivalent to the matrix expression

$$[T]x = y, \qquad \text{where } x = [U] = (x_1, x_2, x_3)^{\mathrm{T}}, \qquad y = [T(U)] = (y_1, y_2, y_3)^{\mathrm{T}}.$$

From our work, the preceding $[T]$ is constructed by

$$[T] = [A_1\ A_2\ A_3], \qquad \text{where } A_j = [T(u_j)] = (a_{1j}, a_{2j}, a_{3j})^{\mathrm{T}};$$

i.e., $T(u_j) = a_{1j} u_1 + a_{2j} u_2 + a_{3j} u_3$. The numbers a_{ij} are the components of the stress tensor in the given coordinate system. For example, $T(u_1) = a_{11} u_1 + a_{21} u_2 + a_{31} u_3$ is the stress vector across the x_2, x_3 plane, oriented by u_1. The coordinate of $T(u_1)$ along u_1 is a_{11}, and is called a *normal stress*. The coordinates along components orthogonal to u_1 are a_{21} and a_{31}, and are called *shear stresses*.

Remark. Construction of a linear operator from a matrix.   The process of constructing an operator from a matrix can be reversed: Given a matrix A and two fixed ordered bases $\beta = \{e_k\}_{k=1}^n$ and $\gamma = \{f_k\}_{k=1}^m$, define an operator $T : V \to W$ by the relationship

$$T(x) = y \Leftrightarrow A[x] = [y].$$

It is easy to check that T is linear and $[T] = A$.

EXAMPLE 6. Let $V = \mathbf{T}_1$ with ordered basis $\{1, \cos(t), \sin(t)\}$. Define $T : V \to V$ by $T(x) = y$ iff $A[x] = [y]$, where

$$A = \begin{bmatrix} 2 & -1 & 1 \\ 3 & 0 & 1 \\ 0 & 5 & 9 \end{bmatrix}.$$

Find

$$y = T(x), \qquad \text{if } x = c_1 1 + c_2 \cos(t) + c_3 \sin(t).$$

Here $[x] = c = (c_1, c_2, c_3)^{\mathrm{T}}$. Then $d = Ac = (2c_1 - c_2 + c_3, 3c_1 + c_3, 5c_2 + 9c_3)^{\mathrm{T}}$. To find $T(x)$, put the coefficients back on the correct members of the basis:

$$y = T(x) = d_1 1 + d_2 \cos(t) + d_3 \sin(t)$$

$$= (2c_1 - c_2 + c_3)1 + (3c_1 + c_3) \cos(t) + (5c_2 + 9c_3) \sin(t).$$

Observe that forming the matrix of T is a way to associate a linear operator $T : V \to W$ with a matrix $[T]$ when bases for V and W are fixed. This correspondence is an analog of the coordinate map that makes a correspondence between a vector x and its coordinate column $[x]$.

THEOREM 4. Let β and γ be bases for V and W, respectively. Let $L(V, W)$ be the set of all linear operators $T : V \rightarrow W$, and let $\mathbf{F}^{m,n}$ be the set of all $m \times n$ matrices with scalar entries. Then $L(V, W)$ is a vector space, and the mapping from $L(V, W)$ to $\mathbf{F}^{m,n}$, taking an operator T to its matrix $[T] = [T]_{\beta,\gamma}$ with respect to β, γ, is

(a) One to one and onto.

(b) Linear: $[T + S] = [T] + [S]$, $[\alpha T] = \alpha[T]$.

That is, the mapping $T \rightarrow [T]$ is an isomorphism from $L(V, W)$ to $\mathbf{F}^{m,n}$.

Proof. $L(V, W)$ is a vector space because it is a subspace of the vector space of all functions $V \rightarrow W$. (a) The mapping is onto by the fact that, given A, there is a linear operator T, with $[T] = A$. The mapping is one to one since a linear transformation is completely determined by its values on a basis. (b) Linearity is clear from the method of construction.

The next theorem determines the matrix of the composition ST and the matrix of T^{-1}, when it exists. For matrix operators $T(x) = Ax$ and $S(y) = By$, we have (assuming correct dimensions): $ST(x) = S(T(x)) = B(Ax) = BAx$. Thus ST is generated by BA. A similar statement is valid in general for ST.

THEOREM 5. When ST is well defined, all the relevant vector spaces are finite dimensional, and bases are fixed in the respective vector spaces, then

(a) $[I] = I$

(b) $[ST] = [S][T]$

(c) T^{-1} exists iff $[T]^{-1}$ exists, and when T^{-1} exists, $[T^{-1}] = [T]^{-1}$.

Proof. Statement (a) is clear from the construction. Proof of (b): Let $Tx = y$ and $Sy = z$, which is equivalent to $[T][x] = [y]$ and $[S][y] = [z]$, according to Theorem 3. Eliminating y and $[y]$ yields $STx = z$, which is equivalent to $[S][T][x] = [z]$. But by Theorem 3, $STx = z$ is equivalent to $[ST][x] = [z]$, which implies that $[ST] = [S][T]$ because the correspondence between operators and matrices is one to one. Proof of (c): If T^{-1} exists, then $[T][T^{-1}] = [TT^{-1}] = [I] = I$, so that $[T^{-1}] = [T]^{-1}$. Now suppose that $[T]^{-1}$ exists. Define $S : W \rightarrow V$ by $Sx = y$ iff $[T]^{-1}[x] = [y]$, so that $[S] = [T]^{-1}$. But now, $STx = y$ iff $[y] = [ST][x] = [S][T][x] = I[x] = [x]$, which implies that $STx = x$ for all x in V; hence $ST = I$. Similarly $TS = I$, so $S = T^{-1}$.

EXAMPLE 7. COMPOSITIONS OF ROTATIONS. Multiplication of a complex number z by $e^{i\theta}$ rotates z counterclockwise through an angle θ. Identify a complex number $z = x + iy$ with a column

$$z = \begin{bmatrix} x \\ y \end{bmatrix}.$$

Thus

$$e^{i\theta} = \begin{bmatrix} \cos(\theta) \\ \sin(\theta) \end{bmatrix},$$

and a counterclockwise rotation through an angle θ is given by

$$T_\theta(z) = e^{i\theta}z = \begin{bmatrix} x\cos(\theta) - y\sin(\theta) \\ y\sin(\theta) + x\cos(\theta) \end{bmatrix} = \begin{bmatrix} \cos(\theta) & -\sin(\theta) \\ \sin(\theta) & \cos(\theta) \end{bmatrix} \begin{bmatrix} x \\ y \end{bmatrix}.$$

Hence

$$[T_\theta] = \begin{bmatrix} \cos(\theta) & -\sin(\theta) \\ \sin(\theta) & \cos(\theta) \end{bmatrix}$$

in the standard basis, by Example 2.

Rotating through $-\theta$ undoes the operation; i.e., $T_{-\theta}T_\theta = I$. We easily confirm that $[T_{-\theta}][T_\theta] = I$.

If T_ϕ is the counterclockwise rotation through an angle ϕ, then $T_\theta T_\phi = T_{\theta+\phi}$ is the counterclockwise rotation through an angle $\theta + \phi$. But

$$[T_\theta][T_\phi] = \begin{bmatrix} \cos(\theta) & -\sin(\theta) \\ \sin(\theta) & \cos(\theta) \end{bmatrix}\begin{bmatrix} \cos(\phi) & -\sin(\phi) \\ \sin(\phi) & \cos(\phi) \end{bmatrix}$$

$$= \begin{bmatrix} \cos(\theta)\cos(\phi) - \sin(\theta)\sin(\phi) & -\cos(\theta)\sin(\phi) - \cos(\phi)\sin(\theta) \\ \cos(\phi)\sin(\theta) + \cos(\theta)\sin(\phi) & -\sin(\phi)\sin(\theta) + \cos(\phi)\cos(\theta) \end{bmatrix}$$

$$= \begin{bmatrix} \cos(\theta + \phi) & -\sin(\theta + \phi) \\ \sin(\theta + \phi) & \cos(\theta + \phi) \end{bmatrix} = [T_{\theta+\phi}],$$

which illustrates part (2) of Theorem 5.

EXAMPLE 8. Let $V = \mathbf{P}_2$, and define $T : V \to V$ by

$$(Tp)(x) = \frac{1}{2}\int_{-1}^{1}(15t^2 + 3xt + 6x^2)p(t)\,dt.$$

Construct T^{-1} and compute $T^{-1}(t^2)$.

Using the basis $\{1, t, t^2\}$ yields

$$[T] = \begin{bmatrix} 5 & 0 & 3 \\ 0 & 1 & 0 \\ 6 & 0 & 2 \end{bmatrix}, \qquad \text{whence } [T]^{-1} = \begin{bmatrix} -1/4 & 0 & 3/8 \\ 0 & 1 & 0 \\ 3/4 & 0 & -5/8 \end{bmatrix}.$$

According to the correspondence between matrices and operators, $T^{-1}(c_0 + c_1 t + c_2 t^2) = d_0 + d_1 t + d_2 t^2$ iff $[T]^{-1}c = d$. Compute

$$[T]^{-1} = \begin{bmatrix} c_0 \\ c_1 \\ c_2 \end{bmatrix} = \begin{bmatrix} (-c_0/4) + (3c_2/8) \\ c_1 \\ (3c_0/4) - (5c_2/8) \end{bmatrix}.$$

Now reattach the coefficients to the proper basis vectors:

$$T^{-1}(c_0 + c_1 t + c_2 t^2) = (-c_0/4 + 3c_2/8)1 + c_1 t + (3c_0/4 - 5c_2/8)t^2.$$

In particular, $T^{-1}(t^2) = \frac{1}{8}(3 - 5t^2)$, by taking $c_0 = c_1 = 0$ and $c_2 = 1$, or using the fact that the third column of $[T]^{-1}$ is the coordinate column for $T^{-1}(t^2)$. Check this calculation by showing that $T[\frac{1}{8}(3 - 5t^2)] = t^2$ directly.

EXAMPLE 9. Given f continuous on $x \geq 0$, define the numbers

$$x_k = \int_0^{k+1} f(t)\,dt, \qquad k = 0, 1, \ldots.$$

Is it possible to recover f from these numbers? Let us answer this when $f \in \mathbf{P}_2$. Define $T : \mathbf{P}_2 \to \mathbf{R}^3$ by

$$T(f) = \left[\int_0^1 f(t)\, dt, \int_0^2 f(t)\, dt, \int_0^3 f(t)\, dt\right]^{\mathrm{T}} = (x_0, x_1, x_2)^{\mathrm{T}}.$$

Clearly T is linear. Use the basis $\beta = \{1, t, t^2\}$ for $\mathbf{P}_2$ and the basis $\delta = \{\delta_1, \delta_2, \delta_3\}$ for $\mathbf{R}^3$. A computation yields

$$T(1) = \begin{bmatrix} 1 \\ 2 \\ 3 \end{bmatrix}, \qquad T(t) = \begin{bmatrix} 1/2 \\ 2 \\ 9/2 \end{bmatrix}, \qquad T(t^2) = \begin{bmatrix} 1/3 \\ 8/3 \\ 9 \end{bmatrix}.$$

Since $x = [x]_\delta$ in $\mathbf{F}^n$,

$$[T] = [[T(1)]_\delta, [T(t)]_\delta, [T(t^2)]_\delta] = \begin{bmatrix} 1 & 1/2 & 1/3 \\ 2 & 2 & 8/3 \\ 3 & 9/2 & 9 \end{bmatrix}.$$

A row reduction (or other test) shows that $[T]$ is nonsingular. This implies that $[T]^{-1}$ exists, and so T^{-1} exists. Thus for each $x \in \mathbf{R}^3$, there is exactly one $f \in \mathbf{P}_2$ with $T(f) = x$ so that f is uniquely determined by x_0, x_1, x_2. To recover f from x, compute

$$[T]^{-1} = \begin{bmatrix} 3 & -3/2 & 1/3 \\ -5 & 4 & -1 \\ 3/2 & -3/2 & 1/2 \end{bmatrix}.$$

For example, to find f for which $T(f) = (0, 0, 1)^{\mathrm{T}} = \delta_2$, compute $[T]^{-1}\delta_2 = (1/3, -1, 1/2)^{\mathrm{T}} = [f]_\beta$, which gives $f(t) = 1/3 - t + 1/2 t^2$.

The effect of a change of basis on the matrix of a linear operator is discussed next. We discuss only the case $V = W$, which is our main interest. This theorem will have wide use in subsequent chapters.

THEOREM 6. Let $T : V \to V$ be a linear operator, and let $\beta = \{e_k\}_{k=1}^n$ and $\gamma = \{f_k\}_{k=1}^n$ be two bases for V. Suppose the change of basis matrix from β to γ is $P = [p_{ij}]$. Then

$$[T]_\gamma = P^{-1}[T]_\beta P. \tag{5}$$

Proof. Two proofs are given. The first proof uses the coordinate map and the relation between coordinate columns $[z]_\beta = P[z]_\gamma$ for all z. Let $T(x) = y$. Then

$$[y]_\beta = [T]_\beta[x]_\beta = [T]_\beta P[x]_\gamma, \qquad \text{and} \qquad [y]_\beta = P[y]_\gamma = P[T]_\gamma[x]_\gamma.$$

Hence $[T]_\beta P[x]_\gamma = P[T]_\gamma[x]_\gamma \ \forall [x]_\gamma$. Since this is true for all $[x]_\gamma$, it must be that $[T]_\beta P = P[T]_\gamma$, which yields (6).

For the second proof, let $A = [T]_\beta$ and $B = [T]_\gamma$. From the definitions of A and B and from the change of basis relation $f_j = \sum_{i=1}^n p_{ij} e_i$, it follows that

$$T(f_k) = \sum_{j=1}^n b_{jk} f_j = \sum_{j=1}^n b_{jk}\left\{\sum_{i=1}^n p_{ij} e_i\right\} = \sum_{i=1}^n \left\{\sum_{j=1}^n b_{jk} p_{ij}\right\} e_i, \qquad \text{and}$$

$$T(f_k) = T\left(\sum_{j=1}^n p_{jk} e_j\right) = \sum_{j=1}^n p_{jk} T(e_j) = \sum_{j=1}^n p_{jk}\left\{\sum_{i=1}^n a_{ij} e_i\right\} = \sum_{i=1}^n \left\{\sum_{j=1}^n p_{ji} a_{ij}\right\} e_i.$$

By the linear independence of $\{e_k\}_{k=1}^n$,

$$\sum_{j=1}^n b_{jk} p_{ij} = \sum_{j=1}^n p_{ji} a_{ij} \qquad \text{for } i = 1, 2, \ldots, n.$$

According to the definition of matrix multiplication, this is exactly $PB = AP$, which is equivalent to (5).

DEFINITION. Two $n \times n$ matrices are *similar* if and only if there is a nonsingular matrix P such that $P^{-1}AP = B$.

By the preceding theorem, two matrices are similar if and only if they are matrices of the same operator, but with respect to two different bases. The next example elaborates on this idea. It gives the matrix of a matrix operator when the basis is not the standard basis $\{\delta_1, \ldots, \delta_n\}$.

COROLLARY 7. Let $T(x) = Ax$ be a matrix operator on $\mathbf{F}^n$. Let $\{E_1, E_2, \ldots, E_n\}$ be a basis for $\mathbf{F}^n$. Then

$$[T]_{\{E_1, \cdots, E_n\}} = P^{-1}AP, \qquad \text{where } P = [E_1 \ E_2 \ \ldots \ E_n]. \tag{6}$$

The old basis is the standard basis $\{\delta_1, \delta_2, \ldots, \delta_n\}$, and the new basis is $\{E_1, E_2, \ldots, E_n\}$. Proposition 4 of Section 1.5 showed that the change of basis matrix is $P = [E_1 \ E_2 \cdots E_n]$. Example 2 showed that the matrix of T with respect to the standard basis is A. The matrix B of T with respect to $\{E_1, E_2, \ldots, E_n\}$ is therefore $B = P^{-1}AP$, by Theorem 6.

EXAMPLE 10. Let $T : \mathbf{R}^2 \to \mathbf{R}^2$ be defined by $T(x) = Ax$, with

$$A = \begin{bmatrix} 1 & 2 \\ -1 & 0 \end{bmatrix}.$$

Let $\{\delta_1, \delta_2\}$ be the old basis and

$$\gamma = \{E_1, E_2\} = \left\{ \begin{bmatrix} 1 \\ 0 \end{bmatrix}, \begin{bmatrix} 1 \\ 1 \end{bmatrix} \right\}$$

be the new basis. Find the matrix B of T with respect to γ by using (6) and also by using the definition of $B = [T]_\gamma$.

To compute the matrix of T with respect to the new basis using the definition, compute the images $T(E_i)$, and then express them as linear combinations of E_1, E_2. Note that $E_2 = \delta_1 + \delta_2$, so $\delta_1 = E_1$ and $\delta_2 = E_2 - E_1$. Thus,

$$T(E_1) = \begin{bmatrix} 1 \\ -1 \end{bmatrix} = \delta_1 - \delta_2 = E_1 - (E_2 - E_1) = 2E_1 - E_2,$$

$$T(E_2) = \begin{bmatrix} 3 \\ -1 \end{bmatrix} = 3\delta_1 - \delta_2 = 3E_1 - (E_2 - E_1) = 4E_1 - E_2.$$

Hence

$$B = \begin{bmatrix} 2 & 4 \\ -1 & -1 \end{bmatrix}.$$

To compute B using (6), observe that $P = [E_1, E_2]$, so

$$P^{-1} = \begin{bmatrix} 1 & -1 \\ 0 & 1 \end{bmatrix},$$

and $B = P^{-1}AP$. The answers agree.

Exercises 2.3

1. Let $T : \mathbf{C}^3 \to \mathbf{C}^3$ be defined by

$$(i)\ T(x) = \begin{bmatrix} 1 & 1 & 2 \\ 1 & -1 & 1 \\ 1 & 5 & 4 \end{bmatrix} x \qquad (ii)\ T(x) = \begin{bmatrix} 1 & 1 & 0 \\ 0 & 0 & -1 \\ 0 & -1 & 1 \end{bmatrix} x.$$

For each of these operators,
   (a) Find the matrix A of T with respect to the standard basis.
   (b) Find the matrix B of T with respect to the basis $E_1 = (1, 1, -1)^T, E_2 = (1, 0, 1)^T$, and $E_3 = (-1, 2, 1)^T$.

2. Let $(0, 0)^T, (1, 1)^T, (1, 2)^T$ be the vertices of a triangle T_1 in $\mathbf{R}^2$, and let $(0, 0)^T, (1, -1)^T$, and $(2, 1)^T$ be the vertices of T_2. Find a linear operator $L : \mathbf{R}^2 \to \mathbf{R}^2$ taking T_1 onto T_2, maintaining the order of the vertices.

3. Define $T : \mathbf{P}_2 \to \mathbf{P}_2$ by $Tp(x) = (1 + 2x^2)p''(x) + (1 - 2x)p'(x) + p(x) + xp(0)$.
   (a) Find the matrix A of T with respect to the basis $\{1, x, x^2\}$.
   (b) Find the matrix B of T with respect to the basis $\{1, 1 + x, 1 + x + x^2\}$.
   (c) Find $N(T)$ and its dimension. Find the rank of T.
   (d) Is T one to one? onto? invertible? Explain.
   (e) Solve, if possible, $Tp(x) = (1 + x)^2$.

4. Define $T : \mathbf{P}_2 \to \mathbf{P}_2$ by $Tp(x) = \frac{1}{2}\int_{-1}^{1}\{3 + 3(t + x) + 6xt + 3x^2\}p(t)\,dt$.
   (a) Find the matrix A of T with respect to $\{1, x, x^2\}$.
   (b) Find $N(T)$ and its dimension. Find the rank of T and a basis for $R(T)$.
   (c) Is T one to one? onto? invertible? Explain.
   (d) Is $x^2 \in R(T)$?
   (e) Find the matrix of T^2 in the given basis.
   (f) Find $T^2(1)$.

5. Let $\Delta : \mathbf{P}^4 \to \mathbf{P}^3$ be defined by $\Delta p(x) = p(x + 1) - p(x)$.
   (a) Find the matrix A of Δ with respect to the bases $\{1, x, \ldots, x^4\}$ and $\{1, x, x^2, x^3\}$ of $\mathbf{P}^4, \mathbf{P}^3$.
   (b) Find the matrix B of Δ with respect to the bases

$$\{1, x, x(x - 1), x(x - 1)(x - 2), x(x - 1)(x - 2)(x - 3)\} \qquad \text{and}$$
$$\{1, x, x(x - 1), x(x - 1)(x - 2)\}.$$

   (c) Find the null space, range, and rank of Δ. Is Δ one to one? onto? invertible? Explain.

6. Define $T : \mathbf{R}^3 \to \mathbf{R}^3$ by $T(x) = 2(x^Te_1)e_1 - (x^Te_2)e_2 + (x^Te_3)e_3$, where $e_1 = (0, 1, 0)^T$, $e_2 = (1, 0, 1)^T$, and $e_3 = (1, 2, 1)^T$.
   (a) Find the matrix of T with respect to the standard basis.
   (b) Find a basis for $N(T)$ and $R(T)$.
   (c) Is T one to one? onto? Does T^{-1} exist? Explain.

7. Define $T : \mathbf{C}^{2,2} \to \mathbf{C}^{2,2}$ by

$$T\left(\begin{bmatrix} a & b \\ c & d \end{bmatrix}\right) = \begin{bmatrix} a + b + c & a - c \\ c - d & a + d \end{bmatrix}. \tag{*}$$

   (a) Show that T^{-1} exists, before doing (b) or (c).
   (b) Find the matrix of T with respect to the basis $\{\delta_1\delta_1^T, \delta_1\delta_2^T, \delta_2\delta_1^T, \delta_2\delta_2^T\}$.
   (c) Find an expression for T^{-1} in the same form as (*); i.e., express T^{-1} explicitly in terms of a, b, c, d.

8. Define $T : \mathbf{P}_2 \rightarrow \mathbf{P}_2$ by $T p(x) = \frac{1}{2}\int_{-1}^{1}[3 + 6tx - 15t^2 x^2]p(t)\,dt$.
   (a) Find the matrix A of T with respect to the basis $\{1, t, t^2\}$.
   (b) Find the null space of T, the rank of T, and a basis for $R(T)$.
   (c) Show that T^{-1} exists, and find an expression for $T^{-1}(a + bt + ct^2)$.
   (d) Find p so that $T(p) = (1 + t)^2$.
   (e) Find q so that $T^2(q) = t^2$.

9. Define $T : \mathbf{T}_1 \rightarrow \mathbf{T}_1$ by $Tx = x'' + x' + 2x - x(0)$.
   (a) Show that T is invertible.
   (b) Find $T^{-1}(a + b\cos(t) + c\sin(t))$ as a linear combination of $1, \cos(t)$, and $\sin(t)$.
   (c) Find $p \in \mathbf{T}_1$ so that $T^2(p) = \sin(t)$.

10. Let

$$B = \begin{bmatrix} 1 & -1 & 0 \\ 1 & 1 & 1 \\ 1 & 3 & 2 \end{bmatrix},$$

   and let V be the subspace of $\mathbf{R}^3$: $V = \{x : x_1 - 2x_2 + x_3 = 0\}$. Define T on V by $T(x) = Bx$, $x \in V$.
   (a) Show that $T : V \rightarrow V$.
   (b) Show that $\{E_1 = (2, 1, 0)^\mathrm{T}, E_2 = (1, 0, -1)^\mathrm{T}\}$ is a basis for V.
   (c) Find the matrix A of T with respect to the basis $\{E_1, E_2\}$.
   (d) Is $T : V \rightarrow V$ invertible? Explain.

11. Let $\mathbf{w}$ be a fixed (geometric) nonzero vector. Let V be the three-dimensional euclidean space of calculus. Define a map $T : V \rightarrow V$ by $T(\mathbf{r}) = \mathbf{w} \times \mathbf{r}$, where $\times$ is the usual cross product.
   (a) Find the matrix A of T in the basis $\mathbf{i}_1 = \mathbf{i}, \mathbf{i}_2 = \mathbf{j}, \mathbf{i}_3 = \mathbf{k}$ (a right-handed system).
   (b) Show that A is *skew-symmetric*, i.e., that $A^\mathrm{T} = -A$.
   (c) Is T invertible?

12. The trailing moving average of length h of a function f is defined by

$$T f(x) = \frac{1}{h} \int_{0}^{h} f(x - t)\,dt.$$

   Analysis of the stock market involves the use of such averages. Can a function be recovered from its trailing moving average?
   Take the domain of T to be $\mathbf{P}_n$. Show that $T : \mathbf{P}_n \rightarrow \mathbf{P}_n$ is linear. Show that, for every $g \in \mathbf{P}_n$, there is a unique $f \in \mathbf{P}_n$ so that $T(f) = g$. Thus, on $\mathbf{P}_n$, functions are characterized by their trailing moving average.

13. Proposition 1 asserts that a linear operator is characterized by its values on a basis. Given a basis $\{e_k\}_{k=1}^{n}$ for $\mathbf{F}^n$ and arbitrary $\{v_k\}_{k=1}^{n}$ in $\mathbf{F}^n$, find the unique linear operator T for which $T(e_k) = v_k$, where $k = 1, 2, \ldots, n$. Find the matrix of T in the standard basis specifically.

14. Suppose that $\beta = \{e_k\}_{k=1}^{n}$ is a basis for V, and $\{f_k\}_{k=1}^{n} \subset V$. Define $T : V \rightarrow V$ by $T(\sum_{i=1}^{n} c_i e_i) = \sum_{i=1}^{n} c_i f_i$. Find the matrix of T with respect to the basis β.

Inner Product Spaces

Inner products appear naturally in a variety of physical and mathematical situations. Many examples of such situations will be presented in this chapter and in later parts of the book. A familiar example of an inner product is the dot product between geometric vectors. Many important practical and theoretical problems may be attacked and solved using the theory related to inner products.

3.1
PRELIMINARIES

This section begins the study of inner products with definitions and a few examples.

DEFINITION. $\langle , \rangle$ is an inner product on the complex vector space V iff for every $x, y \in V$, $\langle x, y \rangle$ is a complex number, and the following properties hold for every x, y, z in V and every α, β in $\mathbf{C}$:

(a) $\langle \alpha x + \beta y, z \rangle = \alpha \langle x, z \rangle + \beta \langle y, z \rangle$ (*Linear* in the first variable)
(b) $\langle y, x \rangle = \overline{\langle x, y \rangle}$ (*Hermitian*)
(c) $\langle x, x \rangle \geq 0$ (*Positive*)
(d) $\langle x, x \rangle = 0 \Leftrightarrow x = 0$ (*Definite*)

From (a) and (b), it follows that $\langle , \rangle$ is *conjugate linear* in the second variable:

$$\langle z, \alpha x + \beta y \rangle = \overline{\alpha} \langle z, x \rangle + \overline{\beta} \langle z, y \rangle \qquad \forall x, y, z \text{ in } V \text{ and } \alpha, \beta \text{ in } \mathbf{C}.$$

Note that $\langle x, 0 \rangle = 0$ for all x, since $\langle x, 0 \rangle = \langle x, 0 \cdot 0 \rangle = \overline{0} \langle x, 0 \rangle = 0$.

$\langle , \rangle$ is an *inner product on the real vector space* V iff properties (a) through (d) hold with the word "complex" replaced by the word "real." In this case, the conjugate signs vanish automatically, since $\overline{\alpha} = \alpha$ for all real α, so that the same set of axioms is valid with the conjugate signs erased. Property (b) is then one of *symmetry:* $\langle x, y \rangle = \langle y, x \rangle$. Also, $\langle , \rangle$ is linear in each variable.

In this chapter, $\mathbf{F}$ is used to represent $\mathbf{R}$ or $\mathbf{C}$. Statements are made as if the scalar field were $\mathbf{C}$, and the correct statement for real scalars will follow automatically from $\text{Re}(\alpha) = \alpha$ and $\bar{\alpha} = \alpha$, when α is real. For finite fields, some of the ideas break down, so such fields are not considered here. Following are a few examples of inner products that occur frequently in applications. Often, the verification of (a)–(d) is straightforward and is not presented.

EXAMPLE 1. $V = \mathbf{F}^n$.

(a) $\langle x, y \rangle = \bar{y}^{\mathrm{T}} x = \sum_{i=1}^{n} x_i \bar{y}_i$. This is called the *standard inner product* or *Euclidean inner product* on $\mathbf{F}^n$. In $\mathbf{R}^3$, it is the usual dot product.

(b) $\langle x, y \rangle = \sum_{i=1}^{n} x_i \bar{y}_i w_i$, where $\{w_k\}_{k=1}^n$ is a fixed sequence of positive numbers, called *weights*. This inner product is called a *weighted inner product*.

(c) $\langle x, y \rangle = y^{\mathrm{H}} B^{\mathrm{H}} B x$, where B is a nonsingular $n \times n$ matrix. Recall that $A^{\mathrm{H}} = \bar{A}^{\mathrm{T}}$ (See Section 0.4).

EXAMPLE 2. FUNCTION SPACES. In (a) and (b), $w(t)$ is a fixed function that is positive, integrable, and continuous on (a, b) and is called a *weight function*. In (c), the w_i are fixed positive numbers called *weights*.

(a) $V = C[a, b]$, $\langle x, y \rangle = \int_a^b x(t)\overline{y(t)}w(t)\,dt$
(b) $V = C^1[a, b]$, $\langle x, y \rangle = \int_a^b x'(t)\overline{y'(t)}w(t)\,dt + x(a)\overline{y(a)}$
(c) $V = \mathcal{F}(\{x_i\}_1^n)$, $\langle f, g \rangle = \sum_{i=1}^{n} f(x_i)\overline{g(x_i)}w_i$

To show that (a) is an inner product, verify the axioms using familiar properties of the integral:

$$\langle \alpha f + \beta g, h \rangle = \int_a^b \{\alpha f(t) + \beta g(t)\}\overline{h(t)}\,dt = \alpha \int_a^b f(t)\overline{h(t)}\,dt + \beta \int_a^b g(t)\overline{h(t)}\,dt$$
$$= \alpha \langle f, h \rangle + \beta \langle g, h \rangle$$

$$\langle f, g \rangle = \int_a^b f(t)\overline{g(t)}\,dt = \int_a^b \overline{\overline{f(t)}g(t)}\,dt = \overline{\langle g, f \rangle}$$

$$\langle f, f \rangle = \int_a^b |f(t)|^2\,dt \geq 0 \quad \text{since } |f(t)|^2 \geq 0 \ \forall t \in [a, b].$$

For the last property, use a theorem from calculus that states that if ϕ is continuous and nonnegative and $\int_a^b \phi(t)\,dt = 0$, then $\phi(t) = 0 \ \forall t \in [a, b]$. Hence

$$\langle f, f \rangle = 0 \Rightarrow |f(t)|^2 = 0 \quad \forall t \in [a, b] \Rightarrow f(t) = 0 \quad \forall t \in [a, b] \Rightarrow f = 0.$$

Observe that there are many inner products on one vector space. Usually, in a particular mathematical or physical problem, there is a natural choice (which is not always immediately obvious) of the inner product that should be used. For example, in the study of euclidean geometry in $\mathbf{R}^n$, one should use the standard inner product because it furnishes the usual concept of length and angle in $\mathbf{R}^n$. The inner products of Example 1(b) are natural choices for certain curve-fitting or statistical problems. In the theory of Sturm–Liouville differential equations, if the differential equation $x^{-1}(xy')' + y = f, 0 < x \leq 1$ (where the prime denotes d/dx), were being considered, the inner product of Example 2(a) with $w(x) = x, a = 0, b = 1$ is used to facilitate certain types of solutions. This differential equation arises in heat flow and vibration problems expressed in polar or cylindrical coordinates; an example of this is given in Chapter 7. More examples will be presented later.

Geometric Properties of an Inner Product

All inner products have properties very similar to those of the dot product in three-dimensional euclidean space, and may be thought of as carrying the information of length and angle (using the following definitions). One reason that inner products are nice to work with is that the geometric conclusions one can reach by drawing pictures and reasoning in ordinary three-dimensional space usually carry over to any other inner-product space.

> **DEFINITIONS.** The *length* or *norm* of a vector x generated by the inner product $\langle\,,\rangle$ is $\|x\| = \sqrt{\langle x, x \rangle}$. $\|x - y\|$ is also called the *distance* between the vectors x and y.

Next some elementary properties and identities for the norm are given. These are stated for the complex case; the conjugate and Re (real part) symbols disappear for the corresponding properties in the real case.

Properties of the norm generated by an inner product. $\forall x, y \in V$, $\forall$ scalars α, the following properties hold:

(a) $\|x\| > 0$ unless $x = 0$ (A norm is *positive definite*)
(b) $\|\alpha x\| = |\alpha|\,\|x\|$ (A norm is *homogeneous*)
(c) $|\langle x, y \rangle| \le \|x\|\|y\|$ (*Cauchy–Schwarz inequality*)
(d) $\|x \pm y\| \le \|x\| + \|y\|$ (*Triangle inequality*)

These properties are derived as follows: Property (a) follows from the definition of the norm and the positive definite properties of the inner product. Property (b) follows from $\|\alpha x\|^2 = \langle \alpha x, \alpha x \rangle = \alpha\overline{\alpha}\langle x, x \rangle = |\alpha|^2\|x\|^2$. The identity

$$\|x + \alpha y\|^2 = \|x\|^2 + 2\,\mathrm{Re}(\overline{\alpha}\langle x, y \rangle) + |\alpha|^2\,\|y\|^2 \tag{1}$$

will be used to prove (c) and (d). In fact,

$$\begin{aligned}
\|x + \alpha y\|^2 &= \langle x + \alpha y, x + \alpha y \rangle \\
&= \langle x, x \rangle + \overline{\alpha}\langle x, y \rangle + \alpha\langle y, x \rangle + \alpha\overline{\alpha}\langle y, y \rangle \\
&= \|x\|^2 + \overline{\alpha}\langle x, y \rangle + \alpha\langle \overline{x, y} \rangle + |\alpha|^2\,\|y\|^2 \\
&= \|x\|^2 + 2\,\mathrm{Re}(\overline{\alpha}\langle x, y \rangle) + |\alpha|^2\,\|y\|^2.
\end{aligned} \tag{$*$}$$

To prove (c), note that when $y = 0$ there is nothing to prove, since $\langle x, 0 \rangle = 0\ \forall x$. When $y \ne 0$, set $\alpha = -\langle x, y \rangle/(\|y\|^2)$ in (1), which gives

$$0 \le \|x + \alpha y\|^2 = \|x\|^2 - 2|\langle x, y \rangle|^2/\|y\|^2 + |\langle x, y \rangle|^2/\|y\|^2 = \|x\|^2 - |\langle x, y \rangle|^2/\|y\|^2.$$

Simplifying this gives Property (c). To prove (d), note that $\pm\mathrm{Re}(z) \le |z|$ for all complex z. Now put $\alpha = \pm 1$ in (1), use $\pm\mathrm{Re}(z) \le |z|$, and use (c):

$$\begin{aligned}
\|x \pm y\|^2 &= \|x\|^2 \pm 2\,\mathrm{Re}(\langle x, y \rangle) + \|y\|^2 \le \|x\|^2 + 2|\langle x, y \rangle| + \|y\|^2 \\
&\le \|x\|^2 + 2\|x\|\,\|y\| + \|y\|^2 = (\|x\| + \|y\|)^2,
\end{aligned}$$

which establishes (d).

Remark. A *norm* on a vector space is a real-valued function $\|x\|$ of a vector that satisfies (a), (b), and (d): $\|x\| > 0$ if $x \neq 0$, $\|\alpha x\| = |\alpha| \|x\|$, and $\|x + y\| \leq \|x\| + \|y\|$ for all vectors x, y and scalars α. Commonly used norms are the 1-norm and the ∞-norm, which are defined as

$$\|x\|_1 = \sum_{i=1}^n |x_i| \quad \text{and}$$

$$\|x\|_\infty = \max\{|x_i| : i = 1, 2, \ldots, n\} \quad \text{for } x \in \mathbf{C}^n$$

$$\|x\|_1 = \int_a^b |x(t)| \, dt \quad \text{and}$$

$$\|x\|_\infty = \max\{|x(t)| : a \leq t \leq b\} \quad \text{for } x \in C[a, b].$$

These norms are not generated by inner products. (See Exercise 17.)

In a real vector space, define the *angle* θ between two nonzero vectors x, y by

$$\cos(\theta) = \frac{\langle x, y \rangle}{\|x\| \|y\|}, \quad 0 \leq \theta \leq \pi. \tag{2}$$

The Cauchy–Schwarz inequality shows that $-1 \leq \cos(\theta) = (\langle x, y \rangle)/(\|x\| \|y\|) \leq 1$, so that θ is well defined. Putting $\alpha = 1$ in (1) yields $\|x+y\|^2 = \|x\|^2 + 2\langle x, y \rangle + \|y\|^2 = \|x\|^2 + 2\|x\| \|y\| \cos(\theta) + \|y\|^2$, which is the law of cosines from plane geometry. If $V = \mathbf{R}^2$ or $\mathbf{R}^3$ and the standard inner product is used, θ is the angle that would be measured with a protractor.

> **DEFINITION.** x and y are said to be *orthogonal* or *perpendicular,* denoted $x \perp y$, iff $\langle x, y \rangle = 0$. Note that the concept of angle and orthogonality depends on the inner product being used. In a real vector space with nonzero x and y, $x \perp y \Leftrightarrow \theta = 90°$. In either the real or complex case, the *Pythagorean law*
>
> $$x \perp y \Rightarrow \|x \pm y\|^2 = \|x\|^2 + \|y\|^2$$
>
> holds. This follows immediately from (1).

Exercises 3.1

1. Let $\langle f, g \rangle = f(0)\overline{g(0)} + f(1)\overline{g(1)} + f(2)\overline{g(2)}$. Is $\langle , \rangle$ an inner product on $\mathbf{P}_2$? on $\mathbf{P}_3$?

2. Let $V = \{f : f, f'$ are continuous on $[0, 1]\}$. Show that $\langle f, g \rangle = \int_0^1 f'(t)\overline{g'(t)} \, dt + f(0)\overline{g(0)}$ is an inner product on V.

3. (a) Verify that the $\langle x, y \rangle$ defined in Examples 1(b) and 1(c) are inner products on $\mathbf{F}^n$.
   (b) Verify that the $\langle x, y \rangle$ defined in Examples 2(b) and 2(c) are inner products.

4. Let $V = \{f : f, f', f''$ are continuous on $[-1, 1]\}$ with real scalars. Show that

$$\langle f, g \rangle = \int_{-1}^1 f''(t)g''(t) \, dt + f(-1)g(-1) + f(1)g(1)$$

is an inner product on V.

5. Verify that the following pairs of vectors are orthogonal in the given inner product. Also verify the Pythagorean law, $\|x + y\|^2 = \|x\|^2 + \|y\|^2$.
   (a) $(1, 2, 1)^T, (1, 0, -1)^T$ $\langle x, y \rangle = y^H x$
   (b) $(1, i)^T, (2, -2i)^T$ $\langle x, y \rangle = y^H x$
   (c) $e^{2\pi it}, e^{4\pi it}$ $\langle x, y \rangle = \int_0^1 x(t)\overline{y}(t)\, dt$
   (d) $1, 3t^2 - 1$ $\langle x, y \rangle = \int_0^1 x(t)\overline{y}(t)\, dt$
   (e) $\sin(\pi \log(t)), \sin(2\pi \log(t))$ $\langle x, y \rangle = \int_1^e x(t)\overline{y}(t)t^{-1}\, dt$

6. Find the angle between the given vectors in the indicated inner product.
   (a) $(1, 1, -1)^T, (0, 1, -1)^T$ $\langle x, y \rangle = y^T x$
   (b) $(1, 1, -1)^T, (0, 1, -1)^T$ $\langle x, y \rangle = y^T \text{diag}(1, 2, 1)x$
   (c) t, t^2 $\langle x, y \rangle = \int_0^1 x(t)y(t)\, dy$
   (d) t, t^2 $\langle x, y \rangle = \sum_{i=0}^4 x(i/4)y(i/4)$

7. Let $\langle , \rangle$ be an inner product on V, and let $\alpha > 0$. Define $\langle , \rangle_1$ by $\langle x, y \rangle_1 = \alpha \langle x, y \rangle$. Show that $\langle , \rangle_1$ is an inner product on V. If V is a real vector space, show that both of these inner products yield the same angle for every pair of nonzero vectors x, y.

8. Show that $\|x + y\|^2 + \|x - y\|^2 = 2(\|x\|^2 + \|y\|^2)$, and interpret this equation geometrically in $\mathbf{R}^3$.

9. If $x \in V$ and $\langle x, y \rangle = 0 \,\forall y \in V$, show that $x = 0$.

10. Let B be a nonsingular $n \times n$ matrix. Show that $\langle x, y \rangle = y^H B^H B x$ is an inner product on $\mathbf{C}^n$.

11. Suppose $\langle x, y \rangle = y^H G x$ is an inner product on $\mathbf{C}^n$. Show that G is nonsingular and that $G^H = G$. Matrices satisfying the latter property are called *Hermitian* matrices.

12. The *trace* of an $n \times n$ matrix is defined to be $\text{Tr}(A) = \sum_{i=1}^n a_{ii}$. Show that $\langle A, B \rangle = \text{Tr}(B^H A)$ is an inner product on $\mathbf{C}^{n,n}$. If $A = [A_1 \ A_2 \cdots A_n]$, express $\|A\|^2$ in terms of the norms of the columns A_i in the standard inner product.

13. Show that $|\langle x, y \rangle| = \|x\| \|y\|$ iff x and y are linearly dependent.

14. Show that $\|x + y\| = \|x\| + \|y\|$ iff $x = 0$ or $y = 0$ or $x = cy$ for some $c > 0$.

15. Prove the following generalization of the Pythagorean law: If $e_i \perp e_j$ when $i \neq j$, then $\|\sum_{i=1}^n c_i e_i\|^2 = \sum_{i=1}^n |c_i|^2 \|e_i\|^2$.

16. **Polar identities for inner products.** Let $\| \ \|$ be the norm generated by an inner product $\langle , \rangle$ on V. Show that the following polar identities hold $\forall x, y \in V$:
   (a) Real scalars: $\langle x, y \rangle = \frac{1}{4}\{\|x + y\|^2 - \|x - y\|^2\}$
   (b) Complex scalars: $\langle x, y \rangle = \frac{1}{4}\{\|x + y\|^2 - \|x - y\|2 + i(\|x + iy\|^2 - \|x - iy\|^2)\}$.

17. Show that neither $\| \ \|_1$ nor $\| \ \|_\infty$ is generated by an inner product for both $V = \mathbf{F}^n$ and $C[a, b]$.

3.2
ORTHOGONAL SETS

Orthogonal sets may be used to facilitate computations and to deduce significant theoretical results. They are extensively used in applications of linear algebra, including differential equations, spectral analysis, and scientific computation, and in theoretical work in the sciences. Many orthogonal sets arise naturally in the study of differential equations.

> **DEFINITION.** A set of vectors $\{e_i : i = 1, 2, \ldots\}$ is said to be *orthogonal* with respect to (or in) the inner product $\langle , \rangle$ if and only if all $\langle e_i, e_i \rangle > 0$ and $\langle e_i, e_j \rangle = 0$ when $i \neq j$. The set is said to be *orthonormal* with respect to $\langle , \rangle$ if and only if, in addition, all $\|e_i\| = 1$. For brevity, the words "with respect to the inner product $\langle , \rangle$" will be suppressed if there is no chance of misunderstanding.

If $\{e_i : i = 1, 2, \ldots\}$ is orthogonal, then the set $\{f_i : i = 1, 2, \ldots\}$, where $f_i = \alpha_i e_i$, $\alpha_i \neq 0$, is also an orthogonal set. For example, if $\alpha_i = \sigma_i/\|e_i\|$, where σ_i is a scalar of unit magnitude, then the set $\{f_i : i = 1, 2, \ldots\}$ is an orthonormal set. This kind of construction is called a *normalization* of the set $\{e_i : i = 1, 2, \ldots\}$ and is not unique. For example, if the e_i are in $\mathbf{F}^n$, one may want to normalize by making the entry of each f_i that is largest in magnitude equal to 1 or by making the euclidean lengths of the f_i all equal to 1.

A set that is orthogonal with respect to one inner product is usually not orthogonal with respect to another. For example, consider the inner products of Example 2 of Section 3.1, with $a = -1$ and $b = 1$. Then 1 and t are orthogonal when $w(t) = 1$, but not when $w(t) = 1 + t^2$.

In the following examples, the computations are routine, so the verifications will not be given.

> **EXAMPLE 1.** The standard basis $\{\delta_i : i = 1, 2, \ldots, n\}$ in $\mathbf{F}^n$ is an orthonormal set with respect to the standard inner product.

> **EXAMPLE 2. ORTHOGONAL SETS OF TRIGONOMETRIC FUNCTIONS.** There are a variety of orthogonal sets, and they occur in the study of periodic functions. In each case, the orthogonality is easily established from integral tables or direct integration.

> (*a*) The following sets are orthogonal with respect to the inner product
>
> $$\langle x, y \rangle = \int_0^{2\pi} x(t)\overline{y(t)}\,dt$$
>
> (*i*) $\{e^{ikt} : k = 0, \pm 1, \pm 2, \ldots\}$. Here $\|e_k\|^2 = 2\pi$ for all k.
> (*ii*) $\{1, \cos(kt), \sin(kt) : k = 1, 2, 3, \ldots\}$. Here $\|1\|^2 = 2\pi$, while $\|e_k\|^2 = \pi$ for all other members e_k of the set.
>
> (*b*) The following sets are orthogonal with respect to
>
> $$\langle x, y \rangle = \int_0^{\pi} x(t)y(t)\,dt$$
>
> (*i*) $\{\sin(kx) : k = 1, 2, 3, \ldots\}$. Here $\|e_k\|^2 = \pi/2$ for all k.
> (*ii*) $\{\cos(kt) : k = 0, 1, 2, \ldots\}$. Now $\|e_0\|^2 = \pi$, and $\|e_k\|^2 = \pi/2$ for $k > 0$.

Remark. In the theory of Sturm–Liouville differential equations, a wide variety of orthogonal sets are found. The inner products are of the form (*a*) in Example 2. These orthogonal sets find extensive use in the solution of partial differential equations by means of "separation of variables." (See G. Birkhoff and G. Rota (1969), for an introduction to this topic.)

Remark. A subset of an orthogonal set is also orthogonal.

Computational Advantages of Orthogonal Sets

Orthogonal sets allow substantial computational advantages. Some of these are illustrated in the next theorem. These advantages make certain theoretical and computational endeavors much easier. Many examples will be presented throughout the book.

THEOREM 1. Assume $\{e_1, e_2, \ldots, e_n\}$ is an orthogonal set and x, y are in span $\{e_1, e_2, \ldots, e_n\}$ so that $x = \sum_{i=1}^{n} x_i e_i$ and $y = \sum_{i=1}^{n} y_i e_i$ for some x_i and y_i. Then

(*a*) The coefficients x_i are uniquely given by

$$x_i = \langle x, e_i \rangle / \|e_i\|^2, \qquad i = 1, 2, \ldots, n. \tag{1}$$

(*b*) $\{e_1, e_2, \ldots, e_n\}$ is a linearly independent set.

The following identities, called *Parseval's identities*, hold:

(*c*) $\langle x, y \rangle = \sum_{i=1}^{n} x_i \bar{y}_i \|e_i\|^2$
(*d*) $\|x\|^2 = \sum_{i=1}^{n} |x_i|^2 \|e_i\|^2$

Proof. Proof of (*a*): Using the linearity of $\langle , \rangle$ in the first variable and using $\langle e_i, e_k \rangle = 0$ when $i \neq j$, it follows that

$$\langle x, e_k \rangle = \left\langle \sum_{i=1}^{n} x_i e_i, e_k \right\rangle = \sum_{i=1}^{n} x_i \langle e_i, e_k \rangle = x_k \langle e_k, e_k \rangle = x_k \|e_k\|^2,$$

and (*a*) follows.

For (*b*), the linear independence of $\{e_k\}_{k=1}^{n}$ follows from the uniqueness of the x_i, or the fact that, if $x = 0$ in (1), then all $x_i = 0$. We deduce (*c*) by conjugate linearity and (1), as follows:

$$\langle x, y \rangle = \left\langle x, \sum_{i=1}^{n} y_i e_i \right\rangle = \sum_{i=1}^{n} y_i \langle x, e_i \rangle = \sum_{i=1}^{n} x_i \bar{y}_i \|e_i\|^2.$$

Identity (*d*) follows from (*c*) by setting $x = y$.

Remarks

1. Parseval's identity (*d*) is a generalization of the Pythagorean law: $\|x \pm y\|^2 = \|x\|^2 + \|y\|^2$ when $x \perp y$.
2. Obtaining the coordinates of a vector with respect to a given basis can often be difficult. Statement (*a*) asserts that the coordinates of a vector with respect to an orthogonal basis are directly found by the simple computation (1).

3. Once the coordinates of two vectors are known, the inner products and norms are computed easily as weighted discrete inner products according to Parseval's identities.

EXAMPLE 3. AN ORTHOGONAL BASIS IN R^3: COMPUTATION OF COORDINATES.
Let $V = \mathbf{R}^3$, and let $E_1 = (2, 2, -1)^T$, $E_2 = (-1, 2, 2)^T$, $E_3 = (2, -1, 2)^T$. These form an orthogonal set using the standard inner product. Therefore they are linearly independent and so form a basis for V. Thus to expand $x = (1, 2, 3)^T$ as a linear combination of the E_i, namely, $x = \sum_{i=1}^{n} c_i E_i$, the system of equations $x = \sum_{i=1}^{n} c_i E_i = [E_1 \, E_2 \, E_3]c$ need not be solved. From Theorem 1, $c_i = \langle x, E_i \rangle / \|E_i\|^2$, whence $c_1 = \frac{1}{3}$, $c_2 = 1$, and $c_3 = \frac{2}{3}$.

EXAMPLE 4. APPLICATION OF PARSEVAL'S IDENTITY TO TRIGONOMETRIC SUMS. Using the orthogonal set of Example 2(b),

$$\int_0^\pi \sum_{k=1}^n \frac{1}{k}\cos(kt) \sum_{k=1}^n \frac{1}{k2^k}\cos(kt)\,dt = \left\langle \sum_{k=1}^n \frac{1}{k}\cos(kt), \sum_{k=1}^n \frac{1}{2^k}\cos(kt) \right\rangle$$

$$= \sum_{k=1}^n \frac{1}{2^k}\|\cos(kt)\|^2 = \sum_{k=1}^n \frac{1}{k2^k}\int_0^\pi \cos^2(kt)\,dt$$

$$= \frac{\pi}{2} \sum_{k=1}^n \frac{1}{k2^k}$$

DEFINITION. Given x and an orthogonal set $\{e_1, e_2, \ldots, e_n\}$, the scalars

$$x_i = \frac{\langle x, e_i \rangle}{\|e_i\|^2} \tag{1}$$

are called the *Fourier coefficients* of x with respect to the orthogonal set $\{e_1, e_2, \ldots, e_n\}$. A sum $s = \sum_{i=1}^n x_i e_i$ is called a *Fourier sum* of x, or an *orthogonal sum* of x with respect to the orthogonal set $\{e_1, e_2, \ldots, e_n\}$. Usually, $x \neq s$; we have $x = s$ only when $x \in \text{span}\{e_1, e_2, \ldots, e_n\}$. Later, we will see that Fourier sums turn out to be extremely useful.

Fourier coefficients and Fourier sums are so named because in 1822, the French physicist Jean Baptiste Joseph Fourier applied (orthogonal) trigonometric series, now called Fourier series, to the study of heat flow. His main work on this subject was of seminal importance. Fourier series have been applied widely in the sciences and have been extensively studied in mathematics.

EXAMPLE 5. FOURIER COEFFICIENTS AND SERIES. For the inner product and orthogonal set of Example 2(ai), with $e_k(t) = e^{ikt}$, the numbers

$$x_i = \frac{\langle x, e_k \rangle}{\|e_k\|^2} = \frac{1}{2\pi}\int_0^{2\pi} x(t)e^{-ikt}\,dt$$

are the usual (complex) Fourier coefficients of the function $x(t)$, and the sum $\sum_{k=-n}^n x_k e^{ikt}$ is the nth partial sum of the complex Fourier series of $x(t)$. Using the orthogonal set of Example 2(aii), $\{1, \cos(t), \sin(t), \ldots, \cos(kt), \sin(kt), \ldots\}$, the numbers $\langle x, e_j \rangle / \|e_j\|^2$ are the usual Fourier coefficients, and the Fourier sums are the partial sums $a_0 + \sum_{j=1}^n [a_j \cos(jt) + b_j \sin(jt)]$ of the usual Fourier series of $x(t)$. The orthogonal sets of Example 2(b) arise from the Fourier series of odd and even functions, respectively.

Construction of Orthogonal Sets

Since the computational properties of orthogonal sets are so attractive, a method to construct them is desirable. It turns out that Fourier sums are essential for the construction. The next lemma gives some properties of Fourier sums that will find many uses.

LEMMA 2. If $\{e_1, e_2, \ldots, e_n\}$ is an orthogonal set and $s = \sum_{i=1}^{n} x_i e_i$ is the Fourier sum of x with respect to this set, then $(x - s) \perp e_k$, for $k = 1, 2, 3, \ldots, n$, $(x - s) \perp s$, and

$$\|x - s\|^2 = \|x\|^2 - \|s\|^2 = \|x\|^2 - \sum_{i=1}^{n} |x_i|^2 \|e_i\|^2. \tag{2}$$

Proof. Compute

$$\langle x - s, e_k \rangle = \left\langle x - \sum_{i=1}^{n} x_i e_i, e_k \right\rangle = \langle x, e_k \rangle - \left\langle \sum_{i=1}^{n} x_i e_i, e_k \right\rangle$$

$$= \langle x, e_k \rangle - \sum_{i=1}^{n} x_i \langle e_i, e_k \rangle = \langle x, e_k \rangle - x_k \|e_k\|^2 = 0$$

using the linearity of the inner product, the orthogonality of the e_k, and the definition of Fourier coefficients. Let $r = x - s$. Then

$$\langle r, s \rangle = \left\langle r, \sum_{i=1}^{n} x_i e_i \right\rangle = \sum_{i=1}^{n} x_i \langle r, e_i \rangle = 0$$

i.e., $r \perp s$. Now

$$\|x\|^2 = \|r + s\|^2 = \|r\|^2 + \|s\|^2 = \|r\|^2 + \sum_{i=1}^{n} |x_i|^2 \|e_i\|^2$$

by the Pythagorean law and by Parseval's identity, and so (2) is established.

The Gram–Schmidt process

This process is employed in the proof of the next theorem, and it constructs an orthogonal set having the same span as a given finite set $\{v_1, v_2, \ldots, v_n\}$ of vectors. Thus if $\{v_1, v_2, \ldots, v_n\}$ were a basis for a vector space V, an orthogonal basis for V would be recovered.

THEOREM 3. Suppose $v_1, v_2, \ldots, v_n$ are linearly independent in V and $\langle , \rangle$ is a given inner product on V. Then there is an orthogonal set $\{e_1, e_2, \ldots, e_n\}$ such that

$$\text{span}\{v_1, v_2, \ldots, v_j\} = \text{span}\{e_1, e_2, \ldots, e_j\} \qquad \text{for } j = 1, 2, \ldots, n. \tag{3}$$

Proof. The algorithm for construction is

Step 0: Set $e_1 = v_1$, and compute $\|e_1\|^2$.

Step k: Suppose $e_1, e_2, \ldots, e_k$ have been constructed and $\|e_i\|^2$, $i = 1, 2, \ldots, k$ have been computed. Compute the Fourier sum of v_{k+1} with respect to $\{e_1, e_2, \ldots, e_k\}$:

$$s_{k+1} = \sum_{j=1}^{k} d_j e_j \qquad \text{where } d_j = \frac{\langle v_{k+1}, e_j \rangle}{\|e_j\|^2}, \qquad j = 1, \ldots, k.$$

Set

$$e_{k+1} = v_{k+1} - s_{k+1}, \tag{4}$$

and compute $\|e_{k+1}\|^2$ directly or by

$$\|e_{k+1}\|^2 = \|v_{k+1}\|^2 - \sum_{j=1}^{k} |d_j|^2 \|e_j\|^2. \tag{5}$$

Next we verify that the algorithm produces the correct results. By Lemma 2, e_{k+1} is orthogonal to $e_1, \ldots, e_k$ so that the process produces orthogonal vectors. Hence (5) is a valid calculation of the norm, by Lemma 2. Now assume span$\{e_1, \ldots, e_k\} =$ span$\{v_1, \ldots, v_k\}$ at step k, and show that the construction step gives span$\{e_1, \ldots, e_{k+1}\} =$ span$\{v_1, \ldots, v_{k+1}\}$. Note that $s_{k+1} \in$ span$\{e_1, \ldots, e_k\}$. This and (4) imply that $v_{k+1} = e_{k+1} + s_{k+1} \in$ span$\{e_1, e_2, \ldots, e_{k+1}\}$. But also $s_{k+1} \in$ span$\{v_1, \ldots, v_k\}$, so

$$e_{k+1} = v_{k+1} - s_{k+1} \in \text{span}\{v_1, \ldots, v_k, v_{k+1}\}.$$

Finally, $e_{k+1} \neq 0$, since otherwise, by (4),

$$v_{k+1} = s_{k+1} \in \text{span}\{v_1, \ldots, v_k\},$$

which contradicts the independence of $\{v_1, \ldots, v_{k+1}\}$.

COROLLARY 4. If V is a nontrivial finite dimensional vector space, then for each inner product on V, there is an orthogonal basis.

Remarks

1. Using (5) is often much easier than directly calculating $\|e_{k+1}\|^2$, except in $\mathbf{F}^n$ with the standard inner product, where norms are easy to compute directly.
2. For each k, $\|e_k\|^2 = \langle v_k, e_k \rangle$. This follows from Lemma 2 and (4).
3. If the set $\{v_1, v_2, \ldots, v_n\}$ is linearly dependent, the Gram–Schmidt process will discover that fact. As noted in the preceding proof, if v_{k+1} is a linear combination of $v_1, \ldots, v_k$, then $e_{k+1} = 0$ since span$\{v_1, v_2, \ldots, v_k\} =$ span$\{e_1, e_2, \ldots, e_k\}$. In this case, v_{k+1} could be deleted from the list of v_j and the process continued. In this way, the dependent vectors are cast out as the process proceeds, and an orthogonal basis for span$\{v_1, v_2, \ldots, v_n\}$ is obtained.
4. During the construction of $\{e_1, e_2, \ldots, e_n\}$, we can normalize at each stage, i.e., replace e_{k+1} by $c_{k+1} e_{k+1}$, where $c_{k+1} \neq 0$. This may be done to give an orthonormal set ($c_{k+1} = 1/\|e_{k+1}\|$) or to achieve some other esthetic or computationally desired result.
5. Equation (3) implies that both the direct and inverse change of basis matrices are upper triangular. If the process in Theorem 3 is used, then the change of basis matrices have positive diagonal entries. In fact, in (4) replace $k + 1$ by k so that $s_k = \sum_{i=1}^{k-1} c_{ik} v_i$ and $e_k = v_k - \sum_{i=1}^{k-1} c_{ik} v_i$ for some c_{ik}. Then the change of basis matrix P from $\{v_k\}_{k=1}^{n}$ to $\{e_k\}_{k=1}^{n}$ has $p_{kk} = 1$. In the reverse change of basis, $v_j = \sum_{i=1}^{k} r_{ij} e_i$. The r_{ij} are thus Fourier coefficients, with $r_{jj} = \langle v_j, e_j \rangle / \|e_j\|^2 = 1$, by Remark 2. If the e_k were normalized by replacing e_k with $e_k/\|e_k\|$, then $p_{kk} = 1/\|e_k\|$ and $r_{kk} = \|e_k\|$.

EXAMPLE 6. CONSTRUCTION OF THE LAGUERRE POLYNOMIALS. Let $v_i(t) = t^i$, $i = 0, 1, 2, 3$, and let

$$\langle x, y \rangle = \int_0^\infty x(t)\overline{y(t)}e^{-t}\, dt.$$

Here the Gram–Schmidt process is applied to find orthogonal polynomials e_i, $i = 0, 1, 2, 3$ with respect to the given $\langle\,,\,\rangle$. The classical Laguerre polynomials L_i are related to the e_i by $L_i = (-1)^{i-1}e_i$ (Abramowitz and Stegun, 1972). We use the following fact:

$$\langle t^n, 1 \rangle = \int_0^\infty t^n e^{-t}\, dt = n! \tag{6}$$

Equation (6) follows from integration by parts, or by noting that the integral is $\Gamma(n+1) = n!$. Observe that $\langle t^p, t^q \rangle = \langle t^{p+q}, 1 \rangle = (p + q)!$.

The Gram–Schmidt process is as follows: First, set $k = 0$: $e_0(t) = 1$ and, from (6),

$$\|e_0\|^2 = \langle 1, 1 \rangle = 0! = 1.$$

Next, set $k = 1$: Using (6), find the Fourier coefficient:

$$d_0 = \frac{\langle t, e_0 \rangle}{\|e_0\|^2} = 1.$$

Then, by (4),

$$e_1(t) = t - d_0 e_0(t) = t - 1.$$

By (5) and (6),

$$\|e_1\|^2 = \|t\|^2 - \|1\|^2 = \langle t, t \rangle - 1 = 2! - 1 = 1.$$

Now set $k = 2$: Find the Fourier coefficients (use (6)):

$$d_0 = \frac{\langle t^2, e_0 \rangle}{\|e_0\|^2} = \langle t^2, t \rangle - \langle t^2, 1 \rangle = 3! - 2! = 4, \quad d_1 = \frac{\langle t^2, e_1 \rangle}{\|e_1\|^2} = \langle t^2, 1 \rangle = 2! = 2.$$

Next, using (4),

$$e_2(t) = t^2 - d_1 e_1(t) - d_0 e_0(t) = t^2 - 4(t - 1) + 2 \cdot 1 = t^2 - 4t + 2.$$

From (5) and (6),

$$\|e_2\|^2 = \|t^2\|^2 - 4^2\|e_1\|^2 - 2^2\|e_0\|^2 = 4! - 16 \cdot 1 - 4 \cdot 1 = 4.$$

Set $k = 3$: Find the Fourier coefficients:

$$d_0 = \frac{\langle t^3, e_0 \rangle}{\|e_0\|^2} = \langle t^3, 1 \rangle = 3! = 6,$$

$$d_1 = \frac{\langle t^3, e_1 \rangle}{\|e_1\|^2} = \langle t^3, t \rangle - \langle t^3, 1 \rangle = 4! - 3! = 18,$$

$$d_2 = \frac{\langle t^3, e_2 \rangle}{\|e_2\|^2} = \tfrac{1}{4}(\langle t^3, t^2 \rangle - 4\langle t^3, t \rangle + 2\langle t^3, 1 \rangle) = \tfrac{1}{4}(5! - 4 \cdot 4! + 2 \cdot 3!) = 9.$$

Then by (4),

$$e_3(t) = t^3 - d_0 e_0(t) - d_1 e_1(t) - d_2 e_2(t),$$
$$= t^3 - 9(t^2 - 4t + 2) - 18t - 6 \cdot 1 = t^3 - 9t^2 + 18t - 6.$$

By (5) and (6),

$$\|e_3\|^2 = \|t^3\|^2 - 9^2\|e_2\|^2 - 18^2\|e_1\|^2 - 6^2\|e_0\|^2$$
$$= 6! - 9^2 \cdot 4 - 18^2 \cdot 1 - 6^2 \cdot 1 = 36.$$

The next theorem is just the translation of Theorem 3 into matrix language when the vector space is $\mathbf{F}^m$, the standard inner product is used, and the orthogonal vectors created in the Gram–Schmidt process are normalized to be of unit length. The decomposition of a matrix A asserted in the next theorem is called the *QR factorization*. The Gram–Schmidt process may be used to compute this factorization, but on a computer the results may suffer from excessive propagation of roundoff error. The QR algorithm in numerical packages uses more subtle and stable methods with which to carry out this factorization. The output of the algorithm is often more general, as well. Refer to the user's manual for the software package in question.

QR FACTORIZATION THEOREM. Let A be an $m \times n$ matrix with entries in $\mathbf{F}$ and linearly independent columns. Then there is an $m \times n$ matrix Q and an $n \times n$ matrix R, both having entries in $\mathbf{F}$, such that

1. $A = QR$.
2. $Q^H Q = I_n$.
3. R is a nonsingular upper triangular matrix with $r_{kk} > 0$.
4. The columns of Q are an orthonormal basis for $R(A)$.
5. Q and R are unique.

When $m = n$, Q is unitary, i.e., $QQ^H = Q^H Q = I$.

Proof. Let A_i be the ith column of A, and using the standard inner product in $\mathbf{F}^m$, apply the Gram–Schmidt process as described in Theorem 3 to $\{A_1, A_2, \ldots, A_n\}$ to get $\{E_1, \ldots, E_n\}$. Then set $Q_i = E_i / \|E_i\|$ and $Q = [Q_1 \ Q_2 \cdots Q_n]$. From Theorem 3, $\mathrm{span}\{A_1, \ldots, A_j\} = \mathrm{span}\{Q_1, \ldots, Q_j\}$ for $j = 1, 2, \ldots, n$. Thus

$$A_j = \sum_{i=1}^{j} r_{ij} Q_i, \qquad j = 1, 2, \ldots, n \tag{7}$$

for some r_{ij}. Define $r_{ij} = 0$ if $i > j$, and put $R = [R_1 \cdots R_n] = [r_{ij}]$. By Proposition 1 of Section 0.4,

$$A_j = \sum_{i=1}^{n} r_{ij} Q_i = [Q_1 \ Q_2 \cdots Q_n] \begin{bmatrix} r_{1j} \\ r_{2j} \\ \vdots \\ r_{jj} \\ 0 \end{bmatrix} = QR_j,$$

and so $A = [A_1 \ A_2 \cdots A_n] = [QR_1 \ QR_2 \cdots QR_n] = QR$. To prove statement 2, note that $Q^H Q = [Q_i^H Q_j] = [\langle Q_j, Q_i \rangle] = [\delta_{ij}] = I_n$. Note that $r_{jj} = \|E_j\| > 0$ by the preceding remarks 2 and 5, and so R is nonsingular. That R is nonsingular follows also from the fact that it is a change of basis matrix. If A is real, then the Q_i may be taken to be real, according to the construction in the Gram–Schmidt process. Hence the r_{ij} are also real (cf. (8), which follows). Since $R(A) = \text{span}\{A_1, \ldots, A_n\} = \text{span}\{Q_1, \ldots, Q_n\}$, the columns of Q are an orthonormal basis for $R(A)$. When $m = n$, $Q^H Q = I$, thus also $QQ^H = I$ by Corollary 5 of Section 0.5. Proving the uniqueness of Q and R is left as an exercise. (See Exercise 9.)

Note that (7) is simply the orthogonal expansion of A_j in terms of the Q_i, so that the r_{ij} are Fourier coefficients and may be computed by

$$r_{ij} = \langle A_j, Q_i \rangle = Q_i^H A_j, \tag{8}$$

according to Theorem 1. Thus the jth column of R consists of the Fourier coefficients of the jth column of A with respect to $Q_1, Q_2, \ldots, Q_n$. This may be restated as

$$R_j = Q^H A_j \tag{9}$$

or as

$$R = Q^H A. \tag{10}$$

Thus, after Q has been found, R is easy to obtain.

EXAMPLE 7. Find the QR decomposition for the following matrix.

$$A = \begin{bmatrix} 1 & 2 & 2 \\ 1 & 1 & 2 \\ 0 & 1 & 3 \\ 1 & 0 & -1 \end{bmatrix} = [A_1\ A_2\ A_3]$$

Perform the Gram–Schmidt process: First set $G_1 = A_1$. Next compute

$$G_2 = A_2 - (\langle A_2, A_1 \rangle / \|A_1\|^2) A_1 = A_2 - A_1 = (1, 0, 1, -1)^T.$$

Finally construct

$$G_3 = A_3 - (\langle A_3, A_1 \rangle / \|A_1\|^2) A_1 - (\langle A_3, G_2 \rangle / \|G_2\|^2) G_2$$
$$= A_3 - A_1 - 2G_2 = (-1, 1, 1, 0)^T.$$

The next step is to normalize G_1, G_2, G_3 to obtain the orthonormal basis Q_1, Q_2, Q_3:

$$Q_1 = A_1/\|A_1\| = \left(1/\sqrt{3}\right)(1, 0, 1, -1)^T, \quad Q_2 = G_2/\|G_2\| = \left(1/\sqrt{3}\right)(1, 0, 1, -1)^T,$$

$$Q_3 = \left(1/\sqrt{3}\right)(-1, 1, 1, 0)^T.$$

The nonzero entries of R are simply the Fourier coefficients of the columns of A:

$$r_{11} = \langle A_1, Q_1 \rangle = \sqrt{3}, \qquad r_{12} = \langle A_2, Q_1 \rangle = \sqrt{3}, \qquad r_{22} = \langle A_2, Q_2 \rangle = \sqrt{3},$$
$$r_{13} = \langle A_3, Q_1 \rangle = \sqrt{3}, \qquad r_{23} = \langle A_3, Q_2 \rangle = 2\sqrt{3}, \qquad r_{33} = \langle A_3, Q_3 \rangle = \sqrt{3}.$$

Hence

$$Q = [Q_1\ Q_2\ Q_3] = \frac{1}{\sqrt{3}} \begin{bmatrix} 1 & 1 & -1 \\ 1 & 0 & 1 \\ 0 & 1 & 1 \\ 1 & -1 & 0 \end{bmatrix}, \qquad R = \sqrt{3} \begin{bmatrix} 1 & 1 & 1 \\ 0 & 1 & 2 \\ 0 & 0 & 1 \end{bmatrix}.$$

Exercises 3.2

1. Show that the sets of vectors in Example 2 are orthogonal.

2. Let

$$f_k(x) = \begin{cases} 1 & k/n \leq x < (k+1)/n \\ 0 & \text{otherwise} \end{cases}, \qquad k = 0, 1, \ldots, n-1.$$

Show that $\{f_k\}_{k=0}^{n-1}$ is an orthogonal set in the inner product $\langle f, g \rangle = \int_0^1 f(t)\overline{g}(t)\,dt$.

3. Evaluate the following:

(a) $\displaystyle \int_0^{2\pi} \left| \sum_{k=1}^{n} 2^{-k} e^{ikt} \right|^2 dt$

(b) $\displaystyle \int_0^{\pi} \left(\sum_{k=0}^{n} 2^k \cos(kt) \right)^2 dt$

(c) $\displaystyle \int_0^{\pi} \sum_{k=1}^{n} \frac{\sin(kt)}{k} \sum_{k=1}^{n} \frac{\sin(kt)}{k+1} dt$

4. Find an orthogonal basis for the span of
  (a) $(1, 1, 1)^T, (3, 2, 1)^T, (1, -3, -1)^T$
  (b) $(1, 2, 2)^T, (4, 5, 2)^T, (5, -2, -5)^T$
  (c) $(0, 1, 1, 1)^T, (1, 1, 2, 0)^T, (3, -2, 1, -2)^T$
  (d) $(1, 0, 1, -1)^T, (2, -1, 1, 0)^T, (2, -1, -2, 3)^T, (0, 1, 1, -2)^T$

5. Find an orthogonal basis for the following subspaces of $\mathbf{R}^4$:
  (a) $M = \{x : \sum_{i=1}^{4} x_i = 0\}$
  (b) $M = \{x : x_1 + x_2 - 2x_3 = 0, x_1 + x_2 + x_3 - 3x_4 = 0\}$
  (c) $M = \{x : x_1 + x_2 + x_3 + x_4 = 0, 2x_1 + x_2 = 0\}$

6. Find the QR factorization of the following:

(a) $\begin{bmatrix} 1 & 1 \\ 3 & -1 \end{bmatrix}$ (b) $\begin{bmatrix} 1 & 3 & 4 \\ 1 & 1 & -6 \\ 1 & -1 & 2 \end{bmatrix}$ (c) $\begin{bmatrix} 2 & 1 \\ 2 & 3 \\ -2 & 1 \end{bmatrix}$

(d) $\begin{bmatrix} 1 & 2 & 0 \\ 0 & 1 & 3 \\ 1 & 1 & 3 \\ -1 & 0 & 0 \end{bmatrix}$ (e) $\begin{bmatrix} 2 & 2 & 2 \\ 2 & -4 & 4 \\ 2 & 2 & -2 \\ 2 & -4 & 0 \end{bmatrix}$

7. Let $A = QR$ be the QR factorization of A. Find $QQ^H A$. Is it true that $QQ^H = I$?

8. Suppose $A = PT$, where P has orthogonal columns P_i and T is a nonsingular upper triangular matrix. Find the QR factorization of A.

9. Prove the uniqueness of the QR factorization. That is, if $A = QR = \tilde{Q}\tilde{R}$, where Q, R and $\tilde{Q}, \tilde{R}$ satisfy statements 1–3 of the QR factorization theorem, then $Q = \tilde{Q}$ and $R = \tilde{R}$. (*Hint:* Consider Eq. (7) for $j = 1, \ldots, n$.)

10. Suppose $A = B^H B$, with B nonsingular. Show that there is a lower triangular matrix L with positive entries on the diagonal such that $A = LL^H$. Such a factorization, called a *Cholesky factorization,* is important in numerical analysis.

11. Modify the QR factorization to include the case in which the columns of A need not be linearly independent.

12. Let $E_1 = \begin{bmatrix} 1 \\ 1 \\ 1 \end{bmatrix}$, $E_2 = \begin{bmatrix} 1 \\ 0 \\ -1 \end{bmatrix}$, and $E_3 = \begin{bmatrix} 1 \\ -2 \\ 1 \end{bmatrix}$, and let $\langle\,,\,\rangle$ be the standard inner product.

   (a) Show that the vectors E_k are orthogonal.
   (b) Find a_k, b_k, c_k so that $\delta_k = a_k E_1 + b_k E_2 + c_k E_3$.
   (c) Let $a = (a_1, a_2, a_3)^T$, etc., $F = [a\ b\ c]$, and $E = [E_1\ E_2\ E_3]$. Find the relationship between E and F, and explain why it is so.

13. Let $V = \mathbf{P}_3$ with real scalars. Apply the Gram–Schmidt process to $1, t, t^2, t^3, t^4$ to find an orthogonal basis $p_0, p_1, \ldots, p_4$ for V using each of the following inner products. In each case, find $\|p_k\|^2$,
   (a) $\langle f, g \rangle = f(-2)g(-2) + f(-1)g(-1) + f(0)g(0) + f(1)g(1) + f(2)g(2)$
   (b) $\langle f, g \rangle = \int_{-1}^{1} f(t)g(t)\,dt$
   (c) $\langle f, g \rangle = \int_{-1}^{1} f'(t)g'(t)\,dt + f(0)g(0)$
   (d) $\langle f, g \rangle = \int_{-1}^{1} f''(t)g''(t)\,dt + f(-1)g(-1) + f(1)g(1)$
   Note: The polynomials obtained in (b), if normalized to satisfy $p_n(1) = 1$, are called *Legendre polynomials* (cf. Exercise 15).

14. Let $\langle f, g \rangle = \int_0^1 f(t)g(t)\,dt$, and assume the scalars are real. Find an orthogonal basis for each of the following vector spaces:
   (a) $V = \{p \in \mathbf{P}_2 : p(0) = 0\}$
   (b) $V = \{p \in \mathbf{P}_2 : \int_0^1 p(t)\,dt = 0\}$
   (c) $V = \{p \in \mathbf{P}_3 : p(1) = 0, p(0) = 0\}$
   (d) $V = \{p \in \mathbf{P}_2 : p'(1) = 0\}$

   In Exercises 15 and 16, we state some facts about two sets of orthogonal polynomials without proof, so that they may be used later. See Lebedev (1972) for a thorough discussion of many families of orthogonal polynomials. See also Abramowitz and Stegun (1972) for a vast list of orthogonal polynomials. Davis (1963) also has a good exposition of orthogonal polynomials.

15. **Legendre polynomials:** $\{P_k\}_{k=0}^{\infty}$
   (a) $P_0(x) = 1$, $P_1(x) = x$ and

$$P_{n+1}(x) = \frac{2n + 1}{n + 1} \times P_n(x) - \frac{n}{n + 1} P_{n-1}(x), \qquad \text{if } n \geq 1. \qquad (*)$$

   (b) $\{P_k\}_{k=0}^{\infty}$ is orthogonal in the inner product $\langle f, g \rangle = \int_{-1}^{1} f(x)\overline{g}(x)\,dx$.

   (c) $\|P_n\|^2 = \dfrac{2}{2n + 1}$.

   Compute the first five Legendre polynomials using the recursion $(*)$. Verify that they are orthogonal.

The recursion (∗) is used to evaluate the Legendre polynomials on a computer, since their expressions in powers of x are cumbersome and prone to extensive error numerically, which arises from the large coefficients of alternating sign.

16. **Chebyshev Polynomials** (of the first kind). These polynomials are extensively used in approximations and numerical analysis. Define

$$T_n(x) = \cos(n\theta), \text{ where } x = \cos(\theta).$$

(a) Using trigonometric identities, establish that

$$T_{n+1}(x) = 2xT_n(x) - T_{n-1}(x), \text{ for } n \geq 1.$$

(b) Compute $T_n(x)$ for $n = 0, 1, \ldots, 5$ as a linear combination of powers x^k.
(c) Show that
   (i) T_n is a polynomial of degree n, with leading coefficient 2^{n-1} if $n > 0$.
   (ii) T_n is an odd function if n is odd and an even function if n is even.
(d) Show that $\{T_k\}_{k=0}^{\infty}$ is orthogonal with respect to

$$\langle f, g \rangle = \int_{-1}^{1} f(x)\overline{g}(x)(1 - x^2)^{-1/2} \, dx.$$

(e) Find $\|T_n\|^2$ for $n = 0, 1, 2, \ldots$.
Hint: Substituting $x = \cos(\theta)$ in the inner product gives

$$\langle f, g \rangle = \int_{0}^{\pi} f(\cos(\theta))\overline{g}(\cos(\theta)) \, d\theta.$$

17. Let w be positive and continuous on $[0, 1]$, and let $\{q_k\}_{k=0}^{\infty}$ be a sequence of real polynomials with the degree of $q_k = k$, for each k. Suppose $\int_0^1 q_k(x)q_j(x)w(x) \, dx = 0$ if $k \neq j$. Show that $\int_0^1 x^k q_n(x)w(x) \, dx = 0$ if $n > k$.

18. **Uniqueness of orthogonal polynomials.** Let $\{q_k\}_{k=0}^{\infty}$ and $\{p_k\}_{k=0}^{\infty}$ be orthogonal in the inner product $\langle f, g \rangle = \int_a^b f(x)\overline{g}(x)w(x) \, dx$, and suppose p_k and q_k are real polynomials of degree k, where $k = 0, 1, \ldots$. Show that for each k, p_k and q_k differ only by a nonzero constant multiple.

19. **A recursion to generate a sequence of polynomials orthonormal with respect to an arbitrary inner product.** Because of the computational advantages of orthogonal sets and because they are useful for making approximations, as we will see, it is useful to have a simpler method than Gram–Schmidt for generating them. The algorithm is as follows:
   (a) Given an inner product $\langle , \rangle$,
      set $q_{-1} = p_{-1} = 0$,   (artificial, to start the recursion)
      set $q_0 = 1$, $p_0 = q_0/\|q_0\|$.
   (b) Given $q_{-1}, \ldots, q_n, p_{-1}, \ldots, p_n$, and $n \geq 1$,
      set $q_{n+1}(x) = xp_n(x) - \langle xp_n, p_n \rangle p_n(x) - \|q_n\|p_{n-1}(x)$,
      set $p_{n+1} = q_{n+1}/\|q_{n+1}\|$.
      Show that the resulting sequence $\{p_k\}_{k=0}^{\infty}$ is orthonormal with respect to the given inner product and that degree(p_k) = k.
      (*Hint:* Use mathematical induction. See Davis (1963).)

20. Phase polynomials: An orthogonal set in a discrete inner product. Let $x_k = (2\pi k)/(N+1)$, for $k = 0, 1, \ldots, N$. Put $\langle f, g \rangle = \sum_{k=0}^{N} f(x_k)\overline{g}(x_k)$. Show that the set $\beta = \{e^{ikx}\}_{k=0}^{N}$ is an orthogonal set in this inner product, and compute $\|e^{ikx}\|^2$. A linear combination $p(\omega) = \sum_{k=0}^{N} c_k \omega^k$, where $\omega = e^{ix}$ is called a *phase polynomial* and is important in the theory of the fast Fourier transform. (*Hint:* $\sum_{k=0}^{n} a^k = (1 - a^{n+1})/(1-a)$ if $a \neq 1$.)

21. Let $T : V \to V$ be linear, and let A be the matrix of T in the basis $\beta = \{e_k\}_{k=1}^{n}$. If β is orthonormal, show that $a_{ij} = \langle T(e_j), e_i \rangle$. What if β is merely orthogonal?

22. Prove that A square matrix U is unitary if and only if the rows (transposed) of U are an orthonormal set with respect to the standard inner product in $\mathbf{F}^n$.

23. Assume $\{v_1, v_2, \ldots, v_n\}$ is an arbitrary, linearly independent set. Define $\langle , \rangle$ by $\langle \sum \alpha_i v, \sum \beta_j v_j \rangle = \sum \alpha_i \overline{\beta}_i$. Show that this rule defines an inner product with respect to which $\{v_1, v_2, \ldots, v_n\}$ is orthonormal.

24. Let U be a unitary matrix and $\langle , \rangle$ be the standard inner product. Show that
   (a) $\langle Ux, Uy \rangle = \langle x, y \rangle \ \forall x, y$.
   (b) If U is real and $x, y \in \mathbf{R}^n$, then
       (i) x and Ux have the same length.
       (ii) The angle between x, y is the same as the angle between Ux, Uy.

25. Suppose $A = USV^H$, where U is an $m \times m$ unitary matrix, V is an $n \times n$ unitary matrix, and S is an $m \times n$ matrix with $s_{ii} = \sigma_i > 0$ for $i = 1, 2, \ldots, r \leq n, m$ and all other $s_{ij} = 0$. Show that
   (a) $Ax = \sum_{i=1}^{r} \langle x, V_i \rangle \sigma_i U_i$.
   (b) $\{U_1, \ldots, U_r\}$ is an orthonormal basis for $R(A)$.
   (c) $\{V_{r+1}, \ldots, V_n\}$ is an orthonormal basis for $N(A)$.

26. Suppose P is an $m \times n$ matrix with orthogonal columns P_i.
   (a) Show that $m \geq n$.
   (b) Find $r(P)$.
   (c) Show that $\operatorname{diag}(\|P_1\|^{-2}, \ldots, \|P_n\|^{-2})P^H$ is a left inverse of P.

27. Alternate Gram–Schmidt algorithm. Let $\{v_1, v_2, \ldots, v_n\}$ be linearly independent. Show that the following algorithm produces an orthogonal set $\{e_1, e_2, \ldots, e_n\}$, which satisfies the conclusions of Theorem 3. This algorithm is said to be more stable numerically and less susceptible to round-off error than the Gram–Schmidt algorithm, and it is easier to code. Given $v_1, v_2, \ldots, v_n$,

   1. Set $e_i = v_i/\|v_i\|^2, i = 1, 2, \ldots, n$.
   2. For $i = 1$ to $n - 1$
        For $j = i + 1$ to n
            Set $d_j = \langle e_j, e_i \rangle$.
            Set $e_j = (e_j - d_j e_i)/(1 - d_j^2)$.
        end j loop
      end i loop.

28. Here we provide an example of an orthogonal set arising in differential equations. (Further examples will appear in Chapter 5.)

Let $b > 0$ and c be given, and let $e_n(x) = \sin(\alpha_n x)$, where α_n is the nth positive root of $\tan(\alpha b) = c\alpha$. Show that $\{e_k\}_{k=1}^{\infty}$ is orthogonal in the inner product $\langle f, g \rangle = \int_0^b f(x)g(x)\,dx$. (*Hints:* Observe that $e_n'' = -\alpha_n^2 e_n$. Consider $\langle e_n'', e_m \rangle$ and integrate by parts twice.)

3.3
APPROXIMATION AND ORTHOGONAL PROJECTION

In this section, the basic facts about approximation and projection in inner product spaces are studied. Many examples are given to illustrate the generality of the theory. In later sections, more specialized problems as well as some additional theory will be considered.

Let us pose two apparently different problems. Let V be a vector space with inner product $\langle , \rangle$, M be a nontrivial subspace, and $f \in V$, and suppose $\| \cdot \|$ is the norm generated by the inner product.

The approximation problem. Given f, find $p \in M$ that is "closest" to f; i.e., find $p \in M$ such that

$$\|f - p\| \le \|f - m\| \qquad \forall m \in M. \tag{1}$$

The inequality (1) means that p is the nearest element of M to f, in the sense of the norm generated by $\langle , \rangle$. p is called the *best approximation* to f in M with respect to the $\| \|$ generated by a given inner product. One can interpret $\|f - p\|$ as the *distance between f and M.* The same problem could have been posed if $\| \cdot \|$ were not a norm generated by an inner product.

One can also rephrase the approximation problem as an *optimization problem:* Minimize $\|f - m\|$, where $m \in M$. This means, Find $p \in M$ so that $\|f - p\| \le \|f - m\| \; \forall m \in M$. $m = p$ is the minimizer.

DEFINITION. r is orthogonal to M, or r is *perpendicular to M* if and only if $r \perp m$ for all $m \in M$. This is written as $r \perp M$.

The projection problem. Resolve f into a sum

$$f = s + r, \qquad \text{with } s \in M \text{ and } r \perp M. \tag{2}$$

Statement (2) means f has been resolved into two components, s and r, one in M, and one perpendicular to M. See Fig. 3.1. s is called the *orthogonal projection of f on M,* and r is called the *residual* or *error* vector.

In $\mathbf{R}^3$, the problems become as follows: a point f and a plane M through the origin are given. To solve the approximation problem, find the point p in the plane that is closest to f. To solve the projection problem, find the perpendicular r from f to M (thereby locating $s = f - r$). In $\mathbf{R}^3$, the problems are familiar and known to be equivalent: $p = s$. It is not so clear that they are equivalent if $M = \mathbf{P}_n$ and $\langle f, g \rangle = \int_0^1 f(t)\overline{g}(t)\,dt$, for example. The remarkable fact is that, in an arbitrary vector space, with an arbitrary inner product, the problems are also equivalent. This is shown in Theorem 1.

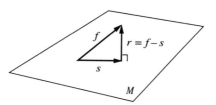

FIGURE 3.1

EXAMPLE 1. The approximation/optimization problem does not always have a solution. For example, let V be the vector space of all Riemann integrable functions and let

$$f(x) = \begin{cases} 0 & x < 0 \\ 1 & 0 \le x \end{cases}, \qquad M = C([-1, 1]), \qquad \langle g, h \rangle = \int_{-1}^{1} g(x)\overline{h}(x)\,dx.$$

It is possible to make continuous functions $m(x)$ very close to the step function $f(x)$ (draw a picture to see this) so that $\|f - m\|^2 = \int_{-1}^{1} |f - m|^2\,dx$ is as small as desired. Thus $\min \|f - m\|$ would have to be equal to 0, if it existed. There is no continuous function m so that $\|f - m\| = 0$.

EXAMPLE 2. PROJECTIONS ON ONE-DIMENSIONAL SUBSPACES. Let $M = \text{span}\{e\} = \{\alpha e : \alpha \in \mathbf{F}\}$ and $f \in V$. Show that the projection of f on M is $s = (\langle f, e \rangle / \|e\|^2)e$.

   c must be found so that $f - ce \perp \alpha e$, for all α, i.e., $0 = \langle f - ce, \alpha e \rangle = \alpha(\langle f, e \rangle - c\langle e, e \rangle) = 0$ for all α. This implies that $c = \langle f, e \rangle / \|e\|^2$ and $s = ce$ is the projection. Note that c is the Fourier coefficient of f relative to e.

Remark. If e is a unit vector in $\mathbf{F}^n$, the projection of f on $\text{span}\{e\}$ using the standard inner product is $s = ee^H f$.

THEOREM 1. For each f, the projection and approximation problems are equivalent; one has a solution if and only if the other does, and the projection p is the same as the best approximation s. The projection is unique, when it exists.

Proof. First assume that the projection problem has a solution $s \in M$, that is, $r = f - s \perp M$. Then for every $m \ne s$, $m \in M$, write $f - m = (f - s) + (s - m) = r + (s - m)$. Since $s - m \in M$, it follows that $r \perp s - m$. Hence

$$\|f - m\|^2 = \|r + (s - m)\|^2 = \|r\|^2 + \|s - m\|^2 > \|r\|^2 = \|f - s\|^2$$

by the Pythagorean law and by the fact that $\|s - m\| > 0$, since $s \ne m$. Therefore s also solves the best approximation problem and is the unique solution of the approximation problem. Now assume the best approximation $p \in M$ exists, and let m in M be arbitrary. Showing that $r = f - p \perp m$ will establish that $s = p$ is the projection. Consider $g(t) = \|f - (p + t\sigma m)\|^2$, where σ is a scalar of unit magnitude and t is a real variable. $g(t) \ge g(0)$ for all t, because $q \equiv p + t\sigma m \in M$, and $\|f - q\|^2$ is minimal when $q = p$, i.e., when $t = 0$. Now choose σ so that $\sigma\langle r, m \rangle = |\langle r, m \rangle|$. Using (1) in Section 3.1 yields $g(t) = \|r\|^2 - 2t\,\text{Re}(\overline{\sigma}\langle r, m \rangle) + |t\sigma|^2\|m\|^2 = \|r\|^2 - 2t|\langle r, m \rangle| + t^2\|m\|^2$. Since g minimizes at $t = 0$, its derivative at $t = 0$ must be zero: $g'(0) = 2|\langle r, m \rangle| = 0$. Hence $r \perp m$ for all $m \in M$, which proves the theorem.

Computation of Projections

Next the existence and computation of projections is discussed. In this section, two fundamental methods for computing the projection are developed. Both methods rely on the fact that the conditions $r = f - s \perp M$ and $s \in M$ characterize the projection. The next lemma establishes a test for the first of these conditions.

LEMMA 2. If $M = \text{span}\{v_1, v_2, \ldots, v_n\}$, then $r \perp M$ if and only if $r \perp v_i$, for $i = 1, \ldots, n$.

Proof. If $r \perp M$, then clearly $r \perp v_i$ for each i. Suppose $r \perp v_i$, for $i = 1, \ldots, n$, and let $m \in M$. Then $m = \sum_{i=1}^{n} \alpha_i v_i$ for some α_i, and $\langle r, m \rangle = \sum_{i=1}^{n} \alpha_i \langle r, v_i \rangle = 0$. Thus $r \perp m$ for all $m \in M$, so that $r \perp M$.

Computation of projections using an orthogonal basis

The next theorem asserts that the projection is easily found if an orthogonal basis is known.

THEOREM 3. Let M be a finite-dimensional subspace of V. Then the projection of f on M exists and is unique for every f in V. If $\{e_1, e_2, \ldots, e_n\}$ is an orthogonal basis for M, then the projection s of f on M is the Fourier sum

$$s = \sum_{i=1}^{n} c_i e_i, \qquad c_i = \frac{\langle f, e_i \rangle}{\|e_i\|^2}, \tag{3}$$

and the length of the residual $f - s$ can be computed by

$$\|f - s\|^2 = \|f\|^2 - \|s\|^2 = \|f\|^2 - \sum_{i=1}^{n} |c_i|^2 \|e_i\|^2. \tag{4}$$

Proof. Lemma 2 of Section 3.2 implies that $f - s \perp e_i$ for all i, and so $f - s \perp M$ by Lemma 2 of Section 3.3. Hence s is the projection of f on M. Equation (4) was proved in Lemma 2 of Section 3.2 and is restated here for emphasis. Uniqueness follows from (3) as well as from Theorem 1.

Remark. Example 2 showed that the projection of f on a one-dimensional subspace M spanned by e is given by $s = ce$, where $c = \langle f, e \rangle / \|e\|^2$. Equation (3) of Theorem 3 then says that the projection of f on $\text{span}\{e_1, e_2, \ldots, e_n\}$ is the sum of the *individual* projections $c_i e_i$ of f on each subspace. This is a rather remarkable fact, and this happens only when the set $\{e_1, e_2, \ldots, e_n\}$ is orthogonal.

Remark. If $\{e_k\}_{k=1}^{M}$ is orthogonal, the projections on the subspaces $M_n = \text{span}\{e_k\}_{k=1}^{n}$ are nested in the following sense: If s_n is the projection on M_n, then the projection on M_{n+1} is $s_{n+1} = s_n + (\langle f, e_{n+1} \rangle / \|e_{n+1}\|^2) e_{n+1}$. That is, just add the next term in the Fourier sum. This makes computation easy. Also note that $\|f - s_{k+1}\|^2 = \|f - s_k\|^2 + |c_{k+1}|^2 \|e_{k+1}\|^2 \leq \|f - s_k\|^2$; i.e., the errors decrease.

Remark. The construction in the Gram–Schmidt process may be restated succinctly: Set $e_1 = v_1$. If $e_1, \ldots, e_k$ have been constructed, then set

$$e_{k+1} = v_{k+1} - (\text{the projection of } v_{k+1} \text{ onto span}\{e_1, \ldots, e_k\}).$$

This follows from Theorem 3 and Eq. (4) of Section 3.2.

EXAMPLE 3. Find the vector in $M = \text{span}\{E_1, E_2, E_3\}$ nearest $f = (2, -2, -2, 6)^T$, where $E_1 = (1, 1, 1, 1)^T$, $E_2 = (1, -1, 0, 0)^T$, and $E_3 = (1, 1, 0, -2)^T$, and the distance from f to M.

Since the E_i are orthogonal, the best approximation is the Fourier sum

$$s = \sum_{i=1}^{3} \frac{\langle f, E_i \rangle}{\|E_i\|^2} E_i = \frac{4}{4} \begin{bmatrix} 1 \\ 1 \\ 1 \\ 1 \end{bmatrix} + \frac{4}{2} \begin{bmatrix} 1 \\ -1 \\ 0 \\ 0 \end{bmatrix} + \frac{-12}{6} \begin{bmatrix} 1 \\ 1 \\ 0 \\ -2 \end{bmatrix} = \begin{bmatrix} 1 \\ -3 \\ 1 \\ 5 \end{bmatrix}.$$

In this example, $\|f - s\|$ is easily calculated directly, without (4). Using (4),

$$\|s\|^2 = 1^2 \|E_1\|^2 + 2^2 \|E_2\|^2 + (-2)^2 \|E_3\|^2 = 4 + 8 + 24 = 36$$

$$\|f - s\|^2 = \|f\|^2 - \|s\|^2 = 48 - 36 = 12,$$

so that the distance from f to M is $\|f - s\| = \sqrt{12}$.

EXAMPLE 4. FOURIER SERIES AND PROJECTIONS ON T_n. Consider the orthogonal sequence $\{e^{ijt} : j = 0, \pm 1, \pm 2, \ldots\}$ and the inner product

$$\langle f, g \rangle = \int_{-\pi}^{\pi} f(t)\overline{g}(t)\, dt.$$

The (complex) Fourier series of a function h is

$$\sum_{k=-\infty}^{\infty} c_k e^{ikt}, \quad \text{where } c_k = \frac{\langle h, e^{ikt} \rangle}{\|e^{ikt}\|^2} = \frac{1}{2\pi} \int_{-\pi}^{\pi} h(t) e^{-ikt}\, dt.$$

The projection of h on T_n, which has an orthogonal basis $\{e^{ijt} : -n \le j \le n\}$ is given by $s = \sum_{k=-n}^{n} c_k e^{ikt}$, which is a partial sum of the series. Moreover, using the fact that projections and approximations are the same, and using (4),

$$\min \int_{-\pi}^{\pi} |h(t) - \sum_{k=-n}^{n} d_k e^{ikt}|^2\, dt = \int_{-\pi}^{\pi} |h(t) - s(t)|^2\, dt = \|f - s\|^2$$

$$= \|f\|^2 - \|s\|^2 = \|f\|^2 - \sum_{k=-n}^{n} |c_k|^2 \|e^{ikt}\|^2$$

$$= \|f\|^2 - 2\pi \sum_{k=-n}^{n} |c_k|^2.$$

For example, if $h(t) = \begin{cases} 0 & x < 0 \\ 1 & x \ge 0 \end{cases}$, then

$$\langle h, e^{ijt} \rangle = \int_{0}^{\pi} e^{-ijt}\, dt = \begin{cases} \pi & j = 0 \\ 0 & j \text{ even and } \ne 0, \\ 2/(ij) & j \text{ odd} \end{cases}$$

using the fact that $\int e^{-ijt}\, dt = (-ij)^{-1} e^{-ijt}$ if $j \ne 0$. Thus

$$s(t) = \frac{1}{2} + \sum_{\text{odd } j=-n}^{n} \frac{e^{ijt}}{\pi i j} = \frac{1}{2} + 2 \sum_{\text{odd } j=1}^{n} \frac{\sin(jt)}{\pi j}$$

where one of Euler's formulae has been used: $e^{i\phi} - e^{-i\phi} = 2i \sin(\phi)$. The (minimal) error is computed by (4):

$$\|h - s\|^2 = \|h\|^2 - \|s\|^2 = \pi - \left[\left(\frac{1}{2} \right)^2 2\pi + \sum_{\text{odd } j=1}^{n} \left(\frac{2}{\pi k} \right)^2 \pi \right]$$

$$= \frac{\pi}{2} - \left(\frac{4}{\pi} \right) \sum_{\text{odd } j=1}^{n} \frac{1}{k^2}.$$

Computation of projections using a spanning set

It may not always be desirable or easy to find an orthogonal basis for a given subspace. The following method yields the projection when only a spanning set is known.

Let $\{v_1, v_2, \ldots, v_n\}$ be a spanning set for M. Since the projection s exists, $s = \sum_{i=1}^{n} c_i v_i$ for certain c_i.

To find the coefficients c_i, consider the following chain of equivalent statements:

$$s \text{ is the projection of } f \text{ on } M \Leftrightarrow$$

$$f - s \perp M \Leftrightarrow$$

$$f - s \perp v_1, \ldots, v_n \Leftrightarrow \tag{5}$$

$$\langle f - s, v_i \rangle = 0, \qquad i = 1, 2, \ldots, n \Leftrightarrow$$

$$\langle f, v_i \rangle - \langle s, v_i \rangle = 0, \qquad i = 1, 2, \ldots, n \Leftrightarrow$$

$$\langle f, v_i \rangle = \langle s, v_i \rangle, \qquad i = 1, 2, \ldots, n \Leftrightarrow$$

$$\sum_{j=1}^{n} c_j \langle v_j, v_i \rangle = \langle f, v_i \rangle, \qquad i = 1, 2, \ldots, n \tag{6}$$

The last system of equations, (6), is traditionally called the *normal equations* for s with respect to the inner product $\langle , \rangle$ and spanning set $\{v_1, \ldots, v_n\}$. Here we also refer to the equivalent statements, (5), as the normal equations since they describe the geometry of projection.

The normal equations (6) may be set in matrix form as

$$Gc = z \tag{6a}$$

where

$$G = [g_{ij}], \qquad g_{ij} = \langle v_j, v_i \rangle, \qquad i, j = 1, 2, \ldots, n$$
$$z = (z_1, z_2, \ldots, z_n)^{\mathrm{T}}, \qquad z_i = \langle f, v_i \rangle, \qquad i = 1, 2, \ldots, n \tag{6b}$$

The matrix G is called the *Gram matrix* of $\{v_k\}_{k=1}^{n}$ with respect to the given inner product.

Remark. Observe that G is Hermitian; i.e., $G^{\mathrm{H}} = G$, since $g_{ij} = \langle v_j, v_i \rangle = \overline{\langle v_i, v_j \rangle} = \overline{g}_{ji}$.

The next theorem summarizes the preceding discussion and describes when the Gram matrix is nonsingular. This result will be sharpened considerably later, in Section 3.6.

THEOREM 4. Let $V = \text{span}\{v_1, v_2, \ldots, v_n\}$, let f be given, and let $s = \sum_{j=1}^{n} c_j v_j$, where $c = (c_1, c_2, \ldots, c_n)^T$. Then the following statements are equivalent:

(a) s is the projection of f on $\text{span}\{v_k\}_{k=1}^{n}$.
(b) c satisfies the normal equations (6) or (6a), (6b).

Further, for every f in V, s exists and there is at least one solution column c for the normal equations (6). The uniqueness of c is equivalent to each of

(c) $\{v_k\}_{k=1}^{n}$ is linearly independent.
(d) the Gram matrix is nonsingular.

Proof. The fact that the normal equations are solvable for every f comes from the fact that the projection exists for every f. If there is more than one solution for c, then s has two expansions $s = \sum_{j=1}^{n} c_j v_j$ with different coefficients, and so $\{v_k\}_{k=1}^{n}$ must be linearly dependent. On the other hand, if $\{v_k\}_{k=1}^{n}$ is linearly dependent, then s has more than one expansion, c is not unique, and so (6) has more than one solution. Further, G is singular iff (6) has more than one solution by Theorem 9 of Section 2.2 or by Theorem 4 of Section 0.5.

COROLLARY 5. If the Gram matrix of $\{v_i\}_{i=1}^{n}$ with respect to one inner product is nonsingular, then for every inner product, the Gram matrix of $\{v_k\}_{k=1}^{n}$ is nonsingular.

Proof. The linear independence of $\{v_k\}_{k=1}^{n}$ does not depend on the notion of inner product.

EXAMPLE 5. PROJECTION ON A PLANE IN $\mathbf{R}^3$ USING THE NORMAL EQUATIONS. Let M be the subspace of $\mathbf{R}^3$ described by $x_1 - x_2 - x_3 = 0$, and take the inner product to be the standard one. Find the projection of $f = (2, 3, 2)^T$ on M.

Using the methods of Section 2.2, a basis for M is $\{v_1, v_2\} = \{(1, 1, 0)^T, (1, 0, 1)^T\}$. Writing $s = a_1 v_1 + a_2 v_2$ and requiring $f - s \perp v_1, v_2$ yields the normal equations

$$\langle v_1, v_1 \rangle a_1 + \langle v_2, v_1 \rangle a_2 = \langle f, v_1 \rangle, \qquad \text{i.e., } 2a_1 + 1a_2 = 5$$
$$\langle v_1, v_2 \rangle a_1 + \langle v_2, v_2 \rangle a_2 = \langle f, v_2 \rangle, \qquad \text{i.e., } 1a_1 + 2a_2 = 4.$$

From this, $a_1 = 2$ and $a_2 = 1$, so that the projection is $s = 2v_1 + 1v_2 = (3, 2, 1)^T$. As a check, we can compute that $f - s$ is indeed orthogonal to v_1, v_2, or we can find an orthogonal basis for $\text{span}\{v_1, v_2\}$ and compute the projection by finding the Fourier sum.

EXAMPLE 6. DISCRETE WEIGHTED APPROXIMATION BY POLYNOMIALS. Find a quadratic polynomial approximating the following data, where $k = 1, 2, \ldots, 5$:

x_k	-1	0	1	2	3
y_k	0.5	0.1	0.5	1	2

Make an unbiased fit, and make another that fits more closely at the ends. More specifically, find a quadratic $p(x)$ that minimizes, for given w_i,

$$\sum_{i=1}^{5} [f(x_i) - p(x_i)]^2 w_i, \qquad \text{where } f(x_i) = y_i.$$

A discrete (real) inner product such that $\|f - p\|^2 = \sum_{i=1}^{5} [f(x_i) - p(x_i)]^2 w_i$ is given by

$$\langle h, g \rangle = \sum_{i=1}^{5} h(x_i)g(x_i)w_i.$$

Let $v_j(x) = x^j$, for $j = 0, 1, 2$, and let $p(x) = \sum_{i=0}^{2} c_i x^i$.
For an unbiased fit, take all the $w_i = 1$. In this case,

$$\langle h, g \rangle = h(-1)g(-1) + h(0)g(0) + h(1)g(1) + h(2)g(2) + h(3)g(3).$$

The normal equations (6) are $\sum_{j=0}^{2} c_j \langle v_j, v_i \rangle = \langle f, v_i \rangle$. For example,

$$\langle v_0, v_0 \rangle = 1 + 1 + 1 + 1 + 1 = 5$$
$$\langle v_1, v_0 \rangle = -1 \cdot 1 + 0 \cdot 1 + 1 \cdot 1 + 2 \cdot 1 + 3 \cdot 1 = 5$$
$$\langle v_2, v_0 \rangle = (-1)^2 \cdot 1 + 0^2 \cdot 1 + 1^2 \cdot 1 + 2^2 \cdot 1 + 3^2 \cdot 1 = 15$$
$$\langle f, v_0 \rangle = 0.5 \cdot 1 + 0.1 \cdot 1 + 0.5 \cdot 1 + 1 \cdot 1 + 2 \cdot 1 = 4.1$$

and so on, yielding the matrix form (6a, b)

$$Gc = [\langle v_j, v_i \rangle]c = \begin{bmatrix} \langle f, v_0 \rangle \\ \langle f, v_1 \rangle \\ \langle f, v_2 \rangle \end{bmatrix} = z.$$

Specifically,

$$\begin{bmatrix} 5 & 5 & 15 \\ 5 & 15 & 35 \\ 15 & 35 & 99 \end{bmatrix} c = \begin{bmatrix} 4.1 \\ 8 \\ 23 \end{bmatrix}.$$

The solution (to four significant digits) is

$$c = (0.2229, -0.02429, 0.2071)^{\mathrm{T}},$$

and the best-fitting quadratic is $p(t) = c_0 + c_1 t + c_2 t^2$.
The pointwise errors are as follows:

x_k	-1	0	1	2	3
$y_k - p(x_k)$	0.0457	-0.1229	0.0943	-0.0029	-0.0143

In order to make the fit closer at $x_k = -1, 3$, the weights w_1, w_5 must be larger than the other weights. For the sake of discussion, let us take $w_1 = w_5 = 10$, and $w_2 = w_3 = w_4 = 1$. The normal equations become

$$\begin{bmatrix} 23 & 23 & 105 \\ 23 & 105 & 269 \\ 105 & 269 & 837 \end{bmatrix} c = \begin{bmatrix} 26.6 \\ 57.5 \\ 189.5 \end{bmatrix}.$$

The solution (to four significant digits) is

$$c = (0.2377, -0.04575, 0.2113)^{\mathrm{T}},$$

and the best-fitting quadratic is $p(t) = p(t) = c_0 + c_1 t + c_2 t^2$.

The pointwise errors are indeed reduced at the ends:

x_k	-1	0	1	2	3
$y_k - p(x_k)$	0.0053	-0.1377	0.0968	0.0087	-0.0020

Remark. If the data are taken from experiments, here is a rule of thumb to determine the weights, w_i: Each w_i should be proportional to the reciprocal of the estimated error at the point x_i. Small differences in the w_i will have little effect.

EXAMPLE 7. Find the orthogonal projection s of $|t|$ on $\mathbf{P}_2$ using the real inner product $\langle f, g \rangle = \int_{-1}^{1} f(t) g(t) dt$.

Equations (6a) and (6b) will be used together with $v_i = t^i$, where $i = 0, 1, 2$. From

$$g_{ij} = \int_{-1}^{1} t^j t^i \, dt, \qquad z_j = \int_{-1}^{1} |t| t^j \, dt, \qquad i, j = 0, 1, 2,$$

and from the supposition that $p = c_0 + c_1 t c_2 t^2$, the normal equations $Gc = z$ are found to be

$$\begin{bmatrix} 2 & 0 & 2/3 \\ 0 & 2/3 & 0 \\ 2/3 & 0 & 2/5 \end{bmatrix} c = \begin{bmatrix} 1 \\ 0 \\ 1/2 \end{bmatrix}.$$

Thus $x = \frac{1}{16}(3, 0, 15)^{\mathrm{T}}$. The projection is $s(t) = \sum_{j=0}^{2} c_j v_j = (3 + t^2)/16$. Note that $s = p$ also minimizes $\int_{-1}^{1} [|t| - p(t)]^2 \, dt, \ p \in \mathbf{P}_2$.

EXAMPLE 8. A PROJECTION PROBLEM WITH LINEARLY DEPENDENT v_i. Let V be the vector space of all real functions continuous on $[-1, 1]$, and let M be the subspace spanned by $v_1(t) = 1$, $v_2(t) = t$, $v_3(t) = 1 + t$. The inner product on V is $\langle f, g \rangle = \int_{-1}^{1} fg \, dt$. Let $f(t) = t^2$. Find the projection of f on M.

The Gram matrix and z are

$$G = \begin{bmatrix} \langle v_1, v_1 \rangle & \langle v_2, v_1 \rangle & \langle v_3, v_1 \rangle \\ \langle v_2, v_1 \rangle & \langle v_2, v_2 \rangle & \langle v_2, v_3 \rangle \\ \langle v_3, v_1 \rangle & \langle v_3, v_2 \rangle & \langle v_3, v_3 \rangle \end{bmatrix} = \begin{bmatrix} 3 & 0 & 3 \\ 0 & 1 & 1 \\ 3 & 1 & 4 \end{bmatrix}, \qquad z = \begin{bmatrix} \langle f, v_1 \rangle \\ \langle f, v_2 \rangle \\ \langle f, v_3 \rangle \end{bmatrix} = \begin{bmatrix} 1 \\ 0 \\ 1 \end{bmatrix}.$$

Solving $Gx = z$ yields the equations

$$x_1 + x_3 = \tfrac{1}{3}, \qquad x_2 + x_3 = 0.$$

Taking x_3 as the free variable, the unique best approximator (projection) on M is

$$s(t) = (\tfrac{1}{3} - x_3) \cdot 1 - x_3 t + x_3(1 + t) = \tfrac{1}{3}.$$

One can also solve this problem by constructing an orthogonal basis for M using the Gram–Schmidt process (casting out dependent vectors), and then forming the Fourier sum.

Projection Operators and Orthogonal Complements

The projection problem, i.e., the resolution of $f = s + r$, with $s \in M$ and $r \perp M$, has now been solved when M is finite-dimensional. Next we formulate a terse and elegant geometric description of this fact and investigate how the projection s depends on f. Three definitions are needed.

> **DEFINITION.** If the projection s exists and is unique for every $f \in V$, then the *projection operator* $\prod_M : V \to V$ is defined by $s = \prod_M(f)$. $\prod$ will be written for $\prod_M$ when confusion is unlikely.

> **DEFINITION.** $M^\perp$ is the set of all vectors orthogonal to M and is called the *orthogonal complement* of M, or M *perp*.

It is easy to verify that $M^\perp$ is a subspace of V. ($M^\perp$ will be discussed in more detail in Section 3.5.) For example, in $\mathbf{R}^3$, if M is the x-y plane, then $M^\perp$ is the z axis, whereas if M is the x axis, then $M^\perp$ is the y-z plane.

> **DEFINITION.** If M_1 and M_2 are two subspaces of V with the property that every $x \in V$ may be written uniquely as $x = x_1 + x_2$ with $x_i \in M_i$, then V is called the *direct sum* of M_1 and M_2. This is written as $V = M_1 \oplus M_2$.

Now it is possible to summarize the theoretical aspects of the preceding work very succinctly.

> **THEOREM 6.** Let M be a finite-dimensional subspace of V, and let $\langle\,,\,\rangle$ be an inner product on V. Then $V = M \oplus M^\perp$.

Proof. The conclusion merely states that each f in V has the unique representation as $f = s + r$, with $s \in M$ and $r \in M^\perp$. This was proved in Theorems 1 and 3.

The fact that Fourier sums are projections allows us to deduce some useful theoretical results.

The next theorem puts Theorems 1 and 3 in the context of projection operators.

> **THEOREM 7.** Let M be a finite-dimensional subspace of an inner product space V. Then the projection operator $\prod = \prod_M : V \to V$ exists and is linear:
>
> $$\prod(\alpha f + \beta g) = \alpha \prod(f) + \beta \prod(g) \qquad \forall f, g \in V, \alpha, \beta \in \mathbf{F}.$$

Moreover

$$\prod(f) = \sum_{i=1}^n \frac{\langle f, e_i \rangle}{\|e_i\|^2} e_i, \qquad \text{where } \{e_k\}_{k=1}^n \text{ is an orthogonal basis for } M. \tag{7}$$

Proof. Equation (7) has already been established as Eq. (3). Thus $s = \prod(f)$ is well-defined for each f in V. Using (7), we obtain

$$\prod(\alpha f + \beta g) = \sum_{i=1}^n \frac{\langle \alpha f + \beta g, e_i \rangle}{\|e_i\|^2} e_i = \sum_{i=1}^n \frac{\alpha \langle f, e_i \rangle + \langle \beta g, e_i \rangle}{\|e_i\|^2} e_i$$

$$= \alpha \prod(f) + \beta \prod(g),$$

which shows that $\prod$ is linear.

The linearity of $\prod$ is perhaps surprising. For example, it is not intuitively obvious why the best approximation of a sum is the sum of the best approximations. If a norm were used that is not generated by an inner product, then $\prod$ would not be linear. The next example applies the ideas of Theorem 7 to find a closed form for the projection operator in $\mathbf{R}^3$. The idea generalizes to $\mathbf{F}^n$, and this generalization is discussed in Section 3.4A.

EXAMPLE 9. Let M be the plane spanned by $v_1 = (1, -1, 0)^{\mathrm{T}}$ and $v_2 = (0, 1, -1)^{\mathrm{T}}$ in $\mathbf{R}^3$. Find the projection operator on M and its matrix in the standard basis.

Using Gram–Schmidt, $v_2 - (\langle v_2, v_1 \rangle / \|v_1\|^2) v_1 = \frac{1}{2}(1, 1, -2)$, so $E_1 = v_1$, and $E_2 = (1, 1, -2)^{\mathrm{T}}$ are an orthogonal basis for M. By (7), the projection of an arbitrary f on M is

$$s = \prod(f) = \frac{\langle f, E_1 \rangle}{\|E_1\|^2} E_1 + \frac{\langle f, E_2 \rangle}{\|E_2\|^2} E_2. \tag{$*$}$$

To find the projection operator, rewrite $(*)$ as follows:

$$\prod(f) = s = \frac{E_1^{\mathrm{T}} f}{\|E_1\|^2} E_1 + \frac{E_2^{\mathrm{T}} f}{\|E_2\|^2} E_2 = E_1 \frac{E_1^{\mathrm{T}} f}{\|E_1\|^2} + E_2 \frac{E_2^{\mathrm{T}} f}{\|E_2\|^2}$$

$$= \left(\frac{E_1 E_1^{\mathrm{T}}}{\|E_1\|^2} + \frac{E_2 E_2^{\mathrm{T}}}{\|E_2\|^2} \right) f = \frac{1}{3} \begin{bmatrix} 2 & -1 & -1 \\ -1 & 2 & -1 \\ -1 & -1 & 2 \end{bmatrix} f = Af.$$

That is, the projection of f is found by the matrix multiplication Af. Thus $[\prod] = A$.

Exercises 3.3

1. Let $M = \mathrm{span}\{v_1, v_2, v_3\}$, where $v_1 = (1, 1, 0, 1)^{\mathrm{T}}$, $v_2 = (3, 1, 2, -1)^{\mathrm{T}}$, and $v_3 = (1, 2, 2, 0)^{\mathrm{T}}$.
   (a) For each of the following f, find the projection of f on M two ways: by use of an orthogonal basis form M and by the normal equations. Use the standard inner product.
      (i) $f = (0, 2, 1, 1)^{\mathrm{T}}$
      (ii) $f = (4, 0, 1, 2)^{\mathrm{T}}$
      (iii) $f = (0, 1, -1, -1)^{\mathrm{T}}$
   (b) Find the matrix of the projection operator in the standard basis.

2. Let $M = \mathrm{span}\{v_1, v_2, v_3\}$, where $v_1 = (1, 1, 1, 1)^{\mathrm{T}}$, $v_2 = (2, 0, 1, 1)^{\mathrm{T}}$, and $v_3 = (2, 4, 3, -1)^{\mathrm{T}}$.
   (a) For each of the following f, find the projection of f on M two ways: by use of an orthogonal basis form M and by the normal equations. Use the standard inner product.
      (i) $f = (1, 1, -2, 0)^{\mathrm{T}}$
      (ii) $f = (3, 1, 2, -2)^{\mathrm{T}}$
      (iii) $f = (5, 1, 0, -2)^{\mathrm{T}}$
   (b) Find the matrix of the projection operator in the standard basis.

3. Find a linear combination s of $v_1 = (1, 0, -2, 1)^{\mathrm{T}}$, $v_2 = (2, 1, -2, 0)^{\mathrm{T}}$, and $v_3 = (1, -1, -1, 3)^{\mathrm{T}}$ for which $\|f - s\|$ is minimal, where $f = (2, 3, 2, 2)^{\mathrm{T}}$. Use the standard inner product.

4. Let $M = \{x : \sum_{i=1}^{4} x_i = 0\} \subset \mathbf{R}^4$. Use the standard inner product.

(a) For each f, find the projection of f on M.

    (i) $f = (1, 1, 1, 1)^T$

    (ii) $f = (1, -2, -1, 2)^T$

    (iii) $f = (1, 1, 0, -1)^T$

    (iv) $f = (2, 4, 2, 4)^T$

(b) Find the matrix of the projection operator in the standard basis.

5. Let $M = \{x : x_1 + x_2 + x_4 = 0, x_1 + x_3 - x_4 = 0, x \in \mathbf{R}^4\}$.

(a) Find the distance from f to M for each of the following, and find the point in M that is closest to f.

    (i) $f = (2, 1, 1, 0)^T$

    (ii) $f = (0, 2, 4, 1)^T$

(b) Find the matrix of the projection operator.

6. Let M be the null space of

$$A = \begin{bmatrix} 1 & 1 & 1 & 1 \\ 1 & 1 & 2 & 0 \end{bmatrix},$$

and let $b = (1, 1, 1, 1)^T$ and $c = (1, 0, 0, 0)^T$.

(a) Find the projections of b and c on M.

(b) Find the matrix of the projection operator in the standard basis.

7. Let $M = R(A)$, where

$$A = \begin{bmatrix} 1 & 2 & 3 \\ 0 & 0 & -2 \\ 1 & 1 & 1 \\ -1 & 0 & 1 \end{bmatrix}.$$

(a) Find an orthogonal basis for M.

(b) Find the projection of each of the following on M.

    (i) $f = (1, 1, 1, 1)^T$

    (ii) $f = (2, 1, 1, 0)^T$

    (iii) $f = (-1, 0, 2, 1)^T$

(c) Find the matrix of the projection operator $\prod : \mathbf{R}^4 \to \mathbf{R}^4$.

8. Let M be the range of

$$A = \begin{bmatrix} 1 & 2 & 0 \\ 0 & 1 & 1 \\ 1 & 1 & 1 \\ -1 & 0 & 0 \end{bmatrix},$$

and let $y = (0, 2, 4, 1)^T$.

(a) Find the projection of y on M using the normal equations.

(b) Find an orthogonal basis for M, and find the projection of y using orthogonal expansions.

(c) Find the matrix of the projection operator in the standard basis.

9. Give $\mathbf{R}^4$ the inner product $\langle x, y \rangle = 2x_1y_1 + 2x_2y_2 + x_3y_3 + x_4y_4$. Let M be the span of $(1, 0, 1, 1)^{\mathrm{T}}$ and $(3, 2, -1, -1)^{\mathrm{T}}$, and let $\prod$ denote the projection operator. Find a matrix A so that $\prod(x) = Ax \ \forall x \in \mathbf{R}^4$.

10. Let $\langle f, g \rangle = \int_{-1}^{1} f\bar{g}\,dx$. Let $h(x) = 1$. For each of the following vector spaces M, find the $p \in M$ that minimizes $\|h - q\|^2$, $q \in M$.
   (a) $M = \{q : q \in \mathbf{P}_3, p(0) = 0\}$
   (b) $M = \{q : q \in \mathbf{P}_3, p(0) = p(1) = 0\}$
   (c) $M = \{q : q \in \mathbf{P}_2, \int_{-1}^{1} p(x)\,dx = 0\}$
   (d) $M = \{q : q \in \mathbf{P}_3, p'(0) = 0\}$

11. Let $h(x) = |x|^3$. For each of the inner products that follow, find the $p \in \mathbf{P}_2$ that minimizes $\|h - p\|$.
   (a) $\langle f, g \rangle = \int_{-1}^{1} f(x)\bar{g}(x)\,dx$. Use the normal equations and an orthogonal set.
   (b) $\langle f, g \rangle = \int_{-1}^{1} f(x)\bar{g}(x)(1 - x^2)^{-1/2}\,dx$. Use Chebyshev polynomials and the substitution $x = \cos(\theta)$ to evaluate the integrals that arise.
   (c) $\langle f, g \rangle = \sum_{i=-2}^{2} f(i/2)\bar{g}(i/2)$. Use either method.
   (d) Graph the best approximations and the data points and compare them. Graph the error $f - p$ in each case. Does the discrete inner product do as well as the integral inner product?

12. Let $h(x) = x + |x|$. For each of the inner products that follow, find the $p \in \mathbf{P}_2$ that minimizes $\|h - p\|$.
   (a) $\langle f, g \rangle = \int_{-1}^{1} f(x)\bar{g}(x)\,dx$. Use the normal equations and an orthogonal set.
   (b) $\langle f, g \rangle = \int_{i=-1}^{1} f(x)\bar{g}(x)(1 - x^2)^{-1/2}\,dx$. Use Chebyshev polynomials and the substitution $x = \cos(\theta)$ to evaluate the integrals that arise.
   (c) $\langle f, g \rangle = \sum_{i=-2}^{2} f(i/2)\bar{g}(i/2)$. Use either method.
   (d) Graph the best approximations and the data points and compare them. Graph the error $f - p$ in each case. Does the discrete inner product do as well as the integral inner product?

13. Approximation of $f(x) = e^x$.
   (a) At $x_k = -1 + k/2$, for $k = 0, 1, \ldots, 4$, the values of e^x are approximately $y_k = 0.37, 0.61, 1, 1.65,$ and 2.72, respectively. Find a quadratic $q(x)$ that minimizes $\sum_{i=0}^{4}[y_i - q(x_i)]^2$.
   (b) Find a quadratic $p(x)$ that minimizes $\int_{-1}^{1}(e_x - p(x))^2\,dx$.
   (c) Compare these quadratics by looking at the errors at the x_k and plotting their graphs for $-1 \le x \le 1$. Plot also the errors $e^x - p(x)$, $e^{-x} - q(x)$.

14. Let $M = \mathrm{span}\{1, \cos(\pi t), \sin(\pi t)\}$, and let $h(t) = t$.
   (a) Find the projection of h on M in each of the inner products.
      (i) $\langle f, g \rangle = \int_0^1 f(t)\bar{g}(t)\,dt$
      (ii) $\langle f, g \rangle = \sum_{i=0}^{n} f(i/4)\bar{g}(i/4)$
   (b) Compare the projections.

15. Let $\langle f, g \rangle = \int_0^{\pi} f(x)\bar{g}(x)\,dx$. Find the linear combination p of $\{\sin(kx) : k = 1, 2, \ldots, n\}$ that minimizes $\int_0^{\pi}[1 - \sum_{k=1}^{n} c_k \sin(kx)]^2\,dx$. Find the minimal value of this integral to four significant figures when $n = 5, 10$.

16. Let $\langle f, g \rangle = \int_0^\pi f(t)\overline{g}(t)\,dt$. For $n = 5, 10$, find the best approximation t_n to x in span$\{1, \cos(x), \ldots, \cos(nx)\}$. Compute the error $\|x - t_n\|$ to four significant digits.

17. Let $f_k(x) = \begin{cases} 1 & (k-1)/n \leq x \leq k/n \\ 0 & \text{otherwise} \end{cases}$, $k = 1, \ldots, n-1$,

$f_n(x) = \begin{cases} 1 & 1 - 1/n \leq x \leq 1 \\ 0 & \text{otherwise} \end{cases}$,   and $\langle f, g \rangle = \int_0^1 f(t)\overline{g}(t)\,dt$.

   (a) Find the best approximation $p(x)$ to $f(x)$ in span$\{f_1, \ldots, f_n\}$ for

      (i) $f(x) = x$

      (ii) $f(x) = e^x$

   (b) For $n = 5$, graph the best approximations together with $f(x)$. What happens as n increases?

18. Let $A_1, A_2, \ldots, A_n \in \mathbf{F}^m$, and let $A = [A_1 \ A_2 \ \cdots \ A_n] = [b_1 \ b_2 \ \cdots \ b_m]^\mathsf{T}$. Show that
   (a) $x \perp A_1, A_2, \ldots, A_n \Leftrightarrow A^H x = 0$
   (b) $x \perp R(A) \Leftrightarrow A^H x = 0$
   (c) If A and x are real, then $x \in N(A) \Leftrightarrow x \perp b_1, b_2, \ldots, b_m$.

19. Let $\{e_k\}_{k=1}^n$ be an orthogonal set in $\mathbf{F}^m$, and let $M = \mathrm{span}\{e_k\}_{k=1}^n$. Show that the matrix of the projection operator in the standard basis is $[\Pi] = \sum_{i=1}^n (e_k e_k{}^H / \|e_k\|^2)$.

20. Suppose $\{A_k\}_{k=1}^n$ is a linearly independent set in $\mathbf{F}^m$, and $M = \mathrm{span}\{A_k\}_{k=1}^n$. Let s be the projection of f on M. Show that
   (a) The normal equations are $A^H A c = A^H f$.
   (b) $s = Ac$.
   (c) In the standard basis, $[\Pi] = A(A^H A)^{-1} A^H$.

21. Define the angle θ between two vectors x, y in a real vector space by $\cos(\theta) = (\langle x, y \rangle / \|x\| \, \|y\|)$. Let s be the projection of f on M, and let θ be the angle between f and s. Show that
   (a) $\langle f, s \rangle = \|s\|^2$
   (b) $\cos(\theta) = \|s\| / \|f\|$
   (c) $\sin(\theta) = \|f - s\| / \|f\|$

22. Let $\langle \, , \rangle$ be an inner product, let $\alpha > 0$, and define an inner product $\langle \, , \rangle_1$ by $\langle x, y \rangle_1 = \alpha \langle x, y \rangle$. If $\dim(M) < \infty$ and $f \in V$, show that $\langle \, , \rangle$ and $\langle \, , \rangle_1$ give the same projection s on M and the same relative error.

23. Discrete trigonometric approximation. This exercise involves ideas found in the theory of the FFT (fast Fourier transform). Let $x_k = 2\pi k/(N+1)$, where $k = 0, 1, \ldots, N$. Suppose N is even. Define the discrete inner product $\langle f, g \rangle = \sum_{k=0}^N f(x_k)\overline{g}(x_k)$.
   (a) Show that the set of functions $\{e^{imx} : m = -N/2, \ldots, 0, 1, \ldots, N/2\}$ is orthogonal with respect to the given $\langle \, , \rangle$. Find the norms of these functions.
   (Hint: $\sum_{j=0}^n a^j = (a^{n+1} - 1)/(a - 1)$ if $a \neq 1$.)
   (b) Show that the trigonometric polynomial p of degree $r \leq N/2$ minimizing the discrete norm $\|f - p\|$ is given by

$$p_r(x) = \sum_{k=-r}^r c_k e^{ikx}, \qquad \text{where } c_k = \frac{\langle f, e^{ikx} \rangle}{\|e^{ikx}\|^2}. \qquad (*)$$

(c) Show that, if $r = N/2$ and p_r is given by (*), then $p(x_k) = f(x_k)$, for $k = 0, 1, \ldots, N$; i.e., p_r interpolates f at the nodes x_k.

Note that obtaining p_{r+1} involves adding just one term to p_r and that $\|f - p_0\| \geq \|f - p_1\| \geq \cdots \geq \|f - p_{N/2}\| = 0$, so that the approximations improve until they are exact.

24. Let V be finite-dimensional, and let M be a nontrivial subspace of V. Define $N = \{x : x \in V, x \perp M\}$. Show that

(a) N is a subspace of M.

(b) If s is the projection of f on M, then $r = f - s$ is the projection of f on N.

3.4
APPLICATIONS OF PROJECTION THEORY

In the first subsection that follows, projections in $\mathbf{F}^m$ will be considered. Several alternative viewpoints of the projection/approximation problems will be presented. In subsequent subsections, the theory developed in the first subsection will be applied in different contexts.

3.4A
PROJECTIONS IN $\mathbf{F}^m$

First, the standard (euclidean) inner product is considered.

Let the subspace be $M = \text{span}\{A_1 \cdots A_n\}$. For a given $y \in \mathbf{F}^m$, the equivalent projection and approximation problems are, respectively,

(a) Find the orthogonal projection s of y on M.

(b) Find $v \in M$ for which $\|y - v\|$ is minimal.

The approximation problem is often rephrased in different ways, depending on the way it arises in problems. Let $A = [A_1 \cdots A_n]$. Then $v \in M$ iff $v = \sum_{i=1}^{n} x_i A_i = Ax$ for some x; in other words, $M = R(A)$. Thus (b) is the same as

(b') Find x for which $\|y - Ax\|$ is minimal.

Another viewpoint is that of finding the best solution of a possibly inconsistent system of equations $\sum_{j=1}^{n} a_{ij} x_j = y_i$. Let $r_i \equiv y_i - \sum_{j=1}^{n} a_{ij} x_j$. r_i is called the *residual* or *error* in the ith equation. Thus $r = y - Ax$. The goal is to

(b'') Minimize $\sum_{i=1}^{n} |r_i|^2$ with respect to x.

But

$$\sum_{i=1}^{n} |r_i|^2 = \sum_{i=1}^{n} |(Ax)_i - y_i|^2 = \|y - Ax\|^2 = \|r\|^2,$$

so that (b'') is merely a restatement of (b'). The name *least squares minimization* is often used in reference to the formulation (b'').

There may be subtle differences in the quantities of interest, however. Let $s = Ax = \sum_{i=1}^{n} x_i A_i$ be the projection and the minimizing v. In some problems, just the projection s is required, and the coefficients x are not of much significance. For example, one may desire the nearest point to a plane, and the basis $\{A_k\}_{k=1}^{n}$ in which the point s is represented is of no matter. In other problems, the coefficients x are of paramount interest, and s may have no particular meaning. Examples of the latter situation are given in the next sections.

The normal equations for the problem are

$$A^H Ax = A^H y. \tag{1}$$

In fact, from (6a, b) of Section 3.3,

$$G = [\langle A_j, A_i \rangle] = [A_i^H A_j] = A^H A,$$

$$f = (\langle y, A_1 \rangle, \ldots, \langle y, A_n \rangle)^T = (A_1^H y, \ldots, A_n^H y)^T = A^H y.$$

Applying the theory of the previous section, the following theorem is obtained.

THEOREM 1. Let $A_i \in \mathbf{F}^m$, where $i = 1, 2, \ldots, n$, and let y be given. Let $M = \mathrm{span}\{A_1, A_2, \ldots, A_n\}$, and set $A = [A_1 \ A_2 \ \cdots \ A_n]$. Let $\langle , \rangle$ be the standard inner product on $\mathbf{F}^m$. Then each of the following statements is equivalent:

(a) $s = Ax$ is the projection of f on M.
(b) $z = x$ minimizes $\|y - Az\|$, where $z \in \mathbf{F}^n$.
(c) x satisfies the normal equations $A^H Ax = A^H y$.

Further, the normal equations always have at least one solution. The projection $s = Ax$ is unique, and x is unique iff the columns of A are linearly independent.

There are three quantities of possible interest: the coefficients x, the projection $s = Ax$, and the projection operator $\prod$. We may want the projection operator to be found as a matrix operator; i.e., a matrix B is sought such that $\prod(y) = By \ \forall y \in \mathbf{F}^m$, which means that $B = [\prod]$ in the standard basis.

Suppose A has n linearly independent columns, so that the Gram matrix $A^H A$ is nonsingular. Some nice expressions for these three quantities are derived next. Finding x is easy (in theory), since the Gram matrix is nonsingular:

$$x = (A^H A)^{-1} A^H y. \tag{2}$$

Now the projection $s = Ax = \prod(y)$ is given by

$$s = \prod(y) = A(A^H A)^{-1} A^H y, \tag{3}$$

so that

$$[\prod] = A(A^H A)^{-1} A^H. \tag{4}$$

Application of the QR decomposition

For work on a computer with larger matrices, $(A^H A)^{-1}$ may be difficult to obtain with good accuracy since often Gram matrices are ill conditioned, i.e., computationally troublesome. Thus the preceding formulae may not offer good numerical solutions. The next method is more attractive in such cases because there are algorithms for computers that accomplish the QR decomposition accurately. Suppose the QR decomposition of A has been computed. The goal is to express x, s, and $\prod$ in terms of Q and R. Recall that the columns of Q are an orthonormal basis for $R(A)$. Let $Q = [E_1\, E_2 \cdots E_n]$. The projection is a Fourier sum so that

$$s = \prod(y) = \sum_{i=1}^{n}(E_i{}^H y)E_i = \sum_{i=1}^{n} E_i(E_i{}^H y) = \left(\sum_{i=1}^{n} E_i E_i{}^H\right)y = (QQ^H)y. \quad (5)$$

Thus

$$\left[\prod\right] = \sum_{i=1}^{n} E_i E_i{}^H = QQ^H. \quad (6)$$

Now let us find x. s is given by (5) and by $s = Ax = QRx$. Multiplying $s = QRx = QQ^H y$ on the left by Q^H gives $Q^H QRx = Rx = Q^H QQ^H y = Q^H y$. Thus

$$x = R^{-1}Q^H y \quad (7)$$

solves the normal equations. In practice, for numerical computations with large matrices, (7) is not used. Rather, $z = Q^H y$ is computed, and then $Rx = z$ is solved using back substitution. Back substitution is easy since R is upper triangular.

In case the columns A_i are linearly dependent, there are infinitely many solutions x to the normal equations. One of these could be selected as the standard solution, e.g., the one of minimal norm. These ideas are pursued extensively in Section 8.2.

Weighted inner products in $\mathbf{F}^m$

Sometimes a weighted inner product,

$$\langle x, y \rangle = \sum_{i=1}^{m} x_i \bar{y}_i w_i = y^H W x, \qquad \text{where } W = \text{diag}(w_1, \ldots, w_n), \quad (8)$$

is appropriate. The next sections show more examples of the utility of these inner products. In the context of the weighted inner product, the approximation problem is to minimize

$$\|y - Ax\|^2 = \sum_{i=1}^{m} |y_i - (Ax)_i|^2 w_i$$

with respect to x. The normal equations (6a) and (6b) of Section 3.3 assert that $s = \sum_{i=1}^{n} x_i A_i = Ax$ is the projection iff x satisfies

$$z = Gx,$$

where

$$z = (\langle y, A_1 \rangle, \langle y, A_2 \rangle, \ldots, \langle y, A_n \rangle)^T = (A_1{}^H W y, \ldots, A_n{}^H W y)^T = A^H W y$$

and

$$G = [\langle A_j, A_i \rangle] = [A_i^H W A_j] = [A_i^H (WA)_j] = A^H W A.$$

Thus the normal equations become

$$A^H W A x = A^H W y \tag{9}$$

where $s = Ax$ is the projection on $M = R(A)$. s is unique, but x is unique iff the columns of A are linearly independent. It is left as an exercise to find the projection operator as a matrix operator when the columns of A are linearly independent and when an orthonormal basis for $R(A)$ is known. (See Exercise 10.)

Another approach is to make the change of coordinates $\tilde{y}_{ii} = \sqrt{w_i} y_i$. This will reduce the projection problem to one involving the standard inner product. This can simplify computations in some situations. For example, if a software package has only routines that solve the least squares problem in the standard inner product, then this approach allows us to solve problems involving weighted inner products. (See Exercise 11.)

Exercises 3.4A

1. "Solve" the following systems, in the sense of minimizing $\|y - Ax\|$ (standard inner product). Find the projection operator two ways, if possible.

(a) $\begin{bmatrix} 0 & 2 & 2 \\ 1 & -1 & 1 \\ 1 & 1 & 3 \\ 1 & 3 & 2 \end{bmatrix} x = \begin{bmatrix} 2 \\ 1 \\ 1 \\ 1 \end{bmatrix}$
(b) $\begin{bmatrix} 1 & 0 & 2 \\ 1 & 1 & 1 \\ 1 & 2 & 0 \end{bmatrix} x = \begin{bmatrix} 1 \\ 1 \\ 2 \end{bmatrix}$
(c) $\begin{bmatrix} 1 & 1 & 2 \\ 1 & 0 & 1 \\ 1 & -1 & 0 \end{bmatrix} x = \begin{bmatrix} 2 \\ 2 \\ 3 \end{bmatrix}$

(d) $\begin{bmatrix} 1 & 1 & 1 \\ 1 & 0 & -1 \\ 1 & 1 & 2 \\ 1 & 0 & 0 \end{bmatrix} x = \begin{bmatrix} 1 \\ 0 \\ 0 \\ 0 \end{bmatrix}$
(e) $\begin{bmatrix} 1 & 3 \\ 2 & 1 \\ 2 & 2 \\ 0 & -2 \end{bmatrix} x = \begin{bmatrix} 1 \\ 0 \\ 0 \\ 1 \end{bmatrix}$

2. (a) Find x such that $\|y - Ax\|$ is minimal (standard $\langle , \rangle$), where

$$A = \begin{bmatrix} 0 & 1 & 2 \\ 1 & 2 & -3 \\ 1 & 3 & -1 \\ 1 & 1 & -2 \end{bmatrix} \quad \text{and} \quad (i)\ y = \begin{bmatrix} 1 \\ 0 \\ 0 \\ 0 \end{bmatrix} \quad (ii)\ y = \begin{bmatrix} 2 \\ 0 \\ 2 \\ 1 \end{bmatrix} \quad (iii)\ y = \begin{bmatrix} 1 \\ 1 \\ -1 \\ 0 \end{bmatrix}$$

(b) Find the matrix of the projection operator in the standard basis.

3. For each of the following subspaces M, find the projection operator (i.e., its matrix in the standard basis). The standard inner product is used. Find the projection operator two ways, if possible.
(a) $M = \text{span}\{V_1, V_2, V_3\}$, where $V_1 = (1, 1, 0, 1)^T$, $V_2 = (3, 1, 2, -1)^T$, and $V_3 = (1, 2, 2, 0)^T$.
(b) $M = \text{span}\{V_1, V_2\}$, where $V_1 = (1, 0, 1, -1)^T$, and $V_2 = (2, 0, 1, 0)^T$.
(c) $M = \{x : \sum_{i=1}^{3} x_i = 0, x \in \mathbf{R}^3\}$.

4. Let y and $A_i \in \mathbf{R}^4$, $i = 1, 2, 3, 4$. The goal is to minimize $\|y - Ax\|^2$. Suppose $A_2 = \alpha A_1 + \beta A_3$ and $\{A_1, A_3, A_4\}$ is linearly independent.

(a) Reduce the problem to a system of equations of the form $B^T B w = B^T y$, with $B = [A_1 \; A_3 \; A_4]$. Can you generalize?

(b) If $y = pA_1 + qA_3 + rA_4$, what is the solution of the original minimization problem?

5. Let M be a nontrivial finite-dimensional subspace of $\mathbf{R}^n$, and let $P = [\prod_M]$ in the standard basis. Show that

(a) $Px = x \; \forall x \in M$

(b) $R(P) = M$

(c) $Px = 0 \Leftrightarrow x \perp m \; \forall m \in M$

(d) $P^2 = P$

(e) $P^H = P$

6. Let P be an $n \times n$ matrix satisfying $P^2 = P$ and $P^H = P$. Let $M = R(P)$. Show that Px is the projection of x on M.

7. Suppose M is a nontrivial subspace of $\mathbf{R}^3$ and

$$[\prod_M] = \frac{1}{3}\begin{bmatrix} 2 & 1 & -1 \\ 1 & 2 & 1 \\ -1 & 1 & 2 \end{bmatrix}.$$

Find M and verify that this matrix (operator) projects onto M.

8. For each of the following matrices P, determine whether P is the matrix of a projection operator. If so, find the subspace M onto which P projects.

(a) $\frac{1}{3}\begin{bmatrix} 2 & 1 & 1 \\ 1 & 2 & 1 \\ -1 & 1 & 2 \end{bmatrix}$ (b) $\frac{1}{3}\begin{bmatrix} 2 & 1 & 1 \\ 1 & 2 & 1 \\ 1 & 1 & 2 \end{bmatrix}$

(c) $\frac{1}{2}\begin{bmatrix} 1 & 0 & 1 \\ 0 & 0 & 0 \\ 1 & 0 & 1 \end{bmatrix}$ (d) $\frac{1}{3}\begin{bmatrix} 1 & -1 & 1 \\ -1 & 1 & -1 \\ 1 & -1 & 1 \end{bmatrix}$

9. Suppose $\{E_k\}_{k=1}^r$ is an orthogonal basis for M in $\mathbf{C}^n$, and set $E = [E_1 \; E_2 \; \cdots \; E_r]$. Let $N = \text{diag}(\|E_1\|^2 \; \|E_2\|^2 \; \cdots \; \|E_r\|^2)$. Show that

(a) $E^H E = N$

(b) $[\prod_M] = E^H N^{-1} E$

10. Consider the weighted inner product $\langle x, y \rangle = \sum_{i=1}^n x_i \bar{y}_i w_i$, and let $M = \text{span}\{A_1, \ldots, A_n\}$, where A is an $m \times n$ matrix. In each of the following cases, find the projection operator as a matrix operator.

(a) $A_1, \ldots, A_n$ are linearly independent.

(b) $E_1, \ldots, E_r$ is an orthogonal basis for M.

11. **Reduction of a weighted least-squares problem to an unweighted one.** Minimize $\|y - Ax\|^2$, where $\| \; \|$ is generated by $\langle x, y \rangle = \sum_{i=1}^m x_i \bar{y}_i w_i$. Here $A = [A_1 \; A_2 \; \cdots \; A_n]$ is an $m \times n$ matrix with rank n.

Hints: Set $\eta_i = \sqrt{w_i} y_i$, $w = \text{diag}(w_1, w_2, \ldots, w_n)$ and define a matrix B by $B = W^{1/2} A$. Show that $\|y - Ax\|$ is minimal iff $\|\eta - Bx\|_{\text{std}}$ is minimal.

12. **Necessary conditions for a minimum using calculus.** Let $G(x) = \frac{1}{2}\|y - Ax\|^2$, where y, A, and x are real. When G is minimal, $\partial G/\partial x_i = 0 \; \forall i$. Compute these equations. See

if they agree with the characterization of the minimizing x just derived. What can be said about the complex case?

13. **Overdetermined systems "$Ax = y$".** Let A be an $m \times n$ matrix with $r(A) = n < m$. Show that
    (a) $Ax = y$ is inconsistent for some y.
    (b) $B = (A^H A)^{-1} A^H$ is a left inverse of A, i.e., $BA = I$.
    (c) If B is restricted to $R(A)$, then B is also a right inverse of A, i.e., $y = Ax \Rightarrow ABy = y$.

14. Suppose x_0 satisfies the normal equations for a given y: $A^H A x_0 = A^H y$. Show that $\|Ax - y\|^2 = \|y - Ax_0\|^2 + \|A(x - x_0)\|^2$ for every $x \in \mathbf{F}^n$.

15. Let $b \neq 0$ and y be given columns in $\mathbf{F}^n$. Show that x_0 minimizes $\|y - Ax\|^2 + (b^H x - 1)^2$ iff x_0 satisfies $A^H y + b = (A^H A + b b^H)x$.

16. Suppose $z \perp Ae_1, \ldots, Ae_m$. Show that $A^H z \perp e_1, \ldots, e_m$.

17. Suppose $\{e_k\}_{k=1}^n$ is an orthogonal basis for $\mathbf{F}^n$ and $z \perp Ae_2, \ldots, Ae_n$. Show that $A^H z = ce_1$ for some scalar c.

3.4B
WEIGHTED LINEAR REGRESSION: DATA FITTING

An example initiates this discussion.

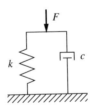

EXAMPLE 1. We are given a spring-dashpot system anchored at one end, as shown in the figure. In many situations, the relation

$$F = cv + kx$$

holds, where F is the applied force, v the velocity of the free end, k the spring constant, and c the damping coefficient. Presented with such a system, we may want to determine the damping coefficient c and the spring constant k by experiment; we can impose a force and measure the displacement and velocity at a given instant of time. Let F_i, v_i, and x_i be the force, velocity, and displacement, respectively, measured in the ith experiment, where $i = 1, 2, \ldots, m$. If the experiments are exactly measured, then

$$F_i = cv_i + kx_i, \qquad 1 = 1, 2, \ldots, m, \tag{1}$$

Clearly two such experiments would be sufficient to determine c and k. When experimental errors exist, two experiments may yield an inconsistent system, or a system that gives highly inaccurate values for c and k. We may want to perform several experiments to obtain better values. The resulting system of equations is overdetermined and most likely inconsistent.

There is a way to salvage the information. Let us suppose that associated with each experiment is a positive confidence weight, w_i, large when the experiment has little error and small when much error is present. One rule of thumb is to use for w_i the reciprocal of the estimated experimental error in the ith experiment. Another rule of thumb is to use for w_i the reciprocal of the variance of the ith experiment. The w_i may be scaled, if desired. For example, replace w_i by w_i/W, where $W = \sum w_i$, so that $\sum w_i = 1$. Scaling will not affect the relative error or angles between vectors.

One way to salvage the information is to give up on equality in (1) and ask for values of c and k that minimize

$$D \equiv \sum_{i=1}^{m} [F_i - (cv_i + kx_i)]^2 w_i.$$

The quantity

$$r_i = F_i - (cv_i + kx_i) \qquad (*)$$

is the residual of the ith equation in (1), and D is then the sum of the squares of the residuals. The problem may be posed as a best approximation problem (cf. Section 3.3). Let $f = (F_1, F_2, \ldots, F_m)^T$, $v = (v_1, v_2, \ldots, v_m)^T$, and $x = (x_1, x_2, \ldots, x_m)^T$. Rewrite (1) exactly, using (*), in the matrix form

$$f = [v \ x] \begin{bmatrix} c \\ k \end{bmatrix} + r. \qquad (2)$$

Then

$$D = \sum_{i=1}^{m} r_i^2 = \|r\|^2 = \left\| f - [v \ x] \begin{bmatrix} c \\ k \end{bmatrix} \right\|^2 = \|f - (cv + kx)\|^2,$$

where $\| \ \|$ is the norm generated by the inner product

$$\langle x, y \rangle = \langle x, y \rangle_w = \sum_{i=1}^{m} x_i y_i w_i. \qquad (3)$$

Thus the problem of minimizing D with respect to c and k has been reduced to finding the best approximation to f from the set of all vectors of the form $cv + kx$. But this is equivalent to the statement that c and k must satisfy the normal equations

$$\langle f, v \rangle = c\langle v, v \rangle + k\langle x, v \rangle$$
$$\langle f, x \rangle = c\langle v, x \rangle + k\langle x, x \rangle, \qquad (4)$$

using the weighted inner product. An appropriate measure of the error is the relative error

$$\frac{\|f - s\|}{\|f\|} = \frac{\sqrt{D}}{\|f\|} \qquad (5)$$

where $s = cv + kx$ is the best approximation/projection, and D is computed at the minimum. The relative error is independent of the scaling of the weights w_i. In this problem, the coefficients c and k are of interest, and not the projection $s = cv + kx$. s is used only to obtain the error and otherwise plays no significant role.

Experimental accuracy and the choice of weights

Now we illustrate the preceding discussion with specific numbers, which are contrived to show how well the approximation works and to show the effect of the choice of weights. To set up the demonstration, $c = 3$ and $k = 5$ were taken, five values of v and x were specified, and the exact forces $F_e = 3v + 5x$ were computed for each of the five values. Then unequal known errors are introduced into these forces to get the measured forces, F. The normal equations are solved with two choices of weights: with all weights equal to 1 and with weights approximately proportional to the reciprocals of the errors.

Experiment number	v	x	F	Error $= F - F_e$
1	1	1	8.01	0.01
2	0	3	14.95	−0.05
3	2	1	11.02	0.02
4	−2	4	14.4	0.4
5	−1	3	12.5	0.5

The normal equations are

$$10c - 8k = -11.25$$
$$-8c + 36k = 158.98$$

giving the solution (to four significant digits) $c = 2.929$, $k = 5.067$, with relative error $= 1.268 \times 10^{-2}$.

If the experimenters had suspected accuracy of ± 0.02 in the first three experiments and an accuracy of ± 0.5 in the last two, then weights proportional to the reciprocals of these accuracies would be $w_i = 25$, for $i = 1, 2, 3$, and $w_i = 1$, for $i = 4, 5$. In this case, the inner product is

$$\langle x, y \rangle = 25(x_1 y_1 + x_2 y_2 + x_3 y_3) + x_4 y_4 + x_5 y_5,$$

and the normal equations are

$$130c + 64k = 709.95$$
$$64c + 300k = 1692.1.$$

The solutions, to four significant digits, are $c = 2.999$ and $k = 5.000$ with relative error $= 6.767 \times 10^{-3}$.

Two points are worth noting. First, there is a smoothing effect in both cases; c and k are more accurate than the errors in F. Second, a good estimate of the errors provides superior accuracy through the determination of appropriate weights.

General weighted linear regression

Now the preceding ideas are expressed in a general setting. Assume that the variable y is related to variables x_i linearly: $y = \sum_{i=1}^{n} a_i x_i$. Suppose that $a_1, a_2, \ldots, a_n$ and y are to be measured experimentally m times in order to determine the x_i. Let y_i and a_{ij} denote the value of y and a_j measured in the ith experiment, and define the residuals r_i by

$$r_i = y_i - \sum_{i=1}^{n} a_{ij} x_j, \qquad i = 1, 2, \ldots, n, \qquad (6)$$

which may be expressed in matrix form

$$r = y - Ax, \qquad A = [a_{ij}]. \tag{7}$$

Assume positive weights w_i have been chosen. As in the preceding discusssion, the w_i could be the reciprocals of the estimated errors or the variances of the experiments. The task is to minimize

$$D \equiv \sum_{i=1}^{n} |r_i|^2 w_i = \|r\|^2 = \|y - Ax\|^2 = \left\| y - \sum_{j=1}^{n} x_j A_j \right\|^2 = \sum_{i=1}^{m} |y_i - (Ax)_i|^2 w_i$$

over all possible scalars x_i, where $\| \ \|$ is the norm generated by the inner product

$$\langle u, v \rangle = \sum_{i=1}^{m} u_i \bar{v}_i w_i = v^H W u, \qquad W = \text{diag}(w_1, \ldots, w_m). \tag{8}$$

In this inner product, the normal equations become

$$A^H W y = A^H W A x, \tag{9}$$

as was demonstrated in the previous section.

Exercises 3.4B

1. A force of 10 units is applied ten times to a spring-dashpot system obeying the law $F = kx + cv$. The following measurements were recorded:

Trial	1	2	3	4	5	6	7	8	9	10
x	1.1	1.3	1.6	1.2	0.9	1.7	1.2	0.8	2.5	2.2
v	2.0	1.9	1.7	1.9	2.0	1.4	1.7	2.3	1.5	1.6

(a) Find the best values of c and k that fit this data, assuming all experiments were performed with equal accuracy. Also determine the relative error.

(b) It has come to your attention that your lab assistant was tired in the last five experiments, and the accuracy may be ten times worse than the accuracy at the other points. Recompute the approximate values of c and k on this basis, and find the relative error.

2. A vehicle moves unpowered on a horizontal road through a distance L and is subjected to linear frictional forces. You want to determine what its initial kinetic energy E must be in order to exit at the distance L with a prescribed velocity v. The equation is

$$E = \tfrac{1}{2}mv^2 + WL\mu, \tag{*}$$

where m is the "effective" mass (which includes the mass and the inertial properties of the rotating parts), W is the weight of the vehicle, and μ is the net coefficient of friction. $WL\mu$ is the work done against friction, and (*) asserts conservation of energy. W is known, but m and μ are not. Some experiments are done to estimate the solution of the problem: The vehicle is run four times with known E, and the velocity v at the end will be measured. The results (in SI units) are as follows, with E in joules and v in m/sec:

E_i	86,000	97,000	163,000	240,000
v_i	4	6	13	18

Here $W = 7840$ newtons and $L = 200$ meters. Do a least-squares fit to (*) using the data. As a by-product, find m and μ. For $v = 10, 20$, find the initial energy E that is needed for these exit velocities.

3. A radioactive material decays as $m = Me^{-\lambda t}$, where M is the original mass in kg and m is the mass at time t. Measurements find the following masses at the indicated times (in minutes):

t_i	10	20	30	40	50
m_i	1.6	1.25	0.9	0.8	0.62

Estimate the original mass and the half-life of the material. Examine the sizes of the residuals, throw out bad data points (if any), and re-estimate the initial mass and half-life. (*Hint:* Consider $\log(m)$.)

4. Three linear amplifiers are connected in a circuit in such a way that the output of the circuit is the sum of the outputs of the amplifiers. If g_i is the gain of amplifier i and s_i is the ith signal, then the output of that amplifier is $g_i s_i$ and the output of the system is $y = g_1 s_1 + g_2 s_2 + g_3 s_3$. In order to determine the gains of the individual amplifiers, signals were measured, and the output y was measured. The following data were recorded:

s_1	s_2	s_3	y
0.5	0.5	0.5	4.8
1.0	0.5	2.0	13.7
0.5	1.0	0.5	6.2
2.0	0.1	0.3	5.6
1.0	0.2	0.4	4.8

Estimate the gain of each amplifier. Find the relative error in the approximation.

3.4C
DISCRETE LEAST-SQUARES CURVE FITTING WITH POLYNOMIALS

Suppose we have $m + 1$ points of data, (x_k, y_k), where $k = 0, \ldots, m$ with distinct x_k, and the task is to fit a polynomial to this data. We want to have a good approximation at the data $p(x_i) \approx y_i$, $i = 0, \ldots, m$. Furthermore, the polynomial should have a smooth graph. We may also want to have the approximation $p(x_i) \approx y_i$ be better at some specified points than at others.

If the degree of the polynomial is allowed to be m, one can always choose the polynomial so that equality holds at each point x_i: $p(x_i) = y_i$ (interpolation). However, if the data have some error, often the interpolating polynomial will oscillate badly in some part of the region of interest. Remarkably, this may also happen if the data are generated by a smooth function f, i.e., if $y_i = f(x_i)$. To counter this phenomenon, the degree n of the polynomial is taken to be less than m, and the data are approximated rather than interpolated. The process is described after the next example.

An example due to Runge is as follows: Take $m + 1$ equispaced points $x_k = -1 + (2k/m)$ in $[-1, 1]$, and put $y_k = f(x_k)$, where $f(x) = (1 + 25x^2)^{-1}$. When these data are interpolated, the interpolating polynomial will show undesirable oscillation in $[-1, 1]$ near ± 1, even for $n = 10$, and the problem worsens with increasing n. In order to smooth out these undesirable effects, we forgo an exact fit and require the degree n to be less than m. See Fig. 3.2.

In Fig. 3.2, the dotted line is $f(x) = (1 + 25x^2)^{-1}$. In Fig. 3.2a, the solid line is the polynomial of degree 10 interpolating f at 11 equidistant points. In part b, the solid line is the polynomial of degree 10 approximating f at 21 equispaced points. Note the difference in scales on the y axis. The approximation is done in the sense described as follows, with equal weights.

The desired approximation can be accomplished by solving the following problem: Given the data (x_i, y_i) with real x_i, the weights $w_i > 0$, where $i = 0, 1, 2, \ldots, m$, and an integer n, find the polynomial $p(x)$ of degree not exceeding n such that

$$D \equiv \sum_{i=0}^{m} |y_i - p(x_i)|^2 w_i \tag{1}$$

is minimized among all polynomials in $\mathbf{P}_n$.

The points $\{x_0, x_1, \ldots, x_m\}$ are sometimes called *nodes.* In practice, n and the weights are chosen. Large weights w_j (relative to the other weights) produce closer approximations $y_j \approx p(x_j)$ at the corresponding nodes. Small variations in the weights will have little effect. Small n will produce graphs that appear smoother and have fewer regions of high slope and curvature, but that fit the data less well.

Define a discrete, weighted, inner product on the vector space of all functions on $S = \{x_0, x_1, \ldots, x_m\}$ by

$$\langle h, g \rangle = \sum_{i=0}^{m} h(x_i)\overline{g(x_i)}w_i, \tag{2}$$

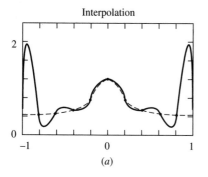
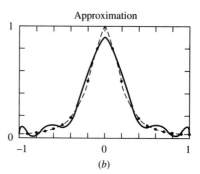

FIGURE 3.2

so that $\|g\|^2 = \sum_{i=0}^m |g(x_i)|^2 w_i$ is the norm generated by this inner product. If $y_i = f(x_i)$, for $i = 0, \ldots, m$, then $D = \|f - p\|^2$, and the problem of minimizing D is a best approximation problem with subspace $M = \mathbf{P}_n$ and the inner product given by (2). The basis $\{1, x, \ldots, x^n\}$ for $\mathbf{P}_n$ is used. If the x_i were not near the origin, then the basis $\{1, x - a, \ldots, (x - a)^n\}$ with a near the center of the x_i would be better for numerical work. Alternatively, the data could be shifted and scaled so as to be near 0. The projection $p(x) = \sum_{j=0}^n c_j x^j$ is characterized by the equations

$$f - p \perp x^k, \qquad k = 0, 1, \ldots, n,$$

which is equivalent to the normal equations (cf. (6) and (6a,b) in Section 3.3)

$$\langle f, x^k \rangle = \sum_{j=0}^n c_j \langle x^j, x^k \rangle, \qquad k = 0, 1, \ldots, n. \tag{3}$$

Carrying the computations further yields (since the x_k are real)

$$\langle x^j, x^k \rangle = \sum_{i=0}^m x_i^{j+k} w_i = \langle x^{j+k}, 1 \rangle$$

$$\langle f, x^k \rangle = \sum_{i=0}^m y_i x_i^k w_i$$

so that the normal equations (3) become

$$\sum_{i=0}^m y_i x_i^k w_i = \sum_{j=0}^m c_j \langle x^{j+k}, 1 \rangle, \qquad k = 0, 1, \ldots, n. \tag{4}$$

Thus the entries $g_{kj} = \langle x^j, x^k \rangle = \langle x^{j+k}, 1 \rangle$ in the Gram matrix follow a pattern, and only $2n + 1$ entries are distinct.

The problem can be converted to one involving matrices and column spaces as follows. When this is done, the results pertaining to projections in column spaces become available, and the minimization can easily be transferred to algorithms designed for column spaces. The trick is to cast D in the form

$$D \equiv \|f - p\|^2 = \|\|y - Ac\|\|^2,$$

where $\|\|\cdot\|\|$ denotes a norm generated by an inner product on $\mathbf{F}^{m+1}$. To do this, let $y = (f(x_0), f(x_1), \ldots, f(x_m))^T$, and $c = (c_0, c_1, \ldots, c_n)^T$. A will be an $(m+1) \times (n+1)$ matrix. Observe that

$$f(x_i) - p(x_i) = y_i - \sum_{j=0}^m c_j x_i^j = y_i - [1 \ x_i \ x_i^2 \ \cdots \ x_i^n] c.$$

Thus, if the matrix A has rows a_i given by

$$a_i = [1 \ x_i \ x_i^2 \ \cdots \ x_i^n], \tag{5}$$

then

$$(Ac)_i = a_i c = [1 \ x_i \ x_i^2 \ \cdots \ x_i^n] c = p(x_i)$$

and

$$Ac = \begin{bmatrix} p(x_0) \\ p(x_1) \\ \vdots \\ p(x_n) \end{bmatrix}. \tag{6}$$

If the suggestive (but abusive) notation

$$1 = \begin{bmatrix} 1 \\ 1 \\ \vdots \\ 1 \end{bmatrix}, \quad x = \begin{bmatrix} x_0 \\ x_1 \\ \vdots \\ x_m \end{bmatrix}, \quad x^j = \begin{bmatrix} x_0^j \\ x_1^j \\ \vdots \\ x_m^j \end{bmatrix}, \quad p(x) = \begin{bmatrix} p(x_0) \\ p(x_1) \\ \vdots \\ p(x_m) \end{bmatrix} \tag{7}$$

is adopted, then $x^j = A_j$, so that

$$A = [1 \ x \ x^2 \ \cdots \ x^n] = \begin{bmatrix} 1 & x_0 & x_0^2 & \cdots & x_0^n \\ \vdots & \vdots & \vdots & \vdots & \vdots \\ 1 & x_m & x_m^2 & \cdots & x_m^n \end{bmatrix}, \tag{8}$$

and (6) can be written as

$$Ac = p(x) = \sum_{i=0}^{n} c_i x^i. \tag{9}$$

A is called the *Vandermonde matrix*. Hence

$$D \equiv \sum_{i=0}^{m} |f(x_i) - p(x_i)|^2 w_i = \sum_{i=0}^{m} |y_i - (Ac)_i|^2 w_i = |||y - Ac|||^2,$$

and the inner product on $\mathbf{F}^{m+1}$ generating $||| \cdot |||$ is

$$\langle z, u \rangle = \sum_{i=0}^{m} z_i \bar{u}_i w_i = u^H W z, \qquad W = \text{diag}(w_0, \ldots, w_m). \tag{10}$$

The normal equations (see (8) in Section 3.4A) are

$$Gc = z, \qquad \text{where } z = A^H W y, \text{ and } G = A^H W A,$$

which agree with (4).

EXAMPLE 1. HIGHER WEIGHTS PRODUCE BETTER LOCAL APPROXIMATION.
Let $f(x) = \cos(\pi x/6)$, and let $x_k = -3 + k$, for $k = 0, \ldots, 6$. Find $p \in \mathbf{P}_2$ best approximating f in the norms generated by the discrete real inner products $\langle f, g \rangle = \sum_{i=1}^{6} f(x_i)g(x_i)w_i$ for

(a) Uniform weights, $w_i = 1$
(b) Weights larger at the ends, $w_i = 1 + 33|x_i|$

Compare the errors at these points.

Here,

$$A = \begin{bmatrix} 1 & -3 & 9 \\ 1 & -2 & 4 \\ 1 & -1 & 1 \\ 1 & 0 & 0 \\ 1 & 1 & 1 \\ 1 & 2 & 4 \\ 1 & 3 & 9 \end{bmatrix}, \qquad y = \begin{bmatrix} 0 \\ 1/2 \\ \sqrt{3}/2 \\ 1 \\ \sqrt{3}/2 \\ 1/2 \\ 0 \end{bmatrix}.$$

For uniform weights,

$$G = A^H W A = \begin{bmatrix} 7 & 0 & 28 \\ 0 & 28 & 0 \\ 28 & 0 & 196 \end{bmatrix}, \quad z = A^H W y = \begin{bmatrix} 3.7321 \\ 0 \\ 5.7321 \end{bmatrix}, \quad c = \begin{bmatrix} .9711 \\ 0 \\ -0.1095 \end{bmatrix},$$

whereas for the nonuniform weights,

$$G = A^H W A = \begin{bmatrix} 403 & 0 & 2404 \\ 0 & 2404 & 0 \\ 2404 & 0 & 18412 \end{bmatrix},$$

$$z = A^H W y = \begin{bmatrix} 126.29 \\ 0 \\ 326.89 \end{bmatrix}, \quad c = \begin{bmatrix} 0.9449 \\ 0 \\ -0.1046 \end{bmatrix}.$$

The errors are (since f and p are even, only the errors for $x_i \geq 0$ are listed)

x	Uniform $f(x) - p(x)$	Nonuniform $f(x) - p(x)$
0	0.0289	0.0551
1	-0.0044	0.0267
2	-0.0332	-0.0224
3	0.0142	0.0057

As expected, the larger weights at the ends force a better approximation there, but yield a bigger error in the center.

EXAMPLE 2. DISCRETE VERSUS CONTINUOUS INNER PRODUCTS. Let $f(x) = e^x$, $0 \leq x \leq 1$, and consider the real inner products

$$\langle f, g \rangle = \int_0^1 f(t)g(t)\, dt, \qquad \langle f, g \rangle_d = \sum_{i=0}^m f(x_i)g(x_i), \qquad x_i = \frac{i}{m}.$$

Compare the approximation of f by polynomials of degree 3 using each of these inner products with $m = 10$.

These inner products are defined on different vector spaces, but e^x belongs to both of them. In particular, these vector spaces are $C([-1, 1])$ and $\mathcal{F}(\{i/m : i = 0, 1, \ldots, m\})$. The point of this example is to illustrate that discrete inner products can produce good results in the interval containing the x_i. In fact, since $1/m\langle f, g \rangle_d \to \langle f, g \rangle$ as $m \to \infty$, as one can see from considering appropriate Riemann sums, and since multiplication of an inner product by a positive scalar does not change the projection operator, one expects that the projections in each inner product will be approximately the same polynomial. Thus if integrals are hard to evaluate, a discrete inner product can be a convenient artifice by which to avoid their calculation. A discrete inner product that produces a better approximation to $\langle f, g \rangle$ is $\langle f, g \rangle' = \frac{1}{2} f(x_0)g(x_0) + \sum_{i=1}^{m} f(x_i)g(x_i) + \frac{1}{2} f(x_m)g(x_m)$, which is based on the trapezoidal approximation to the integral. An even better discrete inner product would be based on Simpson's approximation to the integral.

First consider the integral inner product $\langle \, , \, \rangle$. The Gram matrix H has entries

$$h_{ij} = \langle t_j, t_i \rangle = \int_0^1 t^i t^j \, dt = \frac{1}{i + j + 1}$$

and is called a *Hilbert matrix*. $n \times n$ Hilbert matrices are notoriously difficult to invert accurately, for large n. For $n = 3$ (cubic approximation),

$$H = \begin{bmatrix} 1 & 1/2 & 1/3 & 1/4 \\ 1/2 & 1/3 & 1/4 & 1/5 \\ 1/3 & 1/4 & 1/5 & 1/6 \\ 1/4 & 1/5 & 1/6 & 1/7 \end{bmatrix}.$$

Since

$$H^{-1} = \begin{bmatrix} 16 & -120 & 240 & -140 \\ -120 & 1200 & -2700 & 1680 \\ 240 & -2700 & 6480 & -4200 \\ -140 & 1680 & -4200 & 2800 \end{bmatrix}$$

has large entries with alternating signs, a small error in z can produce a large error in the coefficients $c = H^{-1}z$, which means that the coefficients are sensitive to small errors in z. The system $Hc = z$ is said to be *ill conditioned*. The situation only gets worse as n gets larger. The entries of z are $z_i = \int_0^1 x^i e^x \, dx$, so that $z = (e - 1, 1, e - 2, 6 - 2e)^T$. Thus $c = H^{-1}z = (0.9991, 1.0183, 0.4212, 0.2786)^T$, and the polynomial of best fit is $p(x) = 0.9991 + 1.0183x + 0.4212x^2 + 0.2786x^3$.

For the discrete inner product, $c_d = (0.9995, 1.0162, 0.4229, 0.2792)^T$, and the polynomial of best fit is $q(x) = 0.9995 + 1.0162x + 0.4229x^2 + 0.2792x^3$; this shows that the polynomials of best fit are very close, judging by their coefficients. On the interval $[0, 1]$, the two polynomials agree to within 5.4×10^{-4}, and $p(x)$ agrees with e^x to within 1.06×10^{-3}.

EXAMPLE 3. DATA APPROXIMATION BY POLYNOMIALS. The total personal income in the United States from 1960 to 1990 in trillions of dollars is reported to be

Year	1960	1965	1970	1975	1980	1985	1990
$trn	0.402	0.407	.811	1.265	2.165	3.325	4.680

Fit the data by polynomials, and attempt to estimate the total personal income in 1995 and in 2000.

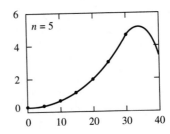

FIGURE 3.3

Let y_i be the personal income in trillions of dollars, and let x_i be the time elapsed since 1960 at the ith data point. Try successive polynomial fits of degrees $1, 2, \ldots, 6$, using uniform weights. Let p_n be the best approximation of degree n. For example, for $n = 1$,

$$G = A^T A = \begin{bmatrix} 7 & 105 \\ 105 & 2275 \end{bmatrix}, \qquad z = \begin{bmatrix} 13.055 \\ 295.945 \end{bmatrix}, \qquad c = \begin{bmatrix} -0.2804 \\ 0.1430 \end{bmatrix},$$

so that $p_1(x) = -0.2804 + 0.1430x$ (c_i is rounded to four sigificant digits). This predicts $p_1(35) = 4.726$, $p_1(40) = 5.441$ trillions of dollars in income. The following table gives the relative errors, and the predicted incomes, (to three significant digits):

n	$\|f - p_n\| / \|f\|$	$x = 35$	$x = 40$
1	0.197	4.73	5.44
2	0.0156	6.36	8.30
3	0.0155	6.36	8.31
4	0.0150	6.23	7.84
5	0.0105	5.25	3.19
6	0	9.02	26.7

Note that the relative errors have a big drop between $n = 1$ and 2, but vary little after that, until interpolation occurs at $n = 6$. This indicates that $n = 2$ is probably the best fit. The graphs for $n = 2$ and 5 are shown in Fig. 3.3. The graphs for $n = 3, 4$ are very similar to the $n = 2$ graph.

Exercises 3.4C

1. Let $f(x) = \frac{1}{2}(x + |x|) = \begin{cases} x & x \geq 0 \\ 0 & x < 0 \end{cases}$. For each of P_3 and P_4, find the projection p on f using the following:
   (a) Legendre polynomials and $\langle f, g \rangle = \int_{-1}^{1} f(x)\bar{g}(x)\,dx$
   (b) Chebyshev polynomials and $\langle f, g \rangle = \int_{-1}^{1} f(x)\bar{g}(x)(1 - x^2)^{-1/2}\,dx$
   (c) Normal equations and $\langle f, g \rangle = \sum_{k=-3}^{3} f(k/3)\bar{g}(k/3)\,dx$
   (d) Graph each of these projections and graph f on $[-1, 1]$. Graph the residual $f - p$ in each case. What is the effect of the weight functions? How does the approximation produced by the discrete inner product compare to the approximation produced by the continous inner product?

2. **Polynomial fitting with nonuniform weights.**

In this numerical example, the behavior of the approximation in a controlled setting is shown. The problem is constructed as follows: Begin with the polynomial $g(x) = 1 + x + \frac{1}{2}x^2$, take the x_i to be the seven points $\pm 3, \pm 2, \pm 1, 0$, and introduce some errors in the $g(x_i)$ values to get the y_i. Specifically,

x_k	-3	-2	-1	0	1	2	3	
y_k	2.51	0.98	0.45	1	3	5.02	8.9	
$y_k - g(x_k)$	0.01	-0.02	-0.05	0	0.5	0.02	0.4	(Errors)

One goal is to see how well the coefficients are recovered. Another is to examine the closeness of fit.

In each of the following cases, find the best approximation $p = c_0 + c_1 x + c_2 x^2$, using the relevant inner product. Compute the errors $f(x_i) - p(x_i)$ at each point and the relative error $\|f - p\| / \|p\|$. Compare the results. How well are the coefficients determined?

(a) Take all the weights to be 1.

(b) Suppose the errors at $x = 1, 3$ were suspected to be on the order of 0.5, with the other errors on the order of 0.02. Introduce a weighted inner product using the rule of thumb suggested.

3. Approximate $\sin(\pi x)$ on $[0, 1]$ by polynomials in $\mathbf{P}_n$ using the inner products

$$\langle f, g \rangle = \int_0^1 f(t)g(t)\,dt, \quad \langle f, g \rangle_d = \sum_{i=0}^{m} f(x_i)g(x_i), \quad x_i = \frac{i}{m}.$$

Try several n and m. Compare the coefficients, graphs, and graphs of the residuals $\sin(\pi x) - p(x)$, where p is the best approximation.

4. Repeat Exercise 3 for $\sqrt{x}$.

5. Approximate $f(x) = (1 + e^{-16x})^{-1}$ on $[-1, 1]$ by polynomials of degree not exceeding n, using the inner product $\langle f, g \rangle_d = \sum_{i=-m}^{m} f(x_i)g(x_i)$, where $x_i = i/m$, with $m = 10$. Compute the relative errors and plot the results for $n = 1, \ldots, 20$.

6. The total labor force in the United States (in millions) between 1930 and 1990 is given in the following table:

1930	1935	1940	1945	1950	1955	1960	1965	1970	1975	1980	1985	1990
29.4	27.1	32.4	40.4	45.2	50.7	54.2	60.8	70.9	76.9	90.6	94.5	103.9

Using uniform weights, approximate the data by polynomials of degree $n = 1, 2, \ldots$. Compute the relative errors and predict the labor force in 1995 and 2000. Plot the graphs to see how well the approximants fit the data.

7. The stock market highs are given for even years as follows:

1970	1972	1974	1976	1978	1980	1982	1984	1986	1988	1990
842	1036	892	1015	908	1000	1071	1287	1956	2184	3000

Using uniform weights, approximate the data by polynomials of degree $n = 1, 2, \ldots$. Compute the relative errors and predict the stock market highs in 1995 and 2000. Plot the graphs to see how well the approximants fit the data.

8. In a polynomial curve-fitting problem involving $\mathbf{P}_n$ and a discrete inner product $\langle h, g \rangle = \sum_{i=0}^{m} h(x_i)\overline{g}(x_i)w_i$, show that the Gram matrix $[\langle x^j, x^i \rangle]$ is singular iff there is a polynomial $q(x)$ of degree n or less having the nodes x_i among its zeros.

9. Let V be the Vandermond matrix for a set of points $\{x_k\}_{k=1}^{m}$. Describe the null space of V in terms of polynomials.

10. A point moves on a line according to $x = A\sin(\omega t + \phi)$. The angular frequency is known to be $\omega = 2$. The goal is to find the amplitude A and phase ϕ approximately. To this end, x is measured at five different times. The data are as follows:

t_i	0	0.2	1.5	2.0	3.5
x_i	-1.6	0.25	2.4	-2.4	1.8

(Hint: $\sin(\omega t + \phi) = \sin(\omega t)\cos(\phi) + \cos(\omega t)\sin(\phi)$. Reduce the problem to approximating x by $a \cdot \cos(\omega t) + b \cdot \sin(\omega t)$ with a discrete inner product. Find a, b, and from these recover A, ϕ.)

11. **Discrete approximation using arbitrary functions.** Let $\{f_k\}_{k=1}^{n}$ be a given set of functions, and let $S = \{x_i\}_{i=1}^{m}$ be a given set of distinct points. Define a polynomial in the f_k to be a linear combination $p = \sum_{k=1}^{n} c_k f_k$. The goal is to develop a system of equations that describes the solution of the problem

$$\text{Minimize } \sum_{i=1}^{m} |h(x_i) - p(x_i)|^2, \quad p \in \text{span}\{f_k\}_{k=1}^{n}.$$

The relevant inner product is $\langle f, g \rangle = \sum_{i=1}^{m} f(x_i)\overline{g}(x_i)$.
(a) Find a system of equations to solve to obtain the c_k, which in turn yields the minimizing p. Write this in terms of the matrix

$$F = \begin{bmatrix} f_1(x_1) & f_2(x_1) & \cdots & f_n(x_1) \\ f_1(x_2) & f_2(x_2) & \cdots & f_n(x_2) \\ \vdots & \vdots & \ddots & \vdots \\ f_1(x_m) & f_2(x_m) & \cdots & f_n(x_m) \end{bmatrix}.$$

Develop notation similar to that used in (7), (8), (9).
(b) Show that the Gram matrix for this problem is singular iff there is a $p = \sum_{k=1}^{n} c_k f_k$ with not all $c_k = 0$ such that $p(x_i) = 0$ for $i = 1, \ldots, m$.
(c) Extend the results in part b to handle this weighted minimization problem:

$$\text{Minimize } \sum_{i=1}^{m} |h(x_i) - p(x_i)|^2 w_i, \quad p \in \text{span}\{f_k\}_{k=1}^{n}.$$

Note: $\{f_k\}_{k=1}^{n}$ is said to be a Chebyshev system or *unisolvent* on I iff for every set $\{x_k\}_{k=1}^{n}$ of distinct points in I, F is nonsingular. Such systems play an important role in approximation theory. $\{1, \sin(x), \cos(x), \ldots, \sin(nx), \cos(nx)\}$ is a Chebyshev system on $[a, a + 2\pi)$, and $\{1, x, \ldots, x^n\}$ is a Chebyshev system on any interval.

12. A mix of n radioactive materials has mass given by $m = \sum_{i=1}^{n} m_i \exp(-\lambda_i t)$, where m_i is the original mass of material i. We have $\lambda_i = \log(2)/h_i$, where h_i is the half life of material i. The radiation emitted is measured and is proportional to the mass of the material present. We want to find the relative amounts of the materials in the mix. Readings are taken at fixed times (in minutes) and divided by the reading at $t = 0$, so that $y = \sum_{i=1}^{n} c_i \exp(-\lambda_i t)$ is the normalized reading at time t, and c_i is the fraction of material i in the mix. In this example, $n = 4$, and the half lives are known to be 7, 12, 30, and 48 minutes. From the following data, estimate the c_i using uniform weights.

t	0	10	20	30	40	50	60	70	80	90	100
y	1	0.67	0.50	0.38	0.31	0.25	0.21	0.18	0.15	0.13	0.11

Also, plot the graph of y versus t, using the estimated c_i, and compute the relative error in the approximation.

13. A helicopter cross section has been digitized to yield pairs (x_k, y_k). It is symmetric about the x axis (which is the vertical axis of the chopper). The goal is to fit the data with periodic functions of the polar coordinate angle θ in order to describe the curve between the data points. Thus the x data are approximated by $x(\theta) = \sum_{k=1}^{n} a_k \cos(k\theta)$, and the y data, by $y(\theta) = \sum_{k=1}^{n} b_k \sin(k\theta)$. The data are as follows:

x	4.07	4.06	4.06	4.03	3.84	3.50	2.82	2.05	1.12	0.17	−0.41
y	0.00	0.38	0.81	1.20	1.51	1.65	1.87	2.08	2.35	2.61	2.77

x	−1.05	−1.65	−2.27	−3.09	−3.83	−4.44	−4.56	−4.66	−4.68	−4.67	−4.65
y	2.81	2.80	2.79	2.78	2.71	2.44	1.97	1.33	0.80	0.29	0.00

Try to fit this data for various n, using uniform weights. Find the relative errors. Be sure to plot $(x(\theta), y(\theta))$ and $(x_k y_k)$ to see what happens between the data points and how well the helicopter shape is captured. Investigate what happens when n is near its maximal value.

14. Approximate e^x on $[0, 1]$ using polynomials of degree less than or equal to n and using $\langle f, g \rangle = \int_0^1 f(t)\bar{g}(t)\,dt$ for $n = 1, 2, \ldots$. What computational problems arise? For which n?

15. Suppose a curve in the complex plane is given by $f(t) = u(t) + iv(t)$ and you want to approximate it by a cubic $h(t) = \sum_{k=0}^{3} c_k t^k$ by a discrete unweighted fit at $t_0, t_1, \ldots, t_m$. This can be done by minimizing $\|f - h\|^2$ with complex c_k or by approximating u, v separately: Minimize $\|u - p\|^2$, where $u \in \mathbf{P}_3$, and $\|v - q\|^2$, where $q \in \mathbf{P}_3$, and then use $p + iq$ to approximate f. Which method is better? Which is easier?

3.5
ORTHOGONAL COMPLEMENTS

As we have seen, the orthogonal complement of a subspace M is a subspace that is closely related to the projection problem. Theorem 6 of Section 3.3 gives the precise relationship. The study of the properties of this subspace will provide new tools that can be applied to projections. As an application of these ideas, in this section, inner products are connected with systems of equations, and the orthogonal complement is used to yield further information about systems of equations and the properties of matrices. First, some definitions are restated.

DEFINITION. For an arbitrary set M, $M^\perp$ is the set of all $x \perp M$ and is called the *orthogonal complement of M*, pronounced "M perp."

$M^\perp$ is a subspace. In fact, if x, $y \in M^\perp$, then for every $m \in M$, $\langle \alpha x + \beta y, m \rangle = \alpha \langle x, m \rangle + \beta \langle y, m \rangle = \alpha 0 + \beta 0 = 0$, so $\alpha x + \beta y \in M^\perp$. For example, if M is a plane in $\mathbf{R}^3$, then $M^\perp$ is a line through the origin perpendicular to this plane.

DEFINITION. Let M_1 and M_2 be subspaces of V. If each vector f in V can be written uniquely as $f = x + y$ with $x \in M_1$ and $y \in M_2$, V is called a *direct sum* of M_1 and M_2, and this fact is denoted by $V = M_1 \oplus M_2$.

THEOREM 1. Let M be a finite-dimensional subspace of V. Then

(a) $V = M \oplus M^\perp$
(b) $(M^\perp)^\perp = M$

If, in addition, V is finite-dimensional, then

(c) $\dim(V) = \dim(M) + \dim(M^\perp)$

Proof

(a) By Theorem 3 of Section 3.3 and the definitions of projection and $M^\perp$, each $f \in V$ has the unique decomposition $f = s + r$, with $s \in M$ and $r \in M^\perp$.
(b) Note that $m \in M$ implies $m \perp M^\perp$ by the definition of $M^\perp$, so $M \subset M^{\perp\perp}$. Let $f \in M^{\perp\perp}$, and write $f = s + r$, with $s \in M$ and $r \in M^\perp$. Since $f \perp M^\perp$, $0 = \langle f, r \rangle = \langle s, r \rangle + \langle r, r \rangle = \|r\|^2$, so $f = s$. Thus $M^{\perp\perp} \subset M$.
(c) Let $\{e_1, \ldots, e_p\}$ be an orthonormal basis for M, and let $\{e_{p+1}, \ldots, e_n\}$ be an orthonormal basis for $M^\perp$. By showing that $\beta = \{e_1, e_2, \ldots, e_n\}$ is a basis for V, (c) is proved. It is sufficient to show β spans V, since β is orthogonal and therefore linearly independent. Given f, find s and r as in (a), and then expand them: $s = \sum_{i=1}^{p} d_i e_i$, and $r = \sum_{i=p+1}^{n} d_i e_i$. But then $f = \sum_{i=1}^{n} d_i e_i$, which proves (c).

Remark. If $\dim(M) = \infty$, then $M^{\perp\perp} \neq M$ is possible. For example, let $V = C([0, 1])$, M be the set of all polynomials, and let $\langle f, g \rangle = \int_0^1 f(t)\overline{g}(t)\, dt$. It turns out that $M^\perp = \{0\}$. The reason for this is that every continuous function may be uniformly approximated arbitrarily well by polynomials. Thus $M^{\perp\perp} = V \neq M$.

EXAMPLE 1. In $\mathbf{R}^4$, let M be the set of x satisfying $x_1 + x_2 + x_3 = 0$ and $x_1 - x_3 + 2x_4 = 0$. Find a basis for $M^\perp$ and find $\dim(M)$.
   $M = \{x : x \perp E_1, E_2\}$, where $E_1 = (1, 1, 1, 0)^{\mathsf{T}}$ and $E_2 = (1, 0, -1, 2)^{\mathsf{T}}$. Thus $M = \mathrm{span}\{E_1, E_2\}^\perp$, so $M^\perp = \mathrm{span}\{E_1, E_2\}^{\perp\perp} = \mathrm{span}\{E_1, E_2\}$. Since $\{E_1, E_2\}$ is linearly independent, it is a basis for $M^\perp$ and $\dim(M^\perp) = 2$. By Theorem 1, $\dim(M) = 4 - 2 = 2$.
   Note that the problem can be restated as follows: Find a basis for $N(A)^\perp$, where

$$A = \begin{bmatrix} 1 & 1 & 1 & 0 \\ 1 & 0 & -1 & 2 \end{bmatrix}.$$

EXAMPLE 2. Let $M = \text{span}\{A_1, A_2\}$ with $A_1 = (1, 0, 1, -1)^T$, $A_2 = (-1, 0, 2, 1)^T$. Find a basis for $M^{\perp}$.

Consider the following equivalent statements: $x \in M^{\perp} \Leftrightarrow x \perp A_1, A_2 \Leftrightarrow \langle x, A_i \rangle = A_i^H x = 0$ (for $i = 1, 2$). Thus $M^{\perp}$ is the set of solutions of

$$[A_1, A_2]^H x = \begin{bmatrix} 1 & 0 & 1 & -1 \\ -1 & 0 & 2 & 1 \end{bmatrix} x = 0.$$

Since $\dim(M) = 2$, $\dim(M^{\perp}) = 4 - 2 = 2$. Using the usual method, a basis for the solutions of this system is $\{B_1 = (1, 0, 0, 1)^T, B_2 = (0, 0, 1, 0)^T\}$. Alternatively, following the proof of Theorem 1, an orthogonal basis $\{E_k\}_{k=1}^4$ for $\mathbf{R}^4$ could have been constructed such that $\text{span}\{E_1, E_2\} = \text{span}\{A_1, A_2\}$, in which case $\{E_3, E_4\}$ would be a basis for $M^{\perp}$.

Note that this problem may be restated as follows: Find a basis for $R(A)^{\perp}$, where $A = [A_1\ A_2]$.

EXAMPLE 3. PROJECTION ON A PLANE IN $\mathbf{R}^3$. Let $V = \mathbf{R}^3$, use the standard inner product, and let $M = \{x : x_1 + x_2 + x_3 = 0\}$. Find the projection of $g = (-1, 1, 1)^T$ on M.

Note that M is two-dimensional since it is the orthogonal complement of the one-dimensional space spanned by $E_3 = (1, 1, 1)^T$. The projection can be calculated using a basis for M, but the ideas of Theorem 1 can also be used. The projection of f on the span of E_3 is $r = (E_3^T g / \|E_3\|^2) E_3 = \frac{1}{3}(1, 1, 1)^T$. Hence $s = g - r = \frac{1}{3}(-4, 2, 2)^T$ is the projection of g on M, since $f - s = r \perp M$.

COROLLARY 2. Let M be a finite-dimensional subspace of V, and let an inner product $\langle , \rangle$ be given.
   (a) If the projection of f on M is p, then the projection of f on $M^{\perp}$ is $f - p$.
   (b) If the projection of f on $M^{\perp}$ is q, then the projection of f on M is $f - q$.
   The proof is left as an exercise. (See Exercise 11.)

Next, we make a connection between inner products and the product Ax. This allows theoretical considerations based on the theory of inner product spaces and orthogonal complements to be applied to matrix theory, especially to the range and null space of a matrix. The inner product will always be the standard one. To illustrate the idea, consider a real system of equations $\sum_j a_{ij} x_j = 0$. Let A have rows a_i^T. Then $Ax = 0$ is the same as the system of equations $a_i^T x = \langle x, a_i \rangle = 0$, $i = 1, 2, \ldots, n$. Thus we have the following chain of equivalent statements: x is in the null space of $A \Leftrightarrow x$ is orthogonal to the rows (transposed) of $A \Leftrightarrow x$ is orthogonal to the columns of $A^T \Leftrightarrow x$ is orthogonal to the span of the columns of A^T. But the range of A^T is the span of its columns. All this information is contained in the terse statement $N(A) = [R(A^T)]^{\perp}$. The next theorem could be proved this way, but a more sophisticated route is taken because it provides another worthwhile technique that generalizes to operators and arbitrary inner products.

First, note that the following identity holds, in which A is an $m \times n$ matrix:

$$\langle Ax, y \rangle = y^H A x = (A^H y)^H x = \langle x, A^H y \rangle \quad \forall x \in \mathbf{F}^n, y \in \mathbf{F}^m. \tag{1}$$

THEOREM 3. Let A be an $m \times n$ matrix. $\langle , \rangle$ is the standard inner product on $\mathbf{F}^n$. Then

$$N(A) = R(A^H)^{\perp} \tag{2}$$
$$R(A) = N(A^H)^{\perp} \tag{3}$$
$$r(A^H) = r(A) \tag{4}$$

Proof. From (1) and various definitions, the following sequence of equivalent statements result:

$$x \in N(A) \Leftrightarrow Ax = 0 \Leftrightarrow \langle Ax, y \rangle = \langle x, A^H y \rangle = 0 \ \forall y$$
$$\Leftrightarrow \langle x, z \rangle = 0 \ \forall z \in R(A^H) \Leftrightarrow x \in R(A^H)^\perp.$$

Thus (2) is established. To prove (3), use (1), (2), and Theorem 1:

$$N(A^H)^\perp = \{R(A^{HH})^\perp\}^\perp = R(A).$$

To prove (4),

$$\begin{aligned}
n - r(A^H) &= n - \dim[R(A^H)] && \text{(definition of rank)} \\
&= \dim[R(A^H)^\perp] && \text{(Theorem 1)} \\
&= \dim[N(A)] && \text{(Equation (2))} \\
&= n - \dim[R(A)] && \text{(rank and nullity theorem)} \\
&= n - r(A).
\end{aligned}$$

EXAMPLE 4. Illustrate the first part of Theorem 3 in the real case. Let

$$A = \begin{bmatrix} a_1^T \\ a_2^T \end{bmatrix} = \begin{bmatrix} 1 & 0 & 1 \\ 1 & 1 & -1 \end{bmatrix}.$$

Then $Ax = 0 \Leftrightarrow x_1 + x_3 = a_1^T x = \langle a_1, x \rangle = 0$, and $x_1 + x_1 - x_3 = a_2^T x = \langle a_2, x \rangle = 0$. That is, $x \perp a_1, a_2$. Since $A^T = [a_1, a_2]$ and x is perpendicular to every linear combination of a_1, a_2, therefore x is perpendicular to every $y \in R(A^T)$. On the other hand, if $x \perp y \in R(A^T)$ for every such y, then $x \perp a_1, a_2$, so $x \in N(A)$.

The next corollary results from Theorem 3 and some standard manipulations. As a challenge, try to unravel the geometry so tersely stated in these equations.

COROLLARY 4. For an $m \times n$ matrix,

(a) $N(A^H) = R(A)^\perp$
(b) $R(A^H) = N(A)^\perp$

Proof. Use the rule $B^{HH} = B$ and Theorem 3.

The next theorem has essentially been done in Section 2.2. It is reproved here as an application of the concept of orthogonality.

DEFINITION. The *row rank* of a matrix A is the dimension of the vector space spanned by the rows of A.

Intuitively, the row rank is the number of linearly independent rows of A.

THEOREM 5. Let A be an $m \times n$ matrix.

(a) $A, \overline{A}, A^T$, and A^H have the same rank.
(b) $r(A) = $ row rank of A.

Proof. Proof of (a): From Theorem 3, $r(A) = r(A^H)$. If A has rows a_i^T, then A^T has columns a_i, and clearly the row rank of A and the rank of A^T are the same. A^H has columns $\overline{a_i}$, and the number of linearly independent columns among $\{\overline{a_k}\}_{k=1}^m$ is the same as the number of independent columns among $\{a_k\}_{k=1}^m$. Thus A^T and A^H have the same rank. The proof of $r(A) = r(\overline{A})$ is similar.

Statement (b) is a consequence of $r(A) = r(A^T)$.

For systems of equations $Ax = y$ that may be inconsistent, the next theorem provides a criterion by which one can test consistency. The Fredholm alternative theorem extends to certain infinite dimensional inner product spaces called Hilbert spaces, which are extremely important in the theory of integral and differential equations.

THE FREDHOLM ALTERNATIVE THEOREM. Let A be an $m \times n$ matrix, and let $y \in \mathbf{F}^m$. Then exactly one of the following is true:

(a) $Ax = y$ is consistent.
(b) There exists z such that $A^H z = 0$ and $z^H y \neq 0$.

Proof. Suppose $Ax = y$ is consistent; then $y \in R(A)$ and, by (3), $y \in N(A^H)^\perp$, i.e., $y \perp z$ for all z such that $A^H z = 0$. This means (b) does not hold. Now suppose (b) holds. This means that there is a $z \in N(A^H)$ so that z is not orthogonal to y; i.e., $z^H y \neq 0$. And this means that $y \notin N(A^H)^\perp$, so by (3), $y \notin R(A)$; therefore $Ax = y$ is inconsistent.

EXAMPLE 5. Suppose A is an $n \times n$ real symmetric matrix of rank $n - 1$. Suppose also that $Au = 0$, where $u = (1, 1, \ldots, 1)^T$. Show that $Ax = y$ has a solution iff $\sum_{i=1}^n y_i = 0$.

By the last theorem, $y \in R(A)$ iff (b) is false; i.e., if $z \in N(A^T) = N(A)$, then $z^T y = 0$. Since $N(A)$ is one-dimensional, $z = \alpha u$ for some α. Thus $y \in R(A)$ iff $u^T y = \sum_{i=1}^n y_i = 0$.

A second important theorem comes from another simple identity:

$$\|Ax\|^2 = (Ax)^H Ax = x^H A^H Ax. \tag{5}$$

Some of the results of the next theorem are not intuitively obvious. For example, it is not immediately clear that A and $A^H A$ should have the same null space. Part (c) says that the number of linearly independent columns of A is the same as the rank of the Gram matrix of these columns, which is plausible and pleasing, but not obvious.

THEOREM 6. Let A be an $m \times n$ matrix. Then

(a) $N(A) = N(A^H A)$, and $N(A^H) = N(AA^H)$
(b) $R(A) = R(AA^H)$, and $R(A^H) = R(A^H A)$
(c) $r(A) = r(AA^H) = r(A^H A)$

Proof. Proof of (a): If $x \in N(A)$, then $A^H Ax = A^H 0 = 0$, so $x \in N(A^H A)$. If $x \in N(A^H A)$, then $0 = A^H Ax$, so $\|Ax\|^2 = 0$ by (4), and thus $Ax = 0$. For the second statement, replace A by A^H. To prove (b), use (a) and Corollary 4: $R(A) = [N(A^H)]^\perp = [N(AA^H)]^\perp = R([AA^H]^H) = R(AA^H)$. The proof of (c) follows from (b) and the definition of rank: $r(A) = \dim(R(A)) = \dim(R(AA^H)) = r(AA^H) = r(A^H) = r(A^H A)$.

Exercises 3.5

1. In each of the following cases, write $x = s + r$, where s is in the given subspace M and r belongs to the orthogonal complement of M.
   (a) $x = (4, -1, -3, 4)^T$, $M = \text{span}\{(1, 1, 1, 1)^T, (1, 2, 2, -1)^T, (1, 0, 0, 3)^T\}$
   (b) $x = (2, 1, -7, 6)^T$, $M = \text{span}\{(1, 1, 0, 1)^T, (3, -1, 2, 1)^T, (1, 2, -1, 3)^T\}$

2. Let $M = \text{span}\{(1, 1, 1, 1)^T, (1, 2, 2, -1)^T\}$.
   (a) Find an orthogonal basis for $M^\perp$.
   (b) Find the projection of δ_1 on M and on $M^\perp$.

3. Let $M = \text{span}(t, t^2)$ in $\mathbf{P}_3$, and let $\langle f, g \rangle = \int_{-1}^{1} f(t)\overline{g}(t)\,dt$.
   (a) Find an orthogonal basis for $M^\perp$.
   (b) Find the projection of 1 on M and on $M^\perp$.

4. Let $M = \{p : \int_{-1}^{1} p(t)\,dt = 0,\ p \in \mathbf{P}_3\}$ and let $\langle f, g \rangle = \int_{-1}^{1} f(t)\overline{g}(t)\,dt$.
   (a) Find an orthogonal basis for $M^\perp$.
   (b) Find the projection of $f(t) = (1 + t)^2$ on M and on $M^\perp$.

5. Let $A = \begin{bmatrix} 1 & 2 & 2 & 4 \\ 2 & 4 & -1 & -2 \\ 2 & -1 & -4 & 2 \end{bmatrix}$.

   (a) Find the rank of A and the dimension of $N(A^H)$.
   (b) Is $\{(1, 0, 0)^T, (1, 1, 0)^T, (1, 1, 1)^T\}$ a basis for $N(A^H)^\perp$? Explain.

6. Let $A = \begin{bmatrix} 1 & 1 & 1 & 1 \\ 1 & 0 & -2 & -1 \\ 1 & 2 & 4 & 3 \end{bmatrix}$.

   (a) Find a basis for $N(A^H)$.
   (b) Find a basis for $N(A^H)^\perp$.
   (c) Find vectors b_i so that $y \perp b_i = 0$ implies $Ax = y$ is solvable.
   (d) Find a basis for the span of the columns of A.

7. Let $A = \begin{bmatrix} 1 & -1 & 1 \\ 1 & 1 & 1 \\ 1 & 0 & -2 \\ 1 & 0 & 0 \end{bmatrix}$.

   (a) Show that the columns of A are orthogonal.
   (b) Find a basis for $N(A^H)$.
   (c) Find a basis for $N(A^H)^\perp$.
   (d) Find vectors b_i so that $y \perp b_i = 0$ implies $Ax = y$ is consistent.
   (e) Find a basis for the span of the columns of A.

8. Let $A = \begin{bmatrix} 1 & 2 & 3 & 0 \\ 2 & -1 & -4 & 5 \\ 2 & 1 & 0 & 3 \\ 1 & 2 & 3 & 0 \end{bmatrix}$.

   (a) Find a basis for $N(A^H)$.
   (b) Find a basis for $N(A^H)^\perp$.
   (c) Find vectors b_i so that $y \perp b_i = 0$ implies $Ax = y$ is solvable.

9. Let $A = \begin{bmatrix} 1 & -2 & 3 \\ 2 & 3 & -1 \\ 4 & -1 & 5 \end{bmatrix}$.

(a) Find the rank of A, and infer the dimension of $N(A^H)$ and of $N(A^H)^{\perp}$.

(b) Find a column b such that $Ax = y$ is consistent iff $\langle y, b \rangle = 0$.

(c) Use the test in part (b) to decide whether $Ax = (0, 1, 1)^T$ is solvable, and check your answer by attempting to solve the system directly.

10. The following matrices are given:

$$(i) \begin{bmatrix} 1 & 1 & -1 \\ 1 & 0 & 0 \\ -1 & 0 & 0 \end{bmatrix} \quad (ii) \begin{bmatrix} 1 & -1 & -1 \\ -1 & 1 & 1 \\ -1 & 1 & 1 \end{bmatrix} \quad (iii) \begin{bmatrix} 2 & -1 & 0 \\ -1 & 1 & 1 \\ 0 & 1 & 2 \end{bmatrix}$$

For each matrix,

(a) Find vectors b_i so that $y \perp b_i$ iff $Ax = y$ is consistent.

(b) Find orthogonal bases for $R(A)$ and $N(A)$.

11. Prove Corollary 2: Let M be a subspace of the finite-dimensional space V, and let an inner product $\langle , \rangle$ be given. Show that

(a) If the projection of f on M is p, then the projection of f on $M^{\perp}$ is $f - p$.

(b) If the projection of f on $M^{\perp}$ is q, then the projection of f on M is $f - q$.

12. Let M be a finite-dimensional subspace of V, and let $\prod$ be the projection operator onto M. Show that

(a) $\prod(f) = f \Leftrightarrow f \in M$

(b) $R(\prod) = M$

(c) $N(\prod) = M^{\perp}$

(d) $I - \prod$ is the projection operator onto $M^{\perp}$

(e) $[\prod]^H = [\prod]$

13. Let A be a square matrix.

(a) Show that $N(A)$ and $N(A^H)$ have the same dimension. Is this true if A is not square?

(b) Show that A is nonsingular iff A^H is nonsingular.

14. Let $A = [A_1 \ A_2 \ \cdots \ A_n]$ be an $m \times n$ matrix of rank n, and let $M = \text{span}\{A_1, A_2, \ldots, A_n\}$. Find the matrix of the projection operator onto $M^{\perp}$ in the standard basis.

15. Let $A = $ be an $m \times n$ matrix of rank m, and let $M = N(A)$. Find the matrices of the projection operators onto M and onto $M^{\perp}$ in the standard basis.

16. If A is an $m \times n$ matrix with rank r, find $\dim(N(A^T))$.

17. Let A be an $m \times n$ matrix with rank r. Find the dimensions of the following subspaces:

(a) $R(A)$　　　　(b) $N(A)$　　　　(c) $R(A^H)$　　　　(d) $N(A^H)$

(e) $R(A^H A)$　　(f) $N(A^H A)$　　(g) $R(AA^H)$　　(h) $N(AA^H)$

18. Let A be an $m \times n$ matrix with rank r. Find the dimensions of the following subspaces:

(a) $R(A)^{\perp}$　　　(b) $N(A)^{\perp}$　　　(c) $R(A^H)^{\perp}$　　　(d) $N(A^H)^{\perp}$

(e) $R(A^H A)^{\perp}$　(f) $N(A^H A)^{\perp}$　(g) $R(AA^H)^{\perp}$　(h) $N(AA^H)^{\perp}$

19. Let A be an $m \times n$ matrix. Show that

(a) Each $x \in \mathbf{F}^n$ may be written uniquely as $x = u + v$, with $u \in N(A)$ and $v \in R(A^H)$.

(b) Each $y \in \mathbf{F}^m$ may be written uniquely as $y = w + z$, with $w \in N(A^H)$ and $z \in R(A)$.

20. **"Solving" underdetermined systems.** Suppose A is an $m \times n$ matrix with rank $m < n$. Thus for each $y \in \mathbf{F}^m$, $Ax = y$ has infinitely many solutions.
  (a) Show that $A^H(AA^H)^{-1}$ is a right inverse of A.
      Let $x_0 = A^H(AA^H)^{-1}y$. Prove the following statements, which assert that x_0 is the solution of $Ax = y$ with minimal length.
  (b) x_0 is a solution of $Ax = y$.
  (c) $x_0 \perp N(A)$.
  (d) If x_1 is another solution of $Ax = y$ and $x_1 \neq x_0$, then x_0 is the projection of x_1 on $N(A)^{\perp}$, and $\|x_1\| > \|x_0\|$.
      $A^H(AA^H)^{-1}$ is called the *Moore–Penrose inverse* of A. For an arbitrary matrix, the Moore–Penrose inverse, or *pseudo-inverse,* of A is the matrix A^+, which satisfies the following statements:
      (i) A^+A and AA^+ are Hermitian.
      (ii) $AA^+A = A$.
      (iii) $A^+AA^+ = A^+$.
      A^+ is unique; it accomplishes the following remarkable feat: $x = A^+y$ is the minimizer of $\|Ax - y\|$ of minimal length. In Section 8.2, following the discussion of the singular value decomposition, A^+ will be constructed for an arbitrary matrix A.

21. **Constrained minimization problems.**
  (a) Consider this problem:
      Minimize $\|y - Ax\|^2$ subject to $b^H x = 1$.
      Show that the solution x satisfies $A^H Ax = A^H y - \lambda b$ and $b^H x = 1$ for some scalar λ.
      Hints: First reduce the problem to one with the constraint $b^H z = 0$, as follows: Pick an x_0 so that $b^H x_0 = 1$, and set $z = x - x_0$ and $w = y - Ax_0$. Then the problem becomes this: Minimize $\|w - Az\|^2$ subject to $b^H z = 0$. Next let $u \in \text{span}(b)^{\perp}$, and show that a minimizing z satisfies $\langle Au, w - Az \rangle = 0$. Then deduce that $A^H w - A^H Az = \lambda b$. For a second method, apply Lagrange multipliers.
  (b) Generalize (a) to the minimization of $\|Ax - y\|^2$ subject to p linearly independent constraints $b_i{}^H x = c_i$.

3.6
THE GRAM MATRIX AND ORTHOGONAL CHANGE OF BASIS

The Gram matrix arose naturally in the study of projections when a nonorthogonal basis was used, and it played a role in the normal equations. In this section, it reappears in the matrix representation of an inner product. Here we also discuss the following problem: Suppose $\{e_k\}_{k=1}^n$ is an orthonormal basis for V. Under what conditions is a new basis $f_j = \sum_{i=1}^n p_{ij}e_i$ also orthonormal?

Matrix Representation of Inner Products

Let $\beta = \{e_1, e_2, \dots, e_n\}$ be a spanning set for V. The computation of an inner product can be reduced to a matrix multiplication involving coordinate columns as follows: Let $x = \sum_{i=1}^n c_i e_i$ and $y = \sum_{i=1}^n d_i e_i$ so that $[x] = c$ and $[y] = d$. Then

$$\langle x, y \rangle = \left\langle \sum_{j=1}^{n} c_j e_j, \sum_{i=1}^{n} d_i e_i \right\rangle = \sum_{j=1}^{n} c_j \left\langle e_j, \sum_{i=1}^{n} d_i e_i \right\rangle = \sum_{i;j=1}^{n} c_j \overline{d_i} \langle e_j, e_i \rangle$$

$$= \sum_{i=1}^{n} \left(\sum_{j=1}^{n} c_j \langle e_j, e_i \rangle \right) \overline{d_i} = d^H G c,$$

where the matrix G is defined by $g_{ij} = \langle e_j, e_i \rangle$. G is the *Gram matrix* of the set β with respect to the given inner product. Thus once the Gram matrix has been computed, the inner products may be obtained by the matrix multiplication

$$\langle x, y \rangle = [y]^H G [x]. \tag{1}$$

EXAMPLE 1. In $\mathbf{F}^n$, using the standard basis and inner product, $G = I$. In fact, $g_{ij} = \delta_j{}^H \delta_i = \delta_{ij}$.

EXAMPLE 2. THE HILBERT MATRIX. This is the Gram matrix of $1, t, \ldots, t^n$ using $\langle f, g \rangle = \int_0^1 f(t)\overline{g(t)}\, dt$:

$$H = \left[\int_0^1 t^{j+i}\, dt \right] = \begin{bmatrix} 1 & 1/2 & 1/3 & \cdots & 1/n \\ 1/2 & 1/3 & 1/4 & \cdots & 1/(n+1) \\ \cdots & \cdots & \cdots & \cdots & \cdots \\ 1/n & 1/(n+1) & 1/(n+2) & \cdots & 1/(2n+1) \end{bmatrix}.$$

If $p(t) = \sum_{i=0}^{n} c_i t^i$ and $q(t) = \sum_{i=0}^{n} d_i t^i$, then (1) asserts that $\int_0^1 p(t)\overline{q}(t)\, dt = d^H H c$, which is easy to check.

In finite element programs, x and y belong to a certain function space on which the inner product is defined by an integral. To avoid computing the integrals $\langle x, y \rangle$ every time they arise, we can compute the Gram matrix once, accurately, and then use (1) to compute $\langle x, y \rangle$. Following is a simple example of this that will appear in some later applications.

EXAMPLE 3. Let $x_k = k/n$, where $k = 0, 1, \ldots, n$, and let $V = \operatorname{span}\{f_1, \ldots, f_{n-1}\}$ where f_i is the hat function defined in Example 6 of Section 1.3. Thus V is the set of all piecewise linear continuous functions on $[0, 1]$ that vanish at 0 and 1. Let $\langle f, g \rangle = \int_0^1 f(t)\overline{g(t)}\, dt$. Find the Gram matrix G of $\{f_1, \ldots, f_{n-1}\}$. If $f \in V$ and $f(x_k) = y_k \in \mathbf{R}$, compute $\int_0^1 f(t)\, dt$ and $\int_0^1 f^2(t)\, dt$ using G and (1).

Solution. In fact,

$$f_i(t) = \begin{cases} nt - (i-1) & x_{i-1} \le t \le x_i \\ -nt + (i+1) & x_i \le t \le x_{i+1} \\ 0 & \text{otherwise} \end{cases}.$$

Thus

$$g_{ij} = \int_0^1 f_i(t) f_j(t)\, dt = \int_{x_{i-1}}^{x_i} f_i(t) f_j(t)\, dt + \int_{x_i}^{x_{i+1}} f_i(t) f_j(t)\, dt = \frac{1}{6n} \begin{cases} 1 & |i \pm j| = 1 \\ 4 & i = j \\ 0 & \text{otherwise} \end{cases}.$$

In other words, $G = 1/(6n)\,\mathrm{Trid}(1, 4, 1)$. By the interpolation property of the hat functions, $f(x_k) = y_k$ implies $f = \sum_{i=1}^{n-1} y_i f_i$, and similarly, $1 = \sum_{i=1}^{n-1} f_i$. Let $u = (1, \ldots, 1)^T$.

From (1):

$$\int_0^1 f(t)\,dy = \langle y, 1 \rangle = y^T G u = \frac{1}{6n}(5y_1 + 6y_2 + \cdots + 6y_{n-2} + 5y_{n-1}),$$

and

$$\int_0^1 f^2(t)\,dt = y^T G y$$

$$= \frac{1}{6n}(4y_1{}^2 + 2y_1 y_2 + 4y_2{}^2 + 2y_2 y_3 + \cdots + 2y_{n-2}y_{n-1} + 4y_{n-1}{}^2).$$

The next proposition has been proved before in the context of projections. A new argument for the criterion that G is nonsingular is presented.

PROPOSITION 1. G is Hermitian, and G is nonsingular iff $\beta = \{e_1, e_2, \ldots, e_n\}$ is linearly independent.

Proof. G is Hermitian since $\langle e_j, e_i \rangle = \overline{\langle e_i, e_j \rangle}$. From (1), $\|x\|^2 = [x]^H G[x]$. If β is linearly independent, then $G[x] = 0$ implies $\|x\|^2 = 0$, which implies $x = 0$, which, in turn, implies $[x] = 0$. Hence G is nonsingular by Theorem 9 of Section 2.2. If G is nonsingular and $\sum_{j=1}^n c_j e_j = 0$, then $0 = \langle \sum_{j=1}^n c_j e_j, \sum_{i=1}^n d_i e_i \rangle = d^H G c \ \forall d$. But then $Gc = 0$, so $c = 0$. Hence β is linearly independent.

Note that the Hilbert matrix and the Gram matrix of Example 3 are nonsingular since they arise from linearly independent functions.

Orthonormal Change of Basis

Suppose the change of basis is given by

$$f_j = \sum_{i=1}^n p_{ij}\, e_i, \qquad j = 1, 2, \ldots, n$$

or equivalently, the change of coordinates is given by

$$[x]_\beta = P[x]_\gamma,$$

where $P = [p_{ij}]$, and where $\beta = \{e_i\}_{i=1}^n$ and $\gamma = \{f_i\}_{i=1}^n$ are bases for V. Suppose β is orthonormal. Then the entries of P are Fourier coefficients:

$$p_{ij} = \langle f_j, e_i \rangle. \tag{2}$$

Next, the condition under which γ is orthonormal is investigated. Parseval's identity gives

$$\langle f_j, f_k \rangle = \sum_{i=1}^n p_{ij}\overline{p_{ik}} = (P^H P)_{kj}. \tag{3}$$

Thus γ is orthonormal iff $P^H P = I = (\delta_{kj})$. This important condition merits a definition, and the preceding discussion is summarized in Theorem 2.

DEFINITION. U is a *unitary* matrix iff U is square and $U^H U = I$. A real unitary matrix is called an *orthogonal* matrix.

Remark. If U is unitary, then $U U^H = I$ and $U^{-1} = U^H$, by Corollary 5 of Section 0.5. Unitary matrices were encountered before in the QR factorization theorem of Section 3.2.

THEOREM 2. If $\beta = \{e_1, e_2, \ldots, e_n\}$ is an orthonormal basis, then $\gamma = \{f_1, f_2, \ldots, f_n\}$ is an orthonormal basis if and only if the change of basis matrix is unitary.

In the exercises, nonorthonormal change of basis is discussed. (See Exercise 8.)

THEOREM 3. Let U be an $n \times n$ matrix. Then U is unitary if and only if the columns of U are an orthonormal set with respect to the standard inner product in $\mathbf{F}^n$.

Proof. The theorem follows immediately from the identity

$$U^H U = [U_i^H \, U_j] = [\langle U_j, U_i \rangle], \qquad \text{where } U = [U_1 \, U_2 \cdots \, U_n]. \tag{4}$$

EXAMPLE 4. DIRECTION COSINES AND UNITARY MATRICES. In $\mathbf{R}^3$, let $\beta = \{\delta_i\}_1^3$ denote the standard basis, and let $\gamma = \{E_i\}_1^3$ be an orthonormal basis. Our goal is to find the geometric meaning of the entries of the unitary change of basis matrix $P = [E_1 \, E_2 \, E_3]$.

Each vector x can be written

$$x = \sum_{i=1}^{3} x_i \delta_i = \sum_{i=1}^{3} y_i E_i = Py, \qquad \text{where } P = [E_1 \, E_2 \, E_3] = [p_{ij}],$$

and P is the unitary change of basis matrix. The coordinates x_i, y_i are easy to obtain, since both expansions are Fourier expansions:

$$y_i = \langle x, E_i \rangle, \qquad \text{and} \qquad x_i = \langle x, \delta_i \rangle.$$

Recall that $x_i = \langle x, \delta_i \rangle = \|x\| \|\delta_i\| \cos(\theta_i) = \|x\| \cos(\theta_i)$ defines the angle θ_i between x and δ_i. The numbers $\cos(\theta_i) = x_i / \|x\|$ are called the *direction cosines* of x. Now apply the preceding idea to $x = E_j$: $p_{ij} = \langle E_j, \delta_i \rangle = \|E_j\| \|\delta_i\| \cos(\theta_{ij}) = \cos(\theta_{ij})$, where θ_{ij} is the angle between δ_i and E_j. Thus P is simply the matrix of direction cosines of the basis $\{E_1, E_2, E_3\}$. Specifically, the jth column of P consists of the direction cosines of E_j with respect to δ_1, δ_2, and δ_3 (the x, y, and z axes), in that order.

The next theorem sharpens the previously proved theorem that a Gram matrix is singular if and only if the vectors defining the Gram matrix are linearly dependent.

THEOREM 4. Let G be the Gram matrix of $\{v_1, v_2, \ldots, v_n\}$. Then the rank of G is the dimension of $\mathrm{span}\{v_1, v_2, \ldots, v_n\}$.

Proof. Let $V = \mathrm{span}\{v_1, v_2, \ldots, v_n\}$. By the definition of the Gram matrix, $g_{ij} = \langle v_j, v_i \rangle$. Let $e_1, \ldots, e_r$ be an orthonormal basis for V. Then $v_j = \sum_{k=1}^{r} b_{kj} e_k$, where $b_{kj} = \langle v_j, e_k \rangle$. Thus

$$g_{ij} = \langle v_j, v_i \rangle = \left\langle \sum_{k=1}^{r} b_{kj} e_k, \sum_{m=1}^{r} b_{mi} e_m \right\rangle = \sum_{k;m=1}^{r} b_{kj} \bar{b}_{mi} \langle e_k, e_m \rangle$$

$$= \sum_{k;m=1}^{r} b_{kj} \bar{b}_{mi} \delta_{km} = \sum_{k=1}^{r} b_{kj} b_{ki},$$

so that $G = B^H B$. Next we show that B has rank r, from which $r(G) = r$ by Theorem 6 of Section 3.5. By showing that for each $y \in \mathbf{F}^r$ there is an x so that $Bx = y$, $R(B) = \mathbf{F}^r$ follows, and thus $r(G) = r$. Let $y = (y_1, \dots, y_r)^T$ be given. Define v by $v = \sum_{i=1}^r y_i e_i$. Since $\{v_k\}_{k=1}^n$ spans, there is an $x = (x_1, x_2, \dots, x_n)^T$ so that $v = \sum_{i=1}^n x_i v_i$. Computing the Fourier coefficients y_j, $y_j = \langle v, e_j \rangle = \langle \sum_{i=1}^n x_i v_i, e_j \rangle = \sum_{i=1}^n b_{ij} x_j$. Thus $Bx = y$, and the proof is finished.

3.6 Exercises

1. Let $V = \mathbf{P}_2$ and $\beta = \{1, t, t^2\}$. The following inner products are given:
   (i) $\int_{-1}^{1} f(t)\overline{g}(t)\, dt$
   (ii) $\int_{0}^{1} f(t)\overline{g}(t)\, dt$
   (iii) $\sum_{k=-2}^{2} f(k/2)\overline{g}(k/2)$
   (iv) $\sum_{k=0}^{4} f(k/4)\overline{g}(k/4)$
   For each of these inner products,
   (a) Find the Gram matrix.
   (b) Verify (1) for $y(t) = 1 - 2t + t^2$, $x(t) = 1 + 2t + t^2$.

2. Let $p_1(t) = 1$, $p_2(t) = e^t$, and $p_3(t) = e^{-t}$. Use the inner product $\int_{0}^{1} f(t)\overline{g}(t)\, dt$.
   (a) Find the Gram matrix of $\{p_1, p_2, p_3\}$.
   (b) Verify (1) for $x = \sinh(t)$, $y = \cosh(t)$.

3. Let $p_k(t) = e^{i(k-2)t}$, $k = 0, 1, \dots, 4$. Use $\langle f, g \rangle = \int_{0}^{\pi} f(t)\overline{g}(t)\, dt$.
   (a) Find the Gram matrix of $\{p_1, p_2, p_3, p_4\}$.
   (b) Verify (1) for $x(t) = \sin(t)$, $y(t) = \cos(2t)$.

4. Let $A_1 = (1, 1, 1)^T$, $A_1 = (1, 0, 1)^T$, $A_3 = (0, 1, 1)^T$.
   (a) Find the Gram matrix of $\{A_1, A_2, A_3\}$.
   (b) Verify (1) for $x = \sum_{i=1}^{3} A_i$, $y = A_1 - A_2 - A_3$.

5. Let $A_1 = (1, 2, 3)^T$, $A_2 = (1, 0, 1)^T$, $A_3 = (-1, 2, 1)^T$. Find the Gram matrix of $\{A_1, A_2, A_3\}$ in the standard inner product and its rank.

6. Let $\{e_1, e_2, e_3\}$ be an orthonormal basis in V. Define f_k by
   $$f_1 = \tfrac{1}{3}(2e_1 - e_2 + 2e_3), \quad f_2 = \tfrac{1}{3}(2e_1 + 2e_2 - e_3), \quad f_3 = \tfrac{1}{3}(-e_1 + 2e_2 + 2e_3).$$
   (a) Show that $\{f_1, f_2, f_3,\}$ is an orthonormal set.
   (b) Find the projection of e_3 on the span of f_1, f_2.
   (c) Write e_2 as a linear combination of the f_j.

7. Let G be the Gram matrix of a linearly independent set $\beta = \{e_k\}_{k=1}^n$. Show that
   (a) G is a diagonal matrix implies that β is orthogonal.
   (b) $G = I$ iff β is orthonormal.

8. **General change of basis and inner product representation**
   (a) Let $f_j = \sum_{i=1}^n p_{ij} e_i$ and $\beta = \{e_i\}_{i=1}^n$, $\gamma = \{f_i\}_{i=1}^n$ be bases for V. Show that
   $$\langle x, y \rangle = [y]_\gamma{}^H H[x]_\gamma, \quad \text{where } H = P^H G P.$$

This connects the Gram matrix H with respect to γ to the Gram matrix G with respect to β.

(b) From part (a), prove that β and γ are orthonormal iff $P^H P = I$.

(c) Investigate what happens when β, γ are both orthogonal, but are not both orthonormal. Consider three cases: β is orthonormal, γ is orthonormal, and neither are orthonormal.

9. A *permutation matrix* is obtained from the identity matrix by rearranging the rows of the identity matrix. Let P be a permutation matrix. Show that $PP^T = I$.

10. Let $\beta = \{e_k\}_{k=1}^n$ be a basis for $\mathbf{R}^n$. Let G be the Gram matrix of this basis in the standard inner product, and let $P = [e_1\ e_2\ \cdots\ e_n]$. Show that
    (a) $[h_{ij}] = [g_{ij}]^{-1}$ exists.
    (b) Define $f_j = \sum_{i=1}^n h_{ji} e_i$. Show that $\langle e_k, f_j \rangle = \delta_{jk}$.
    (c) $\{f_k\}_{k=1}^n$ is a basis for $\mathbf{R}^n$.
    (d) f_j is the transpose of the jth row of P^{-1}.

$\{f_k\}_{k=1}^n$ is called the *basis reciprocal* to $\{e_k\}_{k=1}^n$. It plays an important role in tensor analysis.

11. Let $f(x) = \sin(\pi x)$, and let $M = V$ and $\langle f, g \rangle = \int_0^1 f(t)\overline{g}(t)\,dt$, where V is defined as in Example 3.
    (a) Find the projection of f on M for $n = 5$. Compare this with the interpolant, $g(x) = \sum_{i=1}^{n-1} \sin(\pi x_i) f_i(x)$.
    (b) Do part (a) for larger n.

12. In some situations, it is desirable to use continuous piecewise quadratics. For the interval $[0, 1]$, let $x_k = k/n$, and define

$$f_k(x) = \begin{cases} n^2(x_{k+1} - x)(x - x_{k-1}) & x_{k-1} \leq x \leq x_{k+1} \\ 0 & \text{otherwise} \end{cases}, \qquad k = 1, 2, \ldots, n - 1.$$

(a) Describe $M = \text{span}\{f_1, \ldots, f_{n-1}\}$.

(b) Show that $\{f_1, \ldots, f_{n-1}\}$ has the interpolation property $f_i(x_j) = \delta_{ij}$, where $i, j = 1, \ldots, n - 1$.

(c) Find the Gram matrix G of $\{f_1, \ldots, f_{n-1}\}$ in the inner product $\int_0^1 f(x)\overline{g}(x)\,dx$. Show that G is nonsingular.

(d) If $f \in M$, and $f(x_i) = y_i$, compute $\int_0^1 f(x)\,dx$ and $\int_0^1 f^2(x)\,dx$.

(e) Let $f(x) = \sin(\pi x)$ and $n = 5$. Find the projection of f on M and compare it to the interpolant $g(x) = \sum_{i=1}^{n-1} \sin(\pi x_i) f_i(x)$.

(f) Do part (e) for larger n.

CHAPTER 4

Diagonalizable Linear Operators

The concepts of eigenvalue and eigenvector are introduced in this chapter. First, some of their general properties are studied, and then attention is given to a special class of linear operators known as diagonalizable operators, which have remarkably attractive computational properties. Diagonalizable linear operators occur frequently in applications. (General linear operators will be studied in Chapter 8.) Next, the functions of operators are defined. The ideas in this chapter have extensive applications. Some applications are given in the areas of differential equations, difference equations, matrix theory, and linear algebra. Our main work occurs in nontrivial finite-dimensional vector spaces. All vector spaces are assumed to be of this nature unless otherwise specified.

4.1
EIGENVALUES AND EIGENVECTORS

For a linear operator T, a very important situation occurs when the image $T(e)$ of a nonzero vector e is proportional to e: $T(e) = \lambda e$, for some scalar λ. If this phenomenon occurs on a basis of such vectors, a rather remarkable suite of consequences follows.

Here is an example from mechanics: The equilibrium displacement u of an idealized membrane stretched over a region and fixed on the boundary obeys $\nabla^2 u = f$, where f is proportional to the applied load density and $\nabla^2 = \partial^2/\partial x^2 + \partial^2/\partial y^2$ is the Laplacian. It turns out that those special equilibrium displacements u in which the displacement is proportional to the load density f, i.e., $\nabla^2 u = \lambda u$, form the characteristic shapes, or modes, in which the membrane may vibrate in such a way that the displacement $U(P, t)$ at each time t is always proportional to its maximal displacement: $U(P, t) = \cos(\omega t + \phi)u(P)$. The frequency of oscillation ω is proportional to $\sqrt{|\lambda|}$. Furthermore, all small vibrations of the membrane can be constructed from these modes. Examples of this type will be discussed more later in the study of small vibrations in Chapter 7.

In certain ecological problems, the ith entry, x_i, of $x = (x_1, x_2, \ldots, x_n)^{\mathrm{T}}$ denotes the population of the ith species at a given time, and $T(x)$ is a column whose ith entry denotes the population of the ith species one time unit later. Suppose $e \neq 0$ satisfies the equation $T(e) = \lambda e$, with $\lambda \neq 0$. This means that, after one time unit, each population is magnified (or reduced) by the same number λ. Thus the relative sizes of the populations have not changed with the passage of a single time unit. Such population distributions $e = (e_1, e_2, \ldots, e_n)^{\mathrm{T}}$ may be of special interest, as will be illustrated in Section 4.6.

DEFINITIONS. Let T be a linear operator acting on the vector space V. λ *is an eigenvalue of T acting on V and e is an eigenvector of T in V associated with (or corresponding to) the eigenvalue λ if and only if* $T(e) = \lambda e$, *where* $e \neq 0$ *and* $e \in V$.

The *eigenspace $M(\lambda)$ of the eigenvalue λ* is the set of all eigenvectors associated with λ together with the zero vector: $M(\lambda) = \{x \in V : T(x) = \lambda x\}$.

The set $\sigma_V(T) = \{\lambda \in \mathbf{C} : \lambda$ is an eigenvalue of T on $V\}$ is called the (point) *spectrum of T on V*. The subscript V is suppressed when there is no danger of confusion.

The *eigenvalue problem* for a linear operator T acting on a vector space V is as follows: Find all scalars λ and vectors $0 \neq e \in V$ such that $T(e) = \lambda e$. In other words, find $\sigma(T)$ and every $M(\lambda)$.

Some authors replace the word "eigenvalue" with "proper value" or "characteristic value."

Remark. The spectrum of T and the eigenspaces are uniquely determined by T and V. If the domain V of T is changed, $\sigma(T)$ and the eigenspaces may be radically affected.

Remark. If T is a matrix operator $T(x) = Ax$ (where A is square), and if $Ae = \lambda e$ with $e \neq 0$, then λ and e will also be called an eigenvalue and eigenvector of A, for economy of language.

Remark. The scalars are taken to be complex in the study of the eigenvalue problem. This will avoid some sticky problems of abstract algebra. For example, if the linear operator is defined by $T(x) = Ax$, where A is a real matrix, there is no difficulty in allowing x to be a column of complex numbers. All linear operators on function spaces studied so far apply equally well to real or complex functions. Some results apply to finite fields $\mathbf{F}$, but we will not treat this topic.

EXAMPLE 1. Let

$$A = \begin{bmatrix} 2 & 1 \\ 1 & 2 \end{bmatrix}, \quad e = \begin{bmatrix} 1 \\ 1 \end{bmatrix}, \quad f = \begin{bmatrix} 1 \\ 0 \end{bmatrix}.$$

Are e and f eigenvectors of the matrix operator $T(x) = Ax$?

Since $Ae = 3e$, e is an eigenvector with eigenvalue 3. Af is not a constant multiple of f, so f is not an eigenvector.

EXAMPLE 2. Define $T : \mathbf{P}_n \to \mathbf{P}_n$, by $T(f) = f''$, where $' = d/dx$. Find the eigenvalues, the eigenvectors, and the eigenspaces of T.

$f'' = \lambda f$ has the solution $f(x) = a + bx$, if $\lambda = 0$. Thus $\lambda = 0$ is an eigenvalue with eigenvectors $a + bx \neq 0$ and $M(0) = \text{span}\{1, x\}$. If $\lambda \in \mathbf{C}$ and $\lambda = \beta^2 \neq 0$, then the solution of $f'' = \lambda f$ is $f(x) = Ae^{\beta x} + Be^{-\beta x}$. If $f \neq 0$, then $f \notin \mathbf{P}_n$, and so $\lambda \neq 0$ is not an eigenvalue.

EXAMPLE 3. Define $T : \mathbf{T}_n \to \mathbf{T}_n$ by $T(f) = f''$. Find the eigenvalues, the eigenvectors, and the eigenspaces of T.

Some eigenvalues and eigenvectors can be written down from experience: for $\lambda = 0$, $e(x) = 1$; for $\lambda = -k^2 = (ik)^2$, $e(x) = Ae^{ikx} + Be^{-ikx} = C\cos(kx) + D\sin(kx)$, where $k = 1, 2, \ldots, n$. It is easy to see that these functions are eigenvectors with the given eigenvalues by differentiation. The respective eigenspaces are $M(0) = \text{span}\{1\}$, $M(-k^2) = \text{span}\{e^{ikx}, e^{-ikx}\} = \text{span}\{\cos(kx), \sin(kx)\}$. In Section 4.2, it will be shown by theoretical means that there are no more eigenvectors or eigenvalues. Alternatively, to construct the eigenvalues and vectors, one can solve $T(f) = f'' = \lambda f$, which gives $f(x) = Ae^{\beta x} + Be^{-\beta x}$ if $\lambda = \beta^2 \neq 0$. Then the requirement that $f \in \mathbf{T}_n$ requires that the solution be 2π periodic, which implies that $\beta = ik$, where $k = \pm 1, \pm 2, \ldots$.

Examples 2 and 3 show that changing the domain of the operator has a radical effect on its eigenvalues and eigenvectors.

Next, some elementary consequences of the definitions are given.

PROPOSITION 1

(a) If λ is an eigenvalue, $M(\lambda)$ is a nontrivial subspace of V.
(b) T is one to one on V if and only if 0 is not an eigenvalue of T.
(c) If $T : V \to V$ with $\dim(V) < \infty$, then T^{-1} exists iff 0 is not an eigenvalue of T.

Proof

(a) By definition, $M(\lambda) \neq \{0\}$ and $M(\lambda) \subset V$. Since $M(\lambda) = N(T - \lambda I)$, $M(\lambda)$ is a subspace of V.
(b) Note that λ is an eigenvalue iff $N(T - \lambda I) \neq \{0\}$. Thus 0 is an eigenvalue of T iff $N(T) \neq \{0\}$ iff T is not one to one. (See Proposition 3 of Section 2.1.)
(c) T^{-1} exits iff T is one to one on V by Theorems 3 and 6 of Section 2.2. Now apply part (b).

The consequences of part (a) are worth reiterating: If e is an eigenvector of T associated with the eigenvalue λ, then so is αe, for all scalars $\alpha \neq 0$. Thus an eigenvector may be scaled in any pleasing way by multiplying it by a nonzero constant. If $\{e_1, e_2, \ldots, e_m\}$ are all eigenvectors of T associated with the *same* eigenvalue λ, then so is $\sum_{i=1}^{m} \alpha_i e_i$, as long as this sum is not zero.

An eigenvalue λ is called a *simple eigenvalue of T* on V if and only if $\dim(M(\lambda)) = 1$, and λ is a *degenerate eigenvalue of T* otherwise. When λ is a simple eigenvalue of T, then the eigenvectors associated with λ can only differ by constant multiples, since a basis of $M(\lambda)$ consists of a single nonzero vector. However, when λ is a degenerate eigenvalue and the associated eigenspace is finite-dimensional, then there are infinitely many bases of eigenvectors for this eigenspace. Many physical problems with linear models have only simple eigenvalues.

Theoretical Computation of Eigenvalues and Eigenvectors

As we have seen in some problems, the eigenvalue problem may be solved by working directly with the operator and the vector space. However, for many problems it is better to introduce coordinates into the problem by using a specific basis for V and to represent vectors by columns and the operator by a matrix. The eigenvalue problem is solved for the matrix of the operator, by theoretical or numerical methods, and the results are transferred back to the original vector space.

Next a theoretical procedure for computing eigenvalues and eigenvectors is outlined. The strategies commonly used in computer software for solving the eigenvalue problem will not be discussed. Those strategies involve a good deal of interesting theoretical linear algebra, which takes substantial time to explain and which you will be able to understand after reading this book.

Determinants are needed in the following discussion, and a review of them is presented in the appendix to this chapter.

Let $\{e_k\}_{k=1}^n$ be a basis for V, and construct the matrix A of T with respect to this basis. Recall that $T(x) = y$ iff $Ac = d$, where c and d are the coordinate columns of x and y, respectively, with respect to the chosen basis. Hence $T(e) = \lambda E$ iff $AE = \lambda E$, where E is the coordinate column of e. Thus we have this sequence of equivalent statements:

(i) $T(e) = \lambda e, \; e \neq 0$ $\Leftrightarrow$

(ii) $AE = \lambda E, \; E \neq 0,$ $\Leftrightarrow$

       $(A - \lambda I)E = 0, \; E \neq 0,$ $\Leftrightarrow$

       $A - \lambda I$ is singular, and $E \neq 0$ satisfies $(A - \lambda I)E = 0,$ $\Leftrightarrow$

(iii) $|A - \lambda I| = 0$ and $E \neq 0$ satisfies $(A - \lambda I)E = 0.$

Here $|A| = \det(A)$ is the determinant of the matrix A. Thus the eigenvalue problem (i) for T is equivalent to the eigenvalue problem (ii) for A, and these are in turn equivalent to (iii). To solve (iii), a λ satisfying $|A - \lambda I| = 0$ must be found, and then an $E \neq 0$ for which $(A - \lambda I)E = 0$ must be found. If $T(x) = Ax$ is a matrix operator and the standard basis is used, then $[T] = A$ so that (i) and (ii) are the same (with $e = E$).

Next we discuss the equation $|A - \lambda I| = 0$. The proposition following the definition justifies some of the language in the definition.

DEFINITION. $p(\lambda) \equiv |A - \lambda I|$ is called the *characteristic polynomial* for T. The equation $p(\lambda) = 0$ is called the *secular equation,* or the *characteristic equation* for T.

Thus λ is an eigenvalue for T on V if and only if $p(\lambda) = 0$. Next, some properties of $p(\lambda)$ are investigated.

PROPOSITION 2. If A is an $n \times n$ matrix, then $p(\lambda) = |A - \lambda I|$ is a polynomial in λ of degree n of the form

$$p(\lambda) = (-1)^n \lambda^n + \mathrm{Tr}(-1)^{n-1}\lambda^{n-1} \pm \cdots + |A|. \tag{1}$$

Furthermore, $p(\lambda) = (\lambda_1 - \lambda)(\lambda_2 - \lambda) \cdots (\lambda_n - \lambda)$, and

$$\text{Tr} = \sum_{i=1}^{n} a_{ii} = \sum_{i=1}^{n} \lambda_i \quad \text{(called the } \textit{Trace} \text{ of } A) \tag{2}$$

$$|A| = \prod_{i=1}^{n} \lambda_i. \tag{3}$$

Proof. The fact that the leading coefficient of $p(\lambda)$ is $(-1)^n$ follows easily by induction. From the factorization $p(\lambda) = (\lambda_1 - \lambda)(\lambda_2 - \lambda) \cdots (\lambda_n - \lambda)$, (3) follows by setting $\lambda = 0$ in the definition of $p(\lambda)$. Equation (2) follows from expanding $p(\lambda)$ in powers of λ, and comparing these coefficients with those obtained by expanding $|A - \lambda I|$ in powers of λ using cofactor expansions.

PROPOSITION 3. $p(\lambda)$ is independent of the choice of the basis used to represent T by a matrix. The eigenvalues do not depend on the matrix representing the linear transformation.

Proof. Clearly the eigenvalues depend only on T and V, and therefore are independent of the matrix representation. If another basis were used, then the matrix of T is of the form $B = P^{-1}AP$, and $|B - \lambda I| = |P(A - \lambda I)P^{-1}| = |P| |A - \lambda I| |P|^{-1} = |A - \lambda I|$, so that the characteristic polynomial is the same.

COROLLARY 4. Similar matrices have the same eigenvalues, trace, and determinant.

Proof. Similar matrices represent the same transformation, but in different bases.

Summary of the theoretical procedure for finding eigenvalues and eigenvectors

1. Select your favorite basis, and compute the matrix A of T with respect to that basis.
2. Compute the characteristic polynomial $|A - \lambda I| = p(\lambda)$, and solve the equation $p(\lambda) = 0$ for its distinct roots $\lambda_1, \lambda_2, \ldots$. These are precisely the eigenvalues of T. (See the appendix of this chapter for tips on finding these roots.)
3. For each root $\lambda = \lambda_j$, find a basis for the eigenspaces of A: $M_A(\lambda_j) = N(A - \lambda_j I)$; i.e., find as many linearly independent column vectors $x \neq 0$ as possible that satisfy $(A - \lambda_j I)x = 0$. If $x = (x_1, x_2, \ldots, x_n)^{\mathrm{T}}$ is such a vector, then $v = \sum_{i=1}^{n} x_i e_i \in V$ is an eigenvector of T corresponding to the eigenvalue λ_j. That is, each linearly independent column eigenvector corresponding to λ_j is taken back to V by reattaching coefficients onto the basis.

The next statement is important, and is one reason for requiring the scalars to be complex. It is not true for real scalars, some finite fields, or for infinite-dimensional vector spaces.

THEOREM 5. Let V be a complex vector space with $0 < \dim(V) < \infty$. Then every linear operator $T : V \to V$ has at least one eigenvalue.

Proof. Every polynomial of degree n has at least one complex zero.

If the matrix is real, there is no guarantee that any of the eigenvalues will be real. The next proposition imposes some limits, however.

PROPOSITION 6. Let A be a real $n \times n$ matrix, λ be an eigenvalue of A, and let E be an eigenvector associated with the eigenvalue λ.

(a) If λ is complex, then $\overline{\lambda}$ is an eigenvalue of A and $\overline{E}$ is an associated eigenvector.
(b) If λ is real, then E may be chosen to be in $\mathbf{R}^n$.

Proof

(a) Conjugating both sides of $AE = \lambda E$ gives $A\overline{E} = \overline{\lambda}\overline{E}$.
(b) In solving $(A - \lambda I)E = 0$, complex numbers need not be introduced. For a second proof, take the real and imaginary parts of $AE = \lambda E$ to get $AE_{re} = \lambda E_{re}$ and $AE_{im} = \lambda E_{im}$, where the subscripts re and im refer to the real and imaginary parts of E. Since $E = E_{re} + iE_{im} \neq 0$, one of E_{re}, E_{im} must be nonzero, and thus an eigenvector.

EXAMPLE 4. Let

$$A = \begin{bmatrix} 1 & -1 \\ 1 & 1 \end{bmatrix}.$$

Find the eigenvalues and eigenvectors of A.

From $|A - \lambda I| = (1 - \lambda)^2 + 1$, it follows that the eigenvalues of A are $\lambda_1 = 1 + i$ and $\lambda_2 = \overline{\lambda}_1 = 1 - i$. Solving

$$(A - \lambda_1 I)e = \begin{bmatrix} -i & -1 \\ 1 & -i \end{bmatrix} e = 0$$

yields the eigenvector

$$e = c \begin{bmatrix} 1 \\ -i \end{bmatrix}, \qquad (c \neq 0).$$

Thus the eigenvectors corresponding to λ_2 are of the form

$$\overline{e} = \overline{c} \begin{bmatrix} 1 \\ i \end{bmatrix}.$$

DEFINITIONS. Let $p(\lambda)$ be the characteristic polynomial of T. The *algebraic multiplicity* of an eigenvalue λ_0 is the number of times it appears in the factorization $p(\lambda) = (\lambda_1 - \lambda)(\lambda_2 - \lambda) \cdots (\lambda_n - \lambda)$. The *geometric multiplicity* of λ_0 is the dimension of the eigenspace $M(\lambda_0)$.

The relationship between the algebraic and geometric multiplicities is given in the next theorem, which will not be proved.

THEOREM 7. The algebraic multiplicity and geometric multiplicity of an eigenvalue obey the following relationship:

Geometric multiplicity $\leq$ Algebraic multiplicity.

EXAMPLE 5. If $p(\lambda) = (7 - \lambda)^4(5 - \lambda)(3 - \lambda)^2$, then the spectrum of the operator is $\sigma(T) = \{3, 5, 7\}$, the eigenvalues 3, 5, and 7 have algebraic multiplicities 2, 1, and 4, respectively, and 3, 3, 5, 7, 7, 7, 7 are the eigenvalues listed with their algebraic multiplicities. The geometric multiplicities cannot all be determined from the characteristic polynomial, unless considerably more information is known. All that may be said is this: $\dim(M(1)) \leq 2$, $\dim(M(2)) = 1$, and $\dim(M(7)) \leq 4$.

EXAMPLE 6. Let $V = \mathbf{P}_2$ and $(T(x))(t) \equiv Tx(t) = (1 + t^2)x''(t) + x'(t) + x(t)$, where $' = d/dt$. Find the spectrum and eigenspaces of T.

Solution. One must verify $T : V \to V$, and this can be done in the process of the computations. Pick the ordered basis $\{1, t, t^2\}$, and compute each $T(t^i)$ to find the matrix A of T. If the image under T of each basis member lies in V, then so does the image $T(x)$ of every $x \in V$, by the linearity of T. Hence the construction of A by expressing each $T(t^j)$ as a linear combination of $\{1, t, t^2\}$ can succeed only if $T : V \to V$. From $T(1) = 1$, $T(t) = 1 + t$, and $T(t^2) = 2(1 + t^2) + 2t + t^2$, we have

$$A = [T] = \begin{bmatrix} 1 & 1 & 2 \\ 0 & 1 & 2 \\ 0 & 0 & 3 \end{bmatrix}, \qquad \text{whence } p(\lambda) = |A - \lambda I| = (1 - \lambda)^2(3 - \lambda).$$

Thus $\sigma(T) = \{1, 3\}$ and $\lambda = 1$ has algebraic multiplicity 2. Let $\lambda_1 = 1$ and $\lambda_2 = 3$.
Next, solve $(A - \lambda_1 I)x = 0$: Row-reduction of $A - I$ yields

$$A - I = \begin{bmatrix} 0 & 1 & 2 \\ 0 & 0 & 2 \\ 0 & 0 & 2 \end{bmatrix} \to \begin{bmatrix} 0 & 1 & 0 \\ 0 & 0 & 1 \\ 0 & 0 & 0 \end{bmatrix},$$

and a basis for $N(A - I) = M_A(\lambda_1)$ is $E_1 = (1, 0, 0)^{\mathrm{T}}$. Now solve $(A - \lambda_2 I)x = 0$: Row-reduction of $A - 3I$ yields

$$A - 3I = \begin{bmatrix} -2 & 1 & 2 \\ 0 & -2 & 2 \\ 0 & 0 & 0 \end{bmatrix} \to \begin{bmatrix} -2 & 0 & 3 \\ 0 & 1 & -1 \\ 0 & 0 & 0 \end{bmatrix},$$

which gives $E_2 = (1.5, 1, 1)^{\mathrm{T}}$ as a basis for $N(A - \lambda_2 I) = M_A(\lambda_2)$.
Returning to $\mathbf{P}_2$, the eigenvalue, eigenspace pairs for T are $\lambda_1 = 1$, $M(1) = $ span$\{e_1\}$ with $e_1 = 1$, and $\lambda_2 = 3$, $M(3) = $ span$\{e_2\}$ with $e_2 = 1.5 + t + t^2$. λ_1 has geometric multiplicity 1. Since all the eigenvectors have been found, there is not a basis of eigenvectors for T.

EXAMPLE 7. Let

$$A = \begin{bmatrix} 1 & 2 & 0 & 0 \\ 0 & -1 & 0 & 0 \\ 0 & 0 & 1 & 0 \\ 0 & 0 & -1 & 2 \end{bmatrix}.$$

Find the spectrum and eigenspaces of A. Determine whether A is invertible.
Since $A - \lambda I$ is block diagonal,

$$|A - \lambda I| = \det\left(\begin{bmatrix} 1 - \lambda & 2 \\ 0 & -1 - \lambda \end{bmatrix}\right) \det\left(\begin{bmatrix} 1 - \lambda & 0 \\ -1 & 2 - \lambda \end{bmatrix}\right) = (1 - \lambda)^2(1 + \lambda)(\lambda - 2).$$

Thus $\sigma(A) = \{-1, 1, 2\}$. The eigenvalues $-1, 2$ are simple, and their eigenspaces must have dimension 1, whereas the eigenvalue 1 has algebraic multiplicity 2, and its eigenspace may have dimension 1 or 2. Solving $(A + I)x = 0$ yields $e_1 = (1, -1, 0, 0)^{\mathrm{T}}$, which forms a basis for $M(-1)$. Solving $(A - 2I)x = 0$ yields $e_2 = (0, 0, 1, 0)^{\mathrm{T}}$, which forms a basis for $M(2)$. Solving $(A - I)x = 0$ leads to the equivalent equations $x_2 = 0$ and $x_3 = x_4$, so that x_1 and x_4 are free variables. Thus a basis for $M(1)$ is $\{(1, 0, 0, 0)^{\mathrm{T}}, (0, 0, 1, 1)^{\mathrm{T}}\}$, and the geometric multiplicity of $\lambda = 2$ is 2.

Numerical Computation of Eigenvalues and Eigenvectors

There are many good packages for computing (approximately) the eigenvalues and eigenvectors of matrices. Some of these are even found on handheld calculators. We invite the reader to use them, after doing some problems by hand to see how the theoretical calculations work. Nearly all of the computational problems in this book can be done by hand and have nice answers, albeit some require a bit of tedious work. Relics of a generation past, perhaps, theoretical points are nonetheless illustrated in these problems.

When numerical methods are used, different algorithms or different implementations of the same algorithm can produce different sets of eigenvectors. When an eigenspace is one-dimensional, all algorithms yield eigenvectors proportional to each other (up to round-off error). When an eigenspace is multidimensional, there are infinitely many valid bases for the eigenspace in question, and it can be difficult to reconcile the answers from different methods. A further complication is that some algorithms report certain columns as eigenvectors even when they are not. Thus one has to dedicate some thought to the validity and meaning of the output of the algorithm, unless a priori information can be found that implies that the matrix in question has a basis of eigenvectors. When there is a basis of eigenvectors, the output is usually reliable; however, computational problems beyond the capability of the computer being used always exist.

Computational results can be checked as follows: To check that E is an eigenvector of A, verify that AE is proportional to E: $AE = \lambda E$. If $P = [E_1\ E_2\ \cdots\ E_n]$ is reported to be a matrix of eigenvectors with $\Lambda = \mathrm{diag}(\lambda_1, \lambda_2, \ldots, \lambda_n)$ the matrix of corresponding eigenvalues, verify that $AP = P\Lambda$. (See Exercise 14.)

Exercises 4.1

1. Let

$$A = \begin{bmatrix} 2 & -1 & 1 \\ 1 & 0 & 1 \\ 2 & -2 & 3 \end{bmatrix}.$$

Show that

$$e = \begin{bmatrix} 1 \\ 1 \\ 2 \end{bmatrix}$$

is an eigenvector of A, and find the corresponding eigenvalue.

2. Show that $\sin(nx)$, $\cos(nx)$ are eigenvectors of $T = d^2/dx^2 + cI$.

3. Find the spectrum and eigenspaces for each of the following matrices:

(a) $\begin{bmatrix} 0 & 1 \\ 1 & 0 \end{bmatrix}$ (b) $\begin{bmatrix} 0 & 1 \\ -1 & 0 \end{bmatrix}$ (c) $\begin{bmatrix} 1 & 1 \\ -1 & 3 \end{bmatrix}$ (d) $\begin{bmatrix} 2 & 1 \\ 1 & 2 \end{bmatrix}$ (e) $\begin{bmatrix} 4 & -3 & 1 \\ 1 & 0 & 1 \\ 0 & 0 & 3 \end{bmatrix}$

$$(f) \begin{bmatrix} 0 & 1 & 1 \\ 1 & 0 & 1 \\ 4 & -4 & 3 \end{bmatrix} \quad (g) \begin{bmatrix} 1 & -1 & 0 & 0 \\ -1 & 1 & 0 & 0 \\ 0 & 0 & 1 & 0 \\ 0 & 0 & 1 & 2 \end{bmatrix} \quad (h) \begin{bmatrix} 1 & -1 & 0 & 0 \\ -1 & 0 & 1 & 0 \\ 0 & 1 & 1 & 0 \\ 0 & 0 & 0 & 3 \end{bmatrix}$$

4. $R = \begin{bmatrix} \cos(\theta) & -\sin(\theta) \\ \sin(\theta) & \cos(\theta) \end{bmatrix}$ is the rotation matrix through an angle θ. Find the spectrum and eigenspaces of R.

5. For each of the following operators, find the spectrum and eigenspaces.
   (a) $T : \mathbf{C}^{2,2} \to \mathbf{C}^{2,2}$ $T(A) = A^T$
   (b) $T : \mathbf{C}^{2,2} \to \mathbf{C}^{2,2}$ $T\left(\begin{bmatrix} a & b \\ c & d \end{bmatrix} \right) = \begin{bmatrix} b & 2b - a \\ c & c + 2d \end{bmatrix}$
   (c) $T : \mathbf{P}_2 \to \mathbf{P}_2$ $Tp(x) = p''(x) + p'(x) + p(x) + p(0)$
   (d) $T : \mathbf{T}_1 \to \mathbf{T}_1$ $Tp(x) = p''(x) + p'(x) + p(x) + p(0)$
   (e) $T : \mathbf{P}_2 \to \mathbf{P}_2$ $Tp(x) = \int_{-1}^{1}(x - t)^2 p(t)\, dt - 2p(0)x^2$

6. Let $T(x) = \mathrm{diag}(a_1, a_2, \ldots, a_n)x$ define an operator $\mathbf{C}^n \to \mathbf{C}^n$. Show that δ_i, $i = 1, 2, \ldots, n$ are eigenvectors of T, and find the corresponding eigenvalues. Note that the a_i need not be distinct.

7. Let $T(f) = f''$ act on the vector space of all twice-continuously differentiable functions. Show that every real number is an eigenvalue of T. Can complex numbers be eigenvalues of T?

8. Suppose that $K(x, t) = \sum_{k;j=1}^{n} a_{jk} x^k t^j$, and T is the operator defined by

$$Tf(x) = \int_a^b K(x, t) f(t)\, dt,$$

with a, b finite. Let T act on the space V of all continuous functions on $[a, b]$. Show that the eigenvectors associated with an eigenvalue $\lambda \neq 0$ are polynomials of degree not exceeding n.

9. Show that $e^{i\alpha x}$ is an eigenvector of the finite difference operators $\tilde{\delta}$ and δ^2, which are defined by $\tilde{\delta}f(x) = f(x + h) - f(x - h)$, $\delta^2 f(x) = f(x + h) - 2f(x) + f(x - h)$, where h is a fixed positive number. Find the corresponding eigenvalues.

10. Let $Tf(x) = f'(x)$, and let $T : \mathbf{P}_n \to \mathbf{P}_n$. Find the eigenspaces and spectrum of T.

11. Suppose e is an eigenvector of T with eigenvalue λ. Show that
    (a) e is an eigenvector of T^2, and in general, of T^n. What are the corresponding eigenvalues?
    (b) If T is invertible, show that e is also an eigenvector of T^{-1}. What is the corresponding eigenvalue?

12. (a) Show that A^T has the same spectrum as A.
    (b) How are the spectrums of $\bar{A}$ and A^H related to the spectrum of A?

13. Suppose $P : V \to V$ is *idempotent*; i.e., $P^2 = P$.
   (a) What are the possible eigenvalues and eigenvectors of P?
   (b) Let $M \neq \{0\}$ be a subspace of the finite-dimensional space V, and Π be the orthogonal projection on M. Find the spectrum and eigenspaces of Π.
   (*Hint:* Orthogonal projection operators are idempotent.)

14. Suppose A is a square matrix and $AE_i = \lambda_i E_i$, for $i = 1, 2, \ldots, m$. Let $P = [E_1 \cdots E_m]$. Show that $AP = P\,\mathrm{diag}(\lambda_1, \lambda_2, \ldots, \lambda_m)$.

4.2
LINEAR OPERATORS WITH AN EIGENBASIS

Many operators in engineering, the sciences, and mathematics have a basis of eigenvectors. Such operators have a myriad of computational advantages, and an incredible number of important problems can be solved with them. Most of this chapter, and indeed the next two chapters, will be spent investigating these operators and some of their applications.

> **DEFINITION.** Let $T : V \to V$. *T has an eigenbasis* if and only if there exists a basis $\{e_1, e_2, \ldots, e_n\}$ for V of eigenvectors of T, i.e., $T(e_i) = \lambda_i e_i$. The eigenvalues λ_i are not necessarily distinct. $\{e_1, e_2, \ldots, e_n\}$ is called an *eigenbasis for T*, and the basis $\{e_1, e_2, \ldots, e_n\}$ is said to *diagonalize T*. T is said to be *diagonalizable on V* when it has an eigenbasis. If T is a matrix operator $T(x) = Ax$, then the same language is applied to A. (The reason for the latter terminology is explained in Theorem 1.)

One may think of an eigenbasis as a basis such that, along each "axis" e_i, the operator acts as a pure (possibly complex) "stretching."

> **EXAMPLE 1.** Let $D = \mathrm{diag}(d_1, d_2, \ldots, d_n)$, and let T be the matrix operator $T(x) = Dx$. Then the standard basis is an eigenbasis for T, and the corresponding eigenvalues are d_i, since $T(\delta_i) = D\delta_i = d_i\delta_i$.

There is no requirement that the d_i be distinct in this example. Note that, if $D = I$, then all the eigenvalues are 1.

> **EXAMPLE 2.** Let T be defined by $T(f) = f''$. Then if $n > 1$, T is diagonalizable on $\mathbf{T}_n$ but not on $\mathbf{P}_n$.
> On $\mathbf{T}_n$, T has the eigenbasis $\{e^{ikt}\}_{k=-n}^n$ as well as the eigenbasis $\{1, \cos(x), \sin(x), \ldots, \cos(nx), \sin(nx)\}$ (cf. Example 3 of Section 4.1). On $\mathbf{P}_n$, T has only the eigenspace $M(0) = \mathrm{span}\{1, x\} = N(T)$, and $\{1, x\}$ is not a basis for $\mathbf{P}_n$. (Cf. Example 2 of Section 4.1).

Before we embark on a study to determine which operators are diagonalizable, some striking properties of diagonalizable operators are discussed, as well as some of their potent computational advantages. The next two theorems illustrate these ideas.

> **THEOREM 1.** Let $T : V \to V$, and let V be an n-dimensional space. T has an eigenbasis if and only if there is a basis of V with respect to which the matrix of T is diagonal. More precisely, if $\beta = \{e_1, e_2, \ldots, e_n\}$ is a basis, then
>
> $$[T]_\beta = \mathrm{diag}(\lambda_1, \lambda_2, \ldots, \lambda_n) \Leftrightarrow T(e_k) = \lambda_k e_k, \qquad k = 1, 2, \ldots, n. \tag{1}$$

Proof. This follows immediately from the definition of $[T]_\beta : T(e_k) = \lambda_k e_k$ iff the kth column of $[T]_\beta$ has zero entries in every row but the kth, where it has the entry λ_k.

The next theorem exhibits some of the computational advantages stemming from the existence of an eigenbasis. In earlier chapters, the problems of finding the inverse of an operator, its matrix, and its null space and range were studied. One may also be interested in computing T^n or repeated applications of T on a vector. The next theorem asserts that these entities are easy to find, provided an eigenbasis is known. For example, if T is a matrix operator generated by a large matrix, the null space and range of T, and T^{100}, may not be easy to compute directly but can be easily found if an eigenbasis is known.

Many problems that can be solved when an eigenbasis is known will be illustrated later in this chapter and in the reminder of the book. Furthermore, many theoretical applications are easy for diagonalizable operators. Thus, for many problems, an eigenbasis is the best basis in which to work.

Unfortunately, not all linear operators are diagonalizable, as some of the examples of Section 4.1 show. Nondiagonalizable operators are called *defective operators*. The structure of general linear operators will be undertaken in Chapter 8.

THEOREM 2. Let $\beta = \{e_1, e_2, \ldots, e_n\}$ be an eigenbasis for V, with $T(e_i) = \lambda_i e_i$. Suppose $x = \sum_{i=1}^n x_i e_i$. Then

(a) $T(x) = \sum_{i=1}^n \lambda_i x_i e_i$.
(b) $T^p(x) = \sum_{i=1}^n \lambda_i^p x_i e_i$ for $p = 1, 2, \ldots$.
(c) If all $\lambda_i \neq 0$, then $T^{-1}(x) = \sum_{i=1}^n \lambda_i^{-1} x_i e_i$.
(d) If $\lambda_1, \lambda_2, \ldots, \lambda_r \neq 0$, and $\lambda_i = 0$ for all $i > r$, then
    (i) $R(T) = \text{span}\{e_1, \ldots, e_r\}$.
    (ii) $N(T) = \text{span}\{e_{r+1}, \ldots, e_n\}$.
    (iii) The rank of T is r.

Proof. Let $x = \sum_{i=1}^n x_i e_i$.

(a) $T(x) = \sum_{i=1}^n x_i T(e_i) = \sum_{i=1}^n \lambda_i x_i e_i$, by linearity and by the definition of eigenvalue.

(b) Applying T again, $T^2(\sum_{i=1}^n x_i e_i) = \sum_{i=1}^n \lambda_i x_i T(e_i) = \sum_{i=1}^n \lambda_i^2 x_i e_i$. Again, $T^3(\sum_{i=1}^n \lambda_i x_i e_i) = \sum_{i=1}^n \lambda_i^2 x_i T(e_i) = \sum_{i=1}^n \lambda_i^3 x_i e_i$. The process continues inductively, producing another power of λ_i each time T is applied, and the result follows.

(c) It is sufficient to show that if $y = \sum_{i=1}^n \lambda_i^{-1} x_i e_i$, then $T(y) = x$. But this follows from (a).

(d) Note that $T(x) = \sum_{i=1}^n \lambda_i x_i e_i = \sum_{i=1}^r \lambda_i x_i e_i$. This shows that $\{e_1, \ldots, e_r\}$ is an independent spanning set for the range of T, and (di) follows. Statement (dii) follows from the fact that $T(x) = 0$ implies that $x_i = 0$ for $i = 1, \ldots, r$, by the linear independence of $\{e_1, \ldots, e_r\}$. Hence $\{e_{r+1}, \ldots, e_n\}$ is an independent spanning set for $N(T)$. Statement (diii) follows from (di) and from the definition of rank.

Remark. From statement (c) of Theorem 2 follows the remarkable fact that, if T is diagonalizable and no eigenvalue is zero, then T^{-1} and T^n are also diagonalizable, using the same eigenbasis. This fact will be generalized in the next section.

The next examples show simple applications of Theorem 2.

EXAMPLE 3. Let $T = d/dx$ act on $\mathbf{T}_n = \text{span}\{e^{ikx}\}_{k=-n}^n$. If $f(x) = \sum_{k=-n}^n c_k e^{ikx}$, compute $T^p f$.

Since $T(e^{ikx}) = ik \cdot e^{ikx}$, $\lambda_k = ik$. From (b) of Theorem 2, $T^p f(x) = \sum_{k=-n}^n c_k \lambda_k^p e^{ikx} = \sum_{k=-n}^n c_k(ik)^p e^{ikx}$, which is easily verified by repeated differentiation.

EXAMPLE 4. Let

$$A = \begin{bmatrix} 0.8 & 0.4 \\ 0.2 & 0.6 \end{bmatrix}.$$

Given that

$$e_1 = \begin{bmatrix} 2 \\ 1 \end{bmatrix}, \qquad e_2 = \begin{bmatrix} 1 \\ -1 \end{bmatrix}$$

are eigenvectors of A, and

$$x = \begin{bmatrix} 1 \\ 1 \end{bmatrix},$$

(a) Find $A^n x$ for $n = 1, 2, \ldots$. Does $A^n x$ tend to a limit?
(b) Find $A^{-1} x$.

(a) A calculation shows that $Ae_1 = e_1$ and $Ae_2 = 0.4e_2$, so that $\lambda_1 = 1$, $\lambda_2 = 0.4$. Further, $x = \frac{2}{3}e_1 - \frac{1}{3}e_2$. Thus $A^n x = \frac{2}{3}\lambda_1{}^n e_1 - \frac{1}{3}\lambda_2{}^n e_2 = \frac{2}{3}e_1 - \frac{1}{3}(0.4)^n e_2$. From this it is clear that

$$A^n x \to \begin{bmatrix} 4/3 \\ 2/3 \end{bmatrix} \text{ as } n \to \infty.$$

(b) $A^{-1} x = \frac{2}{3}\lambda_1^{-1} e_1 - \frac{1}{3}\lambda_2^{-1} e_2 = \frac{2}{3}e_1 - \frac{1}{3}\left(\frac{1}{0.4}\right) e_2 = \begin{bmatrix} 0.5 \\ 1.5 \end{bmatrix}.$

The matrix in this example is called a stochastic or probability matrix. Such matrices occur in Markov processes. In Section 4.6, applications involving such matrices will be studied.

Since diagonalizable operators are so computationally attractive, two questions arise: (1) Given a linear operator $T : V \to V$, when is it diagonalizable? (2) How does one find an eigenbasis of a diagonalizable operator? In the next theorem, a simple but not quite satisfying answer is given to the first question. In Chapter 5, a priori criteria that are completely satisfactory in certain areas of science and engineering will be given. A theoretical answer will be given for question 2 in the following discussion. For computations, there exist excellent algorithms that can find an eigenbasis for reasonable matrices.

THEOREM 3. If $\lambda_1, \lambda_2, \ldots, \lambda_m$ are distinct eigenvalues of T and $e_1, e_2, \ldots, e_m$ are associated eigenvectors, then $\{e_1, e_2, \ldots, e_m\}$ is linearly independent. Hence if T is a linear operator on an n-dimensional space V and has n distinct eigenvalues, then T is diagonalizable, and each eigenspace is one-dimensional.

Proof. The first part of this theorem is proved by induction. If $m = 1$, then $e_1 \neq 0$ and the set $\{e_1\}$ is clearly linearly independent. Suppose the theorem is true for m eigenvalues, and let $\{\lambda_1, \lambda_2, \ldots, \lambda_{m+1}\}$ be distinct eigenvalues of T with $T(e_i) = \lambda_i e_i$. Assume

$$0 = \sum_{i=1}^{m+1} c_i e_i. \tag{*}$$

All the c_i must be shown to be 0. Apply T to (*). Then

$$0 = T(0) = \sum_{i=1}^{m+1} c_i T(e_i) = \sum_{i=1}^{m+1} c_i \lambda_i e_i. \tag{**}$$

Multiply (*) by λ_{m+1} and subtract this from (**) to get $\sum_{i=1}^{m}(\lambda_i - \lambda_{m+1})c_i e_i = 0$. Since $\{e_1, e_2, \ldots, e_m\}$ is linearly independent by the induction hypothesis, the coefficients must vanish, so that $c_1 = c_2 = \cdots = c_m = 0$ since $\lambda_i - \lambda_{m+1} \neq 0$. But then the sum in (*) reduces to $c_{m+1} e_{m+1} = 0$, and so $c_{m+1} = 0$ since $e_{m+1} \neq 0$.

The second part of the theorem is proved as follows: Since there are n distinct eigenvalues, each selection of n corresponding eigenvectors must be linearly independent. But n linearly independent vectors in an n-dimensional space must be a basis, by Theorem 6 of Section 1.4, so that T has a basis of eigenvectors and is therefore diagonalizable. The proof that each eigenspace is one-dimensional is left as an exercise.

Remark. Example 1 shows that an operator can have an eigenbasis without the eigenvalues being distinct.

If an operator has an eigenbasis $\{e_1, e_2, \ldots, e_n\}$ and $T(e_j) = \lambda_j e_j$, it is not yet known that every eigenvalue is in the list $\lambda_1, \lambda_2, \ldots, \lambda_n$, nor is it known whether there are more eigenvectors. The next proposition says that, in this case, every eigenvalue is known, and that the eigenspaces are also known.

PROPOSITION 4. Suppose $T : V \to V$ has an eigenbasis $\beta = \{e_1, e_2, \ldots, e_n\}$ and $T(e_j) = \lambda_j e_j$, $j = 1, 2, \ldots, n$.

(a) If $\lambda \in \sigma(T)$, then $\lambda = \lambda_k$ for some k.
(b) If $T(e) = \lambda e$, with $e \neq 0$, then e is a linear combination of those e_i in β such that $\lambda_i = \lambda$; i.e., $M(\lambda) = \text{span}\{e_i : \lambda_i = \lambda\}$.

Proof. Suppose $T(e) = \lambda e$, where $e \neq 0$.

(a) If $\lambda \neq \lambda_i$ for $i = 1, 2, \ldots, n$, then $e_1, e_2, \ldots, e_n, e$ are linearly independent by Theorem 3, which is impossible since $\dim(V) = n$.
(b) Since β is a basis, $e = \sum_{i=1}^{n} c_i e_i$. But

$$T(e) = \sum_{i=1}^{n} c_i \lambda_i e_i = \lambda e = \lambda \sum_{i=1}^{n} c_i e_i \Rightarrow \sum_{i=1}^{n}(\lambda - \lambda_i)c_i e_i = 0 \Rightarrow c_i = 0,$$

when $\lambda \neq \lambda_i$, since β is a basis. This proves (b).

EXAMPLE 5. Let $T : \mathbf{T}_n \to \mathbf{T}_n$ be defined by $T(f) = f''$. Then $\sigma(T) = \{0, -1, -4, \ldots, -n^2\}$, and $M(0) = \{1\}$ and $M(-k^2) = \{Ae^{ikx} + Be^{-ikx}\}$.

On $\mathbf{T}_n$, T has the eigenbasis $\{e^{ikx} : k = -n, \ldots, 0, \ldots, n\}$, and $T(e^{ikx}) = -k^2 e^{ikx}$. Apply Proposition 4. Cf. Example 3 of Section 4.1.

EXAMPLE 6. Define $T : \mathbf{C}^{2,2} \to \mathbf{C}^{2,2}$ by $T(A) = A^{\mathsf{T}}$. Find $\sigma(T)$ and the eigenspaces of T.

Some thought leads to the following observations:

$$T(e_i) = e_i \quad \text{for } e_1 = \begin{bmatrix} 1 & 0 \\ 0 & 0 \end{bmatrix}, \quad e_2 = \begin{bmatrix} 0 & 1 \\ 1 & 0 \end{bmatrix}, \quad e_3 = \begin{bmatrix} 0 & 0 \\ 0 & 1 \end{bmatrix}$$

and

$$T(e_4) = -e_4 \quad \text{for } e_4 = \begin{bmatrix} 0 & 1 \\ -1 & 0 \end{bmatrix}.$$

Clearly $\{e_1, e_2, e_3, e_4\}$ is a basis of eigenvectors for $\mathbf{C}^{2,2}$. Thus $\sigma(T) = \{-1, 1\}$, $M(1) = \text{span}\{e_1, e_2, e_3\}$, and $M(-1) = \text{span}\{e_4\}$. In other words, every 2×2 symmetric matrix $(A^T = A)$ is a linear combination of e_1, e_2, and e_3, and every 2×2 skew symmetric matrix $(A^T = -A)$ is a multiple of e_4.

EXAMPLE 7. Let $V = \mathbf{T}_1$, and define T by $(Ty)(x) \equiv Ty(x) = y(x + \pi/4)$.

(a) Determine whether T is diagonalizable, and solve the eigenvalue problem for T acting on V.

(b) Determine the rank of T and whether T is invertible.

Solution

(a) Use the ordered basis $\{1, \cos x, \sin x\}$. Then

$$T(1) = 1$$

$$T(\cos)(x) = \cos(x + \pi/4) = \frac{1}{\sqrt{2}}[\cos(x) - \sin(x)]$$

$$T(\sin)(x) = \sin(x + \pi/4) = \frac{1}{\sqrt{2}}[\cos(x) + \sin(x)]$$

where the addition formulae for sin and cos have been used. That $T : V \to V$ follows now from the linearity of T. Thus

$$A = [T] = \begin{bmatrix} 1 & 0 & 0 \\ 0 & 1/\sqrt{2} & 1/\sqrt{2} \\ 0 & -1/\sqrt{2} & 1/\sqrt{2} \end{bmatrix},$$

so $p(\lambda) = (\lambda - 1)[(\lambda - 1/\sqrt{2})^2 + \frac{1}{2}]$. Hence $\lambda_1 = 1$, $\lambda_2 = (1 + i)/\sqrt{2}$, $\lambda_3 = (1 - i)/\sqrt{2}$ (the indexing is arbitrary), and $\sigma(T) = \{1, (1 + i)/\sqrt{2}, (1 - i)/\sqrt{2}\}$. Since the eigenvalues are distinct, there is a basis of eigenvectors, so the operator is diagonalizable, and each eigenspace is one-dimensional. Solving $(A - \lambda_j I)x = 0$ for each j yields the following for λ_1, $x = E_1 = (1, 0, 0)^T$; for λ_2, $x = E_2 = (0, 1, i)^T$; and for $\lambda_3 = \overline{\lambda_2}$, $x = E_3 = \overline{E_2} = (0, 1, -i)^T$. Reattaching the coefficients to the basis gives the following eigenvalues and eigenvectors:

$$\lambda_1 = 1, \qquad e_1 = 1$$

$$\lambda_2 = \frac{(1 + i)}{\sqrt{2}}, \qquad e_2 = \cos(x) + i \cdot \sin(x) = e^{ix}$$

$$\lambda_3 = \frac{(1 - i)}{\sqrt{2}}, \qquad e_3 = \cos(x) - i \cdot \sin(x) = e^{-ix}.$$

To check the solution, compute each $T(e_k)$ directly:

$$T(e_1) = 1 = \lambda_1 e_1, \qquad T(e_2) = e^{i\pi/4} e^{ix} = \lambda_2 e_2,$$

$$T(e_3) = e^{-i\pi/4} e^{-ix} = \lambda_3 e_3. \tag{$**$}$$

If the basis $\{1, e^{ix}, e^{-ix}\}$ were used to start with, then (**) would have been observed in the process of attempting to find $[T]$. No further work need be done. The calculations in (**) imply that $\{1, e^{ix}, e^{-ix}\}$ is an eigenbasis, and the eigenvalues are the multipliers λ_i. Since we have an eigenbasis, all of the eigenvalues and eigenvectors are known.

(b) Since all $\lambda_i \neq 0$, T has rank three and is invertible by Proposition 1 of Section 4.1. Clearly, $T^{-1}p(x) = p(x - \pi/4)$.

The next example is a special case of a *Fredholm integral operator:* $Tf(x) = \int_a^b K(x, t)f(t)\, dt$. $K(x, t)$ is called the *kernel* of the operator (cf. Example 7 of Section 2.1). For a discussion of such operators, see, for example, R. Courant and D. Hilbert, 1953.

EXAMPLE 8. On $V = P_2$, T is defined by

$$T p(x) = \frac{3}{8} \int_{-1}^{1} [1 + 4(x + t) + 5(x^2 + t^2) - 15x^2 t^2] p(t)\, dt. \tag{\#}$$

(a) Show that T is diagonalizable, and find an eigenbasis.
(b) Compute $T^3(t)$ three different ways.
(c) Show that T is invertible, and find $T^{-1}(t)$.

Solution

(a) Pick the basis $\{1, t, t^2\}$. While constructing the matrix, observe that each $T(t^i)$, where $i = 0, 1, 2$ is in V, so that $T : V \to V$. We have

$$(T1)(x) = 2 + 3x, \qquad (Tt)(x) = 1, \qquad (Tt^2) = 1 + x - x^2, \qquad \text{whence}$$

$$A = [T] = \begin{bmatrix} 2 & 1 & 1 \\ 3 & 0 & 1 \\ 0 & 0 & -1 \end{bmatrix}, \qquad \text{and} \qquad p(\lambda) = |A - \lambda I| = -(\lambda + 1)^2(\lambda - 3).$$

Thus $\sigma(T) = \{-1, 3\}$ and $\lambda = -1$ have algebraic multiplicity 2. Set $\lambda_1 = -1$ and $\lambda_2 = 3$. Solving $(A - \lambda_1 I)x = 0$ leads to $3x_1 + x_2 + x_3 = 0$, which has the general solution $x = [-(x_2 + x_3)/3, x_2, x_3]^T$. Hence the null space of $A - \lambda_1 I$ has a basis $E_1 = (-1, 3, 0)^T$ and $E_2 = (-1, 0, 3)^T$. The corresponding eigenvectors of T are $e_1 = 3t - 1$, $e_2 = 3t^2 - 1$. Solving $(A - \lambda_2 I)x = 0$ yields a one-dimensional null space spanned by $E_3 = (1, 1, 0)^T$. The corresponding eigenvector of T is $e_3 = t + 1$. Thus T has the eigenspaces $M(-1) = \text{span}\{e_1, e_2\}$ and $M(3) = \text{span}\{e_3\}$. The two-dimensional eigenspace $M(-1)$ has infinitely many bases. Any one of them could have been used. Now relabel the eigenvalues so that $T(e_i) = \lambda_i e_i$, i.e., so that $\lambda_1 = -1$, $\lambda_2 = -1$, and $\lambda_3 = 3$ (keeping the e_i as labeled) in anticipation of using Theorem 2 of Section 4.2.

(b) $T^p(f)$ can be computed three ways: (1) by applying T to fp times using (\#), (2) by computing $[T]^p[f]$ and reattaching coefficients on the basis $1, t, t^2$, and (3) by using the rule $T^p(\sum_{i=1}^n x_i e_i) = \sum_{i=1}^n \lambda_i^p x_i e_i$ derived in Theorem 2 of Section 4.2. For $p = 3$, each of these is tractable, but for large p, only the last is feasible.

Using method 1, if $f(t) = t$, then by (\#), $Tf(x) = 1$, $T^2 f(x) = (T1)(x) = 2 + 3x$, and $T^3 f(x) = [T(2 + 3t)](x) = 7 + 6x$. Using method 2, a computation gives

$$A^3 = \begin{bmatrix} 20 & 7 & 7 \\ 21 & 6 & 7 \\ 0 & 0 & -1 \end{bmatrix}, \qquad \text{so } A^3[f] = \begin{bmatrix} 7 \\ 6 \\ 0 \end{bmatrix},$$

whence $T^3(f) = 7(1) + 6t + 0t^2$, which agrees with the result of method 1. For method 3, note that $f = t = \frac{1}{4}(e_1 + e_3)$. Hence

$$T^3(f) = \lambda_1^3 \tfrac{1}{4} e_1 + \lambda_2^3 0 e_1 + \lambda_3^3 \tfrac{1}{4} e_3 = -\tfrac{1}{4} e_1 + 27 \tfrac{1}{4} e_3 = \tfrac{1}{4}\{27(t + 1) - (3t - 1)\} = 6t + 7.$$

(c) Since no $\lambda_i = 0$, thus $N(T) = \{0\}$, so T is one to one and therefore invertible by Theorem 6 of Section 2.2 or by Proposition 1 of Section 4.1. Alternatively, $[T]$ is nonsingular, so T^{-1} exists by Theorem 5 of Section 2.3. Now $T^{-1}(f) = \lambda_1^{-1}\frac{1}{4}e_1 + \lambda_2^{-1}0e_1 + \lambda_3^{-1}e_3 = \frac{1}{3}(1 - 2t)$. This can be checked by finding $[T]^{-1}$, computing $[T]^{-1}\delta_2$, and putting the entries of this column on the correct powers of t.

The next eigenvalue problem is important in a variety of places, especially in numerical analysis. Examples of its use will be shown later.

EXAMPLE 9. THE EIGENVALUES AND EIGENVECTORS OF TRIDIAGONAL MATRICES. Consider the following $n \times n$ matrix:

$$A = \begin{bmatrix} b & c & & & & 0 \\ a & b & c & & & \\ & a & b & c & & \\ & & \cdots & \cdots & & \\ & & & a & b & c \\ 0 & & & & a & b \end{bmatrix} \equiv \mathrm{Trid}(a, b, c).$$

In other words, $a_{i,i-1} = a$, $a_{i,i} = b$, $a_{i,i+1} = c$, and all other entries are equal to zero. If $a, c \neq 0$, show that the eigenvalues and eigenvectors of A are

$$\lambda_k = b + 2c\rho\cos\left(\frac{k\pi}{n+1}\right), \tag{2a}$$

$$e_k = [\rho\sin(kx_1), \rho^2\sin(kx_2), \ldots, \rho^n\sin(kx_n)]^T, \qquad k = 1, 2, \ldots, n, \tag{2b}$$

where $\rho = \sqrt{a/c}$, $x_j = j\pi/(n+1)$, and $j = 1, 2, \ldots, n$.

Writing out the system of equations $Ae = \lambda e$ yields

$$be_1 + ce_2 = \lambda e_1$$

$$ae_{k-1} + be_k + ce_{k+1} = \lambda e_k, \qquad k = 2, 3, \ldots, n-1 \tag{3}$$

$$ae_{n-1} + be_n = \lambda e_n.$$

Put $e_0 = 0$ and $e_{n+1} = 0$; then (3) holds for $i = 1, 2, \ldots, n$. Thus (3) is a finite difference equation with the boundary conditions $e_i = 0$, $i = 0, n+1$. Finite difference equations with constant coefficients can often be solved by using a trial solution $e_k = z^k$, just as differential equations with constant coefficients can be solved by trial solutions $e^{rx} = (e^r)^x$. Putting $e_k = z^k$ in (3) and cancelling z^{k-1} gives the characteristic equation

$$a + (b - \lambda)z + cz^2 = 0. \tag{4}$$

Since $c \neq 0$, let z_1, z_2 be the roots of (4). Let $N = n + 1$ for convenience. Thus z_1^k and z_2^k are solutions of (3). Since equation (3) is linear, $Az_1^k + Bz_2^k = e_k$ must also be a solution of (4). Applying the boundary conditions at $k = 0$, N gives $A + B = 0$ and $Az_1^N + Bz_2^N = A(z_1^N - z_2^N) = 0$. Since $e \neq 0$, $(z_1/z_2)^N = 1$, and hence $z_1/z_2 = e^{2\pi i p/N} = r$, for $p = 1, 2, \ldots, n$. Since z_1 and z_2 are roots of (4), $a + (b - \lambda)z + cz^2 = c(z - z_1)(z - z_2)$, so that $z_1 z_2 = a/c$ and $z_1 + z_2 = (\lambda - b)/c$. Thus $z_1 z_2 = z_2^2 r = a/c$, so that $z_2 = \sqrt{a/c}\,e^{-i\pi p/N}$, and $z_1 = z_2 r = \sqrt{a/c}\,e^{i\pi p/N}$. The $\sqrt{\ }$ can be taken to be either value, but it must be the same in every expression. Now the eigenvalues and eigenvectors can be found. Taking $A = (2i)^{-1}$ and a fixed p, we have

$$\lambda = b + c(z_1 + z_2) = b + c\sqrt{a/c}(e^{i\pi p/N} + e^{-i\pi p/N}) = b + 2c\sqrt{a/c}\cos(\pi p/N),$$

$$e_k = A(z_1^k - z_2^k) = A(a/c)^{k/2}(e^{ik\pi p/N} - e^{-ik\pi p/N}) = (a/c)^{k/2}\sin(k\pi p/N),$$

where $k = 1, 2, \ldots, n$. Setting $x_k = \pi k/N$, the eigenvalues and corresponding eigenvectors are

$$\lambda_p = b + 2c \sqrt{a/c} \cos\left(\frac{\pi p}{n+1}\right), \qquad p = 1, 2, \ldots, n$$

$$e_p = [\rho \sin(px_1), \rho^2 \sin(px_2), \ldots, \rho^n \sin(px_n)]^T, \qquad \rho = \sqrt{a/c}.$$

Thus A has n distinct eigenvalues so that the eigenvectors are linearly independent and thus a basis for $\mathbf{C}^n$.

In Example 6 of Section 4.1, the algebraic multiplicity of the eigenvalue 1 was 2, but the geometric multiplicity was 1. This kind of discrepancy cannot happen for diagonalizable operators.

THEOREM 5. Let $T : V \to V$ be a linear operator on the n-dimensional vector space V. Then the algebraic multiplicity of each eigenvalue is the same as its geometric multiplicity iff T is diagonalizable on V.

Proof. Suppose T is diagonalizable. The characteristic polynomial is independent of the basis used, so let us use an eigenbasis $\beta = \{e_1, e_2, \ldots, e_n\}$ with $T(e_i) = \lambda_i e_i$. Then $A = [T]_\beta = \mathrm{diag}(\lambda_1, \ldots, \lambda_n)$, and $p(\lambda) = \prod_{i=1}^{n}(\lambda_i - \lambda)$. For example, suppose $\lambda_1 = \lambda_2 = \cdots \lambda_p$ were listed exactly p times on the diagonal of A, so that λ_1 has algebraic multiplicity p. Let e be an arbitrary eigenvector in $M(\lambda_1)$, so that $T(e) = \lambda_1 e$, where $e \neq 0$. Suppose $e = \sum_{i=1}^{n} c_i e_i$. Then

$$0 = T(e) - \lambda_1 e = \sum_{i=1}^{n}(\lambda_i - \lambda)c_i e_i = \sum_{i=p+1}^{n}(\lambda_i - \lambda)c_i e_i.$$

By the linear independence of $\{e_k\}_{k=1}^{n}$, each $(\lambda_i - \lambda_1)c_i = 0$, and therefore $c_i = 0$ for $i = p+1, \ldots, n$. Thus e is a linear combination of the linearly independent set $\{e_1, e_2, \ldots, e_p\}$. Hence $\{e_1, e_2, \ldots, e_p\}$ is a basis for $M(\lambda_1)$, and the theorem follows. If the algebraic multiplicity agrees with the geometric multiplicity for each eigenvalue, then there are n independent eigenvectors for T. These eigenvectors must be a basis, since their number equals the dimension of V.

EXAMPLE 10. DIAGONALIZATION OF A SQUARE MATRIX. Given an $n \times n$ matrix A, find if possible a matrix P so that $A = P\Lambda P^{-1}$, where $\Lambda = \mathrm{diag}(\lambda_1, \ldots, \lambda_n)$.

Two solutions to this problem are given, one using the concepts of operators and bases and one using matrix equations.

Vector space method. Define $T(x) = Ax$. Then T is a linear operator $\mathbf{C}^n \to \mathbf{C}^n$ whose matrix with respect to the standard basis is A, by Example 2 of Section 2.3. In Corollary 7 of Section 2.3, it was shown that the matrix of T with respect to a basis $\{E_1, \ldots, E_n\}$ is $P^{-1}AP$, where $P = [E_1 \; E_2 \cdots E_n]$. The condition $A = P\Lambda P^{-1}$ is equivalent to $P^{-1}AP = \Lambda$, which means, by Theorem 1 of Section 4.2, that the basis $\{E_1, \ldots, E_n\}$ is an eigenbasis with $T(E_i) = AE_i = \lambda_i E_i$. Thus the columns of P are eigenvectors, and the entries on the diagonal are the corresponding eigenvalues. Note that P and Λ exist iff T (or A) is diagonalizable.

Matrix method. We unscramble the algebraic statement $A = P\Lambda P^{-1}$. $A = P\Lambda P^{-1} \Leftrightarrow AP = P\Lambda$, with P nonsingular. Let $P = [E_1 \; E_2 \; \cdots \; E_n]$. Then we have $AP = [AE_1 \; AE_2 \cdots AE_n]$, and $P\Lambda = P[\lambda_1 \delta_1, \lambda_2 \delta_2, \ldots, \lambda_n \delta_n] = [\lambda_1 P\delta_1 \; \lambda_2 P\delta_2 \cdots \lambda_n P\delta_n] = [\lambda_1 E_1 \; \lambda_2 E_2 \cdots \lambda_n E_n]$. Equating columns, it is evident that $AP = P\Lambda$ is equivalent to

$AE_i = \lambda_i E_i$, where $i = 1, 2, \ldots, n$. P will be invertible iff $\{E_1, \ldots, E_n\}$ is a basis. Hence P and Λ exist iff A is diagonalizable.

These considerations are summarized in the next theorem.

THEOREM 6. There exist P and $\Lambda = \text{diag}(\lambda_1, \ldots, \lambda_n)$ such that $A = P\Lambda P^{-1}$ if and only if A is diagonalizable. When such P exists, its columns are a basis of eigenvectors of A, and the λ_i are the corresponding eigenvalues: $P = [E_1 \, E_2 \, \cdots \, E_n]$, and $AE_i = \lambda_i E_i$.
   P and Λ are almost unique. The spectrum and the eigenspaces are unique. Each eigenvalue is repeated as many times as the dimension of the corresponding eigenspace. The indexing of the eigenvalues is arbitrary. For each eigenspace, the basis that is used to form the columns of P is arbitrary.

EXAMPLE 11. Let

$$A = \begin{bmatrix} 0 & 1 & 0 \\ -1 & 0 & 1 \\ 0 & -1 & 0 \end{bmatrix}.$$

Find P so that $P^{-1}AP$ is diagonal, if possible.
   At this point, there are no theorems to tell us, before the computations begin, whether this is possible. By Theorem 6, the eigenvalue problem for A must be solved. We find $p(\lambda) = |A - \lambda I| = -\lambda(\lambda^2 + 2)$, so that $\sigma(A) = \{0, i\sqrt{2}, -i\sqrt{2}\}$. Since A has three distinct eigenvalues, A must be diagonalizable, by Theorem 3. For $\lambda_1 = 0$, solving $(A - \lambda_1 I)x = 0$ yields $x = E_1 = (1, 0, 1)^T$. For $\lambda_2 = i\sqrt{2}$, solving $(A - \lambda_2 I)x = 0$ gives $E_2 = (1, i\sqrt{2}, -1)^T$. Using Proposition 6 of Section 4.1, $\lambda_3 = \bar{\lambda}_2$ and $E_3 = \bar{E}_2 = (1, -i\sqrt{2}, -1)^T$ for the third linearly independent eigenvector. Thus $P = [E_1 E_2 E_3]$ implies that $P^{-1}AP = \text{diag}(0, i\sqrt{2}, -i\sqrt{2})$.

Exercises 4.2

1. $R = \begin{bmatrix} \cos(\theta) & -\sin(\theta) \\ \sin(\theta) & \cos(\theta) \end{bmatrix}$ is the rotation matrix through an angle θ.
   (a) Find an eigenbasis $\{e_1, e_2\}$ for R.
   (b) Compute $R^n x$ if $x = ae_1 + be_2$.
   (c) Compute $R^n x$ if $x = (1, 2)^T$.

2. Let $A = \begin{bmatrix} 0 & 1 & -1 \\ -1 & 2 & -1 \\ -2 & 2 & -1 \end{bmatrix}$ and $e_1 = \begin{bmatrix} 1 \\ 0 \\ -1 \end{bmatrix}$, $e_2 = \begin{bmatrix} 1 \\ 1 \\ 0 \end{bmatrix}$, $e_3 = \begin{bmatrix} 1 \\ 1 \\ 2 \end{bmatrix}$.
   (a) Show that $\{e_1, e_2, e_3\}$ is an eigenbasis for A.
   (b) Find the rank of A.
   (c) Does A^{-1} exist? Explain.
   (d) Find $A^{100}x$, where $x = e_1 + 2e_2 + 3e_3$.
   (e) Find $A^{101}\delta_1$.

3. Let $v_1 = (1, 1, 0, 1)^T$, $v_2 = (1, 0, 1, -1)^T$, $v_3 = (-1, 1, 1, 0)^T$, and $v_4 = (0, -1, 1, 1)^T$. Define a linear operator $T : \mathbf{R}^4 \to \mathbf{R}^4$ by $T(v_1) = v_1$, $T(v_2) = -v_2$, $T(v_3) = v_3$, $T(v_4) = 2v_4$.

(a) Find $T^{-1}(\delta_1)$ and $T^2(\delta_1)$.

(b) Find $T^{-1}(\delta_4)$ and $T^2(\delta_4)$. (*Hint:* $\{v_1, v_2, v_3, v_4\}$ is orthogonal.)

(c) Find the matrix of T^2 in the standard basis. (There are two ways to do this.)

4. Find, if possible, a basis of eigenvectors for the following matrices.

(a) $\begin{bmatrix} 0.1 & 0.5 \\ 0.9 & 0.5 \end{bmatrix}$ (b) $\begin{bmatrix} 0 & -1 & 1 \\ 1 & 2 & -1 \\ -3 & -3 & 4 \end{bmatrix}$ (c) $\begin{bmatrix} 1 & 2 & 0 & 0 \\ 0 & -1 & 0 & 0 \\ 0 & 0 & 1 & 0 \\ 0 & 0 & -1 & 2 \end{bmatrix}$.

5. Find P and Λ so that $P^{-1}AP = \Lambda$ is diagonal, for the following matrices A:

(a) $\begin{bmatrix} 1 & 1 & -1 \\ 1 & 1 & 1 \\ -1 & 1 & 1 \end{bmatrix}$ (b) $\begin{bmatrix} 3 & -1 \\ 2 & 0 \end{bmatrix}$ (c) $\begin{bmatrix} 2 & 2 & 2 \\ -2 & -1 & 1 \\ 2 & 1 & -1 \end{bmatrix}$ (d) $\begin{bmatrix} 0 & 1 \\ -2 & 3 \end{bmatrix}$.

6. Find the general form of a diagonalizable 2×2 matrix having a single eigenvalue $\lambda = \alpha$. Then deduce the requirements in order that a 2×2 matrix be defective.

7. On geometric 3-space, define $T(\mathbf{x}) = \mathbf{w} \times \mathbf{x}$, where $\mathbf{w}$ is a fixed real nonzero vector and $\times$ is the usual cross product.

(a) Find the eigenvalues of T and show that T is diagonalizable.

(b) Is there a real vector $\mathbf{x}$ such that $\mathbf{w} \times \mathbf{x}$ is parallel to $\mathbf{x}$?

(c) If $\mathbf{w} = \mathbf{k}$, find a basis of eigenvectors for T.

8. Let $A = \begin{bmatrix} 0 & 1 \\ -1 & 0 \end{bmatrix}$.

(a) Find an eigenbasis $\{e_1, e_2\}$ for A.

(b) If $x = ae_1 + be_2$, find $A^{101}x$ without computing A^{101}.

(c) Work out A^n by matrix multiplication, and verify part (b).

9. Define $T : \mathbf{T}_1 \rightarrow \mathbf{T}_1$ by $Tf(x) = \int_0^\pi f(x - t)\, dt$.

(a) Find a basis of eigenvectors for T and the corresponding eigenvalues.

(b) Show that T^{-1} exists.

(c) Find $T^{-1}(f)$ and $T^4(f)$, where $f(x) = \cos(x)$.

10. Let $T : \mathbf{P}_2 \rightarrow \mathbf{P}_2$ be defined by $Tp(x) = \frac{1}{2}x(x - 1)p''(x) + xp'(x) + p(x) + x^2p'(0)$.

(a) Find the matrix A of T with respect to the basis $\{1, x, x^2\}$.

(b) Find a basis of eigenvectors for T.

(c) Show that T is invertible, and find $T^{-1}(1 + x + x^2)$.

11. For the following operators $T : \mathbf{C}^{2,2} \rightarrow \mathbf{C}^{2,2}$, find the eigenspaces and spectrum of T. Is T diagonalizable? If T is diagonalizable, find the rank and null space of T.

(a) $T(A) = 2A + A^\mathsf{T}$

(b) $T(A) = \frac{1}{2}(A + A^\mathsf{T})$

(c) $T(A) = \frac{1}{2}(A - A^\mathsf{T})$

(d) $T\left(\begin{bmatrix} a & b \\ c & d \end{bmatrix}\right) = \begin{bmatrix} 2a - 2b & a - b \\ -d & 2c + 3d \end{bmatrix}$.

12. Let $T : V \to V$ be linear, and let $\{f_1, f_2, f_3\}$ be an orthonormal basis for V with respect to $\langle , \rangle$. Suppose $T(f_1) = f_2 + f_3$, $T^2(f_1) = f_3$, and $T^3(f_1) = -2f_1 + 3f_2 + 3f_3$. Find
   (a) the matrix of T with respect to this basis.
   (b) the eigenvalues and eigenvectors of T; is T diagonalizable?
   (c) $\langle T(f_2), f_3 \rangle$, $\langle f_2, T(f_3) \rangle$.

13. In each of the following problems, a matrix A and a subspace V of $\mathbf{R}^3$ are given. The same set of questions is to be answered in each case.

   (a) $A = \begin{bmatrix} 1 & -1 & 2 \\ -2 & 5 & -2 \\ 1 & -1 & 0 \end{bmatrix}$, $V = \{x : x_1 - 2x_2 + x_3 = 0\}$

   (b) $A = \begin{bmatrix} 1 & 1 & 2 \\ 2 & 0 & 0 \\ 1 & -1 & 0 \end{bmatrix}$, $V = \{x : x_1 - x_2 + x_3 = 0\}$

   (c) $A = \begin{bmatrix} 7 & -1 & 0 \\ 3 & 3 & 0 \\ -2 & 0 & 2 \end{bmatrix}$, $V = \{x : x_1 - x_2 = 0\}$

   The questions follow:
   (i) Let $T(x) = Ax$. Show that $T : V \to V$.
   (ii) Find the eigenvectors and eigenvalues of T acting on V. Is T diagonalizable?

14. Consider $A = \begin{bmatrix} 0 & r_1 & r_2 \\ a_1 & 0 & 0 \\ 0 & a_2 & 0 \end{bmatrix}$, with $0 < r_2, r_2 < 1, 0 < a_1, a_2 \le 1$. Let $\lambda_1, \lambda_2, \lambda_3$ be the eigenvalues of A.
   Show that
   (a) A has an eigenvalue $\lambda_1 > 0$.
   (b) The eigenvalues obey $|\lambda_1 \lambda_2 \lambda_3| < 1$.
   (c) λ_2, λ_3 lie in the left half plane.
   (d) A is diagonalizable unless $p(\lambda)$ has a double root, in which case it is defective.

15. Let $T : \mathbf{T}_n \to \mathbf{T}_n$ be defined by $T(f) = f'' + 2f$.
   (a) Show that T has the eigenbasis $\{e^{ikt} : k = -n, \ldots, 0, \ldots, n\}$, and find the eigenvalues of T.
   (b) Compute $T^{10}(\sin(2t) + \cos(t))$.
   (c) Find $T^{-1}(\sin(t) + \cos(3t))$

16. The finite difference operators $\tilde{\delta}$ and δ^2 are defined by the following ($h > 0$ is fixed):
   $$\tilde{\delta} f(x) = f(x + h) - f(x - h), \qquad \delta^2 f(x) = f(x + h) - 2f(x) + f(x - h).$$
   (a) Show that $\tilde{\delta}$, δ^2, and $\delta^2 + I$ are diagonalizable on $\mathbf{T}_n$.
   (b) Find $\tilde{\delta}^3(\sin(2x) + \cos(x))$.
   (c) Find $(\delta^2 + 5I)^{-1}(\cos^2(x) - 2\sin(x))$.

17. Let V be the set of polynomials p of degree not exceeding 4 such that $p(0) = p(1) = 0$, and let $T(p) = p'$, where $' = d/dx$.
   (a) Find the dimension of V.
   (b) Find all of the eigenvalues of T acting on V.
   (c) Explain how this example relates to Theorem 4.

18. Suppose $P : V \to V$ is an orthogonal projection operator onto the subspace $M \neq \{0\}$, $\dim(V) = n$. Show that P is diagonalizable, and find a basis of eigenvectors for P and the corresponding eigenvalues.

19. Let $T : V \to V$ be diagonalizable with eigenbasis $\{e_k\}_{k=1}^n$ and $T(e_i) = \lambda_i e_i$ $\forall i$.
    (a) Find an eigenbasis for T^m, and find the corresponding eigenvalues, for $m = 2, 3, \ldots$.
    (b) Show that $r(T) = r(T^n)$ for $n = 2, 3, \ldots$. What can be said of $N(T^n)$?
    (c) If no eigenvalue of T is zero, show that T^{-1} is diagonalizable. Find an eigenbasis for T^{-1} and the corresponding eigenvalues.

20. Let T be a diagonalizable operator on the n-dimensional vector space V. Suppose there is an integer $p > 1$ such that $T^p = 0$. Show that $T = 0$. Give an example of a nonzero 2×2 matrix $A \neq 0$ such that $A^2 = 0$.

21. Suppose that $\{e_k\}_{k=1}^n$ is an eigenbasis of V for the operator T with $T(e_k) = \lambda_k e_k$, $k = 1, 2, \ldots, n$. Show that if $\mu \notin \sigma(T)$ and $x = \sum_{i=1}^n x_i e_i$, then

$$(T - \mu I)^{-1}(x) = \sum_{i=1}^n x_i e_i.$$

22. Let E and F be nonzero n-tuples, and set $P = EF^H$.
    (a) Find the rank of P.
    (b) Determine when P is diagonalizable, and in that case find an eigenbasis for P. (Hint: Consider $Px = \lambda x$.)

23. Suppose A is an $n \times n$ matrix and B is the matrix of the operator generated by A in the basis $\{P_1, \ldots, P_n\}$. Suppose B has an eigenbasis $\{E_1, \ldots, E_n\}$ and $BE_i = \lambda_i E_i$. Show that A has the same eigenvalues, and find an eigenbasis for A.

24. Suppose A is diagonalizable with eigenbasis $\{e_k\}_{k=1}^n$, $Ae_i = \lambda_i e_i$, and B is similar to A: $B = SAS^{-1}$. Show that B is diagonalizable, and find an eigenbasis for B.

25. Suppose A and B are $n \times n$ matrices, each having the same eigenbasis $\{E_k\}_{k=1}^n$. Show that AB and BA are diagonalizable and that $AB = BA$.

26. Suppose A is diagonalizable and $A = P\Lambda P^{-1}$, where $\Lambda = \mathrm{diag}(\lambda_1, \ldots, \lambda_n)$. Show that $A^m = P\Lambda^m P^{-1}$.

27. (a) Find $\lim_{n \to \infty} \begin{bmatrix} 1 & x/n \\ -x/n & 1 \end{bmatrix}$.

    (b) Find $\lim_{n \to \infty} \begin{bmatrix} 1 & x/n \\ x/n & 1 \end{bmatrix}$.

    (Hint: Use the previous exercise.)

28. Let A be a square matrix.
    (a) Show that if A is diagonalizable, then A^T and A^H are diagonalizable.
    (b) If $\{E_k\}_{k=1}^n$ is an eigenbasis for A and $AE_k = \lambda_k E_k$, find an eigenbasis for A^T and A^H, and find their corresponding eigenvalues.

29. Let A be diagonalizable with a real eigenbasis $E_1, \ldots, E_n$, where $AE_i = \lambda_i E_i$. Let $P = [E_1 \; E_2 \; \cdots \; E_n]$ and suppose P^{-1} has rows F_i^T; i.e., $P^{-1} = \begin{bmatrix} F_1^T \\ \vdots \\ F_n^T \end{bmatrix}$.

Show the following:
(a) $F_i^T A = \lambda_i F_i^T$ (Thus the F_i^T are called left eigenvectors of A.)
(b) Each F_i is an eigenvector of A^T.
(c) $\langle F_i, E_j \rangle = \delta_{ij}$ (Orthogonality of left and right eigenvectors)
(d) $A = \sum_{i=1}^{n} \lambda_i E_i F_i^T$ (The spectral resolution of A by diads)
(e) $I = \sum_{i=1}^{n} E_i F_i^T$ (The spectral resolution of I by diads)
(f) A^T is diagonalizable.
(g) Which of parts (a)–(f) hold when the eigenbasis for A contains complex eigenvectors?

30. A matrix is said to be a *probability matrix* iff every entry is nonnegative and the sum of the entries of each column is 1. Let A be a probability matrix, and let $\mathbf{1} = (1, 1, \ldots, 1)^T$. Show that
(a) A^T has $\mathbf{1}$ as an eigenvector.
(b) A^p is a probability matrix for $p = 1, 2, \ldots$.
(c) Assume that A is diagonalizable and that $\dim(M(1)) = 1$. Show that there is an eigenbasis $\{E_1, \ldots, E_n\}$ for A such that

$$\sum_{i=1}^{n} E_{ij} = \begin{cases} 1 & j = 1 \\ 0 & j > 1 \end{cases},$$

where $E_j = (E_{1j}, E_{2j}, \ldots, E_{nj})^T$.

31. Suppose $B = \begin{bmatrix} A & 0 \\ b & 1 \end{bmatrix}$, where A is an $n \times n$ matrix, b is a row, and 0 is a column of length n.

Suppose A is diagonalizable and no eigenvalue of A is 1. Show that B is diagonalizable, and that, from an eigenbasis for A, an eigenbasis for B can be constructed (with the addition of one more suitable vector). (*Hint:* Use block multiplication.) If $\lambda = 1$ is an eigenvalue of A, give a condition in order that the same result holds.

32. Investigate the eigenvalue problem for $\mathrm{Trid}(a, b, c)$ for the cases $a = 0 \neq c$ and $a \neq 0 = c$.

33. Another way to compute an inverse. Let A be an $n \times n$ matrix, and $\beta = \{E_k\}_{k=1}^{n}$ be an eigenbasis for A, with $AE_i = \lambda_i E_i$. Assume that all $\lambda_i \neq 0$ and that β is orthogonal with respect to the standard inner product. To find $A^{-1}y = x$, we must solve $Ax = y$. Suppose $x = \sum_{i=1}^{n} \alpha_i E_i$ and $y = \sum_{i=1}^{n} \beta_i E_i$. Show that
(a) $\alpha_i = \beta_i / \lambda_i$
(b) $\alpha_i = E_i^H y / (\lambda_i \|E_i\|^2)$
(c) $x = \sum_{i=1}^{n} E_i^H y / (\lambda_i \|E_i\|^2) E_i = A^{-1}y$
(d) $\sum_{i=1}^{n} E_i E_i^H / (\lambda_i \|E_i\|^2) = A^{-1}$
(e) $P \Lambda^{-1} N^{-1} P^H = A^{-1}$, where $P = [E_1 \; E_2 \; \cdots \; E_n]$, $\Lambda = \mathrm{diag}(\lambda_1, \ldots, \lambda_n)$, and $N = \mathrm{diag}(\|E_1\|^2, \ldots, \|E_n\|^2)$
(f) Find an independent way to verify part (e).

4.3
FUNCTIONS OF DIAGONALIZABLE OPERATORS

Two competing methods are used to define a function of a diagonalizable operator. One is more or less traditional, and another uses eigenbases.

The traditional method goes as follows: Powers of T are easy to define: $T^2(x) = TT(x) = T(T(x))$, $T^3(x) = TTT(x) = T(T(T(x)))$, and T^n is simply T applied n times. Set $T^0 = I$. Now polynomials in T can easily be defined: If $p(x) = \sum_{i=0}^n c_i x^i$, then $p(T) = \sum_{i=0}^n c_i T^i$. Next we define rational functions of T: If $r(x) = p(x)/q(x)$, where p and q are polynomials, define $r(T) = p(T)q(T)^{-1}$, provided that $q(T)$ has an inverse. A competing definition is $r(T) = q(T)^{-1}p(T)$. Observe that $r(T)$ may be tedious to calculate if p and q have high degree. Transcendental functions are a bit more difficult. For example, e^T is defined by $e^T = \sum_{k=0}^\infty (1/k!)T^k$. However, the infinite sum must be a limit, and the sense in which the limit is to be taken as well as the convergence of the sum should be discussed. For example, what is $e^{d/dx}$? (Try it.) What is e^T if T is an integral operator? What is e^T if T is a projection? If T is an arbitrary square matrix, it is not hard to show that e^T is a matrix whose entries are found by taking the limit of the entries of $\sum_{k=0}^n T^k/k!$ as $n \to \infty$. (See Section 8.4.) However, even in this case, more problems arise that are discussed later in this section. In general, if $f(x) = \sum_{i=0}^\infty a_i x^i$, then $f(T) = \sum_{i=0}^\infty a_i T^i$. If f has radius of convergence R, for which operators T will $f(T)$ converge? A partial answer to this will be indicated in the exercises. (See Exercise 16.) The defining infinite sum cannot always be found in closed form and can be very difficult to obtain accurately on computers. However, the series expression is very useful for certain theoretical calculations. Another drawback of the traditional method is that it is not so obvious how to define $f(T)$ for other functions, such as $\sqrt{T}$, $|T|$ (absolute value of T), or $\log(T)$.

If an eigenbasis is known, it turns out to be easy to define functions of diagonalizable operators, using the idea that a linear operator is completely determined by its values on a basis (see Proposition 1 of Section 2.3). To motivate the definition, consider $f(x) = x^k$, where $k = 1, 2, \ldots$. If $x = \sum_{i=1}^n x_i e_i$ and $T(e_i) = \lambda_i e_i$, then by Theorem 2 of Section 4.2,

$$f(T)(x) \equiv T^k(x) = \sum_{i=1}^n x_i \lambda_i^{\,k} e_i = \sum_{i=1}^n x_i f(\lambda_i)e_i. \tag{*}$$

This idea is generalized to formulate the next definition.

DEFINITION. Suppose T is a diagonalizable operator on V with eigenbasis $\{e_k\}_{k=1}^n$, $T(e_i) = \lambda_i e_i$, and suppose f is an arbitrary function whose domain of definition includes the spectrum of T. The *linear operator* $f(T)$ is defined by

$$f(T)\left(\sum_{i=1}^n x_i e_i\right) = \sum_{i=1}^n f(\lambda_i)x_i e_i. \tag{1}$$

The definition is unambiguous since a linear operator is completely specified by its action on one basis. The definition is equivalent to specifying that $f(T)$ is linear and

$$f(T)(e_i) = f(\lambda_i)e_i, \qquad i = 1, 2, \ldots, n. \tag{2}$$

An immediate and important consequence of the definition is stated in the next theorem.

THEOREM 1. A function of a diagonalizable operator is diagonalizable. More specifically, if $\{e_k\}_{k=1}^n$ is an eigenbasis for T on V with corresponding eigenvalues $\{\lambda_k\}_{k=1}^n$, and if $f(T)$ is defined, then $f(T)$ has the *same* eigenbasis as T, but with corresponding eigenvalues $f(\lambda_i)$, $i = 1, 2, \ldots, n$. Furthermore,

$$[f(T)]_{\{e_1,\ldots,e_n\}} = \text{diag}(f(\lambda_1), f(\lambda_2), \ldots, f(\lambda_n)). \tag{3}$$

The next task is to verify that the eigenbasis definition agrees with the traditional definition when they both apply. This fact is verified here for polynomials in T and with the series definition of e^T. The verification for rational functions or other power series is left to the exercises. (See Exercise 15.)

If $p(x) = \sum_{k=0}^m c_k x^k$ and $x = \sum_{i=1}^n x_i e_i$, then, using $(*)$ and the definition (1),

$$p(T)(x) \equiv \sum_{i=1}^n p(\lambda_i) x_i e_i = \sum_{i=1}^n \left(\sum_{k=0}^m c_k \lambda_i^k \right) x_i e_i = \sum_{k=0}^m c_k \left(\sum_{i=1}^n \lambda_i^k x_i e_i \right)$$

$$= \sum_{k=0}^n c_k T^k(x).$$

Thus $p(T) = \sum_{k=0}^n c_k T^k$, which is the traditional definition. For e^T, if $x = \sum_{i=1}^n x_i e_i$, then (formally)

$$e^T(x) \equiv \sum_{i=1}^n e^{\lambda_i} c_i e_i = \sum_{i=1}^n \left(\sum_{k=0}^\infty \frac{\lambda_i^k}{k!} \right) c_i e_i = \sum_{k=0}^\infty \frac{1}{k!} \left(\sum_{i=1}^n \lambda_i^k c_i e_i \right)$$

$$= \sum_{k=0}^\infty \frac{1}{k!} T^k(x).$$

Hence $e^T = \sum_{k=0}^\infty T^k/k!$, which is the traditional definition.

Notice that there is a tradeoff between the theoretical and computational aspects of the competing definitions of functions of T. The traditional method is valid for all operators but is limited to a smaller class of functions $f(x)$, since it is not always clear how to define $f(T)$ for certain functions. The traditional method has certain theoretical advantages, a few of which will be illustrated. With this method, $f(T)$ can be very hard to compute for complicated functions; even e^T can be troublesome on a computer when T is a square matrix. Some reasons for this are that the powers T^n can be difficult to compute for large matrices or they may be very inaccurately computed if the matrix T is badly scaled. See Golub and Van Loan (1989). The eigenbasis definition is restricted to diagonalizable operators and, for computational work, requires the calculation of an eigenbasis before anything else can be done. But once the eigenbasis is known, it is possible to tell whether or not $f(T)$ is defined and to compute it easily. Fortunately, now excellent and stable algorithms exist that compute an eigenbasis for a wide range of matrices, even for some fairly large ones. In addition, many theoretical advantages will follow from the definition involving eigenbases that would be harder to obtain from the traditional definition.

A more complicated formula similar to (1) exists for nondiagonalizable operators. It is based on Hermite polynomial interpolation and depends on quantities that are nearly impossible to compute, except in special cases. See N. Dunford and J. T. Schwartz (1958), for a discussion of this.

EXAMPLE 1. Some formal examples of the definition of $f(T)$ follow, assuming that $\{e_k\}_{k=1}^n$ is an eigenbasis for V and $T(e_i) = \lambda_i e_i$.

(a) $|T|(\sum_{i=1}^n x_i e_i:) = \sum_{i=1}^n |\lambda_i| x_i e_i$

(b) $\sqrt{T}(\sum_{i=1}^n x_i e_i) = \sum_{i=1}^n \sqrt{\lambda_i} x_i e_i$, where the branch of the square root must be specified. Clearly $\sqrt{T}\sqrt{T} = T$, no matter how the square root branch is defined. If one uses only real roots, then $\sqrt{T}$ is not defined when a negative real number or a complex number is in the spectrum of T.

(c)

$$(1 - T + 3T^3)(1 + T^2)^{-1}\left(\sum_{i=1}^n x_i e_i:\right) = \sum_{i=1}^n \frac{1 - \lambda_i + 3\lambda_i^2}{1 + \lambda_i^2} x_i e_i,$$

provided that $\pm i$ are not in the spectrum of T.

The next few examples use some of the examples of Section 4.2. The definition of functions of diagonalizable operators is used to find the various functions of the operator.

EXAMPLE 2. Define $T : \mathbf{P}_2 \to \mathbf{P}_2$ by

$$T p(x) = \frac{3}{8}\int_{-1}^1 [1 + 4(x + t) + 5(x^2 + t^2) - 15x^2 t^2]p(t)\, dt. \tag{#}$$

Find T^{-1}, 2^T, and T^{10}, and apply them to the function t^2.

In Example 8 of Section 4.2, it was found that the eigenvalues of T are -1, 3 and that $M(-1) = \text{span}\{3t - 1, 3t^2 - 1\}$, $M(3) = \text{sp}\{t + 1\}$. Let $e_0 = 3t - 1$, $e_1 = 3t^2 - 1$, and $e_2 = t + 1$, so that $T(e_i) = \lambda_i e_i$, with $\lambda_0 = -1$, $\lambda_1 = -1$, and $\lambda_2 = 3$. The operators $T^{-1}, 2^T$, and T^{10} are not found in closed form, but rules are provided by which to compute them. Suppose $p = c_0 e_0 + c_1 e_1 + c_2 e_2$. Since no eigenvalue of T is zero, T^{-1} exists and is given by

$$T^{-1}(p) = \lambda_0^{-1} c_0 e_0 + \lambda_1^{-1} c_1 e_1 + \lambda_2^{-1} c_2 e_2 = -(c_0 e_0 + c_1 e_1) + \tfrac{1}{3} c_2 e_2. \tag{*}$$

(1) was used with $f(x) = 1/x$. See also Theorem 2 of Section 4.2. For example, to find $T^{-1}(t^2)$, t^2 must be expressed in terms of the e_i: Find c_i so that

$$t^2 = c_0 e_0 + c_1 e_1 + c_2 e_2 = c_0(3t - 1) + c_1(3t^2 - 1) + c_2(t + 1).$$

These are $c_0 = -\tfrac{1}{12}$, $c_1 = \tfrac{1}{3}$, and $c_2 = \tfrac{1}{4}$; and now (*) may be used to obtain $T^{-1}(t^2) = \tfrac{1}{3}(1 + t) - t^2$. Next,

$$2^T(p) = 2^{\lambda_0} c_0 e_0 + 2^{\lambda_1} c_1 e_1 + 2^{\lambda_2} c_2 e_2 = \tfrac{1}{2}(c_0 e_0 + c_1 e_1) + 8 c_2 e_2.$$

Thus $2^T(t^2) = -\tfrac{1}{24}(3t - 1) + \tfrac{1}{6}(3t^2 - 1) + 2(t + 1)$.

Finally, for T^{10}, use $f(x) = x^{10}$. Then

$$T^{10}(p) = \lambda_0^{10} c_0 e_0 + \lambda_1^{10} c_1 e_1 + \lambda_2^{10} c_2 e_2 = c_0 e_0 + c_1 e_1 + c_2 3^{10} e_2,$$

so that

$$T^{10}(t^2) = -\tfrac{1}{12}(3t - 1) + \tfrac{1}{3}(3t^2 - 1) + \tfrac{1}{4}3^{10}(t + 1).$$

To compute this using (#) directly, one would have to apply T ten times, i.e., compute $p_1 = T(t^2)$, $p_2 = T(p_1)$, $p_3 = T(p_2)$, ..., $p_{10} = T(p_9) = T^{10}(t^2)$.

EXAMPLE 3. Let $V = \mathbf{T}_n$ and define T by $Tf = f''$, where $' = d/dx = D$. Find e^T.

See Example 2 of Section 4.2. An eigenbasis is $\{1, \cos(jx), \sin(jx) : j = 1, 2, \ldots, n\}$. The eigenvector 1 has the eigenvalue $\lambda_0 = 0$, whereas the eigenvectors $\cos(jx)$, $\sin(jx)$ correspond to the eigenvalue $\lambda_j = -j^2$. Thus if $\phi(x) = a_0 + \sum_{i=1}^n [a_k \cos(kx) + b_k \sin(kx)]$, then, writing $T = D^2$,

$$(e^{D^2}\phi)(x) = e_0 a_0 + \sum_{k=1}^n e^{-k^2}(a_k \cos(kx) + b_k \sin(kx))$$

Functions of Matrices

When the operator is a matrix operator $T(x) = Ax$, a more compact formula can be found for the expression (1) of the definition of $f(T)$, provided that the standard basis is used. One must be sure to distinguish between the operator, which is independent of basis, and the matrix of the operator in a particular basis, which changes every time the basis is changed.

DEFINITION. Let $T(x) = Ax$, where A is an $n \times n$ matrix, and let δ be the standard basis $\delta = \{\delta_k\}_{k=1}^n$. If T (i.e., A) is diagonalizable, then $f(A)$ is defined to be the matrix of $f(T)$ in δ:

$$f(A) = [f(T)]_\delta. \tag{4}$$

According to previous work (cf. Section 2.3), this is equivalent to

$$f(A)x = f(T)(x) \ \forall x \in \mathbf{F}^n. \tag{5}$$

Thus, conversely, $f(T)$ is the matrix operator generated by $f(A)$. From the definition of $f(T)$ and from (5), if $\{E_k\}_{k=1}^n$ is an eigenbasis for A, and with $AE_i = \lambda_i E_i$, then $f(A)$ obeys

$$f(A)E_i = f(\lambda_i)E_i, \tag{6a}$$

$$f(A)\left(\sum_{i=1}^n c_i E_i\right) = \sum_{i=1}^n f(\lambda_i)c_i E_i, \tag{6b}$$

since $f(A)E_i = f(T)(E_i) = f(\lambda_i)E_i$.

EXAMPLE 4. If $D = \text{diag}(d_1, d_2, \ldots, d_n)$, and if all $f(d_i)$ are defined, then

$$f(D) = \text{diag}(f(d_1), f(d_2), \ldots, f(d_n)) \tag{7}$$

Proof. Let $T(x) = Dx \ \forall x$. Since $D\delta_i = d_i\delta_i$, the standard basis is an eigenbasis with corresponding eigenvalues d_i. From (6),

$$f(D)x = f(D)\left(\sum_{i=1}^n x_i \delta_i\right) = \sum_{i=1}^n f(d_i)x_i\delta_i = \text{diag}(f(d_1), f(d_2), \ldots, f(d_n))x.$$

Thus $f(D) = \text{diag}(f(d_1), f(d_2), \ldots, f(d_n))$. Alternatively, Theorem 1 of Section 4.2 immediately yields $f(D) \equiv [f(T)]_\delta = \text{diag}(f(d_1), f(d_2), \ldots, f(d_n))$ since $\delta = \{\delta_k\}_{k=1}^n$ is an eigenbasis.

Remark. Equation (3) of Theorem 1 may now be more succinctly stated as

$$[f(T)]_e = f(\Lambda), \qquad e = \{e_1, e_2, \ldots, e_n\}. \tag{3}$$

Now $f(A)x$ can be computed according to (6b). However, one often wants a compact formula for $f(A)$. The next theorem provides this and illustrates several techniques.

THEOREM 2. Let A be a diagonalizable $n \times n$ matrix, let $\beta = \{E_i : i = 1, 2, \ldots, n\}$ be an eigenbasis of A, with $AE_i = \lambda_i E_i$, and suppose all $f(\lambda_i)$ are defined. Let $\Lambda = \text{diag}(\lambda_1, \lambda_2, \ldots, \lambda_n)$, and let $P = [E_1 \ E_2 \ \cdots \ E_n]$. Then

$$f(A) = Pf(\Lambda)P^{-1}. \tag{8}$$

Proof. First we give a proof using matrix identities. Since $x = \sum_{i=1}^{n} y_i E_i = Py$,

$$f(A)x = f(T)\left(\sum_{i=1}^{n} y_i E_i\right) = \sum_{i=1}^{n} f(\lambda_i)y_i E_i = [f(\lambda_1)E_1, f(\lambda_2)E_2, \ldots, f(\lambda_n)E_n]y$$

$$= [E_1 \ E_2 \ \cdots \ E_n] \, \text{diag}(f(\lambda_1), f(\lambda_2), \ldots, f(\lambda_n))y = Pf(\Lambda)P^{-1}x.$$

Since this is valid for every x, (8) follows.

A second proof using matrix identities is

$$f(A)P = f(A)[E_1 \ E_2 \ \cdots \ E_n] = [f(A)E_1 \ f(A)E_2 \cdots f(A)E_n]$$

$$= [f(\lambda_1)E_1 \ f(\lambda_2)E_2 \cdots f(\lambda_n)E_n]$$

$$= [E_1 \ E_2 \ \cdots \ E_n]\text{diag}(f(\lambda_1), f(\lambda_2), \ldots, f(\lambda_n))$$

$$= Pf(\Lambda).$$

Another proof uses Theorem 1, which implies that $[f(T)]_\beta = f(\Lambda)$. The change of basis formula for linear operators on $\mathbf{F}^n$ says $f(\Lambda) = P^{-1}f(A)P$, which yields (8).

Finally, (8) also follows from Theorem 6 of Section 4.2 if A in that theorem is replaced by $f(A)$ and Λ is replaced by $f(\Lambda)$. These replacements are valid, by Theorem 1 and Example 4.

EXAMPLE 5. Let

$$A = \begin{bmatrix} 0 & 1 \\ -2 & 3 \end{bmatrix}.$$

Find $(A-1)^{2,000}$, e^A, and $(3A - 5I)(1 + 4A - 4A^2 + A^3)^{-1}$.

The characteristic polynomial of A is $p(\lambda) = (\lambda-1)(\lambda-2)$. Let $\lambda_1 = 1$ and $\lambda_2 = 2$. The corresponding eigenvectors are $E_1 = (1, 1)^\mathsf{T}$ and $E_2 = (1, 2)^\mathsf{T}$. Hence

$$P = \begin{bmatrix} 1 & 1 \\ 1 & 2 \end{bmatrix}, \qquad P^{-1} = \begin{bmatrix} 2 & -1 \\ -1 & 1 \end{bmatrix}, \qquad \Lambda = \begin{bmatrix} 1 & 0 \\ 0 & 2 \end{bmatrix}.$$

Then

$$(A - I)^{2,000} = P(\Lambda - I)^{2,000}P^{-1} = P \, \text{diag}(0^{2,000}, 1^{2,000})P^{-1} = \begin{bmatrix} -1 & 1 \\ -2 & 2 \end{bmatrix}.$$

$$e^A = Pe^\Lambda P^{-1} = P \, \text{diag}(e, e^2)P^{-1} = \begin{bmatrix} 2e - e^2 & e^2 - e \\ 2e - 2e^2 & 2e^2 - e \end{bmatrix},$$

and, setting $r(x) = (3x - 5)/(1 + 4x - 4x^2 + x^3)$,

$$(3A - 5I)(1 + 4A - 4A^2 + A^3)^{-1} = Pr(\Lambda)P^{-1} = P\operatorname{diag}(r(1), r(2))P^{-1}$$

$$= P\operatorname{diag}(-1, 1)P^{-1} = \begin{bmatrix} -3 & 2 \\ -4 & 3 \end{bmatrix}.$$

The next corollary allows the computation of $f(T)$ from its matrix representation in arbitrary vector spaces.

COROLLARY 3. Let $T : V \to V$ be diagonalizable, and let $A = [T]$ in some basis $\gamma = \{f_1, f_2, \ldots, f_n\}$. Then, in the same basis,

$$[f(T)] = f(A). \tag{9}$$

Proof. The proof is left as an exercise. (See Exercise 13.)

EXAMPLE 6. $T : \mathbf{P}_2 \to \mathbf{P}_2$ is defined by

$$Tp(x) = \frac{3}{8} \int_{-1}^{1} [1 + 4(x + t) + 5(x^2 + t^2) - 15x^2t^2]p(t)\, dt.$$

Find $\sqrt{3T + 7I}(h)$, where $h(t) = (t + 1)^2$.

In Example 8 of Section 4.2, the matrix A of T with respect to the basis $\{1, t, t^2\}$ was constructed, and a basis of eigenvectors for A was found to be $E_1 = (-1, 3, 0)^T$, $E_2 = (-1, 0, 3)^T$, and $E_3 = (1, 1, 0)^T$, with corresponding eigenvalues -1, -1, and 3. The change of basis matrix is $P = [E_1 \; E_2 \; E_3]$ and $\Lambda = \operatorname{diag}(-1, -1, 3)$. For $f(x) = \sqrt{3x + 7}$, $f(\Lambda) = \operatorname{diag}(2, 2, 4)$, so that

$$f(A) = [f(T)] = Pf(\Lambda)P^{-1} = \begin{bmatrix} 3.5 & 0.5 & 0.5 \\ 1.5 & 2.5 & 0.5 \\ 0 & 0 & 2 \end{bmatrix}.$$

Since $[h] = (1, 2, 1)^T$ and $[f(T)][h] = (5, 7, 2)^T = [f(T)h]$, it follows that $f(T)(h) = 5 + 7t + 2t^2$.

General Properties of Functions of Diagonalizable Operators

Now let us turn to some useful general facts about functions of diagonalizable operators. Some of them are surprising.

THEOREM 4. Suppose $T : V \to V$ is a linear diagonalizable operator, where V is a nontrivial finite-dimensional vector space. Suppose that $f(T)$ and $g(T)$ are well defined functions of T. Then

(a) if $f(\lambda) = g(\lambda)$ for all λ in the spectrum of T, then $f(T) = g(T)$.
(b) $f(T)$ and $g(T)$ commute: $f(T)g(T) = g(T)f(T)$.

Proof. Let $\{e_k\}_{k=1}^{n}$ be a basis, and let $T(e_i) = \lambda_i e_i$.

(a) Since $f(\lambda_i) = g(\lambda_i)$ for all i, $g(T)(\sum_{i=1}^{n} x_i e_i) = \sum_{i=1}^{n} g(\lambda_i)x_i e_i = \sum_{i=1}^{n} f(\lambda_i)x_i e_i = f(T)(\sum_{i=1}^{n} x_i e_i)$ for all x in V. Hence $f(T) = g(T)$.

(b) $f(T)g(T)(\sum_{i=1}^{n} x_i e_i) = f(T)(\sum_{i=1}^{n} x_i g(\lambda_i)e_i) = \sum_{i=1}^{n} x_i f(\lambda_i)g(\lambda_i)e_i$. Similarly, $g(T)f(T)(\sum_{i=1}^{n} x_i e_i) = \sum_{i=1}^{n} x_i f(\lambda_i)g(\lambda_i)e_i$. Thus $f(T)g(T)x = g(T)f(T)x \; \forall x$, whence $f(T)g(T) = g(T)f(T)$.

The next theorem is a quirk of finite dimensionality and the interpolation properties of polynomials. Sometimes it can be quite useful. It also gives us the remarkable fact that there are no transcendental functions of diagonalizable operators on finite-dimensional vector spaces—there are only polynomial functions of such operators. This theorem also implies that if the spectrum of an operator is known, $f(T)$ may be computed without the knowledge of an eigenbasis. A similar but more complicated theorem is valid for nondiagonalizable operators.

THEOREM 5. Let T be a diagonalizable operator on the nontrivial finite-dimensional vector space V. Then every function of T is a polynomial in T. In fact, if the spectrum of T has m distinct points, then for every $f(T)$ that is well defined, there is a polynomial p of degree not exceeding $m - 1$ such that $f(T) = p(T)$.

Proof. Let $\sigma = \{\lambda_1, \lambda_2, \ldots, \lambda_m\}$ be the spectrum of T, and suppose f is defined on σ. The Lagrange interpolation theorem asserts that there is a polynomial of degree not exceeding $m - 1$, such that $f(\lambda_i) = p(\lambda_i)$, for $i = 1, 2, \ldots, m$. Hence $f(T) = p(T)$, by Theorem 4, part (a).

Remark. For defective operators, it is still true that $f(T)$ and $g(T)$ commute when f and g are rational functions for which $f(T)$ and $g(T)$ are defined. See Exercise 15.

EXAMPLE 7. Let

$$A = \begin{bmatrix} 0 & 1 \\ -2 & 3 \end{bmatrix}.$$

Find polynomials p and q such that $A^{-1} = p(A)$ and $2^A = q(A)$.

This matrix was considered in Example 5. There it was found that $\lambda_1 = 1$ and $\lambda_2 = 2$. According to Theorem 5, every function of A is actually a linear polynomial in A: $f(A) = \alpha I + \beta A$. According to Theorem 4, if $f(\lambda) = \alpha + \beta\lambda$ for each λ in the spectrum of A, then $f(A) = \alpha I + \beta A$. Thus to find α and β, one must solve the equations $f(\lambda_1) = \alpha + \beta\lambda_1$, $f(\lambda_2) = \alpha + \beta\lambda_2$. First consider $A^{-1} = f(A)$, where $f(x) = x^{-1}$. The equations $1 = \alpha + \beta$, $\frac{1}{2} = \alpha + 2\beta$ have the solution $\alpha = \frac{3}{2}$, $\beta = -\frac{1}{2}$, so $A^{-1} = \frac{1}{2}(3 - A)$, which is easily checked by multiplication. For $f(x) = 2^x$, the equations are $2 = \alpha + \beta$, $2^2 = \alpha + 2\beta$, and the solution is $\alpha = 0$, $\beta = 2$. Hence $2^A = 2A$. Theorem 2 could have been used to find each $f(A)$, but then $P = [E_1 \ E_2]$ would be needed.

If T is diagonalizable, there are many functions for which $f(T) = 0$: It only requires that $f(\lambda) = 0$ for all $\lambda \in \sigma(T)$. The next theorem examines this more closely.

THEOREM 6. Let $T : V \to V$ be a diagonalizable linear operator, and let $\dim(V) = n \neq 0, \infty$.

(a) If p is the characteristic polynomial of T, then $p(T) = 0$.
(b) If $\sigma(T) = \{\lambda_1, \lambda_2, \ldots, \lambda_m\}$ and $q(x) = (x - \lambda_1)(x - \lambda_2) \cdots (x - \lambda_m)$, then $q(T) = 0$.

Proof

(a) Since $p(\lambda_i) = 0$ for all eigenvalues of T, by (1) of the definition or by Theorem 4, $p(T) = 0$.
(b) (Note that there are no repetitions in the list of eigenvalues in $\sigma(T)$, so these are the distinct eigenvalues of T.) Since $q(\lambda) = 0$ for all λ in the spectrum, $q(T) = 0$ for the same reasons.

DEFINITION. The polynomial defined in statement (b) of Theorem 6 is called the *minimal polynomial* of the diagonalizable operator T. For defective operators, the minimal polynomial of an operator T is defined to be the polynomial q of smallest degree and leading coefficient 1 such that $q(T) = 0$.

Remark. Part (a) of Theorem 6 is called the Cayley–Hamilton theorem and is true even if the assumption that T be diagonalizable is removed. Part (b) is not true when the operator is not diagonalizable, as a rule. (See the exercises in Section 8.3 for an outline of the proof of the Cayley–Hamilton theorem.)

The procedure in the next example is valid for arbitrary operators.

EXAMPLE 8. REDUCTION OF DEGREE OF POLYNOMIALS OF OPERATORS BY LONG DIVISION. Let $g(x)$ be the given polynomial, and let T be the given operator. Suppose $q(x)$ is the characteristic polynomial or the minimal polynomial of T. By long division by $q(x)$, compute $m(x)$ and $r(x)$, with $r(x)$ of smaller degree than $q(x)$, so that $g(x) = m(x)q(x) + r(x)$. Then, since $q(T) = 0$, $g(T) = m(T)q(T) + r(T) = r(T)$.

EXAMPLE 9. Consider T defined on $\mathbf{P}_2$ by

$$Tp(x) = \frac{3}{8}\int_{-1}^{1}[1 + 4(x + t) + 5(x^2 + t^2) - 15x^2t^2]p(t)\,dt.$$

Reduce T^4 to a lower-degree polynomial in T using the methods of Example 7 and Theorems 4 and 5.

In Example 8 of Section 4.2, it was found that T is diagonalizable and that $\sigma(T) = \{-1, 3\}$, so the minimal polynomial is $q(x) = (x-3)(x+1) = x^2 - 2x - 3$. q can be used to reduce *every* polynomial in the operator T to a linear function of T. For $g(x) = x^4$, long division gives

$$
\begin{array}{r}
x^2 + 2x + 7 \\
x^2 - 2x - 3 \overline{)\, x^4 } \\
\underline{x^4 - 2x^3 - 3x^2} \\
2x^3 + 3x^2 \\
\underline{2x^3 - 4x^2 + 6x} \\
7x^2 - 6x \\
\underline{7x^2 - 14x - 21} \\
20x + 21.
\end{array}
$$

Hence $m(x) = x^2 + 2x + 7$, $r(x) = 20x + 21$, and, by the above comments, $T^4 = 20T + 21I$.

According to Theorems 4 and 5, this could also have been done by solving this interpolation problem: Find $p(x) = ax + b$ such that $p(-1) = -a + b = (-1)^4 = 1$, and $p(3) = 3a + b = (3)^4 = 81$. These equations give $a = 20$, $b = 21$, so that $p(T) = T^4$, which agrees with the answer given by the long-division process.

Exercises 4.3

1. Let $A = \begin{bmatrix} 3 & -1 \\ 2 & 0 \end{bmatrix}$.

Find
  (a) a and b so that $A^{-1} = aI + bA$
  (b) a and b so that $2^A = aI + bA$
  (c) e^{At}
  (d) $(1 + 4A - 6A^2 + 2A^3)^{-1}(A^2 - A - I)$
  (e) $\sqrt{3A - 2I}$

2. Let $A = \begin{bmatrix} 1 & 1 \\ -1 & 1 \end{bmatrix}$.

Find
  (a) A^{12}
  (b) $(A - I)^{102}$
  (c) $e^{\pi(I-A)/2}$

3. Let $A = \begin{bmatrix} -1 & -2 & -2 \\ 1 & 2 & 1 \\ -1 & -1 & 0 \end{bmatrix}$.

Find
  (a) (i) A^{-100} (ii) A^{101} (iii) 2^A
  (b) The minimal polynomial of A
  (c) The expressions for the following in the form $aI + bA$:
      (i) A^{-1} (ii) 2^A (iii) $(A + 1)^4$

4. Suppose T is a linear operator on a three-dimensional space V and has eigenvalues $-1, 0, 2$. Express the following as quadratics in T:
  (a) $(T + 2I)^{-1}$
  (b) 2^T
  (c) $e^{\pi i T}$

5. Let T be a linear operator on $\mathbf{C}^4$ such that $T(v_1) = v_1$, $T(v_2) = 2v_2$, $T(v_3) = -v_3$, and $T(v_4) = v_4$, where $v_1 = (0, 1, -1, 1)^T$, $v_2 = (-1, 1, 1, 0)^T$, $v_3 = (1, 0, 1, 1)^T$, and $v_4 = (1, 1, 0, -1)^T$.
  (a) Show that T is invertible.
  (b) Find $T^{-1}(\delta_2)$.
  (c) Find $e^T(\delta_2)$.
  (d) Find a matrix B so that $T^3(x) = Bx \ \forall x$.

6. The matrix

$$A = \begin{bmatrix} 2 & 1 & 2 \\ -1 & 0 & -2 \\ 1 & 1 & 3 \end{bmatrix}$$

has eigenvectors

$$E_1 = \begin{bmatrix} 1 \\ -1 \\ 0 \end{bmatrix}, \qquad E_2 = \begin{bmatrix} -1 \\ 3 \\ -1 \end{bmatrix}, \qquad E_3 = \begin{bmatrix} 1 \\ -1 \\ 1 \end{bmatrix}.$$

    (a) Find the eigenvalues of A.

    (b) Find $(A - 2I)^{20}$.

    (c) Find $\sqrt{A}$.

    (d) Find $f(t)$ and $g(t)$ so that $e^{At} = f(t)I + g(t)A$.

7. Define $T : \mathbf{P}_2 \to \mathbf{P}_2$ by $(Tp)(x) = (1 - x^2)p''(x) - xp'(x) + 2p(x)$.

    (a) Show that T is diagonalizable.

    (b) Find $2^{T/2}(f)$, where $f(x) = x^2$.

8. Let $T : \mathbf{P}_2 \to \mathbf{P}_2$ be defined by $Tp(x) = \frac{1}{2}x(x - 1)p''(x) + xp'(x) + p(x) + x^2p'(0)$. This operator was considered in Exercise 10 of Section 4.2 and has an eigenbasis $e_1 = 1$, $e_2 = x - x^2$, $e_3 = x + x^2$ with corresponding eigenvalues $\lambda_1 = 1$, $\lambda_2 = 1$, $\lambda_3 = 3$.

    (a) Express T^{-1} and 2^T in the form $aI + bT$.

    (b) Find the eigenvalues and eigenvectors of T^{-1} and 2^T.

    (c) If $p(x) = x^2 + x + 1$, find $2^T(p)$ and $T^{-1}(p)$.

    (d) Find $(T - 2I)^{10}$.

    (e) Find $[2^T]_\beta$, where $\beta = \{1, x, x^2\}$.

9. Define $T : \mathbf{P}_2 \to \mathbf{P}_2$ by $Tp(x) = \frac{1}{2}\int_{-1}^{1}[1 + 3t + 3tx + 3x^2]p(t)\,dt$. Let $f(x) = x$. Find

    (a) The matrix A of T with respect to the basis $1, x, x^2$.

    (b) The eigenvalues and eigenvectors of T.

    (c) A basis for $R(T)$ and $N(T)$.

    (d) $T^4(f)$, $(T + I)^{-1}(f)$, and $(\cos(\pi T))(f)$.

10. For each of the following operators $T : \mathbf{T}_n \to \mathbf{T}_n$, find $e^T(p)$, $p \in \mathbf{T}_n$.

    (a) $T = i\dfrac{d}{dx}$

    (b) $T = \tilde{\mu}$, where $\tilde{\mu}f(x) = \frac{1}{2}[f(x + h) + f(x - h)]$ for fixed $h > 0$.

11. Let $A = \text{trid}(-1, 2, -1)$ be an $n - 1 \times n - 1$ matrix. Find the eigenvalues and eigenvectors of $f(T) = 6n^2T(6I - T)^{-1}$.

12. Let $T : V \to V$ be diagonalizable, and let $\dim(V) < \infty$. Show that no polynomial $p(x)$ of degree less than that of the minimal polynomial can have the property $p(T) = 0$.

13. Prove Corollary 3: Let $T : V \to V$ be linear, and let $\phi = \{f_k\}_{k=1}^{n}$ be a basis for V. Show that $[f(T)]_\phi = f([T]_\phi)$.

14. Let $f(x) = p(x)/q(x)$, where p and q are polynomials. If $q(\lambda) \neq 0$ for all λ in the spectrum of T, then $f(T) = p(T)q(T)^{-1} = q(T)^{-1}p(T)$.

15. Let $f(x)$ and $g(x)$ be rational functions, and let T be a linear operator on V.

    (a) Explain what is meant by the "traditional definition" of $f(T)$.

    (b) Assume that $f(T)$ and $g(T)$ are well defined by the traditional definition. (T need not be diagonalizable.) Show that $f(T)g(T) = g(T)f(T)$ agrees with the definition given in the text when T is diagonalizable.

16. Let $f(x) = \sum_{k=0}^{\infty} a_k x^k$ have radius of convergence R. Suppose T is a diagonalizable operator and all its eigenvalues satisfy $|\lambda_i| < R$. Show that the traditional and eigenbasis definitions of $f(T)$ agree, formally. Where is the restriction $|\lambda_i| < R$ used in the calculations?

17. Find the spectral resolution of $f(A)$ by diads, when A is diagonalizable. (See Exercise 29 in Section 4.2 for the definitions.)

18. Let

$$R_\theta = \begin{bmatrix} \cos(\theta) & -\sin(\theta) \\ \sin(\theta) & \cos(\theta) \end{bmatrix}.$$

Find R_θ^n.

4.4
FIRST-ORDER MATRIX DIFFERENTIAL EQUATIONS

Consider the first-order homogeneous system of differential equations

$$x' = Ax, \qquad ' = \frac{d}{dt}, \tag{1}$$

where A is an $n \times n$ constant matrix and x is a vector function. Observe that when the system (1) is written for each component, it becomes $x_i' = \sum_{j=1}^{n} a_{ij}x_j$. Unless A is diagonal, the equations are "coupled" in the sense that the derivative of x_i depends on x_j when $a_{ij} \neq 0$, making the problem hard to solve. In this section, a method for finding the general solution of (1) when A is diagonalizable is derived. An example of an inhomogeneous system, $x' = Ax + f(t)$, is also considered. A general theory is not developed here. One point to be observed here is the computational desirability of diagonalizable matrices.

First, we consider a method of finding *some* solutions that will point the way to the general solution.

The method of trial solutions

By analogy to a scalar differential equation $x' = ax$, we can look for solutions of (1) of the form $x(t) = e^{\lambda t}c$, where c is a nonzero constant column. Putting this x in (1), we have $x' = \lambda e^{\lambda t}c = e^{\lambda t}Ac$, so that $e^{\lambda t}c$ is a solution of (1) $\Leftrightarrow Ac = \lambda c \Leftrightarrow \lambda$ is an eigenvalue of A and c is a corresponding eigenvector. If we have several such solutions $e^{\lambda_i t}c_i$, where $i = 1, 2, \ldots, r$ we can make a linear combination (also called a superposition) of them $x = \sum_{i=1}^{r} \alpha_i e^{\lambda_i t}c_i$, and the result is also a solution. This is verified by substitution into (1). The question arises: When is such a solution the complete (i.e., general) solution of (1)? The answer is: Whenever the eigenvectors form a basis. This follows from the subsequent calculations. Two different approaches to finding the complete solution of the differential equation are given.

Vector space method of solution

Find a basis of eigenvectors $\{E_i\}_{i=1}^{n}$ with $AE_i = \lambda_i E_i$. Now make a change of basis to the eigenbasis $x = \sum_{i=1}^{n} y_i E_i$. Note that the eigenvectors are constant, and the coordinate functions $y_i = y_i(t)$ must exist, since $\{E_k\}_{k=1}^n$ is a basis. Thus there is no loss of generality by this change of basis. Then

$$x' = \sum_{i=1}^n y_i' E_i = A\left(\sum_{i=1}^n y_i E_i\right) = \sum_{i=1}^n y_i A E_i = \sum_{i=1}^n \lambda_i y_i E_i.$$

Since $\{E_k\}_{k=1}^n$ is linearly independent, the coefficients of the columns E_i must agree, resulting in the *decoupled* system

$$y_i' = \lambda_i y_i, \qquad i = 1, 2, \ldots, n,$$

which has the general solution $y_i = c_i e^{\lambda_i t}$. Thus the general solution of $x' = Ax$ is

$$x = \sum_{i=1}^n c_i e^{\lambda_i t} E_i. \tag{2}$$

If initial conditions are given, i.e., $x(0) = x_0$, then the c_i are computed by solving

$$x_0 = \sum_{i=1}^n c_i E_i = [E_1\ E_2 \cdots E_n]c = Pc. \tag{3}$$

Hence

$$c = P^{-1} x_0. \tag{4}$$

Equations (2) and (3) are quite useful, but can be spruced up into compact notation as follows: Since e^{At} obeys $e^{At} E_i = e^{\lambda_i t} E_i$, (2) becomes (with the use of some matrix identities)

$$x = \sum_{i=1}^n c_i e^{\lambda_i t} E_i = \sum_{i=1}^n c_i e^{At} E_i = e^{At}\left(\sum_{i=1}^n c_i E_i\right) = e^{At} Pc = e^{At} x_0,$$

where $P = [E_1\ E_2 \cdots E_n]$, $x_0 = Pc$, and $c = (c_1, c_2, \ldots, c_n)^T$.
Thus the elegant and compact formula

$$x = e^{At} x_0 \tag{5}$$

for the solution of $x' = Ax$, with initial conditions $x(0) = x_0$, is obtained. e^{At} may be computed by the methods of Section 4.3. For example, employing Theorem 2 of that section yields $e^{At} = Pe^{\Lambda t} P^{-1}$. However, if $x(t)$ must be computed for a large number of t values, using this in (5) is very inefficient.

Matrix method of solution

Since A is diagonalizable, there is a matrix P so that $P^{-1}AP = \Lambda = \mathrm{diag}(\lambda_i)$. Let us make the substitution $x = Py$. This is simply the change of coordinates corresponding to the preceding change of basis. Then

$$x' = Py' = Ax = APy,$$

and left-multiplication of this equation by P^{-1} gives

$$y' = P^{-1}Py' = P^{-1}APy = \Lambda y.$$

Writing this out in components yields (again)

$$y_i' = \lambda_i y_i,$$

with the solution

$$y_i = c_i e^{\lambda_i t}.$$

Observing that $y = (c_1 e^{\lambda_1 t}, \ldots, c_n e^{\lambda_n t})^{\mathrm{T}} = \mathrm{diag}(e^{\lambda_1 t}, \ldots, e^{\lambda_n t})c = e^{\Lambda t}c$, one finds that

$$x = P e^{\Lambda t} c \qquad (6)$$

is the general solution of the differential equation.

Imposing the initial conditions $x(0) = x_0 = PY(0) = Pc$ yields the solution (5) again to the initial condition problem:

$$x = P e^{\Lambda t} P^{-1} x_0 = e^{At} x_0.$$

Thus both methods of solution ultimately lead to the same answer. Of course, to get P, the eigenvalue problem mentioned in the first method must be solved. The identity $x = \sum_{i=1}^{n} y_i E_i = Py$ explicitly shows the connection between the methods. The methods are actually the same, but appear different at first glance.

Remark 1. Observe that terms such as $t^k e^{\lambda t}$ never occur in the solution (2) when A is diagonalizable, even if there are repeated eigenvalues. Such terms will appear in the solution only for defective operators. (See Chapter 8.)

Remark 2. The solution form (5) is most useful for theoretical work. It is valid for nondiagonalizable matrices as well. (See Chapter 8.) Accepting the usual rules of differentiation for e^{At}, one can show that (5) is a solution with the following required initial values:

$$\frac{d}{dt}(e^{At} x_0) = \left(\frac{d}{dt} e^{At}\right) x_0 = A e^{At} x_0 = Ax, \qquad x(0) = e^0 x_0 = I x_0 = x_0.$$

For all square matrices,

$$e^A t = \sum_{k=0}^{\infty} \frac{A^k t^k}{k!}.$$

The series converges for all t and every A. The usual rules of calculus hold.

Remark 3. In some situations, (2) has an advantage over (5) for theoretical work. For example, suppose $\lambda_1 \gg \lambda_i$ for $i = 2, 3, \ldots, n$. Since $e^{\lambda_1 t} \gg e^{\lambda_i t}$ for large enough t, from (2) it follows that $x(t) \approx c_1 e^{\lambda_1 t} E_1$ for large t. This is not readily evident from (5).

Remark 4. If $x(t)$ must be computed for many t values, (2) or its equivalent, (6), should be used. Using these will avoid the repeated calculation of e^{At} required by (5).

EXAMPLE 1. Consider

$$x' = \begin{bmatrix} 3 & -1 & 0 \\ -1 & 2 & -1 \\ 0 & -1 & 3 \end{bmatrix} x = Ax.$$

Let

$$x_0 = \begin{bmatrix} 2 \\ 1 \\ 2 \end{bmatrix}.$$

Find the general solution of the differential equation, and find the solution given the initial value $x(0) = x_0$.

Solution. The eigenvalue problem for A has the solution $\lambda_1 = 1$, $E_1 = (1, 2, 1)^T$; $\lambda_2 = 3$, $E_2 = (1, 0, -1)^T$; $\lambda_3 = 4$, $E_3 = (1, -1, 1)^T$. Consequently the change of basis $x = y_1 E_1 + y_2 E_2 + y_3 E_3$ will decouple the differential equations to

$$y_1' = y_1, \quad y_2' = 3y_2, \quad y_3' = 4y_2.$$

Thus the general solution is given by

$$x = c_1 e^t E_1 + c_2 e^t E_2 + c_3 e^{3t} E_3.$$

Writing this out for each component, the general solution is

$$\begin{aligned} x_1 &= c_1 e^t + c_2 e^{3t} + c_3 e^{4t} \\ x_2 &= 2c_1 e^t \qquad\quad - c_3 e^{4t} \\ x_3 &= c_1 e^t - c_2 e^{3t} + c_3 e^{4t} \end{aligned} \qquad (*)$$

With initial conditions $x(0) = x_0$, the c_i are found from

$$x_0 = c_1 E_1 + c_2 E_2 + c_3 E_3 = Pc, \qquad \text{where } P = [E_1 \; E_2 \; E_3].$$

One can solve $Pc = x_0$ or compute $c = P^{-1} x_0$. Since

$$P^{-1} = \frac{1}{6} \begin{bmatrix} 1 & 2 & 1 \\ 3 & 0 & -3 \\ 2 & -2 & 2 \end{bmatrix},$$

one finds that

$$c = \begin{bmatrix} 1 \\ 0 \\ 1 \end{bmatrix}.$$

In this example, there is an easier way to obtain the c_k. Observe that $\{E_1, E_2, E_3\}$ is an orthogonal set, so that the c_k are Fourier coefficients:

$$c_k = \frac{\langle x_0, E_k \rangle}{\|E_k\|^2}, \qquad \text{whence } c_1 = \frac{6}{6} = 1, \quad c_2 = \frac{0}{2} = 0, \quad c_3 = \frac{3}{3} = 1.$$

The solution for the given initial conditions is then

$$\begin{aligned} x_1 &= e^t + e^{4t} \\ x_2 &= 2e^t - e^{4t} \\ x_3 &= e^t + e^{4t}. \end{aligned}$$

To tidy up the general solution (*) into matrix form, write

$$x = \begin{bmatrix} e^t & e^{3t} & e^{4t} \\ 2e^t & 0 & e^{4t} \\ e^t & -e^{3t} & e^{4t} \end{bmatrix} \begin{bmatrix} c_1 \\ c_2 \\ c_3 \end{bmatrix} = \begin{bmatrix} 1 & 1 & 1 \\ 2 & 0 & -1 \\ 1 & -1 & 1 \end{bmatrix} = \text{diag}(e^t, e^{3t}, e^{4t})c$$

$$= Pe^{\Lambda t}c,$$

with

$$P = [E_1 \ E_2 \ E_3] = \begin{bmatrix} 1 & 1 & 1 \\ 2 & 0 & -1 \\ 1 & -1 & 1 \end{bmatrix}, \qquad \Lambda = \text{diag}(1, 3, 4).$$

To write the general solution in the form of Eq. (5), $x = e^{At}x_0$, compute

$$e^{At} = Pe^{\Lambda t}P^{-1} = \frac{1}{6}\begin{bmatrix} e^t + 4e^{3t} + e^{4t} & e^t - e^{4t} & e^t - 2e^{3t} + e^{4t} \\ 3e^t - e^{4t} & 3e^t + e^{4t} & 3e^t - e^{4t} \\ 2e^t - 4e^{3t} + 2e^{4t} & 2e^t - 2e^{4t} & 2e^t + 4e^{3t} + 2e^{4t} \end{bmatrix}.$$

EXAMPLE 2. Solve the inhomogeneous linear differential equation

$$x' = Ax + f(t), \text{ with initial conditions } x(0) = x_0, \qquad (\#)$$

where

$$A = \begin{bmatrix} 1 & 2 \\ 0 & -1 \end{bmatrix}, \qquad f = \begin{bmatrix} 1 \\ 1 \end{bmatrix}t, \qquad \text{and } x_0 = \begin{bmatrix} 1 \\ 2 \end{bmatrix}.$$

The method employed here carries over to the general case. First solve the eigenvalue problem for A. The eigenvalues are $\lambda_1 = -1$, $\lambda_2 = 1$, and the corresponding eigenvectors are $E_1 = (1, -1)^T$ and $E_2 = (1, 0)^T$. The matrix method will be used, so set

$$P = [E_1 \ E_2] = \begin{bmatrix} 1 & 1 \\ -1 & 0 \end{bmatrix}, \qquad \text{whence } P^{-1} = \begin{bmatrix} 0 & -1 \\ 1 & 1 \end{bmatrix}.$$

Then $PAP^{-1} = \text{diag}(-1, 1) = \Lambda$. Substituting $x = Py$ into (#) yields $Py' = APy + f$. Now multiplying this on the left by P^{-1} gives

$$y' = P^{-1}APy + P^{-1}f = \Lambda y + P^{-1}f = \begin{bmatrix} -y_1 \\ y_2 \end{bmatrix} + \begin{bmatrix} -t \\ 2t \end{bmatrix},$$

and the decoupled equations are

$$y_1' = -y_1 - t$$
$$y_2' = y_2 + 2t.$$

To find a particular solution for each of these equations, try $y_{1,2} = at + b$, and determine a and b. Adding the homogeneous solution to the particular solution gives the general solution to each of the decoupled equations:

$$y_1 = c_1 e^{-t} + 1 - t,$$
$$y_2 = c_2 e^t - 2 - 2t.$$

The general solution of (#) is $x = Py = y_1 E_1 + y_2 E_2$, which gives

$$x_1 = c_1 e^{-t} + c_2 e^t - 1 - 3t,$$
$$x_2 = -c_1 e^{-t} - 1 + t.$$

The initial condition requires that

$$x_0 = Py(0) = P\begin{bmatrix} c_1 + 1 \\ c_2 - 2 \end{bmatrix},$$

so that

$$\begin{bmatrix} c_1 + 1 \\ c_2 - 2 \end{bmatrix} = P^{-1}x_0 = \begin{bmatrix} -2 \\ 3 \end{bmatrix},$$

which yields $c_1 = -3$, $c_2 = 5$. Thus the solution is

$$x = Py = (-3e^{-t} + 1 - t)E_1 + (5e^t - 2 - 2t)E_2 = \begin{bmatrix} 5e^t - 3e^{-t} - 1 - 3t \\ 3e^{-t} - 1 + t \end{bmatrix}.$$

Remark. Converting second-order systems to first-order systems. A second-order system is $x'' + Ax' + Bx = f$, where A, B are $n \times n$ matrices and f is a column that can be made into a first-order system by the following artifice: Set $y = x'$; then the system

$$x' = y$$
$$y' = -Ay - Bx + f$$

can be written as a first-order system in block matrix form as

$$\begin{bmatrix} x \\ y \end{bmatrix}' = \begin{bmatrix} 0 & I \\ -B & -A \end{bmatrix} + \begin{bmatrix} 0 \\ f \end{bmatrix}, \tag{7}$$

or

$$z' = Cz + g,$$

with

$$z = \begin{bmatrix} x \\ y \end{bmatrix}, \qquad C = \begin{bmatrix} 0 & I \\ -B & -A \end{bmatrix}, \qquad g = \begin{bmatrix} 0 \\ f \end{bmatrix}.$$

Then techniques for first-order equations may be used to solve this system. Second-order systems arising in vibration theory will be treated in Chapter 7; better techniques are used for those systems.

Exercises 4.4

1. Solve the following systems of equations $x' = Ax$, where A is the following:

(a) $\begin{bmatrix} 1 & 1 & -1 \\ -1 & 3 & -1 \\ -2 & 2 & 0 \end{bmatrix}$ (b) $\begin{bmatrix} 2 & 2 & 2 \\ -2 & -1 & 1 \\ 2 & 1 & -1 \end{bmatrix}$ (c) $\begin{bmatrix} 0 & 3 & -1 \\ -1 & 4 & -1 \\ 0 & 0 & 1 \end{bmatrix}$

(d) $\begin{bmatrix} i & 1 \\ 1 & i \end{bmatrix}$ (e) $\begin{bmatrix} 0 & 1 & 0 \\ -1 & 0 & 0 \\ 0 & 0 & 1 \end{bmatrix}$ (f) $\begin{bmatrix} 1 & 1 & 0 & 0 \\ 3 & 3 & 0 & 0 \\ 0 & 0 & 0 & 1 \\ 0 & 0 & -1 & 0 \end{bmatrix}$

2. Solve $x' = Ax$, where $x(0) = x_0$, in each of the following cases:

(a) $A = \begin{bmatrix} 2 & -1 & 0 \\ -1 & 1 & 1 \\ 0 & 1 & 2 \end{bmatrix}$, $x_0 = \begin{bmatrix} 1 \\ 1 \\ 1 \end{bmatrix}$ (b) $A = \begin{bmatrix} 1 & 1 & -1 \\ 1 & 1 & 1 \\ -1 & 1 & 1 \end{bmatrix}$, $x_0 = \begin{bmatrix} 1 \\ -2 \\ 0 \end{bmatrix}$

3. Solve $x' = Ax + f$ in each of the following cases:

(a) $A = \begin{bmatrix} 0 & 1 \\ -2 & 3 \end{bmatrix}$, $f = \begin{bmatrix} 1 \\ 0 \end{bmatrix}$, $x(0) = \begin{bmatrix} 0 \\ 1 \end{bmatrix}$

(b) $A = \begin{bmatrix} 1 & 1 \\ 1 & 1 \end{bmatrix}$, $f = 4\begin{bmatrix} t \\ 0 \end{bmatrix}$, $x(0) = \begin{bmatrix} 2 \\ 1 \end{bmatrix}$

(c) $A = \begin{bmatrix} 1 & -1 & 1 \\ -1 & 3 & -1 \\ 1 & -1 & 1 \end{bmatrix}$, $f = \begin{bmatrix} 1 \\ 0 \\ -1 \end{bmatrix}$, $x(0) = 6\begin{bmatrix} 1 \\ 0 \\ 0 \end{bmatrix}$

4. Let A be a diagonalizable matrix with eigenvalues λ_i, and let $x(t)$ be a solution of $\dot{x} = Ax$. Show that
  (a) Every solution $x(t)$ is bounded as $t \to \infty$ iff all $\lambda_i \le 0$.
  (b) Every solution $x(t) \to 0$ as $t \to \infty$ iff all $\lambda_i < 0$.

5. Let A be a diagonalizable matrix. Show that the series definition of e^{At} agrees with the definition of the text, i.e., $\sum_{k=0}^{\infty}(A^k t^k)/(k!) = Pe^{\Lambda t}P^{-1}$. (*Hint:* $A = P\Lambda P^{-1}$.)

6. An ordinary linear differential equation can sometimes be converted to a system $x' = Ax$ and solved by diagonalization of A. For example, consider $y''' + ay'' + by' + cy = f(t)$. Set $x_1 = y$, $x_2 = y'$, $x_3 = y''$. Then the differential equation has the matrix form

$$x' = \begin{bmatrix} x_1 \\ x_2 \\ x_3 \end{bmatrix}' = \begin{bmatrix} 0 & 1 & 0 \\ 0 & 0 & 1 \\ -c & -b & -a \end{bmatrix}\begin{bmatrix} x_1 \\ x_2 \\ x_3 \end{bmatrix} + \begin{bmatrix} 0 \\ 0 \\ f(t) \end{bmatrix} = Ax + g(t).$$

Put the following ordinary differential equations in the preceding form. Determine whether the resulting matrix A is diagonalizable and, if so, solve the differential equation using the methods of this section.
  (a) $y'' - y' - 2y = 1$
  (b) $y''' + y = 0$
  (c) $y'''' + y = 0$
  (d) $y''' - y'' = 0$

4.5
ESTIMATES OF EIGENVALUES: GERSHGORIN'S THEOREMS

The following theorems of Gershgorin are used to estimate eigenvalues. More methods for Hermitian operators will be found in Chapter 5.

  The first theorem is best discussed with the idea of a vector norm. One example of a norm is the norm generated by an inner product: $\|x\| = \sqrt{\langle x, x \rangle}$. In general, a norm $\| \ \|$ is a real-valued function of a vector variable satisfying the following:

$$\|x\| > 0 \text{ if } x \neq 0 \qquad\qquad \text{(Positive definite)} \qquad (1a)$$

$$\|\alpha x\| = |\alpha|\,\|x\| \text{ for all scalars } \alpha \text{ and vectors } x \qquad \text{(Homogeneous)} \qquad (1b)$$

$$\|x + y\| \leq \|x\| + \|y\| \text{ for all } x,\, y. \qquad \text{(Triangle inequality)} \quad (1c)$$

Two commonly used norms on $\mathbf{F}^n$ are

$$\text{The 1-norm:} \qquad \|x\|_1 = \sum_{i=1}^{n} |x_i|$$

$$\text{The } \infty\text{-norm:} \qquad \|x\|_\infty = \max\{|x_i| : i = 1, 2, \ldots, n\}.$$

It is easy to see that these satisfy the three criteria (1a–c).

Now let us show how to estimate the maximum size of eigenvalues. For a particular norm $\|\ldots\|$, suppose a function of matrices $M(A)$ has been found such that

$$\|Ax\| \leq M(A)\|x\| \text{ for all } x \in \mathbf{C}^n. \tag{2}$$

Suppose $Ae = \lambda e$, with $e \neq 0$. Let $f = e/\|e\|$. Then $Af = \lambda f$, and $\|f\| = 1$ by homogeneity. Using (2) gives

$$|\lambda| = |\lambda|\,\|f\| = \|\lambda f\| = \|Af\| \leq M(A)\|f\| = M(A). \tag{3}$$

In other words, no eigenvalue exceeds $M(A)$ in absolute value.

If a good $M(A)$ is found for the 1-norm, then Gershgorin's first theorem will follow as an example. Let $A = [A_1\ A_2 \cdots A_n]$. Then

$$\|Ax\|_1 = \left\| \sum_{j=1}^{n} x_j A_j \right\| \leq \sum_{j=1}^{n} |x_j|\,\|A_j\|_1 \leq \max\{\|A_j\|_1 : j = 1, 2 \ldots, n\} \sum_{j=1}^{n} |x_j|$$

$$= \max\{\|A_j\|_1 : j = 1, 2, \ldots, n\}\|x\|_1.$$

That is, (2) holds with $M(A) = \max\{\|A_j\|_1 : i = 1, 2, \ldots, n\}$. This $M(A)$ is called the *maximum absolute column sum* since $\|A_j\|_1$ is the sum of the absolute values of the column A_j. One can show that no smaller $M(A)$ can satisfy (2) using $\|\ \|_1$. These results are expressed in the next theorem.

GERSHGORIN'S FIRST THEOREM. If λ is an eigenvalue of A, then $|\lambda| \leq \max\{\sum_{i=1}^{n} |a_{ij}| : j = 1, 2, \ldots, n\}$.

COROLLARY 1. If λ is an eigenvalue of A, then $|\lambda| \leq \max\{\sum_{j=1}^{n} |a_{ij}| : i = 1, 2, \ldots, n\}$.

Proof. A^{T} has the same eigenvalues as A. Apply the first theorem to A^{T}.

It is left to the reader to discover a good $M(A)$ for the norm $\|\ \|_\infty$ and for the norm generated by an inner product. (See Exercises 7 and 8.)

EXAMPLE 1. Bound the eigenvalues of

$$A = \begin{bmatrix} 0 & 2 & -1 \\ 1 & 3 & 0 \\ 0 & 1 & 2 \end{bmatrix}.$$

The column sums are 1, 6, and 3, so that $|\lambda| \leq 6$. The row sums are 3, 4, and 3, so that $|\lambda| \leq 4$, which is a better result.

GERSHGORIN'S SECOND THEOREM. Let A be an $n \times n$ matrix, and define the row radii R_i by

$$R_i = \sum_{j=1, \, j\neq i}^{n} |a_{ij}|, \qquad i = 1, 2, \ldots, n. \tag{4}$$

Define the corresponding closed disk in the complex plane as $D_i = \{z : |z - a_{ii}| \leq R_i\}$. Then each eigenvalue λ of A lies in the union of these disks. If k of the disks form a set U disjoint from the other $n - k$ disks, then k eigenvalues (counted with their algebraic multiplicity) lie in U.

Proof. Suppose $Ae = \lambda e$, where $e = (e_1, e_2, \ldots, e_n)^T \neq 0$. Choose k such that $|e_j| \leq |e_k|$ for all j. Then looking at the kth row of the equation $Ae = \lambda e$ gives

$$|(\lambda - a_{kk})e_k| = \left| \sum_{j=1, \, j\neq k}^{n} a_{jk}e_j \right| \leq \sum_{j=1, \, j\neq k}^{n} |a_{jk}| \, |e_j| \leq \left\{ \sum_{j=1, \, j\neq k}^{n} |a_{jk}| \right\} |e_k|.$$

Cancelling $|e_k|$ yields the first part of the assertion. The second part depends on a theorem from complex analysis that says that the zeros of a polynomial continuously depend on the coefficients. Write $A = \mathrm{diag}(a_{11}, \ldots, a_{nn}) + B$, and let $A(t) = \mathrm{diag}(a_{11}, \ldots, a_{nn}) + tB$ and $p_t(\lambda)$ be the characteristic polynomial of $A(t)$. Note that $p_1(\lambda)$ is the characteristic polynomial of A and the coefficients of $p_t(\lambda)$ are continuous functions of t. As t increases toward 1, the zeros $\lambda_1(t), \ldots, \lambda_n(t)$ of $p_t(\lambda)$ (listed with their algebraic multiplicity) move continuously to the eigenvalues of A. According to the first part of the theorem, each $\lambda_i(t)$ lies in the union of the disks $D_i(t) = \{z : |z - a_{ii}| \leq tR_i\}$. For $t = 0$, the eigenvalues are the a_{ii}, and so there are exactly k of them in U. As t increases, the eigenvalues move continuously and cannot "jump" from U to one of the other $n - k$ disks. Thus, at $t = 1$, there are exactly k eigenvalues in U.

Remark. Since A^T has the same eigenvalues as A, Gershgorin's second theorem can sometimes be applied to A^T to get better estimates. This simply converts the row sums R_i in the theorem into analogous column sums.

EXAMPLE 2. Estimate the eigenvalues of A, where

$$A = \begin{bmatrix} 2 & 1 & 0 \\ -1 & 4 & 1 \\ 2 & -1 & 10 \end{bmatrix}.$$

The eigenvalues lie in the union of the disks: $|\lambda - 2| \leq 1, |\lambda - 4| \leq 2, |\lambda - 10| \leq 3$. The last disk is disjoint from the first two, which overlap. Thus there is exactly one eigenvalue in the disk $|\lambda - 10| \leq 3$, and the other two are in the union of the first two disks. If the last theorem is applied to A^T, then the eigenvalues lie in the union of $|\lambda - 2| \leq 3, |\lambda - 4| \leq 2, |\lambda - 10| \leq 1$. Again the first pair of disks overlap but are disjoint from the last. Thus the eigenvalue near 10 satisfies $|\lambda - 10| \leq 1$, an improvement over the analysis using the row radii. Observe that A is nonsingular, since no eigenvalue is 0.

Similar matrices have the same eigenvalues, and this can sometimes be used to improve the bounds obtained in Gershgorin's theorems. The next example illustrates one method of using this idea.

EXAMPLE 3. Estimate the eigenvalues of

$$A = \begin{bmatrix} 10 & 1 & 0 \\ 1 & 10 & 1 \\ 0 & 1 & 0 \end{bmatrix}.$$

Applying Gershgorin's second theorem, the eigenvalues lie in the union of the disks $|\lambda - 10| \leq 1$, $|\lambda - 10| \leq 2$, and $|\lambda - 0| \leq 1$. Thus two eigenvalues lie in the disk $|\lambda - 10| \leq 2$ and one in the disk $|\lambda| \leq 1$. Let $D = \mathrm{diag}(d_1, d_2, d_3)$ and $B = D^{-1}AD$; i.e., let

$$B = \begin{bmatrix} 10 & d_2/d_1 & 0 \\ d_1/d_2 & 10 & d_3/d_2 \\ 0 & d_2/d_3 & 0 \end{bmatrix}.$$

Let us try to improve the bound for the eigenvalue near 0. Take $d_2/d_3 = 1/5$ and $d_1 = d_2$. The Gershgorin disks are $|\lambda - 10| \leq 1$, $|\lambda - 10| \leq 7$, $|\lambda - 0| \leq 1/5$. The last disk is disjoint from the first two, so the eigenvalue near 0 obeys $|\lambda| \leq 1/5$. To sharpen the result for the eigenvalues near 10, take $d_1 = d_2$ and $d_3/d_2 = 1/5$. The Gershgorin disks are now $|\lambda - 10| \leq 1$, $|\lambda - 10| \leq 1.2$, and $|\lambda - 0| \leq 5$, and the last is again disjoint. Thus two eigenvalues lie in the disk $|\lambda - 10| \leq 1.2$.

The following corollary gives a fast a priori test for the invertibility of a certain class of matrices. The type of matrices involved appears frequently in applications.

DEFINITION. An $n \times n$ matrix is *strictly diagonally dominant* iff

$$|a_{ii}| > \sum_{j=1, j \neq i}^{n} |a_{ij}|, \qquad i = 1, 2, \ldots, n. \tag{5}$$

COROLLARY 2. Strictly diagonally dominant matrices are invertible.

Proof. Inequality (5) is equivalent to $|a_{ii}| > R_i$. Gershgorin's second theorem implies that each eigenvalue is in the union of the disks $|\lambda - a_{ii}| \leq R_i$. But $\lambda = 0$ is not in this union since $|a_{ii}| > R_i \; \forall i$. Thus 0 is not an eigenvalue of A, so A is nonsingular.

Example 2 exhibits an example of this corollary.

Exercises 4.5

In Exercises 1–3, you may wish to compute the eigenvalues of the matrices to see how well the estimates work.

1. For the following matrices, estimate the eigenvalues using Gershgorin's second theorem. Sketch the disks in the complex plane that contain the eigenvalues.

(a) $\begin{bmatrix} 7 & 2 & -1 \\ -1 & 10 & 1 \\ -1 & 1 & 6 \end{bmatrix}$ (b) $\begin{bmatrix} 0 & -1 & 1 \\ -1 & 6 & 2 \\ 1 & 2 & 6 \end{bmatrix}$ (c) $\begin{bmatrix} 10 & 0 & -1 & 1 \\ -1 & 12i & -1 & 2 \\ -1 & 3 & 20 & 2 \\ 1 & 2i & 3 & -45 \end{bmatrix}.$

2. Estimate the eigenvalues of the following matrices:

(a) $\begin{bmatrix} 4 & 0.2 & 0 \\ 0.1 & -3 & -0.3 \\ 0 & -0.3 & 5 \end{bmatrix}$
(b) $\begin{bmatrix} 2 & 0.5 & 0.2 \\ 0.4 & 0 & 0 \\ 0 & 0.7 & 0 \end{bmatrix}$
(c) $\begin{bmatrix} 2 & -0.4 & 0 \\ -2.5 & 3 & -0.5 \\ 0 & -2 & 2 \end{bmatrix}$

(d) $\begin{bmatrix} 10 & -5 & 0 \\ -10 & 60 & -5 \\ 0 & -5 & 5 \end{bmatrix}$
(e) $\begin{bmatrix} 0.6 & 0 & 0.2 & 0 \\ 0.1 & 0.8 & 0 & 0 \\ 0 & 0.2 & 0.7 & 0.1 \\ 0.3 & 0 & 0.1 & 0.9 \end{bmatrix}$.

3. Apply Gershgorin's second theorem to the following. Then use the method of Example 3 to sharpen the bounds for the eigenvalues.

(a) $\begin{bmatrix} 1 & 0.01 & 0.02 \\ 0.01 & 2 & 0.01 \\ 0.03 & 0.01 & 4 \end{bmatrix}$
(b) $\begin{bmatrix} 0 & 1 & 0 & 0 \\ 1 & 5 & 1 & 0 \\ 0 & 1 & 20 & 1 \\ 0 & 0 & 1 & 20 \end{bmatrix}$

(c) $\begin{bmatrix} 0 & 0.35 & 0.1 \\ 0.8 & 0 & 0 \\ 0 & 0.6 & 0 \end{bmatrix}$
(d) $\begin{bmatrix} 1 & 10 & 20 \\ 0.5 & 0 & 0 \\ 0 & 0.8 & 0 \end{bmatrix}$.

4. Show that Gershgorin's second theorem implies Gershgorin's first theorem.

5. A *probability matrix* A is one that has nonnegative entries and whose columns all sum to 1. (Cf. Exercise 30 in Section 4.2). Show that the eigenvalues of a probability matrix lie in the unit disk.

6. Let A be given. Find a column x such that $\|x\|_1 = 1$ and $\|Ax\|_1 = \max\{\|A_j\|_1 : j = 1, 2, \ldots, n\}$. This shows that inequality (2) is the best possible in the sense that it becomes an equality for certain vectors $x \neq 0$.

7. The idea behind Gershgorin's first theorem can be extended to the norm $\|x\|_2 \equiv \sqrt{\sum_{i=1}^n |x_i|^2}$ generated by $\langle x, y \rangle = y^H x$ as follows:
(a) Let $A = [A_1 \ A_2 \cdots A_n]$. Justify the following steps:

$$\|Ax\|_2 \leq \sum_{j=1}^n \|A_j\|_2 |x_j| \leq \left[\sum_{j=1}^n \|A_j\|_2^2 \right]^{1/2} \left[\sum_{j=1}^n |x_i|^2 \right]^{1/2}.$$

(b) Deduce that $\|Ax\|^2 \leq M(A)\|x\|_2$, where $M(A) = \sqrt{\sum_{i,j=1}^n |a_{ij}|^2}$.

(c) Deduce that if λ is an eigenvector of A, then $|\lambda| \leq \sqrt{\sum_{i,j=1}^n |a_{ij}|^2}$.

8. Let $\|x\|$ denote the ∞-norm, and let A be an $n \times n$ matrix.
(a) Find an $M(A)$ so that (i) $\|Ax\| \leq M(A)\|x\|$ $\forall x$, and (ii) there is an x with $\|x\| = 1$ so that $\|Ax\| = M(A)$.
(b) State the corresponding version for Gershgorin's first theorem for this $M(A)$.

4.6
APPLICATION TO FINITE DIFFERENCE EQUATIONS

Some examples of finite difference equations and how they arise in modeling problems are presented. Several techniques used to analyze finite difference equations are demonstrated in the process of discussing these examples. (A general theory for the solution or analysis of finite difference equations will not be presented.)

EXAMPLE 1. A certain species lives three years and reproduces in the second and third years. The fraction of organisms that survive from year i to year $i + 1$ is σ_i, and the number of offspring per adult is α during the second year and β during the third year. Find expressions for the population in each age group for an arbitrary year. Determine the behavior of the populations as time becomes large.

Solution. The situation can be represented graphically as follows:

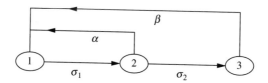

In year n, let P_n, Q_n, R_n be the number of individuals in their first year, second year, and third year, respectively. In the $n + 1$st year, the populations are as follows:

$$P_{n+1} = \alpha Q_n + \beta R_n \qquad \text{(Number in first year)}$$

$$Q_{n+1} = \sigma_1 P_n \qquad \text{(Number in second year)}$$

$$R_{n+1} = \sigma_2 Q_n \qquad \text{(Number in third year)}.$$

This can be written in matrix form:

$$x_{n+1} = \begin{bmatrix} 0 & \alpha & \beta \\ \sigma_1 & 0 & 0 \\ 0 & \sigma_2 & 0 \end{bmatrix} \begin{bmatrix} P_n \\ Q_n \\ R_n \end{bmatrix} = A x_n. \tag{1}$$

Applying this repeatedly yields

$$x_1 = A x_0, \qquad x_2 = A x_1 = A^2 x_0, \qquad x_3 = A x_2 = A A^2 x_0 = A^3 x_0,$$

and, by induction,

$$x_n = A^n x_0, \qquad n = 1, 2, \ldots. \tag{2}$$

If A is diagonalizable and $\{E_1, E_2, E_3\}$ is an eigenbasis with $A E_i = \lambda_i E_i$, then the x_n may be computed in a couple of ways. Let $P = [E_1\ E_2\ E_3]$. According to (6) in Section 4.3, or Theorem 2 of Section 4.2, if $x_0 = \sum_{i=1}^{3} c_i E_i$, then

$$x_n = A^n x_0 = \sum_{i=1}^{3} \lambda_i^n c_i E_i, \tag{3a}$$

where

$$x_0 = \sum_{i=1}^{3} c_i E_i = [E_1 \; E_2 \; E_3]c = Pc. \tag{3b}$$

On the other hand, if $\Lambda = \text{diag}(\lambda_1, \lambda_2, \lambda_3)$, then by (8) in Section 4.3,

$$x_n = A^n x_0 = P\Lambda^n P^{-1} x_0. \tag{4}$$

For the sake of an example, suppose $\alpha = 14$, $\sigma_1 = \frac{1}{2}$, $\beta = 28$, and $\sigma_2 = \frac{3}{7}$; we have a fecund species, perhaps akin to rabbits. The characteristic polynomial of A is $-\lambda^3 + 7\lambda + 6$, and the eigenvalues are $\lambda_1 = 3$, $\lambda_2 = -1$, and $\lambda_3 = -2$. From (3) it follows that

$$x_n = 3^n c_1 E_1 + (-1)^n c_2 E_2 + \left(-\tfrac{1}{2}\right)^n c_3 E_3, \tag{*}$$

and it is clear that a nonzero starting population will lead to populations that get arbitrarily large as long as $c_1 \neq 0$. The condition needed to ensure that $c_1 \neq 0$ is discussed below. Factoring out $\lambda_1 = 3$ gives

$$x_n = 3^n \{ c_1 E_1 + \left(-\tfrac{1}{3}\right)^n E_2 + \left(-\tfrac{1}{6}\right)^n c_3 E_3 \},$$

so that if $c_1 \neq 0$,

$$x_n = (P_n, Q_n, R_n)^{\mathrm{T}} \approx 3^n c_1 E_1 \qquad \text{for large } n. \tag{**}$$

Equation (*) gives us an exact solution, and (**) tells us what the populations are asymptotically. A basis for $M(3)$ is $E_1 = (1, 3, 9)^{\mathrm{T}}$, so by (**), the components P_n, Q_n, R_n of x_n are in the approximate ratio $1 : 3 : 9$ for large n. Note that the eigenvector E_1 has a special significance: It describes the relative sizes of the population for large time, i.e., the limiting population distrubution. The total population T_n at year n is

$$T_n = P_n + Q_n + R_n \approx 3^n c_1 (1 + 3 + 9) \approx 13(3^n)c_1.$$

To find the exact population at a given year, the other eigenvectors are needed, as well as the initial population vector x_0. For λ_2 and λ_3, choose $E_2 = (1, -1, 1)^{\mathrm{T}}$ and $E_3 = (1, -2, 4)^{\mathrm{T}}$, respectively. Then

$$P = \begin{bmatrix} 1 & 1 & 1 \\ 3 & -1 & -2 \\ 9 & 1 & 4 \end{bmatrix} \quad \text{and} \quad P^{-1} = \begin{bmatrix} 0.1 & 0.15 & 0.05 \\ 1.5 & 25 & -0.25 \\ -0.6 & -0.4 & 0.2 \end{bmatrix},$$

so that $c = P^{-1} x_0$. Observe that $c_1 = 0.1 P_0 + 0.15 Q_0 + 0.05 R_0$, so that $c_1 = 0$ only if the initial population is zero. Now suppose that at $n = 0$, there are 1 organism in its first year, 2 in their second year, and 3 in their third year. Then, $c = P^{-1} x_0 = (1.4, -1, 0.6)^{\mathrm{T}}$. For example, in the third year, using (3):

$$x_3 = \sum_{i=1}^{3} \lambda_i^3 c_i E_i = 1.4(3^3)\begin{bmatrix} 1 \\ 3 \\ 9 \end{bmatrix} + (-1)(-1)^3\begin{bmatrix} 1 \\ -1 \\ 1 \end{bmatrix} + 0.06(-2)^3\begin{bmatrix} 1 \\ -2 \\ 4 \end{bmatrix} = \begin{bmatrix} 34 \\ 122 \\ 322 \end{bmatrix}.$$

Or one can use (4), $x_3 = P\Lambda^3 P^{-1} x_0$, to obtain the same result.

Example 1 is a special case of the *Leslie demographic model*. This model divides the population into m disjoint age groups, each of one time period in length.

α_i denotes the number of new offspring from age group i in one time period, and σ_i is the fraction of those in age group i who survive (as breeders) into the next age group in one time step. If $x_i^{(n)}$ is the number of individuals in group i at time n, then

$$x_1^{(n+1)} = \sum_{i=1}^{m} \alpha_i x_i^{(n)}$$

$$x_{i+1}^{(n+1)} = \sigma_i x_i^{(n)}, \qquad i = 1, \ldots, m - 1,$$

which may be written in matrix form as

$$x^{(n+1)} = \begin{bmatrix} \alpha_1 & \alpha_2 & \cdots & & a_m \\ \sigma_1 & & \cdots & & 0 \\ 0 & \sigma_2 & \cdots & & 0 \\ \cdots & & \cdots & & \cdots \\ 0 & 0 & \cdots & \sigma_{m-1} & 0 \end{bmatrix} x^{(n)}, \qquad x^{(n)} = \begin{bmatrix} x_1^{(n)} \\ x_2^{(n)} \\ \vdots \\ x_m^{(n)} \end{bmatrix}. \tag{5}$$

The matrix in (5) is called a *Leslie* matrix.

EXAMPLE 2. Consider a vaccine against disease z. The vaccine immunizes some individuals for a long time, and in others the effect is short. Having the disease also confers immunity, possibly temporarily. The population may be divided into three categories: immune, sick, and at risk (not immune and not sick). Suppose that in one time period, 20 percent of the immune people stay immune, and 80 percent lose their immunity. Of those at risk, 70 percent stay this way, 10 percent get immunized, and 20 percent get sick. Assume that those who get sick all recover and that 40 percent of them become immune and 60 percent do not. Find the distribution of immune, at-risk, and sick people at a given time and in the long term. Investigate such questions as, If the vaccination rate were increased, what would be the long-term effects on the population?

To analyze the problem, we will quantify how the population distribution is affected in one time period. The transitions in one time step can be diagrammed as:

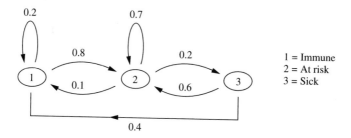

Let $x = (x_1, x_2, x_3)^T$ be the population distribution vector at a given time, where $x_1 = $ number of immune, $x_2 = $ number at risk, and $x_3 = $ number of sick individuals. Let $y = (y_1, y_2, y_3)^T$ be the corresponding populations after one time period. Then

$$\begin{aligned} y_1 &= 0.2x_1 + 0.1x_2 + 0.4x_3 \\ y_2 &= 0.8x_1 + 0.7x_2 + 0.6x_3 \\ y_3 &= \qquad\quad 0.2x_2 \end{aligned} \quad \text{or} \quad y = \begin{bmatrix} 0.2 & 0.1 & 0.4 \\ 0.8 & 0.7 & 0.6 \\ 0 & 0.2 & 0 \end{bmatrix} x = Ax.$$

Suppose x is the column of initial populations. Note that the distribution after two periods is $Ay = AAx = A^2x$, and so on, so that the population distribution after k time periods is A^kx.

Suppose A is diagonalizable: $AE_i = \lambda_iE_i$, $\{E_1, E_2, E_3\}$ is a basis, and $P = [E_1\ E_2\ E_3]$. Then, according to Section 4.4, if $x = \sum_{i=1}^{3} c_iE_i = Pc$, then

$$A^nx = \sum_{i=1}^{3} \lambda_i^n c_iE_i = P\Lambda^nP^{-1}x.$$

Since all entries are nonnegative and the columns sum to 1, Gershgorin's first theorem tells us that all the eigenvalues lie in the unit disk. Thus the populations cannot explode as they did in Example 1. This fits with intuition, since there are no births and deaths in this model; i.e., individuals are conserved. The characteristic polynomial of A is $p(\lambda) = -\lambda^3 + 0.9\lambda^2 + 0.06\lambda + 0.04$. It turns out that $\lambda = \lambda_1 = 1$ is an eigenvalue. If fact, this happens for a great many Markov processes. Division by $\lambda - 1$ gives $p(\lambda) = (1 - \lambda)(\lambda^2 + 0.1\lambda + 0.04)$, so the other two eigenvalues are $\lambda_{2,3} = (-0.1 \pm i\sqrt{0.15})/2$, which are of magnitude less than 1. The matrix is thus diagonalizable, having three distinct eigenvalues. Since $\lambda_i^n \to 0$ as $n \to \infty$ for $i = 2, 3$ we have, for large n,

$$A^nx = c_1E_1 + \lambda_2^nc_2E_2 + \lambda_3^nc_3E_3 \approx c_1E_1, \qquad (\#)$$

assuming $c_1 \neq 0$. Computing E_1 gives $E_1 = (1.125, 5, 1)^T$. Thus, in the long run (if $c_1 \neq 0$), the population ratio becomes

$$x_1 : x_2 : x_3 = 1.125 : 5 : 1.$$

Again, the eigenvector E_1 has special significance. If $c_1 = 0$, then the population vector $A^nx \to 0$. This, however, cannot occur for real populations in this model because no one leaves the system or dies. (*Challenge:* Find a mathematical proof of the fact that $c_1 \neq 0$, for all starting $x \neq 0$.) Using ($\#$), an estimate can be made of the number of steps needed to reach the steady state c_1E_1. Since $|\lambda_2| = |\lambda_3| = 0.2$, the second two terms in ($\#$) will be less than 0.001 of their initial size when $(0.2)^n < 0.001$, and this occurs when $n \geq 5$. The convergence is very fast.

Now let us assume that the immunization rate went up, and column 2 of A becomes $A_2 = (0.8, 0.16, 0.04)^T$. Again it turns out that $\lambda = 1$ is an eigenvalue. This may be seen by computing the determinant of $A - I$ or attempting to solve $(A - I)E = 0$, which indeed gives $E = E_1 = (25.5, 25, 1)^T$. So (assuming A is diagonalizable and $1 > |\lambda_2|, |\lambda_3|$), the eventual population ratios are (approximately)

$$x_1 : x_2 : x_3 = 25.5 : 25 : 1,$$

a dramatic improvement.

Markov processes

Example 2 illustrates a Markov process. In general, there are n classes of objects, and the fraction of the members of class j moving to class i in one time period is a given number a_{ij}. (These "fractions" may be interpreted as probabilities.) Thus, the jth column of $A = [a_{ij}]$ exhibits how group j is redistributed among the groups in one time step. It is assumed that objects are neither created nor destroyed, so that the entries of the column must sum to 1. In the example, $A_1 = (0.2, 0.8, 0)^T$ has entries that are the fractions of the immune people who become immune, at risk, or sick, respectively, in one time period. Let x_i be the number in group i at a particular time and y_i the number of people in group i at the next time. The ith row exhibits the proportion in each group who move to group i, so that $y_i = \sum_{j=1}^{n} a_{ij} x_j$. Thus $y = Ax$ gives the distribution of the number of people in each category after one period, given the initial distribution x. A is called the *transition matrix* for the problem. Markov processes have many applications and have been well studied. For further examples and discussion, see Roberts (1976), Kemeny (1960), Luenberger (1979).

EXAMPLE 3. Consider a production line with three stations. In each time interval, the product moves from station 1 to station 2 to station 3, from which it is shipped. Once during each time period, parts to make c products are supplied to station 1. From station 1 all parts are passed to station 2. At station 2, 52 percent of the products are found to be defective (on the average) and must be returned to station 1 for reprocessing; the rest are passed to station 3. At station 3, 20 percent of the parts are returned to station 1 for reprocessing, and the rest are shipped. The questions to be answered are: Find the long-time behavior of the flow along the line; in particular, find, if possible, the value of c such that 100 products per time period may be shipped. Estimate the number of time steps from a cold start that it takes to attain steady flow in the line.

Solution. The production line may be diagrammed as follows:

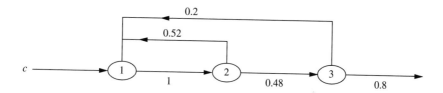

Let $x_i^{(n)}$ denote the number of "products" at station i at t_n. Then

$$x_1^{(n+1)} = 0.52x_2^{(n)} + 0.2x_3^{(n)} + c$$
$$x_2^{(n+1)} = x_1^{(n)}$$
$$x_3^{(n+1)} = 0.48x_2^{(n)}$$

In matrix form,

$$x^{(n+1)} = \begin{bmatrix} 0 & 0.52 & 0.2 \\ 1 & 0 & 0 \\ 0 & 0.48 & 0 \end{bmatrix} x^{(n)} + c \begin{bmatrix} 1 \\ 0 \\ 0 \end{bmatrix} = Ax^{(n)} + c\delta_1.$$

Successively computing $x^{(1)}$, $x^{(2)}$, etc., we obtain

$$x^{(n)} = A^n x^{(0)} + (I + A + A^2 + \cdots + A^{n-1})c\delta_1. \tag{6}$$

Using Gershgorin's second theorem (on the rows of A) to estimate the eigenvalues of A yields $|\lambda| \leq 1$. This can be improved by setting $D = \text{diag}(1, 1.25, 1)$, so that

$$D^{-1}AD = \begin{bmatrix} 0 & 0.65 & 0.2 \\ 0.8 & 0 & 0 \\ 0 & 0.6 & 0 \end{bmatrix},$$

which has the same eigenvalues as A. Applying Gershgorin's theorem again yields $|\lambda| \leq 0.85$. A theorem that will be proved in Chapter 8 asserts that, if all the eigenvalues of a matrix obey $|\lambda| < 1$, then $A_n \to 0$ as $n \to \infty$. (See the exercises for a proof for diagonalizable operators.) To find the behavior of the second term in Eq. (6), note that

$$(I - A)(I + A + A^2 + \cdots + A^{n-1}) = I - A^n.$$

Letting n tend to infinity yields $(I - A)\sum_{i=0}^{\infty} A_i = I$, so that

$$I + A + A^2 + \cdots + A^{n-1} \to \sum_{i=0}^{\infty} A^i = (I - A)^{-1}.$$

Hence $x^{(n)} \to (I - A)^{-1}c\delta_1$. Let $B = [B_1 B_2 B_3] = (I - A)^{-1}$. Then $B\delta_1 = B_1$, so that the limiting distribution (i.e., steady state) is cB_1, where B_1 is the first column of $(I - A)^{-1}$. A computation yields

$$B_1 \approx \begin{bmatrix} 2.604 \\ 2.604 \\ 1.25 \end{bmatrix}.$$

The output of the system at steady state is $0.8x_3 = 0.8 \cdot c \cdot 1.25 = c$, which is what should occur, since nothing was gained or lost along the line. Thus $c = 100$ for 100 units of output in each time interval.

To investigate the rate at which the steady state is approached, the eigenvalues of A are computed and found to be $\lambda_1 = 0.8$, $\lambda_2 = -0.6$, $\lambda_3 = -0.2$. At a cold start, $x^{(0)} = 0$. Let $\delta_1 = d_1 E_1 + d_2 E_2 + d_3 E_3 = Pd$, where $AE_1 = \lambda_i E_i$ and $p = [E_1 E_2 E_3]$. Then

$$
\begin{aligned}
(I + A + A2 + \cdots + A^{n-1})\delta_1 &= d_1 \left(\sum_{i=1}^{n-1} \lambda_1^i \right) E_1 + d_2 \left(\sum_{i=1}^{n-1} \lambda_2^i \right) E_2 + d_3 \left(\sum_{i=1}^{n-1} \lambda_3^i \right) E_3 \\
&= d_1 \frac{1 - \lambda_1^n}{1 - \lambda_1} E_1 + d_2 \frac{1 - \lambda_2^n}{1 - \lambda_2} E_2 + d_3 \frac{1 - \lambda_2^n}{1 - \lambda_2} E_3 \\
&= d_1 \frac{1}{1 - \lambda_1} E_1 + d_2 \frac{1}{1 - \lambda_2} E_2 + d_3 \frac{1}{1 - \lambda_2} E_3 \\
&\quad - d_1 \frac{\lambda_1^n}{1 - \lambda_1} E_1 - d_2 \frac{\lambda_2^n}{1 - \lambda_2} E_2 - d_3 \frac{\lambda_2^n}{1 - \lambda_2} E_3 \\
&= (I - A)^{-1}\delta_1 \\
&\quad - \left\{ d_1 \frac{\lambda_1^n}{1 - \lambda_1} E_1 + d_2 \frac{\lambda_2^n}{1 - \lambda_2} E_2 + d_3 \frac{\lambda_2^n}{1 - \lambda_2} E_3 \right\}.
\end{aligned}
$$

The terms in the braces are the error terms. If $n \geq 28$, then $\lambda_1^n/(1 - \lambda_1) \leq 0.01$ and the other two terms are $\leq 10^{-6}$. Thus, within about 28 steps, the system is within 1 percent of steady state.

Another kind of finite difference equation is common. Here is an example.

EXAMPLE 4. Solve the recursion $a_{n+1} = -\frac{1}{2}a_n + \frac{5}{2}a_{n-1} - a_{n-2}$ with the initial values $a_0 = 0$, $a_1 = 1$, $a_2 = 0$.

The trick is to reduce the problem to a matrix iteration that can be solved with eigenvalue methods. The method is akin to reducing an nth-order ODE to a first-order system. To this end, let $b_n = a_{n+1}, c_n = a_{n+2}$ $\forall n$. Then

$$a_{n+1} = b_n \qquad \text{(definition)}$$

$$b_{n+1} = c_n \qquad \text{(definition)}$$

$$c_{n+1} = -\tfrac{1}{2}c_n + \tfrac{5}{2}b_n - a_n \qquad \text{(the recursion)}.$$

In matrix form, with $x_n = (a_n, b_n, c_n)^T$,

$$x_{n+1} = \begin{bmatrix} 0 & 1 & 0 \\ 0 & 0 & 1 \\ -1 & 2.5 & -0.5 \end{bmatrix} x_n \equiv A x_n.$$

Once again, $x_n = A^n x_0$ gives the formula for the nth iterate. Computing the eigenvalues and eigenvectors of A yields

$$\lambda_1 = -2, \quad E_1 = \begin{bmatrix} 1 \\ -2 \\ 4 \end{bmatrix}; \qquad \lambda_2 = 1, \quad E_2 = \begin{bmatrix} 1 \\ 1 \\ 1 \end{bmatrix}; \qquad \lambda_3 = 0.5, \quad E_3 = \begin{bmatrix} 1 \\ 0.5 \\ 0.25 \end{bmatrix}.$$

With $x_0 = \sum_{i=1}^{3} d_i E_i = Pd$ and $\Lambda = \text{diag}(\lambda_1, \lambda_2, \lambda_n)$, we have

$$x_n = A^n x_0 = \sum_{i=1}^{3} d_i \lambda_i^n E_i = P\Lambda^n P^{-1} x_0.$$

Equating the top entries of each column gives

$$a_n = \lambda_1^n d_1 + \lambda_2^n d_2 + \lambda_3^n d_3 = (-2)^n d_1 + d_2 + (0.5)^n d_3, \quad n = 0, 1, \ldots .$$

The coefficients c_i can be obtained from $x_0 = Pd$ or from the initial conditions at $n = 0, 1, 2$. By either method, $d_1 = -0.2, d_2 = 1, d_3 = -0.8$.

Exercises 4.6

1. **Population control.** In Example 1, the hapless species will outbreed its habitat unless controlled. We will investigate two methods of population control. In each case that works, find the relative sizes of the limiting population.
   (a) If we take out a number of individuals in the their second year in proportion to the number present each year (before they breed), this is the same as reducing σ_2 and is called *proportional harvesting*.
      (i) Find, if possible, a value of σ_1 for which the species neither dies out nor over-populates. All the other parameters are fixed.
      (ii) Find a value of σ_2 that will accomplish the same goal.
   (b) A second method is to control the birth rates α and β.
      (i) Find, if possible, a value of α for which the species neither dies out nor overpop-ulates. Do the same for β.
      (ii) Let $\beta = 2\alpha$ and find a value of α for which the species neither dies out nor overpopulates.

2. In Example 1, suppose $\alpha = \frac{13}{3}, \beta = 4, \sigma_1 = \frac{3}{4}, \sigma_2 = \frac{1}{2}$.
   (a) At year 0, the population distribution is 10 in their first year, 0 in their second year, and 0 in their third year. Find the population distribution in the fifth year and the tenth year.

(b) Estimate the populations for large n. Find the relative sizes in the limiting population distribution.

(c) This species needs to be stabilized. As in Exercise 1, we want the species to neither die nor overpopulate. Is it possible to adjust σ_1 to achieve this? What about σ_2? If it is possible, find the relative sizes in the limiting population distribution.

3. Consider the general problem for Example 1: $0 < \sigma_i \leq 1, \alpha, \beta > 0$. Let $\lambda_i, i = 1, 2, 3$, be the eigenvalues of the matrix A.

(a) Show there is exactly one eigenvalue $\lambda_1 > 0$.

(b) Find a criterion that tests whether or not λ_2, λ_3 are real or complex. Determine when the matrix is diagonalizable.

(c) When λ_2, λ_3 are real, show that $|\lambda_2|, |\lambda_3| < \lambda_1$.

(d) When λ_2, λ_3 are complex, show that they lie in the left half plane.

(e) Show that $\sigma_1 \sigma_2 \beta = \lambda_1 |\lambda_2| |\lambda_3|$.

4. Suppose it is impossible to make the transition from rich to poor or vice versa without being middle-class. (No one is allowed to go away and become a guru or anything else.) In one year, of the rich, two-thirds remain so and one-third become (to their dismay) middle-class. Of the middle-class, one-third stay in this happy state, and one-third become impoverished and one-third rich. Of the poor, two-thirds stay in this predicament and one-third become middle-class. Find an expression for the population at the nth step. If initially there are 1000 poor folks and no one else, what is the population in 5 years? In 10 years? What is the limiting population distribution?

5. Consider the same three economic groups as in Exercise 4, with the following modifications: It is possible to move between any two groups. Of those initially poor, one-third stay that way, one-third become middle class, and one-third become rich. Of those initially middle-class, one-sixth become poor, one-half become rich, and the rest remain as they are. Of those initially rich, one-sixth become impoverished, one-half are relegated to middle-class, and the remainder stay blissfully rich. Answer the same questions posed in Exercise 4.

6. The following data from the U.S. Treasury Department indicate income group mobility from 1979 to 1988.

		Status in 1988					
		Group 1	Group 2	Group 3	Group 4	Group 5	Group 6
	Group 1	47.3%	38.6%	7.7%	3.8%	0.4%	2.2%
	Group 2	5.3	59.4	20.3	9.4	4.4	1.1
Status	Group 3	0.6	34.8	37.5	14.8	9.3	3.1
in	Group 4	0.4	14.6	32.3	33.0	14.0	5.7
1979	Group 5	0.3	10.8	19.5	29.6	29.0	10.9
	Group 6	0.3	14.4	25.3	25.0	20.7	14.2

For example, of those people in group 1 in 1979 (top row), 47.3 percent remained in group 1, 38.6 percent moved to group 2, and 2.2 percent moved to group 6 by 1988. (There are some rounding errors in the table.) The groups are formed according to income: Group 1 is the top 1 percent, group 2 is the top 2 percent to 20 percent, group 3 includes the next

20 percent, group 4 is the middle 20 percent, group 5 is the next poorest 20 percent, and group 6 is the poorest 20 percent. Find the limiting distribution, assuming the transition matrix stays constant, and express the limiting populations as percentages. Estimate the number of steps needed to get within 10 percent of steady state.

7. In the production line problem of Example 3, suppose that all the product is sent from station 1 to station 2, station 2 ships 39 percent back to station 1 for repairs and 35 percent to station 3, and station 3 returns 20 percent to station 1 and ships the rest to customers. Find the input c needed at station 1 to produce 200 units/period for shipment to customers, when production is steady. At steady state, how many units per time period are being scrapped? About how many time steps are needed for the line to get to 1 percent of steady state?

8. A control mechanism is to be added to station 3 of the production line of Example 3. The number of units x_3 at station 3 is compared with a number a, the difference is multiplied by a factor β, and the input of parts to station 1 is $\beta(a - x_3)$ on the next cycle. Specifically, if x_i are the current numbers of units at the stations and $\tilde{x}_1$ is the next number of units at station 1, then $\tilde{x}_1 = r_2 x_2 + r_3 x_3 + \beta(a - x_3)$, where r_2 and r_3 are the fractions being sent back to station 1 by stations 2 and 3. This time station 1 sends 80 percent of its units to station 2, station 2 sends 60 percent of its units to station 3, and $r_2 = 35$ percent, $r_3 = 30$ percent. Station 3 ships all other units. The individual controlling the line knows that the steady output is proportional to a, so an initial value of a will be tried and then adjusted to get the required output. Suppose $\beta = 0.2$. a is given an initial setting of 140. When the line reaches steady state, a must be readjusted so as to produce 50 units/period when the system settles to steady state. Determine the number of time units required for the system to reach steady state (approximately). Find the value of a needed to achieve the required output at the next steady state. Could you have predicted a before the first run? What happens if β is increased to 2?

9. Find the general solution of the following difference equations. Describe the behavior of the solution as $n \to \infty$.
   (a) $a_{n+2} - a_n = 0$
   (b) $a_{n+2} - 0.75a_{n+1} + 0.125a_n = 0$
   (c) $a_{n+2} - 2.5a_{n+1} + a_n = 0$
   (d) $a_{n+2} + \sqrt{2}a_{n+1} + a_n = 0$
   (e) $a_{n+3} - 2.5a_{n+2} + 0.5a_{n+1} + a_n = 0$
   (f) $a_{n+3} + a_n = 0$
   (g) $a_{n+1} = 0.5a_n + x_{n-1} - 0.5a_{n-2}$

10. Show that $x_{n+1} = Ax_n + f$ has the solution $x_n = A^n x_0 + (\sum_{i=0}^{n-1} A^i)f$.

11. Suppose A is diagonalizable and all its eigenvalues obey $|\lambda_i| < 1$. Show
    (a) $A^n \to 0$
    (b) $\sum_{i=0}^{n-1} A^i \to (1 - A)^{-1}$
    (Hint: $A = P\Lambda P^{-1}$.)

12. **The power method for finding an eigenvector and eigenvalue.** Suppose A is $n \times n$ and diagonalizable with an eigenbasis $\{E_k\}_{k=1}^n$ and $AE_i = \lambda_i E_i$. Suppose there is an eigenvalue λ_1 of largest magnitude: $|\lambda_1| > |\lambda_2|, |\lambda_1|, \ldots, |\lambda_n|$. λ_1 is called a *dominant*

eigenvalue. Assume that $x_0 = \sum_{i=1}^{n} c_i E_i$, with $c_1 \neq 0$. Define iteratively: $x_{k+1} = Ax_k$. Show that

(a) For large k, $x_k \approx c_1 \lambda_1^k E_1$ in the sense that the other terms are small in comparison with this one.

(b) An approximate eigenvector of unit length is $x_k/\|x_k\|$, for large enough k.

(c) An approximation of λ_1 is obtained by looking at ratios of entries of z and y: $\lambda_1 \approx z_i/y_i$, where $y = x_k$, $z = x_{k+1}$, and only those $y_i \neq 0$ are considered.

APPENDIX
REVIEW OF DETERMINANTS

The determinant of a square matrix is denoted $\det(A) = |A|$

Computation of Determinants

$$1 \times 1 \text{ matrix:} \qquad \det([a]) = \|[a]\| = a$$

$$2 \times 2 \text{ matrix:} \qquad \det \begin{bmatrix} a & b \\ c & d \end{bmatrix} = ad - bc$$

Two ways to compute determinants for higher-order matrices will be discussed.

Computation by Cofactor Expansions

DEFINITION. The *minor* M_{ij} of a_{ij} in $A = [a_{ij}]$ is the determinant of the submatrix of A obtained by deleting the ith row and jth column. The *cofactor* of a_{ij} is $C_{ij} = (-1)^{i+j} M_{ij}$.

The following theorem connects determinants of matrices of size n to those of size $n - 1$.

THEOREM 1. For each i, j the following hold:

(a) Expansion along the ith row: $|A| = \sum_{k=1}^{n} a_{ik} C_{ik}$

(b) Expansion along the jth column: $|A| = \sum_{k=1}^{n} a_{kj} C_{kj}$

(c) $\sum_{j=1}^{n} a_{kj} C_{ij} = 0$, if $k \neq i$

(d) $\sum_{i=1}^{n} a_{ik} C_{ij} = 0$, if $k \neq j$

For example, expanding along the first row and then the second column:

$$|A| = \begin{vmatrix} 1 & 2 & 3 \\ 4 & 5 & 6 \\ 7 & 8 & 9 \end{vmatrix} = 1 \begin{vmatrix} 5 & 6 \\ 8 & 9 \end{vmatrix} - 2 \begin{vmatrix} 4 & 6 \\ 7 & 9 \end{vmatrix} + 3 \begin{vmatrix} 4 & 5 \\ 7 & 8 \end{vmatrix} = 0$$

$$|A| = -2 \begin{vmatrix} 4 & 6 \\ 7 & 9 \end{vmatrix} + 5 \begin{vmatrix} 1 & 3 \\ 7 & 9 \end{vmatrix} - 8 \begin{vmatrix} 1 & 3 \\ 4 & 6 \end{vmatrix} = 0$$

The next theorem follows easily from the last.

THEOREM. If A is upper triangular or lower triangular, then $|A|$ is the product of the elements on the main diagonal: $|A| = \prod_{i=1}^{n} a_{ii}$.

Computation by Gauss Elimination to Row Echelon Form

The effects of the elementary row operations are given next. In the following, B is obtained from A by the stated elementary row operation.

1. Interchange of rows: $|B| = -|A|$
2. Add a scalar times row j to row i, $i \neq j$: $|B| = |A|$
3. Multiply one row times a scalar c: $|B| = c|A|$

Thus if R is an upper triangular matrix obtained from A using operations 1 and 2,

$$|A| = (-1)^s \prod_{i=1}^{n} r_{ii}, \quad \text{where } s \text{ is the number of row interchanges used.}$$

Other Properties

1. $|AB| = |A||B|$
2. $|\overline{A}| = \overline{|A|}$
3. $|A^{\mathrm{T}}| = |A|$
4. $|A| \neq 0$ iff A is nonsingular, in which case $|A^{-1}| = |A|^{-1}$.
5. When A is nonsingular, $A^{-1} = |A|^{-1} \operatorname{Adj}(A)$, where the *classical adjoint of A* is defined by $\operatorname{Adj}(A) = [C_{ij}]^{\mathrm{T}}$; that is, the classical adjoint is the transpose of the matrix of cofactors of A. In fact, from the preceding theorem on cofactors,

$$\sum_{j=1}^{n} a_{kj} C_{ij} = |A| \delta_{ik} = \sum_{j=1}^{n} a_{kj} (\operatorname{Adj}(A))_{ji}.$$

This formula is useful for calculating the inverse of a small matrix ($n = 2, 3$, e.g.) and in theoretical work. It is often used in elasticity theory.

EXAMPLE

$$\begin{bmatrix} a & b \\ c & d \end{bmatrix}^{-1} = \frac{1}{|A|} \begin{bmatrix} d & -b \\ -c & a \end{bmatrix}$$

The next theorem may be proved using cofactor expansions.

THEOREM If A is a block diagonal matrix, $A = \operatorname{diag}(A_1, A_2, \ldots, A_p)$, where each A_i is a square matrix, then $|A| = \prod_{i=1}^{p} |A_i|$.

EXAMPLE

$$\det\left(\begin{bmatrix} A_1 & 0 \\ 0 & A_2 \end{bmatrix}\right) = |A_1||A_2|$$

6. *Cramer's rule:* The solution of $Ax = y$, when A is nonsingular, is given by $x_i = |A|^{-1}|M_i|$, where M_i is the matrix A with its ith column replaced by y.

   EXAMPLE. Solve $Ax = y$, where

   $$A = \begin{bmatrix} 1 & 2 \\ -1 & 1 \end{bmatrix}, \qquad x = \begin{bmatrix} x_1 \\ x_2 \end{bmatrix}, \qquad y = \begin{bmatrix} 3 \\ 4 \end{bmatrix}.$$

   Solution

   $$|A| = 3, \qquad x_1 = \frac{1}{3}\begin{vmatrix} 3 & 2 \\ 4 & 1 \end{vmatrix} = \frac{-5}{3}, \qquad x_2 = \frac{1}{3}\begin{vmatrix} 1 & 3 \\ -1 & 4 \end{vmatrix} = \frac{7}{3}$$

Tips on Solving for the Zeros of Polynomials

In many of the exercises the author has been thoughtful enough to design problems with nice (integer or simple rational) answers, and this serves to indicate to the student when the correct answer has been found (if it is not in the back of the book). A few facts will be reviewed to help find these solutions.

First, suppose $p(\lambda) = 0$ is a polynomial equation with rational coefficients. Multiply it through by the least common denominator of the coefficients to get an equation with integer coefficients. Suppose a_n is the leading coefficient and a_0 is the constant term. If $\lambda = p/q$ (p, q integers, with no common factors) is a root of the equation, then p must divide a_0 and q must divide a_n. This can severely limit the possibilities. We can make a list of the possible rational zeros. The next step is to evaluate $p(\lambda)$ for each of the numbers in the list. There is no guarantee that the list contains a zero. Once a zero λ_1 has been found, we can do long division to factor the polynomial: $p(\lambda) = (\lambda - \lambda_1)q(\lambda)$, and the process continues on q.

Gershgorin's first theorem in Section 4.5 can limit the possibilities in the list, if $p(\lambda)$ is a characteristic polynomial. The second theorem can also be useful.

Furthermore, Descartes's rule of signs can be helpful. This theorem says that, if $p(\lambda) = \sum_{i=1}^{n} c_i\lambda^i$ with real c_k and $c_n \neq 0$, and there are k sign changes in the coefficients, then p has $n - 2k$ positive zeros for some $k = 0, 1, 2, \ldots, n/2$. For example, $p(\lambda) = \lambda^3 - 5\lambda^2 + 4\lambda + 5$ has two sign changes in the coefficients, so it either has two or no positive zeros. $p(-\lambda) = -\lambda^3 - 5\lambda^2 - 4\lambda + 5$ has one sign change in the coefficients, so it has just one positive zero; i.e., $p(\lambda)$ has exactly one negative zero.

For small problems, Newton's method or the secant method are useful for computing the real eigenvalues. A graphing calculator is a good way to come up with an initial guess for these algorithms. Solvers on modern calculators can find the roots accurately, and these roots (since the answers are "nice") can be checked by long division.

CHAPTER 5

The Structure of Normal Operators

When a vector space has an inner product specified, it is often of interest to ask how the behavior of linear operators relates to the inner product. Quite often there is a kind of "compatibility" between an operator and the inner product that produces dramatic and highly useful structure in the operator. This structure is described in the spectral theorem, which is the central result of this chapter. Furthermore, it is possible to "classify" operators into some important and commonly occurring types. These results will be applied to matrix theory. The Rayleigh quotient and its relation to several important extremal properties of eigenvalues of Hermitian operators will be discussed, as will methods for approximating some of these eigenvalues and eigenvectors. Finally, as a culmination of the theoretical work of the chapter, the Rayleigh–Ritz method of approximating eigenvalues and eigenfunctions will be proved and examples given. Most of the work in this chapter is theoretical; in later chapters applications to some specific physical problems will be given.

Scalars will always be complex in this chapter, and the domain of the operators considered will be nontrivial.

5.1
ADJOINTS AND CLASSIFICATION OF OPERATORS

In order to begin our classification, definitions are needed. These will be somewhat general, in order to yield useful applications.

DEFINITION. Let $T : V \to W$ be linear, let V be a subspace of the vector space W, and let $\langle \, , \rangle$ be an inner product on W. An operator $T^* : V \to W$ satisfying

$$\langle T(x), y \rangle = \langle x, T^*(y) \rangle \quad \text{for all } x \text{ and } y \text{ in } V \tag{1}$$

is called an *adjoint operator of T* with respect to the inner product $\langle \, , \rangle$ *on V*, or simply an *adjoint of T*.

In most cases of interest, the operator T^* can be directly computed and is unique. Later it will be shown that T^* exists when $V = W$ and V is finite-dimensional. However, the general theory in infinite-dimensional spaces has some nontrivial difficulties that will not be discussed here. For example, if T is a differential operator, the domain of T^* is allowed to be different from the domain of T, but is related to it in a specific way. In order to avoid these fine points, the preceding definition is made. The interested reader is referred to N. I. Achiezer and I. M. Glazman (1961), or to any of the other functional analysis books listed in the references. In the exercises, a more general definition of T^* is given. When infinite-dimensional spaces are discussed, it will be assumed that T^* exists and is unique.

The following examples illustrate the construction of T^*. In these examples, the method used is to start with $\langle T(x), y \rangle$ and perform some manipulative tricks to obtain an expression of the form $\langle x, z \rangle$. Then, from $z = T^*(y)$, the operator T^* is found.

EXAMPLE 1. ADJOINTS OF MATRIX OPERATORS WITH THE STANDARD INNER PRODUCT. Let $T(x) = Ax$, where A is a given $n \times n$ matrix, $V = W = \mathbf{C}^n$, and $\langle x, y \rangle = y^H x$ is the standard inner product on V. Find the adjoint T^*.

Compute

$$\langle T(x), y \rangle = \langle Ax, y \rangle = y^H Ax = y^H A^{HH} x = (A^H y)^H x = \langle x, A^H y \rangle = \langle x, T^*(y) \rangle,$$

so that $T^*(y) = A^H y$. Thus T^* is a matrix operator generated by the matrix A^H; i.e., the matrix of T^* with respect to the standard basis is

$$[T^*] = A^H = \overline{A}^{\mathsf{T}}. \tag{2}$$

This is not true with a general inner product on $\mathbf{C}^n$. However, there are some important situations in which (2) generalizes, as will be seen in the discussion of Theorem 1. Chapter 7 will present many important applications that use a nonstandard inner product on $\mathbf{C}^n$.

Adjoints of differential operators are important in the theory of differential equations, especially in the theory of Sturm–Liouville operators, which arise in the solution of partial differential equations. A simple example is given next, with more examples given in the exercises.

EXAMPLE 2. ADJOINTS OF A SECOND-ORDER DIFFERENTIAL OPERATOR. Let $T : V \to W$ be defined by $T(x) = x''$, where $' = d/dt$. V consists of functions $x(t)$ with two continuous derivatives satisfying $x(0) = x(1) = 0$ and $W = C[a, b]$. Find T^* if the inner product is

$$\langle x, y \rangle = \int_0^1 x(t)\overline{y(t)}w(t)\, dt \qquad \text{for} \quad (i)\ w(t) = 1, \quad (ii)\ w(t) = e^t.$$

The trick is to integrate by parts twice and use the fact that x, y vanish at 0 and that x', y' vanish at 1. For (i)

$$\langle T(x), y \rangle = \int_0^1 x'' \overline{y}\, dt = x'\overline{y}\Big|_0^1 - \int_0^1 x'\overline{y}'\, dt = 0 - x\overline{y}'\Big|_0^1 + \int_0^1 x\overline{y}''\, dt$$

$$= \int_0^1 x\overline{y''}\, dt \equiv \langle x, T^*(y) \rangle.$$

Therefore, $T^*(y) = y'' = T(y)$, so that $T^* = T$. Such operators are called *self-adjoint* or *Hermitian*. For (ii)

$$\langle T(x), y \rangle = \int_0^1 x'' \bar{y} e^t \, dt = x' \bar{y} e^t \Big|_0^1 - \int_0^1 x' \bar{y}' e^t \, dt$$

$$= x(\bar{y}' + \bar{y}) e^t \Big|_0^1 + \int_0^1 x(\bar{y}'' + 2\bar{y}' + \bar{y}) e^t \, dt$$

$$= \int_0^1 x \overline{(y'' + 2y' + y)} e^t \, dt \equiv \langle x, T^*(y) \rangle,$$

so that $T^*(y) = y'' + 2y' + y$, and $T^* \neq T$.

EXAMPLE 3. FREDHOLM INTEGRAL OPERATORS. Define T and the inner product by

$$(Tx)(t) = \int_a^b k(t, s)x(s) \, ds \quad \text{with } \langle x, y \rangle = \int_a^b x(u)\overline{y(u)} \, du,$$

find T^*.

k is assumed to be continuous and is called the *kernel of the transform*. If (a, b) is a finite interval, then $V = W = C[a, b]$ will be taken. If (a, b) is infinite, extra conditions must be added to the definition of V in order that the calculations that follow are permissible. These conditions can be quite technical and are suppressed here.

For example, $k(t, s) = e^{-st}$, $a = 0$, $b = \infty$ yields the Laplace transform, whereas $k(t, s) = e^{-its}/\sqrt{2\pi}$, $a = -\infty$, $b = \infty$ yields the normalized Fourier transform. The method used to construct T^* is to interchange the order of integration (Fubini's theorem):

$$\langle Tx, y \rangle = \int_a^b \left\{ \int_a^b k(t, s)x(s) \, ds \right\} \overline{y(t)} \, dt = \int_a^b \int_a^b k(t, s)x(s)\overline{y(t)} \, dt \, ds$$

$$= \int_a^b x(s) \left\{ \int_a^b k(t, s)\overline{y(t)} \, dt \right\} ds = \int_a^b x(s) \left\{ \overline{\int_a^b \overline{k(t, s)}y(t) \, dt} \right\} ds$$

$$= \langle x, T^*y \rangle,$$

where the fact that $\int \bar{f} = \overline{\int f}$ has been used. Thus

$$(T^*y)(s) = \int_a^b \overline{k(t, s)}y(t) \, dt.$$

Interchanging the dummy variables t and s gives

$$(T^*y)(t) = \int_a^b \overline{k(s, t)}y(s) \, ds,$$

which is a Fredholm integral operator with kernel $k^*(t, s) \equiv \overline{k(s, t)}$. This is similar to the rule $[T^*]_{ij} = \overline{[T]}_{ji}$ of Example 1. For the Laplace transform, $k = k^* = e^{-st}$, which implies that $T = T^*$, and so (formally) the Laplace transform is self-adjoint. For the Fourier transform, $k^*(t, s) = e^{its}/\sqrt{2\pi}$, which is the kernel of the inverse Fourier transform, so (formally) $T^* = T^{-1}$, or $T^*T = I$. Such operators are called *unitary operators*.

The next theorem shows that T^* always exists and is unique when V is finite-dimensional and $V = W$. When $\dim(V) = \infty$, more hypotheses are needed to ensure the existence and uniqueness of T^*.

THEOREM 1. If $T : V \to V$ is linear, $\dim(V) < \infty$, and $\langle , \rangle$ is an inner product on V, then the adjoint of T on V with respect to $\langle , \rangle$ exists and is unique. If $\beta = \{e_1, e_2, \ldots, e_n\}$ is an arbitrary basis for V, and G is the Gram matrix of the given inner product with respect to β, then

$$[T^*]_\beta = G^{-1}[T]_\beta^H G. \tag{3}$$

Thus

$$[T^*]_\beta = [T]_\beta^H \quad \text{when } \beta \text{ is orthonormal.} \tag{4}$$

Proof. Suppose $[T]_\beta = A$. From Eq. (1) in Section 3.6,

$$\langle x, y \rangle = [y]^H G [x] \quad \forall x, y \in V \tag{5}$$

where G is the Gram matrix of the basis β and $[x], [y]$ are the coordinate columns of x, y. (The proof is $x = \sum_{j=1}^n c_j e_j, \; y = \sum_{i=1}^n d_i e_i \Rightarrow \langle x, y \rangle = \sum_{i,j=1}^n \overline{c_j d_i} \langle e_j, e_i \rangle = d^H G c = [y]^H G [x]$.) Now assume T^* exists and find $[T^*]_\beta = B$. Since $A[x]$ is the coordinate vector of $T(x)$ and $B[y]$ is the coordinate vector of $T^*(y)$ by Theorem 3 of Section 2.3, Eq. (5) yields

$$\langle Tx, y \rangle = [y]^H G(A[x]) = [y]^H GA[x] \quad \forall x, y$$
$$\langle x, T^*y \rangle = (B[y])^H G[x] = [y]^H B^H G[x] \quad \forall x, y.$$

Therefore, $GA = B^H G$. Taking conjugate transposes, $(GA)^H = A^H G^H = A^H G = (B^H G)^H = G^H B^{HH} = GB$. Thus $B = G^{-1} A^H G$. If the basis is orthonormal, $G = I$ and Eq. (4) follows. Conversely, if S is defined to be that operator whose matrix is $G^{-1} A^H G$, then the preceding steps also show that S satisfies $\langle Tx, y \rangle = \langle x, Sy \rangle$ for all x, y in V, and so $S = T^*$ exists. Since B is unique, so is T^*.

EXAMPLE 4. A MATRIX OPERATOR WITH A WEIGHTED INNER PRODUCT. Let

$$A = \begin{bmatrix} a & b \\ c & d \end{bmatrix}, \quad T(x) = Ax, \quad \text{and} \quad \langle x, y \rangle = w_1 x_1 y_1 + w_2 x_2 y_2.$$

Find the matrix of T^* in the standard basis.

Solution. Compute

$$G = [\langle \delta_j, \delta_i \rangle] = \begin{bmatrix} w_1 & 0 \\ 0 & w_2 \end{bmatrix}.$$

By Eq. (3),

$$[T^*] = G^{-1} A^H G = \begin{bmatrix} \bar{a} & (w_2/w_1)\bar{c} \\ (w_1/w_2)\bar{b} & \bar{d} \end{bmatrix}.$$

Thus $[T^*] \neq [T]$ if $w_1 \neq w_2$.

These examples show clearly that T^* depends heavily on the inner product being used.

The next theorem lists some of the principal algebraic laws for the adjoint. This theorem says that the mapping $T \to T^*$ has the same kind of properties as $A \to A^H$ (cf. Section 0.4).

THEOREM 2. Assume L, T, S are linear operators and assume that their adjoints exist and are unique. Assume also that the following sums and products exist. Then

(a) $(T + S)^* = T^* + S^*$
(b) $(\alpha T)^* = \bar{\alpha} T^*$
(c) $(T^*)^* = T$
(d) $I^* = I$
(e) $(TL)^* = L^* T^*$
(f) If one of the inverses T^{-1} or $(T^*)^{-1}$ exists, then so does the other, and $(T^{-1})^* = (T^*)^{-1}$.

Proof. The proofs are formal manipulations using the definition. Only two will be demonstrated.

(b) $\langle (\alpha T)x, y \rangle = \langle \alpha Tx, y \rangle = \alpha \langle Tx, y \rangle = \alpha \langle x, T^*y \rangle = \langle x, \bar{\alpha} T^*y \rangle \; \forall x, y$. But, by definition of $(\alpha T)^*$, $\langle (\alpha T)x, y \rangle = \langle x, (\alpha T)^*y \rangle \; \forall x, y$. Thereforem $(\alpha T)^* = \bar{\alpha} T^*$.
(c) $\langle T^*x, y \rangle = \langle x, T^{**}y \rangle \; \forall x, y$ by definition of T^{**}. But $\langle T^*x, y \rangle = \overline{\langle y, T^*x \rangle} = \overline{\langle Ty, x \rangle} = \langle x, Ty \rangle \; \forall x, y$. Thus $T^{**} = T$.

Now the definitions for the classification of operators may be given.

DEFINITIONS. Given $T : V \to W$ linear with $V \subset W$ and an inner product $\langle , \rangle$ on W, T is *normal* on V with respect to $\langle , \rangle$ iff T commutes with its adjoint:

$$T^*T = TT^*. \tag{6}$$

T is *Hermitian* or *self-adjoint* on V with respect to $\langle , \rangle$ iff

$$T^* = T. \tag{7}$$

In other words, T is Hermitian iff

$$\langle T(x), y \rangle = \langle x, T(y) \rangle \quad \forall x, y \text{ in } V. \tag{8}$$

T is *skew-Hermitian* on V with respect to $\langle , \rangle$ iff

$$T^* = -T. \tag{9}$$

T is *unitary* on V with respect to $\langle , \rangle$ iff

$$T^* = T^{-1}. \tag{10}$$

One simply says "T is Hermitian" (for example) without reference to the vector space and inner product when there is no danger of confusion. Be warned again that T^* depends heavily on V and the $\langle , \rangle$ selected. Note that Eqs. (6) and (10) require $T, T^* : V \to V$.

Remark. Each Hermitian, skew-Hermitian, and unitary operator is also a normal operator.

The next theorem gives an alternative characterization of unitary operators $T : V \to V$ on finite-dimensional spaces V. Its geometric interpretation is that unitary operators are simply those that preserve length and angle. For example, rigid rotations must preserve length and angle and therefore are unitary transformations.

THEOREM 3. Let $\dim(V) < \infty$ and let $T : V \to V$ be linear. Let $\langle , \rangle$ be an inner product on V. Then the following statements are equivalent:

(i) T is unitary on V with respect to $\langle , \rangle$.
(ii) $T^*T = I$.
(iii) $\langle Tx, Ty \rangle = \langle x, y \rangle \quad \forall x, y$ in V. $\tag{11}$

Proof. We show $(i) \Rightarrow (ii) \Rightarrow (iii) \Rightarrow (i)$. Assume (i) holds. If T is unitary, then $T^* = T^{-1}$ so (ii) holds. From (ii) it follows that $\langle Tx, Ty \rangle = \langle x, T^*Ty \rangle = \langle x, Iy \rangle = \langle x, y \rangle$ for all x, y. On the other hand, if (iii) holds, then $\langle Tx, Ty \rangle = \langle x, T^*Ty \rangle = \langle x, y \rangle$ for all x, y in V. But $\langle x, u \rangle = \langle x, v \rangle$ for all x implies $u = v$; hence $T^*Ty = y = Iy$ for all y in V, which is the same as $T^*T = I$. Since V is finite-dimensional, T^* is also a right inverse of T, and therefore the inverse of T, by Theorem 6 of Section 2.2.

Remark. Equation (11) implies that T preserves length and angle, since it implies that $\|Tx\|^2 = \|x\|^2$ and $\|x\| \|y\| \cos(\theta) = \langle x, y \rangle = \langle Tx, Ty \rangle = \|Tx\| \|Ty\| \cos(\phi)$, where θ is the angle between x and y and ϕ is the angle between Tx and Ty (in a real vector space).

EXAMPLE 5. Let $V = \mathbf{T}_n$, let α be real and define $Tf(x) = f(x + \alpha)$, and take

$$\langle f, g \rangle = \int_0^{2\pi} f(x)\overline{g(x)}\, dx.$$

Show that T is unitary.

$$\langle Tf, Tg \rangle = \int_0^{2\pi} f(x+\alpha)\overline{g(x+\alpha)}\, dx = \int_\alpha^{2\pi+\alpha} f(t)\overline{g(t)}\, dt = \int_0^{2\pi} f(t)\overline{g(t)}\, dt = \langle f, g \rangle.$$

The last step follows because the integrand is 2π-periodic. Hence, the operator is unitary, and $T^* = T^{-1}$.

EXAMPLE 6. Show that unitary matrices generate unitary operators in the standard inner product.

Recall that a unitary matrix U is an $n \times n$ matrix for which $U^H = U^{-1}$. Let $T(x) = Ux$. By Example 1, $T^*(x) = U^H x$. From

$$\langle T(x), T(y) \rangle = \langle Ux, Uy \rangle = (Uy)^H Ux = y^H U^H Ux = y^H x = \langle x, y \rangle$$

it follows that T is a unitary operator.

Remark. Equation (11) holds for the Fourier transform (under the proper hypotheses) since it is a unitary operator: If $\hat{x}$ and $\hat{y}$ are the (normalized) Fourier transforms of x and y, then

$$\langle \hat{x}, \hat{y} \rangle = \int_{-\infty}^{\infty} \hat{x}(w)\overline{\hat{y}(w)}\, dw = \int_{-\infty}^{\infty} x(t)\overline{y(t)}\, dt = \langle x, y \rangle.$$

This is called *Parseval's identity.* It is possible to develop the theory of the Fourier transform so that it is indeed a unitary operator on an appropriate vector space and shares many of the properties discussed here. Many technical details of analysis are needed, however, which cannot be discussed here.

The preceding classifications are formulated in terms of the operator and inner product and are valid in general. It is convenient to have tests for the finite-dimensional case that use a matrix of the operator to determine the classification. The next theorem and corollary address this issue. With a basis that is not orthonormal, the Gram matrix of the inner product plays a crucial role, and the test is cumbersome. When the basis is orthonormal, the tests are much easier.

THEOREM 4. Let $\dim(V) < \infty$, and let $T : V \to V$ be linear. If $\beta = \{e_1, e_2, \ldots, e_n\}$ is a basis for V, $\langle \, , \, \rangle$ is an inner product on V, $A = [T]_\beta$, and G is the Gram matrix of β with respect to $\langle \, , \, \rangle$, then

(a) T is normal $\Leftrightarrow G^{-1}A^H G$ commutes with A
(b) T is Hermitian $\Leftrightarrow GA = A^H G$
(c) T is unitary $\Leftrightarrow G = A^H GA$
(d) T is skew-Hermitian $\Leftrightarrow -GA = A^H G$.

Proof. These all follow from $[T^*]_\beta = G^{-1}A^H G$ (cf. Theorem 1) and the property required of the operator. For example, T is normal $\Leftrightarrow T^*T = TT^* \Leftrightarrow [TT^*] = [T^*T] \Leftrightarrow [T][T^*] = [T^*][T]$. Now use $[T^*] = G^{-1}A^H G$. T is Hermitian $\Leftrightarrow T^* = T \Leftrightarrow [T^*] = [T] \Leftrightarrow G^{-1}A^*G = A \Leftrightarrow GA = A^H G$.

Statements (a)–(d) of the last theorem simplify radically if the basis is orthonormal since, in this case, $G = I$.

COROLLARY 5. Under the same hypotheses as Theorem 4, if $\beta = \{e_1, e_2, \ldots, e_n\}$ is an *orthonormal* basis for V, and $A = [T]_\beta$, then

(a) T is normal $\Leftrightarrow AA^H = A^H A : A^H$ commutes with A
(b) T is Hermitian $\Leftrightarrow A = A^H$
(c) T is unitary $\Leftrightarrow AA^H = A^H A = I$
(d) T is skew-Hermitian $\Leftrightarrow -A = A^H$.

If matrix A is large, Theorem 4 may not be economical or effective to use numerically to establish the classification of T, especially if G^{-1} is needed. On the other hand, Corollary 5 may be impractical to use numerically because of the difficulty in constructing an orthonormal basis rapidly and accurately by numerical methods. Thus it is often worthwhile to be able to classify the operator a priori.

From Example 1 or Corollary 5, the next result follows immediately.

COROLLARY 6. Let $T(x) = Ax$ define a matrix operator on $\mathbf{C}^n$, and use the standard inner product. Then

(a) T is normal $\Leftrightarrow A^H$ commutes with A
(b) T is Hermitian $\Leftrightarrow A = A^H$
(c) T is unitary $\Leftrightarrow AA^H = A^H A$
(d) T is skew-Hermitian $\Leftrightarrow -A = A^H$.

DEFINITION. CLASSIFICATION OF MATRICES. An $n \times n$ matrix A is given the same classification as the operator $T(x) = Ax$ that it generates when the inner product is taken to be the standard one in $\mathbf{C}^n$. When nothing to the contrary is said, the standard inner product will be used in $\mathbf{C}^n$.

Exercises 5.1

1. Define $T : \mathbf{R}^2 \to \mathbf{R}^2$ by $T(x) = Ax$, where

$$A = \begin{bmatrix} 2 & 1 \\ 4 & 1 \end{bmatrix}.$$

(a) Find the matrix of T^* with respect to the inner product $\langle x, y \rangle = x_1 y_1 + x_2 y_2$.
(b) Find the matrix of T^* with respect to the inner product $\langle x, y \rangle = 3x_1 y_1 + x_2 y_2$.

2. Define $T : \mathbf{C}^2 \to \mathbf{C}^2$ by $T(x) = Ax$, where

$$A = \begin{bmatrix} 0 & 1 & 0 \\ 1 & 0 & 0 \\ 0 & 0 & 1 \end{bmatrix}.$$

Classify T relative to the standard inner product on $\mathbf{C}^2$.

3. Let V be the three-dimensional space of "directed line segment" vectors with $\langle \mathbf{x}, \mathbf{y} \rangle = \mathbf{x} \circ \mathbf{y}$ and define $T(\mathbf{x}) = \mathbf{w} \times \mathbf{x}$, where $\mathbf{w}$ is a fixed real unit vector and $\times$ is the cross product. Classify T.

4. Let $T : \mathbf{C}^2 \to \mathbf{C}^2$ be the operator generated by

$$A = \begin{bmatrix} a & b \\ c & d \end{bmatrix}.$$

Let $\langle , \rangle$ be the standard inner product and let $\langle x, y \rangle_1 = x_1 y_1 + 2x_2 y_2$.
(a) Find A that is Hermitian in $\langle , \rangle$ but not in $\langle , \rangle_1$.
(b) Find A that is Hermitian in $\langle , \rangle_1$ but not in $\langle , \rangle$.

5. Let $T(x) = Ax$ be a matrix operator on $\mathbf{C}^n$ and $\langle x, y \rangle = y^H G x$ be an inner product on $\mathbf{C}^n$. Show directly that $T^*(x) = G^{-1} A^H G x$.

6. Let $V = \{f : f, f' \text{ continuous on } [0, b], f(0) = f(b) = 0\}$ and $T(f) = f'$.
(a) If $\langle f, g \rangle = \int_0^b f(t)\overline{g(t)}\, dt$, find T^* and classify T.
(b) If $\langle f, g \rangle = \int_0^b f(t)\overline{g(t)}e^t dt$, find T^*.

7. Let $T(x) = x''$, where $' = d/dt$, and $\langle x, y \rangle = \int_{-\pi}^{\pi} x(t)\overline{y(t)}\, dt$. Find T^* if $T : T_n \to T_n$.

8. Let $T : \mathbf{P}_2 \to \mathbf{P}_2$ be defined by $T(f) = f''$ and $\langle x, y \rangle = \int_{-1}^{1} f(x)\overline{g}(x)\, dx$. Find $T^*(ae_0 + be_1 + ce_2)$ where $e_k(x) = x^k$.

9. V is the vector space of smooth functions satisfying $f(0) = f(1) = 0$. Define inner products by
(a) $\langle f, g \rangle = \int_0^1 f(t)\overline{g(t)}dt$.
(b) $\langle f, g \rangle_1 = \int_0^1 f(t)\overline{g(t)}e^t\, dt$.
For the operator $S(f) = f'' + f'$, find the adjoint and classify the operator, if possible.

10. Let w be a unit vector in $\mathbf{C}^n$, and set $U = I - 2ww^H$. Show
(a) $U^2 = I$.
(b) U is Hermitian.
(c) U is unitary.

11. Let A be an $n \times n$ matrix. Show that
(a) A is Hermitian implies a_{ii} is real.
(b) A is skew-Hermitian implies a_{ii} is pure imaginary.
What can be said if A is real and skew-Hermitian?

12. Let M be a subspace of V, and let $\langle , \rangle$ be an inner product on V. Suppose the projection operator $\prod = \prod_M$ onto M is well defined. This means that for every $f \in V$, the projection $s = \prod(f)$ exists, where $s \in M$ and $f - s \perp M$. Show that

(a) $\prod$ is linear.

(b) If $x \in M$, then $\prod(x) = x$.

(c) $\prod^2 = \prod$.

(d) $\prod(V) = M$.

(e) $\prod$ is Hermitian.

(f) $N(\prod) = M^\perp$.

13. Let $P : V \to V$ be linear. Suppose $P^2 = P$, and P is Hermitian with respect to an inner product $\langle , \rangle$. Define $M = R(P)$. Show that P is the orthogonal projection on $M : P = \prod_M$.

Thus Hermitian linear maps satisfying $P^2 = P$ are called orthogonal *projection operators*.

14. Suppose P is linear and $P^*(I - P) = 0$. Show that P is an orthogonal projection operator.

15. Show that if T is normal, so is $T - \lambda I$.

16. Suppose $T : V \to V$ is linear, and $\{e_k\}_{k=1}^n$ and $\{f_k\}_{k=1}^n$ are orthonormal bases for V with respect to $\langle , \rangle$. Suppose that $T(e_k) = f_k$, $k = 1, 2, \ldots, n$. Show that T is unitary.

17. Show that $\langle T(x), T(y) \rangle = \langle x, y \rangle \; \forall x, y \in V \Leftrightarrow \|T(x)\| = \|x\| \; \forall x \in V$.

18. Let $\{e_1, e_2, \ldots, e_n\}$ be a basis for V, and $T : V \to V$. Show

(a) if $TT^*(e_i) = T^*T(e_i) \; \forall i$, then T is normal.

(b) if $\langle T(e_i), e_j \rangle = \langle e_i, T(e_j) \rangle \; \forall i, j$, then T is Hermitian.

19. Suppose P is linear, $P^2 = P$, and P is Hermitian. Show that $I - 2P$ is unitary.

20. Show that the class of normal operators contains the classes of Hermitian, skew-Hermitian, and unitary operators.

21. Let V be a finite-dimensional vector space with inner product $\langle , \rangle$. Show that the set of all Hermitian linear operators $T : V \to V$ is a real vector space (using the usual operations). Is it a complex vector space? What can you say about the set of all unitary operators? Normal operators?

22. Let T be a linear operator on V, and let $\langle , \rangle$ be an inner product on V. Prove that if M is an invariant subspace of T, i.e., $T(M) \subset M$, then $M^\perp$ is an invariant subspace of T^*.

23. Let $T : V \to V$, $\dim(V) < \infty$, and let $\langle , \rangle$ be an inner product on V. Show that

(a) $N(T) = R(T^*)^\perp$ (b) $N(T^*) = R(T)^\perp$

(c) $N(T)^\perp = R(T^*)$ (d) $N(T^*)^\perp = R(T)$

(e) (*Fredholm alternative*) $T(x) = y$ has a solution iff $y \perp z$ for every z for which $T^*(z) = 0$.

(*Remark:* This generalizes Theorem 3 and Corollary 4 of Section 3.5.)

24. Let w, p, q be fixed real continuous functions on $[a, b]$ with $w > 0$ and p' continuous. Define T by

$$Tf(x) = \frac{1}{w(x)} \frac{d}{dx}(p(x)f'(x)) + q(x)f(x)$$

and define an inner product by $\langle f, g \rangle = \int_a^b f(x)\overline{g}(x)w(x)\,dw$. Show that T is Hermitian in this inner product on each of the vector spaces defined by the following conditions.

(a) $f(a) = f(b) = 0$

(b) $f'(a) = f(b) = 0$

(c) $f'(a) = f'(b) = 0$

(d) $c_1 f(a) + c_2 f'(a) = 0$, $d_1 f(b) + d_2 f'(b) = 0$, where (c_1, c_2) and $(d_1, d_2) \neq 0$ and are real.

25. Let Ω be a bounded domain in $\mathbf{R}^2$, let V_1 be the vector space of functions that satisfy $f = 0$ on the boundary of Ω, and let V_2 be the vector space of functions that satisfy $\partial f / \partial n = 0$ on the boundary of Ω, where $\partial f / \partial n$ is the outward normal derivative on the boundary of Ω. Show that the Laplacian $\Delta = \partial^2/\partial x^2 + \partial^2/\partial y^2$ is Hermitian on V_1 and on V_2 in the inner product $\langle f, g \rangle = \int_\Omega f(p)\overline{g}(p)\,dA$. (*Hint:* Look up Green's identities in vector calculus.)

26. In this exercise a more general definition of T^* will be given, and some of its properties in finite-dimensional spaces will be discussed. This will put much of the discussion in Section 3.5 in a more abstract setting.

DEFINITION. Let $T : V \to W$ be a linear operator, $\langle , \rangle$ an inner product on V, and $(,)$ an inner product on W. The adjoint T^* of T with respect to these inner products is that linear operator $T^* : W \to V$, for which $(T(x), y) = \langle x, T^*(y) \rangle \; \forall x \in V, y \in W$.

If V and W are finite-dimensional, it is not hard to show that T^* exists and is unique. Part $(a)(ii)$ is a step in this direction. More hypotheses are needed to prove this in infinite-dimensional spaces.

(a) (i) Let $V = \mathbf{C}^n$, $W = \mathbf{C}^m$, and both inner products be the standard inner products. Suppose $T(x) = Ax$. Show that $T^*(x) = A^H x$.

(ii) Let $V = \mathbf{C}^n$, $W = \mathbf{C}^m$, and $T(x) = Ax$. Suppose $\langle x, y \rangle = y^H G x$, $(u, v) = v^H \Gamma u$. If $T^*(x) = Bx$, find B.

The next sequence of exercises generalizes the results of Section 3.5. In them, let $T : V \to W$ be a linear operator, $\langle , \rangle$ an inner product on V, and $(,)$ an inner product on W with $\dim(V) = n$, $\dim(W) = m$ (finite). The operational rules of Theorem 2 will hold. Prove the following:

(b) $N(T) = R(T^*)^\perp$

(c) $R(T) = N(T^*)^\perp$

(d) $r(T^*) = r(T)$

(e) $N(T) = N(T^*T)$

(f) $R(T) = R(TT^*)$

(g) $r(T) = r(TT^*) = r(T^*T)$

5.2
THE SPECTRAL THEOREM

If an operator is normal, its eigenvectors and eigenvalues have some startling properties, which we explore in this section. The theory deduced in this section is of enormous importance and has many far-reaching applications, some of which will be explored in later sections of the book.

LEMMA 1. Suppose S is normal. Then $\|S(x)\| = \|S^*(x)\| \; \forall x$ and $N(S) = N(S^*)$.

Proof. $\|S(x)\|^2 = \langle S(x), S(x) \rangle = \langle x, S^*S(x) \rangle = \langle x, SS^*(x) \rangle = \langle S^*(x), S^*(x) \rangle = \|S^*(x)\|^2$. Now suppose $x \in N(S)$; then $S(x) = 0$ and $0 = \|S(x)\| = \|S^*(x)\|$, so that $S^*(x) = 0$ and thus $x \in N(S^*)$. Similarly, $S^*(x) = 0 \Rightarrow S(x) = 0$. Thus $N(S) = N(S^*)$.

The next theorem describes the nature of the eigenvalues of various types of operators, and gives an important property of the eigenvectors.

THEOREM 2. Suppose T is normal. Then eigenvectors in different eigenspaces are orthogonal. Further:

(a) If T is Hermitian, then the eigenvalues of T are real.
(b) If T is skew-Hermitian, then the eigenvalues of T are imaginary.
(c) If T is unitary, then the eigenvalues of T lie on the unit circle in the complex plane.
(d) If T is normal and $T(e) = \lambda e$, then $T^*(e) = \bar{\lambda} e$.

Proof. First (a)–(d) will be proved. Suppose $T(e) = \lambda e$, $e \neq 0$.

(a) If T is Hermitian, then $\langle T(e), e \rangle = \langle \lambda e, e \rangle = \lambda \langle e, e \rangle = \langle e, T^*(e) \rangle = \langle e, \lambda e \rangle = \bar{\lambda} \langle e, e \rangle$. Since $\langle e, e \rangle \neq 0$, $\lambda = \bar{\lambda}$ and λ is real.
(b) This proof is left as an exercise.
(c) Now, if T is unitary, $\langle T(e), e \rangle = \langle \lambda e, e \rangle = \lambda \langle e, e \rangle = \langle e, T^*(e) \rangle = \langle e, T^{-1}(e) \rangle = \langle e, \lambda^{-1} e \rangle = \bar{\lambda}^{-1} \langle e, e \rangle$, so that $\lambda \bar{\lambda} = |\lambda|^2 = 1$.
(d) Let $S = T - \lambda I$. First, show that S is normal. The rules of Theorem 2 of Section 5.1 yield $S^* = T^* - \bar{\lambda} I$. Thus $S^* S = (T^* - \bar{\lambda} I)(T - \lambda I) = T^* T - \lambda T^* - \bar{\lambda} T - \lambda \bar{\lambda} I = TT^* - \lambda T^* - \bar{\lambda} T - \lambda \bar{\lambda} I = SS^*$. Now suppose $Te = \lambda e$. Then $e \in N(S)$ and, by Lemma 1, $e \in N(S^*)$. But this means that $0 = S^*(e) = (T^* - \bar{\lambda} I)(e)$, which yields $T^*(e) = \bar{\lambda} e$.

Finally, suppose that e and f are eigenvectors in different eigenspaces: $T(e) = \lambda e$, $T(f) = \mu f$, with $\lambda \neq \mu$. Then $\lambda \langle e, f \rangle = \langle \lambda e, f \rangle = \langle T(e), f \rangle = \langle e, T^*(f) \rangle = \langle e, \bar{\mu} f \rangle = \mu \langle e, f \rangle$. Since $\lambda \neq \mu$, $\langle e, f \rangle = 0$.

At this point it becomes relevant to ask whether or not the preceding properties characterize the special kinds of operators that are being studied. It turns out that in some sense they do, at least in finite-dimensional spaces.

THEOREM 3. Let $\dim(V) < \infty$, $T : V \to V$, and $\langle , \rangle$ be an inner product on V. Suppose there exists an orthogonal basis of eigenvectors for T. Then T is normal with respect to $\langle , \rangle$. Further,

(a) If all the eigenvalues are real, then T is Hermitian.
(b) If all the eigenvalues are imaginary, then T is skew-Hermitian.
(c) If all the eigenvalues lie on the unit circle, then T is unitary.

Proof. Two proofs will be given. First, let $T(e_i) = \lambda_i e_i$ and $\beta = \{e_k\}_{k=1}^n$ be an orthonormal basis for V with respect to $\langle , \rangle$. Let e_k be an arbitrary vector in β. Since β is a basis, $T^*(e_k) = \sum_{i=1}^n c_i e_i$ for certain c_i. These c_i are Fourier coefficients; therefore, $c_i = \langle T^*(e_k), e_i \rangle = \langle e_k, T(e_i) \rangle = \langle e_k, \lambda_i e_i \rangle = \bar{\lambda}_i \langle e_k, e_i \rangle = \bar{\lambda}_i \delta_{ij}$. Hence

$$T^*(e_k) = \bar{\lambda}_k e_k, \quad k = 1, 2, \ldots, n. \tag{*}$$

Thus $TT^*(e_k) = T(\bar{\lambda}_k e_k) = \bar{\lambda}_k T(e_k) = \bar{\lambda}_k \lambda_k e_k = |\lambda_k|^2 e_k$, using (*). Similarly, $T^* T(e_k) = T^*(\lambda_k e_k) = \lambda_k T^*(e_k) = \lambda_k \bar{\lambda}_k e_k = |\lambda_k|^2 e_k$. Since $TT^* = T^* T$ on a basis, $T^* T = TT^*$ on V, so that T is normal.

(a) If all λ_i are real, then $\bar{\lambda}_i = \lambda_i$, and clearly $T^*(e_k) = T(e_k) = \lambda_k e_k$ for all k, giving $T^* = T$.
(b) The proof is left as an exercise.
(c) If all $|\lambda_i| = 1$ and $x = \sum_{i=1}^n x_i e_i$, then $T^* T x = \sum_{i=1}^n \lambda_i x_i e_i = x$, hence $T^* T = I$, and so $T^* = T^{-1}$ by Theorem 6 of Section 2.2.

The second proof uses the matrix of the operator and Corollary 5 of Section 5.1 as follows. The basis β may be chosen to be orthonormal, so that $G = I$. Since β is an eigenbasis, $[T]_\beta = \Lambda = \text{diag}(\lambda_1, \lambda_2, \ldots, \lambda_n)$, by Theorem 1 of Section 4.2. Then $[T^*]_\beta = \Lambda^H = \overline{\Lambda}$, by Theorem 1 of Section 5.1. Now clearly, $\Lambda\Lambda^H = \text{diag}(|\lambda_i|^2) = \Lambda^H\Lambda$, so that T is normal, by Corollary 5 of Section 5.1. By the same corollary, if all the eigenvalues are real, $\Lambda^H = \Lambda$, and T must be Hermitian; if all the eigenvalues have modulus 1, then $\Lambda\Lambda^H = I$, and hence $TT^* = I$ and T is unitary. The remaining part is left as an exercise.

The assumption that T is diagonalizable in Theorem 3 is a rather strong one. It is natural to ask whether or not there are normal operators on finite-dimensional spaces that are not diagonalizable. The answer to that question is no, according to the following theorems.

The Spectral Theorem and Spectral Resolution

THEOREM 4. Let $\dim(V) < \infty$ and $T : V \to V$ be a linear operator, normal with respect to the inner product $\langle , \rangle$ on V. Let $\sigma(T) = \{\lambda_1, \ldots, \lambda_r\}$ be the distinct eigenvalues of T and $M_i = M(\lambda_i)$ the corresponding eigenspaces. Then every x in V can be written uniquely as

$$x = \sum_{i=1}^{r} v_i \quad \text{for some } v_i \in M_i. \tag{1}$$

Moreover, for each i, v_i is the orthogonal projection of x on M_i.

Proof. Let M be the set of all x in V such that the representation (1) is possible. Note that each $M_i \subset M$ and that M is a subspace, since each M_i is a subspace. Our goal is to show $M = V$. Since each $x \in V$ can be written $x = m + r$ with $m \in M$, $r \in M^\perp$ by Theorem 1 of Section 3.5, it is enough to show $M^\perp = \{0\}$. First let us show that $T : M^\perp \to M^\perp$. Let $r \in M^\perp$; then $r \perp v_i$, $Tv_i = \lambda_i v_i$, and $T^*v_i = \overline{\lambda}_i v_i$ for $i = 1, 2, \ldots, r$, so that

$$\left\langle Tr, \sum_{i=1}^{r} c_i v_i \right\rangle = \sum_{i=1}^{r} \overline{c}_i \langle Tr, v_i \rangle = \sum_{i=1}^{r} \overline{c}_i \langle r, T^*v_i \rangle = \sum_{i=1}^{r} c_i \langle r, \overline{\lambda}_i v_i \rangle$$

$$= \sum_{i=1}^{r} \overline{c}_i \lambda_i \langle r, v_i \rangle = 0.$$

Hence $Tr \in M^\perp$, validating $T : M^\perp \to M^\perp$. Now if $M^\perp \neq \{0\}$, then T has an eigenvector e in $M^\perp$ by Theorem 5 of Section 4.1. Thus $Te = \mu e$, $e \neq 0$, and $e \perp M$. But $\mu \in \sigma(T)$, say $\mu = \lambda_1$, and so $e \in M_1 \subset M$. But $e \perp M$, so $\langle e, e \rangle = 0$, a contradiction to $e \neq 0$. Thus it must be that $M^\perp = \{0\}$. To show that v_i is the projection of x on M_i, it is enough to show that $x - v_i \perp M_i$. But $x - v_i = \sum_{j:j\neq i} v_j$, and since vectors in different eigenspaces are orthogonal, each $v_j \perp M_i$ when $j \neq i$, and so $x - v_i \perp M_i$. The v_i are unique, since projections are unique.

Remark. Equation (1) may be stated more succinctly as (cf. Section 3.5)

$$V = M(\lambda_1) \oplus M(\lambda_2) \oplus \cdots \oplus M(\lambda_r). \tag{1'}$$

Putting the last few results together yields the spectral theorem. This important theorem is very remarkable. It says that every normal operator is diagonalizable, and may even be diagonalized by an orthogonal basis. The spectral theorem has many applications, which will be explored later. A second proof is outlined in the exercises.

THE SPECTRAL THEOREM. Let $\dim(V) < \infty$ and $T : V \to V$. Then T is normal iff T has an orthogonal basis of eigenvectors. Further, when T is normal:

(a) If all the eigenvalues are real, then T is Hermitian.
(b) If all the eigenvalues are imaginary, then T is skew-Hermitian.
(c) If all the eigenvalues lie on the unit circle, then T is unitary.

Proof. If T has an orthogonal basis of eigenvectors, it is normal, by Theorem 3. Now assume T is normal. For each eigenspace $M(\lambda_i)$ of Theorem 4, construct an orthogonal basis. Since eigenvectors in different eigenspaces are orthogonal, the union of these orthogonal bases is an orthogonal basis for V. The rest of the statements are from Theorem 3 and are added for reference.

Remark. The construction of the orthogonal basis in the spectral theorem parallels the following theoretical procedure to be used in finding an orthogonal eigenbasis:

1. First find the spectrum of T, $\sigma(T) = \{\lambda_1, \ldots, \lambda_r\}$, i.e. the distinct roots λ_i of the characteristic equation.
2. Next find each eigenspace $M(\lambda_i) = N(T - \lambda_i I)$. If this eigenspace has dimension greater than 1, find an orthogonal basis for it. Note that the dimension of each eigenspace is equal to the algebraic multiplicity of the eigenvalue, according to Theorem 5 of Section 4.2.
3. Put these together to get an orthogonal eigenbasis for T.

Remarks concerning uniqueness. $\sigma(T)$ and the eigenspaces are uniquely determined by T and V. The indexing of the λ_i is not, and is arbitrary. If all M_i are one-dimensional, the orthogonal eigenbasis is unique except for order and scalar multiples. In this case, an orthonormal basis will be unique up to order and scalar multiples by scalars of unit magnitude. If some M_i has dimension greater than 1, then it has infinitely many orthonormal bases, all of which are connected by unitary change-of-basis matrices.

Surprisingly, an operator satisfying the hypotheses of Theorem 4 can be expressed as a linear combination of projections. In fact, this is simply another way to state the spectral theorem.

THE SPECTRAL RESOLUTION OF A NORMAL OPERATOR. Let T satisfy the hypotheses of the spectral theorem, and let $\prod_i$ be the orthogonal projection of V onto the eigenspace $M(\lambda_i)$. Then

$$I = \sum_{i=1}^{r} \prod_i \qquad \text{(resolution of the identity)} \qquad (2)$$

$$T = \sum_{i=1}^{r} \lambda_i \prod_i \qquad \text{(spectral resolution of } T) \qquad (3)$$

$$\prod_i \prod_j = 0 \quad \text{if } i \neq j. \qquad (4)$$

Proof. Equation (2) means $I(x) = x = \sum_{i=1}^{r} \prod_i(x) = \sum_{i=1}^{r} v_i$, which is Eq. (1). Equation (3) follows from $T(x) = T(\sum_{i=1}^{r} v_i) = \sum_{i=1}^{r} T(v_i) = \sum_{i=1}^{r} \lambda_i v_i = \sum_{i=1}^{r} \lambda_i \prod_i(x)$. Equation (4) follows from the fact that the vectors in $M(\lambda_i)$ and $M(\lambda_j)$ are orthogonal.

Remark. In the infinite-dimensional inner-product spaces known as Hilbert spaces, there are generalizations of the spectral theorem and the spectral resolution of a normal operator. The sums in the spectral resolution become integrals, however, unless extra hypotheses are made on the operator T. The resulting theory is known as *Hilbert–Schmidt* theory and is quite elegant. Many important integral operators satisfy this extra hypothesis, and the resulting spectral theorem is essential in the study of these integral operators. Since certain differential operators have inverses that are integral operators satisfying this extra condition, the theory can also be applied to these differential operators to obtain some extremely useful facts about differential equations. See Liusternik and Sobolev (1961) for example.

EXAMPLE 1. Given

$$V = \mathbf{C}^3, \qquad T(x) = Ax = \begin{bmatrix} 2 & 2 & 1 \\ 2 & 2 & -1 \\ 1 & -1 & -1 \end{bmatrix} x, \qquad \text{standard} \langle , \rangle,$$

classify T, and find an orthogonal basis of eigenvectors.

In the standard basis, $[T] = A$ and $G = I$. Thus $[T^*] = A^H = A = [T]$, and the operator is Hermitian with respect to the standard inner product by Corollary 6 of Section 5.1 and has an orthogonal eigenbasis and real eigenvalues. Computing the characteristic equation, the eigenvalues are $-2, 1, 4$. Thus each eigenspace is one-dimensional and the eigenvectors are automatically orthogonal. Solving $(A - (-2)I)x = 0$ gives the eigenvector $E_1 = (1, -1, -2)^T$; solving $(A - I)x = 0$ gives the eigenvector $E_2 = (1, -1, 1)^T$, and solving $(A - 4I)x = 0$ yields the eigenvector $(1, 1, 0)^T$. The operator cannot be unitary, because not all of the eigenvalues lie on the unit circle. It cannot be skew-Hermitian because none of the eigenvalues lie on the imaginary axis.

EXAMPLE 2. Let $V = \mathbf{P}_2$, and define $T : V \rightarrow V$ and $\langle , \rangle$ on V by

$$Tp(x) = \frac{3}{8} \int_{-1}^{1} [1 + 4(x + t) + 5(x^2 + t^2) - 15x^2 t^2] p(t) \, dt,$$

$$\langle p, q \rangle = \int_{-1}^{1} p(t) \overline{q(t)} \, dt.$$

Classify the operator T, and find an orthogonal eigenbasis.

The operator can be shown to be Hermitian without computation of the eigenvalues and eigenvectors, using Example 3 of Section 5.1. The kernel satisfies $k(t, x) = \overline{k(x, t)}$ and therefore $T^* = T$. Theorem 4 may also be used as follows. T and V were considered in Example 8 of Section 4.2. There it was found that $\lambda_1 = -1$, $M(-1) = \text{span}\{3t - 1, 3t^2 - 1\}$; $\lambda_2 = 3$, $M(3) = \text{span}(t + 1)$. This operator is potentially Hermitian (real λ). A calculation shows that the eigenvectors in different eigenspaces are orthogonal. An orthogonal eigenbasis can be easily constructed for $M(-1)$, so that T has an orthogonal eigenbasis with real eigenvalues and so is Hermitian (and therefore normal) with respect to the given inner product by Theorem 3 or the spectral theorem. Luckily, $\{3t - 1, 3t^2 - 1, t + 1\}$ is an orthogonal eigenbasis.

EXAMPLE 3. Let

$$V = \mathbf{T}_1, \qquad Tf(x) = f(x + \pi/4), \qquad \text{and } \langle f, g \rangle = \int_0^{2\pi} f(t)\overline{g(t)}\, dt.$$

Classify the operator T.

This is another example that has been considered before. There are two ways to classify T. In Example 7 of Section 4.2 the eigenbasis $e_1 = 1$, $e_2 = e^{ix}$, $e_3 = e^{-ix}$ with corresponding eigenvalues $\lambda_1 = 1$, $\lambda_2 = e^{i\pi/4}$, $\lambda_3 = e^{-i\pi/4}$ was found. Since the eigenvectors are orthogonal with respect to the given $\langle , \rangle$, and the eigenvalues lie on the unit circle, the operator is unitary by Theorem 3 or the spectral theorem. In Example 5 of Section 5.1, T was shown to be unitary by showing that $\langle Tf, Tg \rangle = \langle f, g \rangle \ \forall f, g \in V$, and applying Theorem 3 of Section 5.1. This avoids the diagonalization.

Functions of Normal Operators

Since normal operators on finite-dimensional spaces are diagonalizable, functions of them are easy to calculate when an eigenbasis is known, according to the definition in Section 4.3. Questions such as the following arise: What is the classification of $f(T)$? When does $f(T)$ have the same classification as T? These kinds of questions are investigated next.

THEOREM 5. Let $\dim(V) < \infty$, $T : V \to V$ be linear, and $\langle , \rangle$ be an inner product on V. If T is normal on V with respect to $\langle , \rangle$ and $f(T)$ is defined, then $f(T)$ is normal on V with respect to $\langle , \rangle$. Further,

(a) If $f(\lambda)$ is real-valued for all $\lambda \in \sigma(T)$, then $f(T)$ is Hermitian.
(b) If $|f(\lambda)| = 1$ for all $\lambda \in \sigma(T)$, then $f(T)$ is unitary.

In particular, the inverse of a unitary operator is unitary. The inverse of a Hermitian operator is Hermitian, when it exists, and the inverse of a normal operator is normal, when it exists.

Proof. If $\{e_1, e_2, \ldots, e_n\}$ is an orthogonal eigenbasis for T, then $f(T)$ has the same orthogonal eigenbasis but has the corresponding eigenvalues $f(\lambda_i)$ by Theorem 1 of Section 4.3, and so is normal, by Theorem 3 or the spectral theorem. The other statements come from the same theorems, depending on the nature of the $f(\lambda_i)$.

The only change that affects us in computing functions of a normal operator is the necessity of ensuring that the basis is orthogonal, not just an eigenbasis. If all the eigenvalues are simple, i.e., $\dim(M(\lambda)) = 1$, then the eigenvectors are automatically orthogonal and must only be normalized. When an eigenspace is degenerate, an orthogonal basis for it must be found.

EXAMPLE 4. Let $T(x) = Ax$ be an operator on $\mathbf{C}^n$ defined by

$$A = \begin{bmatrix} 2 & 2 & 1 \\ 2 & 2 & -1 \\ 1 & -1 & -1 \end{bmatrix}.$$

Classify e^T and $e^{i\pi T}$.

From Example 1, T is Hermitian with eigenvalues -2, 1, 4. Both e^T and $e^{i\pi T}$ are normal. e^T has eigenvalues e^{-2}, e^1, e^4, which are real, whence e^t is Hermitian. $e^{i\pi T}$ has eigenvalues $e^{-2\pi i} = 1$, $e^{\pi i} = -1$, $e^{4\pi i} = 1$ and is both Hermitian and unitary because its eigenvalues are real and on the unit circle.

Simultaneous Diagonalization of Normal Operators

In future applications, it will be of interest to orthogonally diagonalize more than one normal operator using the same eigenbasis. Examples of the use of this theorem will be given in Chapter 7, in the context of the theory of small oscillations.

THEOREM 6. Let $T, S : V \to V$ be normal operators on the finite-dimensional nontrivial vector space V with respect to the inner product $\langle , \rangle$. Then S and T have a common orthogonal eigenbasis if and only if they commute: $ST = TS$.

Proof. First assume that $\{e_1, e_2, \ldots, e_n\}$ is a basis that is orthogonal with respect to $\langle , \rangle$ and that $Te_i = \lambda_i e_i$, and $Se_i = \mu_i e_i$. For an arbitrary $x = \sum_{i=1}^n x_i e_i$,

$$TS(x) = T(S(x)) = T\left(\sum_{i=1}^n \mu_i x_i e_i\right) = \sum_{i=1}^n \lambda_i \mu_i x_i e_i = ST(x),$$

so that T and S commute.

Now suppose that $ST = TS$. Let $M(\lambda_1)$ be an eigenspace of T. The first step is to show $S : M(\lambda_1) \to M(\lambda_1)$. To see this, let $x \in M(\lambda_1)$. Then $T(x) = \lambda_1 x$, and $T(S(x)) = S(T(x)) = S(\lambda_1 x) = \lambda_1 S(x)$, so that $S(x) \in M(\lambda_1)$. Since $S : M(\lambda_1) \to M(\lambda_1)$ and $M(\lambda_1)$ is finite-dimensional and nontrivial, S has an eigenvector e_1 in $M(\lambda_1)$ by Theorem 5 of Section 4.1. Thus one common eigenvector e_1 has been found. Now let $M_1 = \text{span}\{e_1\}^\perp$, and show that both $T, S : M_1 \to M_1$. In fact, if $x \in M_1$, then $\langle T(x), e_1 \rangle = \langle x, T^*(e_1) \rangle = \langle x, \bar{\lambda}_1 e_1 \rangle = \lambda_1 \langle x, e_1 \rangle = 0$, so that $T(x) \perp e_1$; hence $Tx \in M_1$. Similarly, $S(x) \in M_1$ when $x \in M_1$. Now the same argument as before is made, but with T and S restricted to M_1; i.e., V is replaced by M_1. This yields an e_2 in M_1 that is a common eigenvalue of T and S. e_2 is orthogonal to e_1, since $e \in M_1 = \text{span}\{e_1\}^\perp$. Next set $M_2 = \text{span}\{e_1, e_2\}^\perp$, and show that $T, S : M_2 \to M_2$ in the same way. A common eigenvector e_3 in M_2 of S and T is found. It must be orthogonal to e_1 and e_2 because it is in M_2. This process is continued until the vector space V is exhausted, in $n = \dim(V)$ steps.

The next two propositions give criteria guaranteeing that the simultaneous diagonalization can be done. The first follows from the fact that a function of a diagonalizable operator does not disturb the eigenvectors, and the second is based on the fact that two functions of one diagonalizable operator commute (see Theorem 4 of Section 4.3).

PROPOSITION 7. Let $T : V \to V$ be a normal operator on the finite-dimensional vector space V with respect to the inner product $\langle , \rangle$. If $S = f(T)$ and $\beta = \{e_1, e_2, \ldots, e_n\}$ is an orthogonal eigenbasis for T, then S is normal and β is also an orthogonal eigenbasis for S.

Proof. In fact, $S(e_i) = f(\lambda_i)e_i$, by the definition of $f(T)$ in Section 4.3. S is normal by Theorem 3.

PROPOSITION 8. Let $L : V \to V$ be a normal operator on the finite-dimensional vector space V with respect to the inner product $\langle , \rangle$. If $T = f(L)$, $S = g(L)$, and $\beta =$

$\{e_1, e_2, \ldots, e_n\}$ is an orthogonal eigenbasis for L, then S and T are normal and β is also an orthogonal eigenbasis for T and S.

Proof. Left as an exercise.

Exercises 5.2

1. The following operators obey $T : \mathbf{P}_2 \to \mathbf{P}_2$, and the inner product is $\langle f, g \rangle = \int_{-1}^{1} f(t)\overline{g(t)}\, dt$. Determine whether or not the operator is diagonalizable and classify the operator. If the operator is diagonalizable and normal, find an orthogonal basis of eigenvectors.
   (a) $Tp(x) = \frac{1}{4}\int_{-1}^{1}\{15x^2t^2 - 6xt - 3\}p(t)\,dt$
   (b) $Tp(x) = (x^2 - 1)p''(x) + 3xp'(x) + p(x)$
   (c) $Tp(x) = [(x^2 - 1)p'(x)]' + p(x)$
   (d) $Tp(x) = (a + b) + (a + c)x + (b + c)x^2$, where $p(x) = a + bx + cx^2$
   (e) $Tp(x) = (\frac{3}{2}x^2 - x - \frac{1}{2})p''(x) + p'(x) + p(x)$
   (f) $Tp(x) = xp'(x) + p(x)$

2. For the operators in parts (a), (c), (d), and (f) of the previous exercise, find the matrix B of T^* in the basis $1, x, x^2$.

3. Define $T : \mathbf{R}^4 \to \mathbf{R}^4$ by $T(v_1) = 2v_1, T(v_2) = 2v_2, T(v_3) = -v_3, T(v_4) = -v_4$, where $v_1 = (1, 0, 1, 0)^T$, $v_2 = (1, 1, 0, 0)^T$, $v_3 = (0, 0, 0, 1)^T$, $v_4 = (1, 1, -1, 1)^T$. Let $\mathbf{R}^4$ have the standard inner product. Is T self-adjoint? Explain. If it is, find an orthogonal eigenbasis.

4. Define $T : \mathbf{R}^4 \to \mathbf{R}^4$ by $T(v_1) = -2v_1, T(v_2) = -2v_2, T(v_3) = v_3, T(v_4) = v_4$, where $v_1 = (0, 1, 1, 1)^T$, $v_2 = (1, 1, 2, 0)^T$, $v_3 = (1, -1, 0, 1)^T$, $v_4 = (1, 1, -1, 0)^T$. Let $\mathbf{R}^4$ have the standard inner product.
   (a) Show that T is self-adjoint and find an orthogonal eigenbasis.
   (b) Classify the operators e^T, $e^{i\pi T}$, $i = \sqrt{-1}$, but do not compute them.

5. Let $T : \mathbf{T}_2 \to \mathbf{T}_2$ be defined by $T(f) = f'$, and $\langle f, g \rangle = \int_0^{2\pi} f(t)\overline{g(t)}\, dt$. Let $h(t) = \sin(t) + 2\cos(2t)$. Classify T and compute $e^T(h)$ and $e^{\pi T}(h)$.

6. Let $T : \mathbf{P}_2 \to \mathbf{P}_2$ be defined by
$$(Tf)(x) = \frac{1}{8}\int_{-1}^{1}[7 + 6(x + t) - 6xt - 15(x^2 + t^2) + 45x^2t^2]f(t)\,dt.$$

   Let $\langle f, g \rangle = \int_{-1}^{1} f(t)\overline{g(t)}\, dt$.
   (a) Find an orthogonal eigenbasis for T and classify T.
   (b) Find T^2, T^{-1}, and $\cos(\pi T)$.
   (c) Classify $S = \log(T + 2I)$ and compute $S(h)$, where $h(t) = 1$.

7. Let $T : \mathbf{P}_2 \to \mathbf{P}_2$ be given by
$$Tp(x) = \frac{1}{6}\int_{-1}^{1}[20 + 9xt - 30(x^2 + t^2) + 45x^2t^2]p(t)\,dt$$

and define the inner product by $\langle f, g \rangle = \int_{-1}^{1} f(x)\overline{g}(x)\, dx$. Find the eigenvalues and an orthogonal basis of eigenvectors, if one exists, for each of the following operators and classify the operator.

(a) 2^T (b) $\frac{1}{3}(3I + T - T^2)$ (c) $e^{\pi i T}$ (d) $(T - 2)^{-1}$

8. (a) Show that a skew-Hermitian operator is normal and its eigenvalues are purely imaginary.

   (b) Show that if T has an orthogonal eigenbasis and its eigenvalues are purely imaginary, then T is skew-Hermitian.

9. Prove Proposition 8.

10. Let $T : V \to V$ be Hermitian in $\langle , \rangle$ and $\dim(V) < \infty$. Classify e^T and e^{iT}.

11. Suppose that $T : V \to V$ is a unitary operator whose eigenvalues lie in the set $\{-1, -1\}$ and $\dim(V) < \infty$. Show that T is Hermitian and $T^2 = I$.

12. Let $T : V \to V$ be Hermitian on V with respect to $\langle , \rangle$. Suppose $\langle T(x), x \rangle = 0$ for all $x \in V$. Show $T = 0$.

13. Suppose $T : V \to V$ is normal with respect to $\langle , \rangle$, and that $\{e_k\}_{k=1}^{n}$ is an orthogonal basis for V with $T(e_i) = \lambda_i e_i$. Suppose $f(T)$ is defined and invertible, and put $f_i = f(T)(e_i)$. Show that $\{f_k\}_{k=1}^{n}$ is an orthogonal set.

14. **Cayley transforms.** Let $H : V \to V$ be Hermitian in $\langle , \rangle$ and $\dim(V) < \infty$.
   (a) Show that $I - iH$ is invertible.
   (b) Let $U = (I + iH)(I - iH)^{-1}$. Show that U, the Cayley transform of H, is unitary.
   (c) Show $H = i(I - U)(I + U)^{-1}$.

15. Let $A = I - vv^H$, where $\|v\| = 1, v \in \mathbf{C}^n$.
   (a) Show that $A^2 = A$, and A is Hermitian.
   (b) Find an orthogonal eigenbasis for A. (*Hint:* Consider the equation $Ae = \lambda e, e \neq 0$.)

16. Let $A = I - 2vv^H$, where v is a unit vector in $\mathbf{C}^n$.
   (a) Show that A is Hermitian and unitary.
   (b) Find A^{-1}.
   (c) Find an orthogonal eigenbasis for A.

17. **Spectral resolution using an orthonormal eigenbasis.** Let $T : V \to V$ be normal with respect to $\langle , \rangle$ and let $\{e_k\}_{k=1}^{n}$ be an orthonormal eigenbasis for T with $T(e_i) = \lambda_i e_i$. Show that $T(x) = \sum_{i=1}^{n} \lambda_i \langle x, e_i \rangle e_i$ and $T^*(x) = \sum_{i=1}^{n} \overline{\lambda_i} \langle x, e_i \rangle e_i$ for every $x \in V$.

18. **Further classification of Hermitian operators.** Let $T : V \to V$ be Hermitian in the inner product $\langle , \rangle$. T is *positive definite* iff $\langle T(x), x \rangle > 0$ when $x \neq 0$, and *positive* when $\langle T(x), x \rangle \geq 0\ \forall x$. Now suppose V is finite-dimensional. Show that, for Hermitian T,
   (a) T is positive definite iff all eigenvalues of T are positive.
   (b) T is positive iff all eigenvalues of T are nonnegative.
   (c) T is positive definite iff $\langle T(x), y \rangle$ is an inner product on V.

19. Let $T : V \to V$ be normal in $\langle , \rangle$ and $\dim(V) < \infty$. Suppose T has the spectral resolution $T = \sum_{i=1}^{p} \lambda_i \prod_i$. Let g be a function such that $g(T)$ is defined. Show that $g(T) = \sum_{i=1}^{p} g(\lambda_i) \prod_i$.

20. **Converse to the spectral resolution theorem.** Let $T : V \to V$ and $\dim(V) < \infty$. Suppose $\prod_i, i = 1, 2, \ldots, p$ are orthogonal projection operators and (i) $\prod_i \prod_j = 0$ if $i \neq j$, (ii) $I = \sum_{i=1}^{p} \prod_i$, (iii) $T = \sum_{i=1}^{p} \lambda_i \prod_i$. Show that T is normal.

21. **A second proof of the spectral theorem.** Consider a normal operator $T : V \to V$. First find an eigenvector $e_1 : T(e_1) = \lambda_1 e_1$. Then find an orthogonal basis $\{e_1, v_2, \ldots, v_n\}$ for V. Let $M_1 = \text{span}\{v_k\}_{k=2}^{n} = \text{span}\{e_1\}^{\perp}$. Show that $T : M_1 \to M_1$. Deduce that T has an eigenvector e_2 in $M_1 : T(e_2) = \lambda_2 e_2$, and $e_2 \perp e_1$. Now find an orthogonal basis $\{e_1, e_2, w_3, \ldots, w_n\}$ of V. Put $M_2 = \text{span}\{w_k\}_{k=3}^{n} = \text{span}\{e_1, e_2\}^{\perp}$. Show $T : M_2 \to M_2$. Continue the process. Show that the process stops with an orthogonal basis of eigenvectors.

22. Suppose $T : V \to V$ is normal with respect to $\langle , \rangle$ and $\dim(V) = n < \infty$. Show that $T^* = p(T)$ for some polynomial $p(x)$.

23. Let V be the vector space of all continuous functions on $[0, 1]$ and give V the inner product $\langle f, g \rangle = \int_0^1 f(x)\bar{g}\, dx$. Define T by $Tf(x) = xf(x)$. Show that T is Hermitian but has no eigenvalues.

24. Let $T : V \to V$, $\dim(V) = n < \infty$, and $\langle , \rangle$ be an inner product on V. Show that the following are equivalent:
    (a) T is normal.
    (b) $T = T_1 + iT_2$, where T_1, T_2 are Hermitian and commute.
    (c) $T(x) = \lambda x \Rightarrow T^*(x) = \bar{\lambda} x$.
    (d) There is an orthogonal eigenbasis for T.
    (e) There is an orthogonal basis β in which $[T]_\beta$ is diagonal.
    (f) $T^* = p(T)$ for some polynomial T.
    (g) There exist operators E_k and scalars c_k such that $T = \sum_{n=1}^{n} c_k E_k, I = \sum_{n=1}^{n} E_k,$ $E_k^2 = E_k, E_k^* = E_k, E_i E_j = 0$ if $i \neq j$.

25. Let T be normal on V. Suppose $y \in R(T)$ and $y \in N(T)$. Show $y = 0$.

5.3
APPLICATIONS TO MATRIX THEORY

In $\mathbf{C}^n$ the distinction between the operator and its matrix becomes blurred when the standard basis is used, since then $T(x) = Ax \Leftrightarrow [T] = A$. When the standard inner product is used, the standard basis is orthonormal so that $[T^*] = A^H$ and thus the matrix rules for classifying operators given in Corollary 5 of Section 5.1 become especially simple. By definition, the matrix inherits the classification of the property satisfied by the operator. The following table summarizes the names and conditions.

Let A be an $n \times n$ matrix, let T be the matrix operator generated by $A : T(x) = Ax$, and let $\mathbf{C}^n$ be given the standard inner product.

Classification	Operator property	Matrix property
Normal	$T^*T = TT^*$	$A^H A = AA^H$
Hermitian	$T^* = T$	$A^H = A$
Skew-Hermitian	$T^* = -T$	$A^H = -A$
Unitary	$T^* = T^{-1}$	$A^H A = I$

Three theorems involving matrices that are direct consequences of the spectral theorem will be derived. These theorems will find very useful applications in the next two chapters. The first theorem characterizes normal matrices in terms of their eigenvectors and eigenvalues.

THEOREM 1. Let A be an $n \times n$ matrix. Then A is normal if and only if there is a unitary matrix U and a diagonal matrix Λ such that $U^H A U = \Lambda$.

When $U = [E_1\ E_2\ \cdots\ E_n]$ and $\Lambda = \mathrm{diag}(\lambda_1, \lambda_2, \ldots, \lambda_n)$ exist, the columns of U are an orthonormal eigenbasis for A, and the entries of Λ are the corresponding eigenvalues, i.e.,

$$AE_i = \lambda_i E_i, \tag{1}$$

$$E_i^H E_j = \delta_{ij}, \quad i, j = 1, 2, \ldots, n. \tag{2}$$

Further, when A is normal,

(a) If all the eigenvalues are real, then T is Hermitian.
(b) If all the eigenvalues are imaginary, then T is skew-Hermitian.
(c) If all the eigenvalues lie on the unit circle, then T is unitary.

Proof. In Theorem 6 of Section 4.2, it was shown that if $P = [E_1\ E_2\ \cdots\ E_n]$ and $\Lambda = \mathrm{diag}(\lambda_1, \lambda_2, \ldots, \lambda_n)$, then $P^{-1}AP = \Lambda \Leftrightarrow AE_i = \lambda_i E_i\ \forall i$ and $\beta = \{E_1, E_2, \ldots, E_n\}$ is a basis for $\mathbf{C}^n$. If A is normal, β can be chosen to be an orthogonal basis of eigenvectors. Observe that

$$U^H U = [E_i^H E_j] = [\langle E_j, E_i \rangle] = \mathrm{diag}(\|E_1\|^2, \|E_2\|^2, \ldots, \|E_n\|^2). \tag{*}$$

If the E_i were normalized, then $U^H U = I$, so that $U^{-1} = U^H$ and $U^H A U = \Lambda$. Conversely, suppose $U^H U = I$ and $U^H A U = \Lambda$. Then $U^{-1} = U^H$, and by (*) the columns of U are orthonormal, since $U^H U = I$. Since $U^{-1}AU = \Lambda$, by the quoted theorem, the columns of U are eigenvectors of A, so that A is normal. A can also be shown to be normal by using $A = U \Lambda U^H$, $U^H U = I$, and verifying that $AA^H = A^H A$. The rest of the statements follow from Theorem 3 of Section 5.2 and are added for reference.

Remark. If A is a real symmetric matrix, then U can be taken to be a real unitary matrix. Real unitary matrices are called *orthogonal* matrices. This follows from the fact that, in solving $(A - \lambda_i I)x = 0$ for the eigenvectors, complex numbers do not need to be introduced.

Remark. If the eigenvectors are orthogonal but not necessarily of unit length, then by putting

$$N^2 = \text{diag}(\|E_1\|^2, \ldots, \|E_n\|^2),$$

$U^H U = I$ and $U^H A U = \Lambda$ are replaced by

$$U^H U = N^2, \tag{3a}$$

$$U^H A U = \Lambda N^2 = N^2 \Lambda. \tag{3b}$$

The next theorem specifies the spectral resolution of a normal matrix. A direct matrix theory proof is given.

THEOREM 2. SPECTRAL RESOLUTION OF A NORMAL MATRIX. Let A be a normal matrix and $\{E_k\}_{k=1}^n$ an orthonormal eigenbasis: $AE_i = \lambda_i E_i$. Then

$$A = \sum_{i=1}^n \lambda_i E_i E_i^H, \qquad I = \sum_{i=1}^n E_i E_i^H. \tag{4}$$

Proof. Using the identities of Section 0.4, we have

$$I = UU^H = [E_1\ E_2\ \cdots\ E_n][E_1\ E_2\ \cdots\ E_n]^H = \sum_{i=1}^n E_i E_i^H,$$

$$A = U\Lambda U^H = U[\lambda_1 \delta_1\ \lambda_2 \delta_2\ \cdots\ \lambda_n \delta_n]U^H = [\lambda_1 U\delta_1\ \lambda_2 U\delta_2\ \cdots\ \lambda_n U\delta_n]U^H$$

$$= [\lambda_1 E_1\ \lambda_2 E_2\ \cdots\ \lambda_n E_n][E_1\ E_2\ \cdots\ E_n]^H = \sum_{i=1}^n \lambda_i E_i E_i^H.$$

Remark. $E_i E_i^H$ is the matrix of the orthogonal projection on the one-dimensional space span(E_i). This follows from Example 2 in Section 3.3.

The next example illustrates the use of Theorem 1 and the computational advantage of an orthogonal eigenbasis in the solution of matrix differential equations (cf. Examples 1 and 2 of Section 4.4). A solution involving e^{At} is outlined in the exercises.

EXAMPLE 1. SOLUTION OF AN INHOMOGENEOUS LINEAR SYSTEM OF ODES.
The matrix of the system considered is normal. Solve

$$x' = Ax + f(t), \qquad x(0) = x_0, \tag{*}$$

where

$$A = \begin{bmatrix} 0 & 1 & 0 \\ 1 & 1 & -1 \\ 0 & -1 & 0 \end{bmatrix}, \qquad f(t) = \begin{bmatrix} 4 \\ 6 \\ 4 \end{bmatrix} t, \qquad x_0 = \begin{bmatrix} 0 \\ -1 \\ 4 \end{bmatrix}.$$

A "vector space method" will be employed. The reader may wish to try a "matrix method" using Eq. (1) or (2). In either case, the eigenvalue problem for A must be solved. Since A is Hermitian, an orthogonal eigenbasis may be found by the spectral theorem.

(a) *Solve the eigenvalue problem.* The characteristic polynomial is $|A - \lambda I| = p(\lambda) = -\lambda(\lambda - 2)(\lambda + 1)$, so $\sigma(T) = \{-1, 0, 2\}$. Thus each eigenspace is one-dimensional, and the eigenvectors are automatically orthogonal. For $\lambda_1 = -1$, solving $(A + 1 \cdot I)x = 0$ gives $E_1 = (-1, 1, 1)^T$; for $\lambda_2 = 0$, solving $(A - 0 \cdot I)x = 0$ gives $E_2 = (1, 0, 1)^T$, and finally, for $\lambda_3 = 2$, solving $(A - 2 \cdot I)x = 0$ yields $E_3 = (1, 2, -1)^T$.

(b) **Decouple the differential equations.** Set $x = \sum_{i=1}^{3} y_i E_i$, substitute this into (*), and use the fact that $AE_i = \lambda_i E_i$:

$$x' = \sum_{i=1}^{3} y_i' E_i = A \sum_{i=1}^{3} y_i E_i + f = \sum_{i=1}^{3} \lambda_i y_i E_i + f.$$

Since $\beta = \{E_1, E_2, E_3\}$ is a basis, f has an expansion $f = \sum_{i=1}^{3} g_i E_i$, so the last equation becomes

$$\sum_{i=1}^{3} y_i' E_i = \sum_{i=1}^{3} (\lambda_i y_i + g_i) E_i.$$

The linear independence of the E_i implies that the coefficients must be equal, and thus we obtain the decoupled system

$$y_i' = \lambda_i y_i + g_i.$$

Up to this point, nothing new has occurred. From this point on, the orthogonality of the E_i facilitates the solution substantially. Since β is an orthogonal set, the g_i are Fourier coefficients:

$$g_i = \langle f, E_i \rangle / \|E_i\|^2.$$

Thus the decoupled equations are

$$y_1' = -y_1 + 2t$$
$$y_2' = 4t$$
$$y_3' = 2y_3.$$

(c) **Solve the decoupled differential equations.** A particular solution for $y_1' = -y_1 + 2t$ may be found by the "method of undetermined coefficients." This amounts to setting $y_1 = At + B$ (from experience), putting this in the differential equation, and solving the resulting equation for the constants A, B. This yields $y_1 = 2(t - 1)$. Thus the general solution of the decoupled equations is

$$y_1 = c_1 e^{-t} + 2(t - 1),$$
$$y_2 = c_2 + 2t^2,$$
$$y_3 = c_3 e^{2t}.$$

(d) **Fit the initial conditions.** The vector initial condition is

$$x_0 = \sum_{i=1}^{3} y_i(0) E_i,$$

which is a Fourier expansion, so that each $y_i(0)$ is a Fourier coefficient:

$$y_i(0) = \langle x_0, E_i \rangle / \|E_i\|^2.$$

Computing these and putting $t = 0$ in the general solution yields

$$c_1 - 2 = y_1(0) = 1$$
$$c_2 = y_2(0) = 2$$
$$c_3 = y_3(0) = -1.$$

Thus $c_1 = 3$, $c_2 = 2$, $c_3 = -1$, so that $x = \sum_{i=1}^{3} y_i E_i$ is completely known:

$$x = [3e^{-t} + 2(t-1)]\begin{bmatrix} -1 \\ 1 \\ 1 \end{bmatrix} + [2 + 2t^2]\begin{bmatrix} 1 \\ 0 \\ 1 \end{bmatrix} - e^{2t}\begin{bmatrix} 1 \\ 2 \\ -1 \end{bmatrix} = \begin{bmatrix} -3e^{-t} - 2(t^2 - t + 2) - e^{2t} \\ 3e^{-t} + 2(t-1) \quad -2e^{2t} \\ 3e^{-t} + 2(t+t^2) \quad + e^{2t} \end{bmatrix}.$$

Functions of Normal Matrices

If T is the operator on $\mathbf{C}^n$ generated by the matrix A, according to the results of Section 4.3, $f(A)$ is the matrix of the operator $f(T)$ in the standard basis. Hence the results of Theorem 5 of the previous section apply to the classification of $f(A)$. This yields the next theorem.

THEOREM 3. Let A be an $n \times n$ matrix. If A is normal and $f(A)$ is defined, then $f(A)$ is normal. Further:

(a) If $f(\lambda)$ is real-valued for all $\lambda \in \sigma(A)$, then $f(A)$ is Hermitian.
(b) If $|f(\lambda)| = 1$ for all $\lambda \in \sigma(A)$, then $f(A)$ is unitary.

In particular, the inverse of a unitary matrix is unitary. The inverse of a Hermitian matrix is Hermitian, when it exists, and the inverse of a normal matrix is normal, when it exists.

Computation of functions of a normal matrix

If A is an $n \times n$ normal matrix, and $\{E_k\}_{k=1}^n$ is an orthonormal basis of eigenvectors with $AE_i = \lambda_i E_i$, then by Theorem 2 of Section 4.3,

$$f(A) = Uf(\Lambda)U^{-1} = Uf(\Lambda)U^{H} \tag{5}$$

where $U = [E_1\ E_2 \ \cdots\ E_n]$, $f(\Lambda) = \text{diag}(f(\lambda_1), f(\lambda_2), \ldots, f(\lambda_n))$.

EXAMPLE 2. ROTATIONS IN TWO DIMENSIONS. Let A be given by

$$A = \begin{bmatrix} \cos(\theta) & -\sin(\theta) \\ \sin(\theta) & \cos(\theta) \end{bmatrix} = [A_1\ A_2].$$

A is unitary, and the columns A_j of A are δ_1 and δ_2 rotated counterclockwise by an angle θ. The eigenvalues of A are $(\theta \neq k\pi)$

$$e^{i\theta} = \cos(\theta) + i\sin(\theta), \qquad e^{-i\theta} = \cos(\theta) - i\sin(\theta),$$

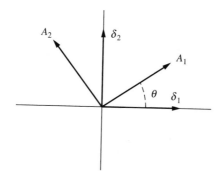

and the corresponding eigenvectors are $E_1 = (1/\sqrt{2})(1, -i)^T$ and $E_2 = (1/\sqrt{2})(1, i)^T$. Thus, with $U = [E_1 \; E_2]$,

$$A^n = U \operatorname{diag}(e^{ni}\theta, e^{-in\theta})U^H = \begin{bmatrix} \cos(n\theta) & -\sin(n\theta) \\ \sin(n\theta) & \cos(n\theta) \end{bmatrix},$$

so that A^n is the same as one rotation through an angle $n\theta$, which makes geometric sense. A^n must be unitary, since $|\lambda_i^n| = 1$ and its eigenvectors are also the orthonormal pair E_1, E_2. This is easily confirmed by direct calculation. If $\theta = 0$, then $A = I$; if $\theta = \pi$, then $A = -I$, and the preceding results are correct, except that the eigenbasis is no longer uniquely determined.

EXAMPLE 3. For the matrix

$$A = \frac{1}{2}\begin{bmatrix} 1 & 0 & -1 \\ 0 & 2 & 0 \\ -1 & 0 & 1 \end{bmatrix},$$

(a) Classify and compute $\sqrt{A}$, e^{At}, 2^A.
(b) Classify $e^{\pi i A}$.

Solution

(a) $|A - \lambda I| = p(\lambda) = -\lambda(\lambda - 1)^2$, so the eigenvalues are $\lambda_1 = 0, \lambda_2 = 1$. Since $\sqrt{\lambda_i}$, $e^{t\lambda_i}$, 2^{λ_i} are real, $\sqrt{A}$, e^{At}, 2^A are Hermitian. For the computations, an orthonormal eigenbasis is needed. $M(0)$ is one-dimensional and $M(1)$ two-dimensional. A unit vector spanning $M(0)$ is $E_1 = (1, 0, 1)^T/\sqrt{2}$, and from the many orthonormal bases for $M(1)$, select $E_2 = (1, -1, -1)^T/\sqrt{3}$ and $E_2 = (1, 2, -1)^T/\sqrt{6}$. A check on computations is to verify that $E_1 \perp E_2, E_3$, since eigenvectors from different eigenspaces must be orthogonal. Set $U = [E_1 \; E_2 \; E_3]$. Then

$$\sqrt{A} = U \operatorname{diag}\left(0, \sqrt{1}, \sqrt{1}\right)U^T = A.$$

Next, since e^{At} has eigenvalues $e^{\lambda_i t}$ and the same eigenbasis,

$$e^{At} = U \operatorname{diag}(1, e^t, e^t)U^T = \begin{bmatrix} (1 - e^t)/2 & 0 & (1 - e^t)/2 \\ 0 & 2e^t/3 & 0 \\ (1 - e^t)/2 & 0 & (1 + e^t)/2 \end{bmatrix}$$

$$2^A = U \operatorname{diag}(1, 2, 2)U^T = \begin{bmatrix} 1.5 & 0 & -0.5 \\ 0 & 2 & 0 \\ -0.5 & 0 & 1.5 \end{bmatrix}.$$

(b) $e^{\pi i A}$ has eigenvalues $e^{\pi i 0} = 1$ and $e^{\pi i 1} = -1$. Hence $e^{\pi i A}$ is Hermitian and unitary.

The Generalized Eigenvalue Problem

The last problem in matrix theory to be considered here arises in the context of differential equations connected with vibration theory and in conjunction with the analysis of quadratic forms. It is useful also in the study of the geometry of surfaces. This problem involves solving another kind of eigenvalue problem, which is defined next. The subsequent theorem asserts the solvability of the problem, in certain circumstances.

DEFINITION. The *generalized eigenvalue problem* for matrices A and G is: Find all λ and E satisfying $AE = \lambda GE$, $E \neq 0$. λ is called a generalized eigenvalue, and E is called a generalized eigenvector corresponding to λ.

THEOREM 4. Given an inner product $\langle x, y \rangle \equiv \langle x, y \rangle_G = y^H G x$ on $\mathbf{F}^n$ and an $n \times n$ Hermitian matrix A, there exist numbers λ_i and columns $E_1, E_2, \ldots, E_n$ orthogonal with respect to $\langle , \rangle_G$ such that $AE_i = \lambda_i GE_i$ $\forall i$.

Proof. G is the Gram matrix of the inner product in the standard basis, so it is Hermitian and nonsingular. $AE_i = \lambda_i GE_i$ is equivalent to $G^{-1}AE_i = \lambda_i E_i$; i.e., the E_i are eigenvectors of $L = G^{-1}A$. To show that such E_i and λ_i exist, it is sufficient to show that the operator $T(x) = Lx$ is Hermitian with respect to $\langle , \rangle_G$, because the spectral theorem then guarantees that an orthogonal eigenbasis exists. But since A and G^{-1} are Hermitian matrices,

$$\langle T(x), y \rangle_G = \langle Lx, y \rangle_G = y^H G(G^{-1}Ax) = y^H Ax,$$

$$\langle x, T(y) \rangle_G = \langle x, Ly \rangle_G = (G^{-1}Ay)^H Gx = y^H A^H (G^{-1})^H Gx = y^H A G^{-1} Gx = y^H Ax.$$

Thus $\langle T(x), y \rangle = \langle x, T(y) \rangle = y^H Ax$, and the proof is finished.

Next, the preceding result is formulated in terms of matrices.

COROLLARY 5. Given an inner product $\langle x, y \rangle = \langle x, y \rangle_G = y^H Gx$ on $\mathbf{F}^n$ and an $n \times n$ Hermitian matrix A, there exist matrices $P = [E_1 \ E_2 \ \cdots \ E_n]$, $\Lambda = \mathrm{diag}(\lambda_1, \ldots, \lambda_n)$ such that

$$P^H G P = I \tag{6}$$

$$P^H A P = \Lambda. \tag{7}$$

In fact, $P = [E_1 \ E_2 \ \cdots \ E_n]$ has columns that are orthonormal with respect to $\langle , \rangle_G$ and satisfy $AE_i = \lambda_i GE_i$ $\forall i$.

Proof. Let $E_1, \ldots, E_n$ be the orthogonal generalized eigenvectors as described in Theorem 4, and normalize them so that $\|E_i\|_G = 1$. Put $P = [E_1 \ E_2 \ \cdots \ E_n]$. Then

$$P^H G P = [E_1 \ E_2 \ \cdots \ E_n]^H [GE_1 \ GE_2 \ \cdots \ GE_n] = [E_j^H GE_i] = [\langle E_i, E_j \rangle_G] = I.$$

Since $AE_j = \lambda_j GE_j$,

$$P^H A P = [E_i^H AE_j] = [\lambda_j E_i^H GE_j] = \mathrm{diag}(\lambda_1 \|E_1\|_G^2, \ldots, \lambda_n \|E_n\|_G^2) = \Lambda.$$

Remark. Given that $\langle x, y \rangle_G$ is an inner product, the converse holds: Eqs. (6) and (7) imply $\{E_1, E_2, \ldots, E_n\}$ is orthonormal with respect to $\langle x, y \rangle_G$ and $AE_i = \lambda_i GE_i$. This is left as an exercise.

Remark. If the eigenvectors are orthogonal but not necessarily of unit length with respect to $\langle , \rangle_G$, then putting

$$N^2 = \mathrm{diag}(\|E_1\|_G^2, \ldots, \|E_n\|_G^2) \tag{8}$$

yields

$$P^H G P = N^2 \tag{9}$$

$$P^H A P = \Lambda N^2 = N^2 \Lambda.$$

Remark. Criteria that imply that $\langle x, y \rangle_G \equiv y^H G x$ is an inner product are explored at length in Chapter 6. See Exercise 14 for a partial discussion.

Computation of the generalized eigenvalues

Theoretically, solving the generalized eigenvalue problem is equivalent to solving the eigenvalue problem for $L \equiv G^{-1}A$. This is usually not a good idea, with a few exceptions. If the work is done by hand, it forces extra computation. If the work is done on a computer, it again forces the calculation of G^{-1}, which can give rise to significant error when G is large or "ill conditioned." The ill conditioning will produce a computed G^{-1} with serious loss of accuracy, so the matrix L will be badly contaminated. In practice, G often arises as the Gram matrix of an inner product on some function space, and these are often ill conditioned. There are good numerical routines for solving the generalized eigenvalue problem, and they do not compute G^{-1}. Computing the generalized eigenvalues via L does make sense if the problem is small or if G^{-1} is easy to compute (e.g., diagonal). Some other approaches are given in the exercises. (See Exercise 36.)

Theoretical computations

The following procedure is primarily for hand calculations or theoretical discussions. Note the parallel between this procedure and the procedure for the usual eigenvalue problem. Observe that

$$AE = \lambda GE, \quad E \neq 0 \Leftrightarrow (A - \lambda G)E = 0, \quad E \neq 0$$

$$\Leftrightarrow |A - \lambda G| = 0 \quad \text{and} \quad 0 \neq E \in N(A - \lambda G).$$

Thus the eigenvalues are exactly the zeros of the characteristic polynomial $p(\lambda) = |A - \lambda G|$, and for each such root, the associated eigenspace is $M(\lambda) = N(A - \lambda G)$. Thus the procedure is

1. Determine the roots of the characteristic equation $|A - \lambda G| = 0$.
2. For each of the distinct eigenvalues λ_i, find the eigenspace $M(\lambda_i) = N(A - \lambda_i G)$. (Recall that for Hermitian operators, the dimension of the eigenspace is the multiplicity of the eigenvalue as a root of the characteristic equation, and that eigenvectors from distinct eigenspaces are automatically orthogonal with respect to $\langle , \rangle_G$.)
3. (a) For each eigenspace of dimension 1, select an eigenvector.
   (b) For each eigenspace with dimension greater than 1, find an orthogonal basis for that eigenspace, e.g., by the Gram–Schmidt process.

The collection of the eigenvectors so found satisfies the requirements of Theorem 4 and Corollary 5.

One example will be presented here, and many more will follow in later chapters.

EXAMPLE 4. Let

$$A = \begin{bmatrix} 3 & 1 & -1 \\ 1 & 4 & -1 \\ -1 & -1 & 7 \end{bmatrix} \quad \text{and} \quad \langle x, y \rangle_G = 2x_1\bar{y}_1 + 3x_2\bar{y}_2 + 6x_3\bar{y}_3,$$

i.e., $G = \text{diag}(2, 3, 6)$.

(a) Find an orthogonal basis of generalized eigenvectors of A and G.

(b) Find matrices P and $\Lambda = \text{diag}(\lambda_1, \ldots, \lambda_n)$ such that $P^H G P = I$ and $P^H A P = \Lambda$.

Solution

(a) A computation shows that $|A - \lambda G| = 36(1 - \lambda)^2(2 - \lambda)$. Thus the eigenspaces $M(1)$ and $M(2)$ have dimension 2 and 1, respectively. Solving $(A - 1 \cdot G)x = 0$ gives $x_1 + x_2 - x_3 = 0$. A basis for this subspace is $E_1 = (0, 1, 1)^T$ and $F_1 = (1, 0, 1)^T$. Orthogonalizing yields $F_1 - \{\langle F_1, E_1 \rangle_G / \|E_1\|_G^2\} E_1 = F_1 - \frac{2}{3} E_1 = \frac{1}{3}(3, -2, 1)^T$. Now multiply this by 3 for aesthetics to get $E_2 = (3, -2, 1)^T$. Finally, E_1, E_2 is an orthogonal basis for the eigenspace $M(1)$. Solving $(A - 2 \cdot G)x = 0$ gives the eigenvector $E_3 = (3, 2, -1)^T$. A check shows $E_3 \perp E_2, E_1$, using $\langle , \rangle_G$. Thus E_1, E_2, E_3 and $\lambda_1 = \lambda_2 = 1, \lambda_3 = 2$ provide a solution for (a). Recall that the eigenspaces are unique, but the particular basis selected for each is not.

(b) Using the results of (a), take $\Lambda = \text{diag}(1, 1, 2)$ and put $e_i = E_i / \|E_i\|_G$ and $P = [e_1 \, e_2 \, e_3]$. Since $\|E_1\|_G^2 = 9, \|E_2\|_G^2 = \|E_3\|_G^2 = 36$, and

$$P = \frac{1}{6} \begin{bmatrix} 0 & 3 & 3 \\ 2 & -2 & 2 \\ 2 & -1 & 1 \end{bmatrix}.$$

Simultaneous generalized eigenvalue problems

The problem is as follows: Given square matrices A, B, and G, find (if possible) a basis $\{E_1, E_2, \ldots, E_n\}$ so that $AE_i = \lambda_i GE_i$, $BE_i = \gamma_i GE_i$. The next theorem gives conditions sufficient for the existence of such a basis. Moreover, under the hypotheses of the theorem, the generalized eigenvectors will be orthogonal with respect to a certain inner product. Examples of the application of this theorem will appear in Chapter 7.

THEOREM 6. An inner product $\langle x, y \rangle = \langle x, y \rangle_G = y^H G x$ on $\mathbf{F}^n$ and $n \times n$ Hermitian matrices A, B are given. Then $AG^{-1}B$ is Hermitian iff there exist numbers λ_i, γ_i and columns $E_1, E_2, \ldots, E_n$ orthogonal with respect to $\langle , \rangle_G$ such that $AE_i = \lambda_i GE_i$ and $BE_i = \gamma_i GE_i$ $\forall i$. Furthermore, when such λ_i, γ_i, and E_i exist, if $P = [E_1 \, E_2 \, \cdots \, E_n]$, $\Lambda = \text{diag}(\lambda_1, \lambda_2, \ldots, \lambda_n)$, $\Gamma = \text{diag}(\gamma_1, \gamma_2, \ldots, \gamma_n)$, and $\{E_1, E_2, \ldots, E_n\}$ is orthonormal in $\langle , \rangle_G$, then

$$P^H G P = I$$
$$P^H A P = \Lambda \tag{10}$$
$$P^H B P = \Gamma.$$

Proof. Let $V = \mathbf{C}^n$, and define $\langle x, y \rangle = y^H G x$, $S = G^{-1}B$, $T = G^{-1}A$. T and S are Hermitian with respect to this $\langle , \rangle$ (cf. Theorem 4). Observe that $ST = G^{-1}BG^{-1}A = TS = G^{-1}AG^{-1}B \Leftrightarrow BG^{-1}A = AG^{-1}B = A^H(G^{-1})^H B^H = (BG^{-1}A)^H$. Now apply Theorem 6 of Section 5.2. Equations (10) follow from the proof of Corollary 5.

There are certain situations where it can be determined in advance that the multiple simultaneous diagonalization may be done without having to check that $AG^{-1}B$ is Hermitian. These follow from Propositions 7 and 8 of Section 5.2 and from the discussion in Section 4.3. The proofs are left for the exercises. (See Exercise 37.) Examples of the use of these theorems will be given in Chapter 7.

THEOREM 7. Let A and G be $n \times n$ Hermitian matrices, and suppose $\langle x, y \rangle_G \equiv y^H G x$ is an inner product. Suppose $\beta = \{E_k\}_{k=1}^n$ is a basis of generalized eigenvectors orthogonal with respect to $\langle , \rangle_G$:

$$AE_i = \lambda_i GE_i, \quad E_i^H GE_j = 0 \quad \text{if } i \neq j, \quad i, j = 1, 2, \ldots, n.$$

If $B = Gf(G^{-1}A)$, then $\{E_k\}_{k=1}^n$ is a basis of generalized eigenvalues for B as well; specifically,

$$BE_i = \gamma_i GE_i \quad \text{with } \gamma_i = f(\lambda_i), \quad i = 1, 2, \ldots, n. \tag{11}$$

THEOREM 8. If G, A, and B are all real-valued functions of a single Hermitian matrix L, and $\langle x, y \rangle_G \equiv y^H G x$ is an inner product, then an eigenbasis $\{E_1, E_2, \ldots, E_n\}$ for L that is orthogonal in the standard inner product will be orthogonal with respect to $\langle , \rangle_G$, and there will exist numbers λ_i, γ_i such that $AE_i = \lambda_i GE_i$ and $BE_i = \gamma_i GE_i$ $\forall i$. Specifically, if $G = g(L)$, $A = a(L)$, $B = b(L)$, then

$$\lambda_i = \frac{a(\alpha_i)}{g(\alpha_i)}, \qquad \gamma_i = \frac{b(\alpha_i)}{g(\alpha_i)}.$$

Exercises 5.3

1. Classify each of the following matrices, and if the matrix A is normal, find a unitary matrix P and a diagonal matrix Λ such that $P^H A P = \Lambda$.

(a) $\begin{bmatrix} 0 & 1 \\ -1 & 0 \end{bmatrix}$ (b) $\dfrac{1}{2}\begin{bmatrix} 1+i & 1-i \\ 1-i & 1+i \end{bmatrix}$ (c) $\dfrac{1}{2}\begin{bmatrix} 1 & \sqrt{3} \\ -\sqrt{3} & 1 \end{bmatrix}$

(d) $\begin{bmatrix} 2+i & 2-i \\ 2-i & 2+i \end{bmatrix}$ (e) $\dfrac{1}{\sqrt{2}}\begin{bmatrix} i & 1 \\ 1 & i \end{bmatrix}$

(f) $\begin{bmatrix} 2 & i & 0 \\ -i & 1 & -i \\ 0 & i & 2 \end{bmatrix}$ (g) $\begin{bmatrix} 1 & 1 & -1 \\ 1 & 1 & 1 \\ -1 & 1 & 1 \end{bmatrix}$ (h) $\dfrac{1}{2}\begin{bmatrix} 1 & 1 & -\sqrt{2} \\ 1 & 1 & \sqrt{2} \\ \sqrt{2} & -\sqrt{2} & 0 \end{bmatrix}$

(i) $\begin{bmatrix} 1 & -1 & 1 \\ -1 & 3 & -1 \\ 1 & -1 & 1 \end{bmatrix}$ (j) $\dfrac{1}{2}\begin{bmatrix} 1 & 0 & i \\ 0 & 2 & 0 \\ -i & 0 & 1 \end{bmatrix}$ (k) $\begin{bmatrix} 1 & 1 & 0 \\ 1 & 0 & 1 \\ 0 & 1 & 1 \end{bmatrix}$ (l) $\begin{bmatrix} 0 & 1 & 0 \\ 1 & 0 & 0 \\ 0 & 0 & 1 \end{bmatrix}$

2. Let $A = \begin{bmatrix} i & 2 \\ -2 & i \end{bmatrix}$, $i = \sqrt{-1}$.

   (a) Classify A and find a unitary matrix U so that $U^H A U$ is diagonal.
   (b) Classify e^A and find a unitary matrix V so that $V^H e^A V$ is diagonal.

3. Let $A = \begin{bmatrix} 3 & -1 \\ -1 & 3 \end{bmatrix}$.

   (a) Classify A and find a unitary matrix U so that $U^H A U$ is diagonal.
   (b) For each of the matrices

$$B = e^{At}, \qquad C = [\tfrac{1}{2}A - I]^{100}, \qquad D = \sqrt{A}, \qquad E = e^{i\pi A/2},$$

$$F = f(A) \quad \text{where } f(x) = \frac{x^7 - 6x^6 + 8x^5 + x - 2}{x^4 - 6x^3 + 8x^2 + x - 3},$$

classify the matrix and find V unitary and Γ diagonal such that $M = V\Gamma V^H$, where $M = B, C, D, E, F$.

(c) Find the matrices C, E, and F.

4. Let $A = \begin{bmatrix} 2 & i \\ -i & 2 \end{bmatrix}$.

   (a) Classify A and find a unitary matrix U so that $U^H A U$ is diagonal.
   (b) For each of the following matrices

   $$B = [A - 2I]^5, \qquad C = e^{i\pi A/2}, \qquad D = (A^2 - 4A + 5I)(A + I)^{-1},$$

   find a unitary matrix V and a diagonal matrix Γ so that $V^H M V = \Gamma$, and classify M where $M = B, C, D$.
   (c) Find D.

5. Let $A = \begin{bmatrix} 1 & -1 & 1 \\ -1 & 3 & -1 \\ 1 & -1 & 1 \end{bmatrix}$. Classify the following matrices.

   (a) $(iI + A)(A - iI)^{-1}$ (b) $i \ln(A + I)$ (c) e^{iA} (d) $e^{i\pi A}$

6. For the following matrix A, find the indicated matrices $f(A)$. Classify $f(A)$. Find a unitary matrix P and a diagonal matrix $\Gamma = \mathrm{diag}(\gamma_1, \gamma_2, \gamma_3)$ such that $P^H f(A) P = \Gamma$:

   $$A = \begin{bmatrix} 1 & 1 & 0 \\ 1 & 0 & 1 \\ 0 & 1 & 1 \end{bmatrix}.$$

   (a) A^{-1} (b) $[\tfrac{1}{3}(4 - A^2)]^{10}$ (c) 2^T (d) $e^{i\pi A}$
   (e) $\sin(\pi A/2)$ (f) $(I + iA)(I - iA)^{-1}$ (g) $|A|$ (h) $\tfrac{i}{3}(I + 3A - A^2)$

7. Solve $x' = \begin{bmatrix} 1 & -1 & 1 \\ -1 & 3 & -1 \\ 1 & -1 & 1 \end{bmatrix} x$, $x(0) = 6 \begin{bmatrix} 1 \\ 0 \\ 0 \end{bmatrix}$.

8. Solve $x' = \begin{bmatrix} 1 & 1 & -1 \\ 1 & 1 & 1 \\ -1 & 1 & 1 \end{bmatrix} x + \begin{bmatrix} 1 \\ 1 \\ 1 \end{bmatrix}$, $x(0) = \begin{bmatrix} 0 \\ 0 \\ 1 \end{bmatrix}$.

9. Solve $x' = \begin{bmatrix} 1 & -1 & 0 \\ -1 & 2 & -1 \\ 0 & -1 & 1 \end{bmatrix} x + \begin{bmatrix} 1 \\ 1 \\ 1 \end{bmatrix}$, $x(0) = 12 \begin{bmatrix} 1 \\ 0 \\ 0 \end{bmatrix}$.

10. **Solutions of $x' = Ax + f(t)$ via e^{At}.** Assume A is diagonalizable. (This restriction will be removed in Chapter 8.) From Chapter 4, $(d/dt)e^{At} = Ae^{At} = e^{At}A$. Look for a solution of $x' = Ax + f$ of the form $x = e^{At}v(t)$. Conclude that $v'(t) = e^{-At}f(t)$. Deduce the solution $x(t) = e^{At}x(0) + e^{At}\int_0^t e^{-Au}f(u)\,du$.

11. Let A be a normal matrix with a basis of orthonormal eigenvectors E_i and corresponding eigenvalues λ_i. Show that $A^H = \sum_{i=1}^{n} \bar{\lambda}_i E_i E_i^H$.

12. Let A have an orthogonal eigenbasis $\{E_k\}_{k=1}^{n}$ and $AE_i = \lambda_i E_i$, $i = 1, 2, \ldots, n$. Show that

$$f(A) = P f(\Lambda) N^{-2} P^H,$$

where $N = \text{diag}(\|E_1\|, \|E_2\|, \ldots, \|E_n\|)$, $P = [E_1\ E_2\ \cdots\ E_n]$, $\Lambda = \text{diag}(\lambda_1, \ldots, \lambda_n)$.

13. Let $L = \text{Trid}(-1, 2, -1)$ be $n \times n$, and suppose $\{e_k\}_{k=1}^{n}$ is an orthogonal eigenbasis for L with $Le_k = \mu_k e_k$. In fact, $\mu_k = 4\sin^2(k\pi/2N)$, $e_k = (\sin(k\pi x_1), \sin(k\pi x_2), \ldots, \sin(k\pi x_n))^T$, where $N = n + 1$, $x_i = i/N$.
   (a) Show that $\{e_k\}_{k=1}^{n}$ is orthogonal in the standard inner product. It can be shown that $\|e_k\|^2 = N/2$.
   (b) Show that $A = \text{Trid}(a, b, a)$ is a linear combination of L and I, and find the eigenvalues of A in terms of the μ_i and an eigenbasis for A.

14. **Further classification of Hermitian matrices.** Let A be Hermitian in the standard inner product. A is *positive definite* iff $\langle Ax, x \rangle > 0$ when $x \neq 0$, and *positive* when $\langle Ax, x \rangle \geq 0\ \forall x$. (Note that a positive or positive definite matrix is always Hermitian by definition.) For Hermitian A, prove (a)–(d).
   (a) A is positive definite iff all eigenvalues of A are positive.
   (b) A is positive iff all eigenvalues of A are nonnegative.
   (c) A is positive definite iff $\langle Ax, y \rangle$ is an inner product on $\mathbf{C}^n$.
   (d) A is positive definite iff $\sum_{i,j=1}^{n} a_{ij} x_j \bar{x}_i > 0$ when $x \neq 0$.
   (e) $\langle x, y \rangle \equiv y^H G x$ is an inner product iff G is positive definite.

15. Let A be a Hermitian positive matrix. Show that there is a matrix B such that $A = B^H B$. If A is positive definite, then B is nonsingular. (*Hint: B* may also be taken to be Hermitian.)

16. Let C be an $n \times n$ matrix. Show that C is Hermitian iff there exist real matrices A, B with A symmetric and B skew-symmetric so that $C = A + iB$. (*Hint:* Consider $C \pm \bar{C}$.)

17. Let H be Hermitian. Show that $U = (I + iH)(I - iH)^{-1}$ is unitary and $H = i(I - U)(I + U)^{-1}$.

18. Let A be a nonsingular Hermitian matrix with $AE_i = \lambda_i E_i$, where $\{E_k\}_{k=1}^{n}$ is an orthogonal set in $\mathbf{C}^n$, using the standard inner product. Show that $\{AE_k\}_{k=1}^{n}$ is also an orthogonal set.

19. Let E be a nonzero vector in $\mathbf{C}^n$ and put $A = EE^H$. Classify A. Find a basis of eigenvectors for A.

20. If U is a unitary matrix, show that $|\det(U)| = 1$.

21. Let U be a 3×3 orthogonal matrix with determinant 1. Show that there is a unit vector e with $Ue = e$.

22. If A is a Hermitian matrix, then $\det(A)$ is real. Prove or disprove this.

23. If U_1 and U_2 are unitary matrices, prove or give a counterexample disproving the following statements:
  (*a*) $U_1 + U_2$ is always unitary.
  (*b*) aU_1 is always unitary, where a is a fixed complex scalar.
  (*c*) $e^{iq}U_1$ is always unitary, where q is a fixed real number.
  (*d*) U_1U_2 is always unitary.

24. If A_1 and A_2 are Hermitian matrices, prove or give a counterexample disproving the following statements:
  (*a*) $A_1 + A_2$ is always Hermitian.
  (*b*) aA_1 is always Hermitian, where a is a fixed complex scalar.
  (*c*) qA_1 is always Hermitian, where q is a fixed real number.
  (*d*) A_1A_2 is always Hermitian.

25. Let A and B be Hermitian $n \times n$ matrices.
  (*a*) If A and B have the same eigenvalues listed according to their algebraic multiplicity, show that there is a unitary matrix U such that $U^{-1}AU = B$.
  (*b*) Does (*a*) remain true if it is assumed only that $\sigma(A) = \sigma(B)$?

26. Suppose A is an $n \times n$ normal matrix with eigenvalues λ_k. Show that the eigenvalues of A^HA are $|\lambda_k|^2$.

27. Suppose A is a real unitary matrix whose eigenvalues are all ± 1. Show that A is Hermitian and $A^2 = I$.

28. A *permutation matrix* P is one whose columns are obtained from the identity matrix by a rearrangement. Show that P is unitary. What can be said about a matrix Q that is obtained from the identity by rearranging rows?

29. Let A be an $n \times n$ matrix. Suppose A has k orthogonal eigenvectors. Show that there is a unitary matrix U such that

$$U^{-1}AU = \begin{bmatrix} \Lambda & B \\ 0 & C \end{bmatrix},$$

where Λ is a $k \times k$ diagonal matrix, and B is $k \times n - k$ and C is $n - k \times n - k$.

30. Suppose A is a normal matrix and $\{E_k\}_{k=1}^n$ is an orthogonal eigenbasis for A with $AE_k = \lambda_k E_k$. Suppose $f(A)$ is defined, and put $F_i = f(A)E_i$. Show $\langle F_i, F_j \rangle = 0$ if $i \neq j$.

31. Solve the generalized eigenvalue problem for the following pairs of matrices, given that $\langle x, y \rangle = y^T Gx$ is an inner product on $\mathbf{R}^n$.

(*a*) $A = \begin{bmatrix} 1 & 1 & -1 \\ 1 & 3 & 1 \\ -1 & 1 & 3 \end{bmatrix}, G = \begin{bmatrix} 2 & 0 & -1 \\ 0 & 3 & 0 \\ -1 & 0 & 2 \end{bmatrix}$

(*b*) $A = \begin{bmatrix} 1 & 1 & -1 \\ 1 & 2 & 1 \\ -1 & 1 & 5 \end{bmatrix}, G = \begin{bmatrix} 2 & 0 & 0 \\ 0 & 3 & 0 \\ 0 & 0 & 6 \end{bmatrix}$

$$(c)\ A = \begin{bmatrix} 6 & 0 & 0 \\ 0 & 3 & -3 \\ 0 & -3 & 3 \end{bmatrix}, G = \begin{bmatrix} 6 & -3 & 0 \\ -3 & 6 & 0 \\ 0 & 0 & 3 \end{bmatrix}$$

$$(d)\ A = \begin{bmatrix} 2 & 0 & 3 \\ 0 & 2 & 1 \\ 3 & 1 & 5 \end{bmatrix}, G = \begin{bmatrix} 2 & 0 & 3 \\ 0 & 3 & 0 \\ 3 & 0 & 6 \end{bmatrix}$$

32. **Simultaneous diagonalization of Hermitian tridiagonal matrices.** Let $L = \mathrm{Trid}(-1, 2, -1)$ be $n \times n$ and $e_k = (\sin(k\pi x_1), \sin(k\pi x_2), \ldots, \sin(k\pi x_n))^{\mathrm{T}}$, $\mu_k = 4\sin^2(k\pi/2N)$, where $N = n + 1$, $x_i = i/N$. It is known that $\{e_k\}_{k=1}^n$ is an orthogonal eigenbasis for L with $Le_k = \mu_k e_k$. (See Exercise 13 for useful information.)
    (a) Find sufficient conditions in terms of u and v, which are independent of n, for $x^{\mathrm{H}} \mathrm{Trid}(u, v, u)x$ to be an inner product on $\mathbf{C}^n$. (*Hint:* Write $x = \sum_{i=1}^n c_i e_i$.)
    (b) Suppose $A = \mathrm{Trid}(a, b, a)$, $G = \mathrm{Trid}(u, v, u)$, and $y^{\mathrm{H}} G x$ is an inner product. Show that $\{e_k\}_{k=1}^n$ solves the generalized eigenvalue problem for A and G. Find the generalized eigenvalues λ_k in $Ae_k = \lambda_k Ge_k$.

33. Suppose $AE_i = \lambda_i GE_i$, $i = 1, 2, \ldots, n$, $P = [E_1\ E_2\ \cdots\ E_n]$, and $\Lambda = \mathrm{diag}(\lambda_1, \lambda_2, \ldots, \lambda_n)$. Show that $AP = GP\Lambda$.

34. Suppose $\{E_1, E_2, \ldots, E_n\}$ is orthonormal in $\langle x, y \rangle = y^{\mathrm{H}} G x$ and $AE_i = \lambda_i GE_i$. Show that
    (a) $A = \sum_{i=1}^n \lambda_i GE_i(GE_i)^{\mathrm{H}}$
    (b) $I = \sum_{i=1}^n GE_i E_i^{\mathrm{H}} = \sum_{i=1}^n E_i(GE_i)^{\mathrm{H}}$

35. Prove the following: Let $P = [E_1\ E_2\ \cdots\ E_n]$, $\Lambda = \mathrm{diag}(\lambda_1, \lambda_2, \ldots, \lambda_n)$, and $N^2 = \mathrm{diag}(\|E_1\|_G^2, \ldots, \|E_n\|_G^2)$. Then $P^{\mathrm{H}} GP = N^2$ and $P^{\mathrm{H}} AP = \Lambda N^2$ imply
    (a) $\{E_1, E_2, \ldots, E_n\}$ is orthogonal with respect to $\langle\ ,\ \rangle_G$.
    (b) $AE_i = \lambda_i GE_i\ \forall i$.

36. **Alternative methods for the solution of $Ae = \lambda Ge$.** These methods reduce the problem to solving the ordinary eigenvalue problem for a Hermitian matrix. The advantage of this on a computer is that routines for solving the ordinary eigenvalue problem for a Hermitian matrix are much sounder numerically than ones for an arbitrary matrix. Let $\langle x, y \rangle \equiv y^{\mathrm{H}} G x$ be an inner product and A Hermitian.
    (a) Let $B = G^{-1/2} A G^{-1/2}$. Show that B is Hermitian. Suppose that $\{f_k\}_{k=1}^n$ is an eigenbasis for B, orthonormal in the standard inner product. Using the f_k, construct a basis $\{e_k\}_{k=1}^n$ satisfying $Ae_k = \lambda_k Ge_k$, $\langle e_k, e_j \rangle_G = \delta_{kj}$. For which G is $G^{-1/2}$ easy to compute?
    (b) In Chapter 6 it will be shown that G can be factored as $G = LL^{\mathrm{H}}$, where L is lower triangular with positive diagonal entries. This is called the *Cholesky decomposition* of G. Use the notation $L^{-\mathrm{H}} = (L^{-1})^{\mathrm{H}}$. Put $B = L^{-1} A L^{-\mathrm{H}}$. Show that B is Hermitian. Suppose that $\{f_k\}_{k=1}^n$ is an eigenbasis for B, orthonormal in the standard inner product. Using the f_k, construct a basis $\{e_k\}_{k=1}^n$ satisfying $Ae_k = \lambda_k Ge_k$, $\langle e_k, e_j \rangle_G = \delta_{kj}$.

37. (a) Prove Theorem 7.
    (b) Prove Theorem 8.

5.4
EXTREMUM PRINCIPLES FOR HERMITIAN OPERATORS

The various minimum/maximum principles discussed in this section play an essential role in estimates of eigenvalues of Hermitian operators. Most of them involve the Rayleigh quotient. For example, in many physical systems the natural frequencies of oscillation ω_i are related to the eigenvalues of a related Hermitian operator by $\omega_i^2 = \lambda_i$. Examples of this are provided in Chapter 7. It is often of practical interest to estimate the natural frequencies of vibration, especially the largest and smallest natural frequencies. In Section 5.6 an extremely useful technique for estimating the eigenvalues and eigenvectors of certain Hermitian operators that relies on the estimates in this section will be discussed.

In this section, V will denote a nontrivial vector space, possibly of infinite dimension, and T will be assumed to be Hermitian on V with respect to the inner product $\langle , \rangle: \langle Tx, y \rangle = \langle x, Ty \rangle \; \forall x, y \in V$.

DEFINITION. The *Rayleigh quotient of T* with respect to $\langle , \rangle$ is

$$R(x) = \frac{\langle Tx, x \rangle}{\|x\|^2} \quad \text{where } x \neq 0.$$

The following lemma embodies a simple principle that will be used often.

LEMMA 1. Let T be Hermitian on V with respect to $\langle , \rangle$. Suppose $\{e_k\}_{k=1}^n$ is orthogonal with respect to $\langle , \rangle$, $T(e_i) = \lambda_i e_i$, and $\lambda_1 \leq \lambda_2 \leq \cdots \leq \lambda_n$. If $0 \neq x = \sum_{i=1}^n x_i e_i$, then

$$\lambda_1 \leq R(x) \leq \lambda_n. \tag{1}$$

Proof. Since $T(e_i) = \lambda_i e_i$ and $x = \sum_{i=1}^n x_i e_i$, $T(x) = \sum_{i=1}^n \lambda_i x_i e_i$. By Parseval's identity,

$$\langle T(x), x \rangle = \left\langle \sum_{i=1}^n \lambda_i x_i e_i, \sum_{i=1}^n x_i e_i \right\rangle = \sum_{i=1}^n \lambda_i |x_i|^2 \|e_i\|^2 \tag{$*$}$$

and

$$\|x\|^2 = \langle x, x \rangle = \left\langle \sum_{i=1}^n x_i e_i, \sum_{i=1}^n x_i e_i \right\rangle = \sum_{i=1}^n |x_i|^2 \|e_i\|^2.$$

If λ_i is replaced in $(*)$ by λ_n, the sum increases since $\lambda_i \leq \lambda_n$. Hence

$$\langle T(x), x \rangle \leq \sum_{i=1}^n \lambda_n |x_i|^2 \|e_i\|^2 = \lambda_n \sum_{i=1}^n |x_i|^2 \|e_i\|^2 = \lambda_n \|x\|^2.$$

Division by $\|x\|^2$ yields $R(x) \leq \lambda_n$. Similarly, replacing each λ_i by λ_1 decreases the sum, and so $R(x) \geq \lambda_1$. ∎

EXAMPLE 1. Let $T(x) = Ax$, where

$$A = \begin{bmatrix} 5 & 4 & 3 \\ 4 & 4 & 4 \\ 3 & 4 & 6 \end{bmatrix}.$$

Estimate the largest and smallest eigenvalues of A using the Rayleigh quotient. Compare these with estimates from Gershgorin's theorems.

Blind guessing and Lemma 1 are used for the Rayleigh quotient estimate. For example, $x = (1, 1, 1)^{\mathrm{T}}$ yields $R(x) = 37/3$, so $\lambda_1 \leq 37/3 \leq \lambda_3$. $x = (-1, 2, -1)^{\mathrm{T}}$ yields $R(x) = 1/6$, and now $\lambda_1 \leq 1/6 < 37/3 \leq \lambda_3$. Gershgorin's first theorem gives us $|\lambda| \leq \max\{\|A_i\|_1\} = 13$. The second theorem says that each eigenvalue is in one of the intervals $|\lambda - 5| \leq 7$, $|\lambda - 4| \leq 8$, $|\lambda - 6| \leq 7$, so the eigenvalues lie between -1 and 13. In summary, $-1 \leq \lambda_1 \leq 1/6$ and $37/3 \leq \lambda_3 \leq 13$. These are good bounds, since $\lambda_1 \approx 0.1287$ and $\lambda_3 \approx 12.36$, as will be demonstrated in the next section.

The result in Lemma 1 can be sharpened to determine when equality holds in the inequality (1) provided V is finite-dimensional.

THEOREM 2. Suppose $T : V \to V$ is Hermitian with respect to $\langle , \rangle$, $\dim(V) < \infty$, $\lambda_1 < \lambda_2 < \cdots \lambda_r$ are the distinct eigenvalues of T, and $R(x)$ is the Rayleigh quotient of T. Then

(a) $\lambda_1 \leq R(x) \leq \lambda_r \quad \forall x \neq 0, \; x \in V$.
(b) $R(x) = \lambda_i$ if $0 \neq x \in M(\lambda_i)$.
(c) $R(x) = \lambda_1$ iff x is an eigenvector associated with λ_1.
(d) $R(x) = \lambda_r$ iff x is an eigenvector associated with λ_r.

Proof

(a) This was proved in Lemma 1.
(b) If $Tx = \lambda_i x$, $x \neq 0$, $R(x) = \langle Tx, x \rangle / \|x\|^2 = \langle \lambda_i x, x \rangle / \|x\|^2 = \lambda_i$.
(c) By Theorem 4 of Section 5.2, $x = \sum_{i=1}^{r} v_i$, uniquely, with $v_i \in M(\lambda_i)$. Since $Tv_i = \lambda_i v_i$, $Tx = \sum_{i=1}^{r} \lambda_i v_i$. The $v_1, \ldots, v_r$ are orthogonal because they lie in different eigenspaces, so that Parseval's identity implies

$$R(x) = \frac{\langle Tx, x \rangle}{\|x\|^2} = \frac{\sum_{i=1}^{r} \lambda_i \|v_i\|^2}{\sum_{i=1}^{r} \|v_i\|^2}.$$

If one or more of the vectors $v_1, \ldots, v_{r-1} \neq 0$, then replacing $\lambda_1, \ldots, \lambda_{r-1}$ by λ_r will cause the numerator in $R(x)$ to strictly increase, so that

$$R(x) < \frac{\sum_{i=1}^{r} \lambda_r \|v_i\|^2}{\sum_{i=1}^{r} \|v_i\|^2} = \lambda_r \frac{\sum_{i=1}^{r} \|v_i\|^2}{\sum_{i=1}^{r} \|v_i\|^2} = \lambda_r.$$

Thus $x \notin M(\lambda_r) \Rightarrow R(x) < \lambda_r$. In other words, $R(x) = \lambda_r$ only if $0 \neq x = v_r \in M(\lambda_r)$.
(d) Similary, $R(x) > \lambda_1$ unless $0 \neq x \in M(\lambda_1)$.

Theorem 2 may be rephrased in terms of constrained minima and maxima. These are of special interest, partly because many physical laws are expressed in terms of minimum principles and partly because many mathematical problems are expressed in terms of determining minima and maxima.

Observe that the Rayleigh quotient is independent of the scaling of x:

$$R(cx) = \frac{\langle Tcx, cx \rangle}{\|cx\|^2} = \frac{c\bar{c}\langle Tx, x \rangle}{|c|^2 \|x\|^2} = R(x) \quad \forall c \neq 0.$$

Thus x can be normalized to have length 1 in the Rayleigh quotient. This observation allows the conversion of Theorem 2 into the next theorem.

THEOREM 3. If $T : V \to V$ is Hermitian with respect to $\langle , \rangle$, $\dim(V) < \infty$, $\lambda_r = \max \sigma(T)$, and $\lambda_1 = \min \sigma(T)$, then,

(a) The maximum of $\langle Tx, x \rangle$ subject to $\|x\| = 1$ is λ_r, and the maximum is attained at x iff $x \in M(\lambda_r)$, $\|x\| = 1$.
(b) The minimum of $\langle Tx, x \rangle$ subject to $\|x\| = 1$ is λ_1, and the minimum is attained at x iff $x \in M(\lambda_1)$, $\|x\| = 1$.

Theorem 2 can be extended to obtain other eigenvalues as the minima of certain constrained minimization problems, as will be seen next. The following theorem is also referred to as a "variational principle" by some authors. Of course, in the next two theorems, the vectors x in the minima could be normalized to $\|x\| = 1$ if desired.

EXTENDED MINIMUM PRINCIPLE. Suppose T is Hermitian on V with respect to $\langle\,,\,\rangle$, and $\{e_k\}_{k=1}^n$ is an orthogonal basis for V, with $Te_i = \lambda_i e_i$, and $\lambda_1 \le \lambda_2 \le \cdots \le \lambda_n$. Then

$$\lambda_1 = \min\{R(x) : x \ne 0\} \tag{2a}$$

$$\lambda_i = \min\{R(x) : x \ne 0,\ x \perp e_1, e_2, \ldots, e_{i-1}\} \quad \text{for } i = 2, 3, \ldots, n. \tag{2b}$$

Proof. Equation (2a) has been proved. Let $x = \sum_{j=1}^n x_j e_j$ satisfy the conditions in Eq. (2b). Since $x_j = \langle x, e_j\rangle/\|e_j\|^2 = 0$ for $j < i$, Parseval's identities yield

$$R(x) = \frac{\sum_{j=i}^n \lambda_j |x_j|^2 \|e_j\|^2}{\sum_{j=i}^n |x_j|^2 \|e_j\|^2} \ge \frac{\lambda_i \sum_{j=i}^n |x_j|^2 \|e_j\|^2}{\sum_{j=i}^n |x_j|^2 \|e_j\|^2} \ge \lambda_i,$$

which follows from replacing each λ_j by the smaller number λ_i, just as in the proof of Lemma 1. If $x = e_i$, then $x \perp e_1, e_2, \ldots, e_{i-1}$ and $R(x) = \lambda_i$. An alternative proof is to set $N = \text{span}\{e_1, e_2, \ldots, e_{i-1}\}$, and notice that $T : N^\perp \to N^\perp$ (i.e., $x \in N^\perp$ implies $T(x) \in N^\perp$). Now apply Theorem 2 to T restricted to $N^\perp$. This implies Eq. (2b) because the restricted operator has the smallest eigenvalue λ_i.

This theorem generalizes to infinite-dimensional spaces when the Hermitian operator T has an "infinite-orthogonal basis $\{e_i\}_1^\infty$ of eigenvectors" and the corresponding eigenvalues satisfy $\lambda_1 \le \lambda_2 \le \cdots \le \lambda_n \to \infty$. To say that $\{e_i\}_1^\infty$ is an "infinite-dimensional orthogonal basis" means that (1) $\{e_i\}_1^\infty$ is orthogonal with respect to $\langle\,,\,\rangle$, and (2) every x has an expansion $x = \sum_{i=1}^\infty x_i e_i$, which is valid in the sense that $\|x - \sum_{i=1}^n x_i e_i\| \to 0$. Sequences $\{e_i\}_1^\infty$ satisfying these properties are also called *complete orthogonal sequences*, and it can be shown that the coefficients x_i are exactly the Fourier coefficients of x: $x_i = \langle x, e_i\rangle/\|e_i\|^2$. In order to ensure the existence of such an $\{e_i\}_1^\infty$, it is not enough that T be Hermitian. The interested reader is invited to look up "compact operators" in a book discussing Hilbert spaces. (See the reference list.)

There is also an extended maximum principle in finite-dimensional spaces.

EXTENDED MAXIMUM PRINCIPLE. Suppose T is Hermitian on V with respect to $\langle\,,\,\rangle$, and $\{e_k\}_{k=1}^n$ is an orthogonal basis, with $Te_i = \lambda_i e_i$ and $\lambda_1 \le \lambda_2 \le \cdots \le \lambda_n$. Then

$$\lambda_n = \max\{R(x) : x \ne 0\} \tag{3a}$$

$$\lambda_i = \max\{R(x) : x \ne 0,\ x \perp e_n, e_{n-1}, \ldots, e_{i+1}\} \quad \text{for } i = 1, 2, \ldots, n-1. \tag{3b}$$

Proof. The proof is left as an exercise.

A natural question that arises is: What if the eigenvectors were replaced by another set of vectors in the extended minimum and maximum principles? The next theorem asserts that the corresponding minimum in Eq. (2b) is no larger than λ_i and the corresponding maximum in Eq. (3b) is no smaller than λ_i. This gives new estimates for the eigenvalues of a Hermitian operator. These estimates are then applied to a problem in matrix theory.

COURANT–FISCHER THEOREM. Let T be Hermitian on V with respect to $\langle\,,\,\rangle$, $\beta = \{e_k\}_{k=1}^n$ be an orthogonal set in V, $T(e_i) = \lambda_i e_i$, and $\lambda_1 \le \lambda_2 \le \cdots \le \lambda_n$.

(a) For arbitrary $f_1, f_2, \ldots, f_{p-1}$, define

$$m(f_1, f_2, \ldots, f_{p-1}) = \min\{R(x) : x \neq 0, x \perp f_1, f_2, \ldots, f_{p-1}\}.$$

Then

$$m(f_1, f_2, \ldots, f_{p-1}) \leq \lambda_p. \tag{4}$$

Equality holds if β is a basis and $f_i = e_i$ for $i = 1, 2, \ldots, p - 1$.

(b) For arbitrary $f_1, f_2, \ldots, f_{n-p}$, define

$$M(f_1, f_2, \ldots, f_{n-p}) = \max\{R(x) : x \neq 0, x \perp f_1, f_2, \ldots, f_{n-p}\}.$$

Then

$$M(f_1, f_2, \ldots, f_{n-p}) \geq \lambda_p. \tag{5}$$

Equality holds if β is a basis and $f_i = e_{i+p}, i = 1, 2, \ldots, n - p$.

When β is a basis, the results can be expressed more succinctly as:

$$\max\{\min\{R(x) : x \neq 0, x \perp f_1, f_2, \ldots, f_{p-1}\} : f_1, f_2, \ldots, f_{p-1}\} = \lambda_p$$

$$\min\{\max\{R(x) : x \neq 0, x \perp f_1, f_2, \ldots, f_{n-p}\} : f_1, f_2, \ldots, f_{n-p}\} = \lambda_p.$$

Proof

(a) Choose $z = \sum_{i=1}^{p} x_i e_i$ so that $x \perp f_1, f_2, \ldots, f_{p-1}, x \neq 0$. To see that this may be done, consider the $p - 1$ linear equations $\langle z, f_j \rangle = \sum_{i=1}^{p} z_i \langle e_i, f_j \rangle = 0, j = 1, \ldots, p - 1$. Since there are more unknowns than equations, there is a nontrivial solution for the z_i, and $z \neq 0$ because $\{e_1, \ldots, e_p\}$ is linearly independent. Observe that $m(f_1, f_2, \ldots, f_{p-1}) \leq R(z)$ by the definition of m, and that $R(z) \leq \lambda_p$ by Lemma 1 (with $n = p$). Thus $m(f_1, f_2, \ldots, f_{p-1}) \leq R(z) \leq \lambda_p$. However, if $f_i = e_i, i = 1, 2, \ldots, p - 1$, then $m(f_1, f_2, \ldots, f_{p-1}) = \lambda_p$ by the extended minimum principle.

(b) The proof is left as an exercise.

The next theorem is an application of the last theorem to matrix theory. It relates the eigenvalues of a Hermitian matrix to those of a special $n - 1 \times n - 1$ submatrix, and is called an interlacing theorem because of the way the eigenvalues are related.

INTERLACING THEOREM. Let A be an $n \times n$ Hermitian matrix and B be the $(n - 1) \times (n - 1)$ submatrix formed by deleting the last row and column of A. Suppose A has eigenvalues $\lambda_1 \leq \lambda_2 \leq \cdots \leq \lambda_n$ and B has eigenvalues $\mu_1 \leq \mu_2 \leq \cdots \leq \mu_{n-1}$. Then $\lambda_i \leq \mu_i \leq \lambda_{i+1}$.

Proof. First $\lambda_1 \leq \mu_1 \leq \lambda_2$ will be shown. Let $y = (x_1, x_2, \ldots, x_{n-1})^T$, and consider the Rayleigh quotients of B and A: $R_1(y) = y^H B y / \|y\|^2$ and $R(x) = x^H A x / \|x\|^2$. Let $\tilde{y} = (x_1, x_2, \ldots, x_{n-1}, 0)^T$, and note that $R_1(y) = R(\tilde{y})$ for all $y \neq 0$. By Lemma 1, $R_1(y) = R(\tilde{y}) \geq \lambda_1$ for all y, so that $\mu_1 = \min\{R_1(y) : y \neq 0\} \geq \lambda_1$ by (2a). Now take $f_1 = \delta_n$ in (4) to find that $\lambda_2 \geq m(\delta_n) = \min\{R(x) : 0 \neq x \perp \delta_n\} = \min\{R_1(y) : 0 \neq y\} = \mu_1$. Next $\lambda_2 \leq \mu_2 \leq \lambda_3$ will be shown. Let e_1 and f_1 denote eigenvectors of A and B corresponding to λ_1 and μ_1, respectively, and let E_1 be obtained from e_1 by setting the last component to 0. Then $\mu_2 \geq \min\{R_1(y) : y \perp E_1\} = \min\{R(\tilde{y}) : \tilde{y} \perp e_1\} \geq \lambda_2$ using (4) and Lemma 1. On the other hand, $\lambda_3 \geq \min\{R(x) : x \perp \delta_n, x \perp \tilde{f}_1\} = \min\{R_1(y) : y \perp f_1\} = \mu_2$, using (4) and the extended minimum principle. The rest of the proof proceeds in the same way.

For an extensive discussion of applications of the Courant–Fischer theorem, the reader is invited to peruse Horn and Johnson (1985).

Exercises 5.4

1. For the following matrices, estimate the eigenvalues using Gershgorin's theorems and the Rayleigh quotient.

    (a) $\begin{bmatrix} 1 & -1 & 1 \\ -1 & 10 & 9 \\ 1 & 9 & 11 \end{bmatrix}$

    (b) $\begin{bmatrix} 3 & -5 & 2 \\ -5 & 9 & -4 \\ 2 & -4 & 3 \end{bmatrix}$

    (c) $\begin{bmatrix} 1 & 1 & 0 \\ 1 & 1 & 1 \\ 0 & 1 & 10 \end{bmatrix}$

    (d) $\begin{bmatrix} 9 & 3 & 7 \\ 3 & 2 & 4 \\ 7 & 4 & 8 \end{bmatrix}$

2. $A = \begin{bmatrix} 6 & 1 \\ -5 & 0 \end{bmatrix}$ has eigenvalues $1, 5$. Find vectors x and y so that $R(x) > 5$ and $R(y) < 1$.

3. Let A be an $n \times n$ Hermitian matrix with an orthogonal eigenbasis $\{e_k\}_{k=1}^n$ with $Ae_i = \lambda_i e_i$ and $\lambda_1 \le \lambda_2 \le \cdots \le \lambda_n$.
    (a) Show that $\lambda_1 \le a_{ii} \le \lambda_n$, $i = 1, 2, \ldots, n$.
    (b) Show that $\lambda_1 \le (1/n) \sum_{i,j=1}^n a_{ij} \le \lambda_n$.

4. Let A be an $n \times n$ Hermitian matrix with columns A_k and eigenvalues λ_k indexed so that $|\lambda_1| \le |\lambda_2| \le \cdots \le |\lambda_n|$. Show
    (a) $\mathrm{Tr}(A^2) = \sum_{i=1}^n \lambda_i^2$
    (b) $\mathrm{Tr}(A^H A) = \sum_{i=1}^n \|A_i\|^2$
    (c) $\lambda_1^2 \le (1/n) \sum_{i=1}^n \|A_i\|^2 \le \lambda_n^2$

5. Let A be an $n \times n$ Hermitian matrix with an orthogonal eigenbasis $\{e_1, e_2, \ldots, e_n\}$ with $Ae_i = \lambda_i e_i$ and $0 \le \lambda_1 \le \lambda_2 \le \cdots \le \lambda_n$. Find the max and min of $\|Ax\|^2/\|x\|^2$ and the vectors at which the max and min are attained.

6. Let A be an $n \times n$ Hermitian matrix with $\lambda_n > |\lambda_i|$ if $i \ne n$. Find

$$\mu \equiv \max \left\{ \frac{1}{\|x\|^2} : \langle Ax, x \rangle = 1 \right\}.$$

7. Let $A = LDL^T$, where L is a real lower triangular matrix, $D = \mathrm{diag}(d_1, \ldots, d_n)$ and $l_{kk} = 1$, $d_k > 0$ for $k = 1, 2, \ldots, n$. Let λ_1, λ_n be the smallest and largest eigenvalues of A, and

$d_b = \max d_i$, $d_s = \min d_i$. Show

(a) $\lambda_1 > 0$.

(b) $\lambda_1 \le d_s/\|y_s\|^2$, where y_s is the solution of $L^T y_s = \delta_s$.

(c) $\lambda_n \ge d_b\|x_b\|^2$, where $x_b = L\delta_b$.

(d) $\kappa \equiv \lambda_n/\lambda_1 \ge d_b/d_s$.

(*Hints:* (a) Consider $R(x)$. (c) Consider the Rayleigh quotient for A^{-1}.) The quantity κ is called the *condition number* of A. It estimates how well $Ax = y$ can be solved numerically. The smaller, the better; about one digit of accuracy may be lost for every power of 10 in κ. To illustrate this, suppose A is Hermitian with $0 < \lambda_1 < \lambda_2 < \cdots < \lambda_n$ with corresponding orthonormal eigenvectors e_i. Consider $Ax = y$. If $y = e_n$, $x = e_n/\lambda_n$. If y is perturbed slightly, $y' = e_n + \alpha e_1$, $Ax' = y'$ has the solution $x' = x + \alpha e_1/\lambda_1$. The relative error in y is $\|y' - y\|/\|y\| = |\alpha|$ (small), while the relative error in x is $\|x' - x\|/\|x\| = \kappa|\alpha| \gg |\alpha|$, if $\kappa \gg 1$.

8. Let A be an $n \times n$ symmetric matrix with $a_{ij} > 0$ $\forall i, j$, and let λ_1 be the largest eigenvalue of A. Show that (i) $\lambda_1 > |\lambda_i|$ if $i \ne 1$, (ii) $M(\lambda_1)$ is one-dimensional, and (iii) there is an eigenvector u with positive components such that $Au = \lambda_1 u$.

   Outline of a proof: Let $Ae = \lambda_1 e$, $e \ne 0$, and $u = (|e_1|, |e_2|, \ldots, |e_n|)^T$. Show

   (a) $Au = \lambda_1 u$

   (b) $\lambda_1 > 0$

   (c) $\lambda_1 > |\lambda_i|$ if $i \ne 1$

   (d) $\dim(M(\lambda_1)) = 1$

   (e) $u_i > 0$ for every i

   Which of (a)–(d) remain true if $A \ne 0$ and only $a_{ij} \ge 0$ is assumed?

9. Let $T : \mathbf{P}_2 \to \mathbf{P}_2$ and an inner product be defined by

$$Tp(x) = \frac{1}{8}\int_{-1}^{1} \{45x^2t^2 - 15(x^2 + t^2) - 6xt + 6(x + t) + 7\}p(t)\,dt$$

$$\langle f, g \rangle = \int_{-1}^{1} f(t)\overline{g(t)}\,dt.$$

   Find the maximum and minimum of $\langle Tp, p \rangle$ subject to $\|p\| = 1$ and the vectors at which the max and min are attained.

10. Let $T : \mathbf{P}_2 \to \mathbf{P}_2$ be defined by

$$Tp(x) = \frac{d}{dx}[(1 - x^2)p'(x)] + 3p(x).$$

   Find the maximum and minimum of

$$\int_{-1}^{1} Tg(t)\overline{g(t)}\,dt \quad \text{subject to} \quad \int_{-1}^{1} |g(t)|^2\,dt = 1, g \in \mathbf{P}_2$$

   and the vectors at which the max and min are attained.

11. Prove the extended maximum principle.

12. Prove part (b) of the Courant–Fischer theorem.

5.5
THE POWER METHOD

In this section, a simple method to approximate some of the eigenvalues and eigenvectors of Hermitian operators on finite-dimensional spaces is presented. At the end of the section a "subspace" method for approximating several eigenvalues and eigenvectors simultaneously will be given.

Estimating the Eigenvalue of Largest Magnitude

Assume $T : V \to V$ is Hermitian with respect to $\langle , \rangle$, $\dim(V) < \infty$, and let $\sigma(T) = \{\lambda_1, \ldots, \lambda_r\}$. Assume that $|\lambda_1| > |\lambda_j|$, if $j > 1$. Such a λ_1 is called the *dominant eigenvalue of T*. Note that, if $T \neq 0$ and all $\lambda_i \geq 0$, then a dominant eigenvalue exists, namely, $\lambda_1 = \max \lambda_i$. When the eigenvalues are of mixed sign and (say) $\lambda_1 = -\lambda_2 \geq |\lambda_i|$ for $i > 2$, then there is no dominant eigenvalue. In many physical problems involving oscillation, all the eigenvalues are nonnegative.

Select a vector x_0. By Theorem 4 of Section 5.2, x_0 has the expansion

$$x_0 = \sum_{i=1}^{r} v_i \quad \text{for some } v_i \in M(\lambda_i).$$

Since $T(v_i) = \lambda_i v_i$, $T(x_0) = \sum_{i=1}^{r} \lambda_i v_i$, and after applying T n times, one obtains (cf. Theorem 2 of Section 4.2)

$$T^n(x_0) = \sum_{i=1}^{n} \lambda_i^n v_i. \tag{*}$$

Assume $v_1 \neq 0$. Geometrically, the condition $|\lambda_1| > |\lambda_j|$ if $j > 1$ means that T produces a larger expansion along the v_1 component of x relative to the expansions along the other components v_i. Repeatedly applying T enhances this effect, so that for large n, the dominant term in the sum (*) is the eigenvector $\lambda_1^n v_1$, since $|\lambda_1|^n \gg |\lambda_i|^n$. Thus for large n,

$$T^n(x_0) \approx \lambda_1^n v_1 \in M(\lambda_1)$$

and
$$R(T^n(x_0)) \approx R(\lambda_1^n v_1) = \lambda_1.$$

This can be made into an iteration as follows: Let $x_{n+1} = T(x_n)$, $n = 0, 1, \ldots$. Then $x_1 = T(x_0)$, $x_2 = TT(x_0) = T^2(x_0), \ldots, x_n = T^n(x_0)$, etc. An algorithm based on this iteration is

Choose x_0 and set $n = 0$.
Given n, x_n:
    Compute $x_{n+1} = T(x_n)$ and $R_{n+1} = R(x_n) = \langle x_{n+1}, x_n \rangle / \|x_n\|^2$.
Print $n + 1$, R_{n+1}, x_{n+1}

Remark. $R_n \leq \lambda_1$ for every n. This follows from Lemma 1 of Section 5.4. Thus the iteration provides better lower bounds for λ_1 as it proceeds.

Convergence and rate of convergence

The following will be established: Usually $R_n \to \lambda_1$, and for a suitable normalization y_n of x_n, $y_n \to v_1 \in M(\lambda_1)$, with the convergence of R_n to λ_1 being substantially faster.

Computing the Rayleigh quotient of x_n gives us

$$R(x_n) = \frac{\langle x_{n+1}, x_n \rangle}{\|x_n\|^2} = \frac{\sum_{i=1}^r \lambda_i^{2n+1} \|v_i\|^2}{\sum_{i=1}^r \lambda_i^{2n} \|v_i\|^2}. \tag{1}$$

If $v_1 \neq 0$, that is, x_0 has a component in the eigenspace $M(\lambda_1)$, then factor out the highest power of λ_1 in the numerator and denominator and set $\theta_i = \lambda_i/\lambda_1$, to get

$$R(x_n) = \lambda_1 \frac{\|v_1\|^2 + \sum_{i>1}(\theta_i)^{2n+1}\|v_i\|^2}{\|v_1\|^2 + \sum_{i>1}(\theta_i)^{2n}\|v_i\|^2} = \lambda_1(1 - c_n\theta^{2n}), \tag{2}$$

where c_n is bounded, and $\theta = \max\{|\theta_i| : i > 1\}$. Thus it is clear that $R(x_n) \to \lambda_1$, since $|\theta_i| \leq \theta < 1$ for all $i > 1$. The "more dominant" λ_1 is, i.e., the smaller θ is, the faster the θ_i^{2n} converge to 0, and the faster $R(x_n)$ approaches λ_1. Equation (2) should be interpreted as "$R(x_n) \to \lambda_1$ with order θ^{2n}."

Now let us do the same kind of calculation with the iterates x_n:

$$x_n = \lambda_1^n \left(v_1 + \sum_{i>1} \theta_i^n v_i \right) = \lambda_1^n(v_1 + \theta^n d_n), \tag{3}$$

where d_n is a bounded sequence of vectors. From this it is clear that

$$x_n \approx \lambda_1^n v_1 \quad \text{for large } n. \tag{4}$$

The sequence x_n may not converge. For example, if $\lambda_1 = 2$, $x_n \approx 2^n v_1$ is unbounded; if $\lambda_1 = -1$, $x_n \approx (-1)^n v_1$ oscillates; and if $\lambda_1 = -2$, $x_n \approx (-2)^n v_1$ oscillates unboundedly. In any case, x_n is approximately a (variable) scalar times v_1 for large n. The language "x_n becomes more and more parallel to v_1", or "span $\{x_n\} \to$ span$\{e_1\}$" will be used to describe this situation. To obtain a convergent sequence of vectors, consider the behavior of the normalized vectors $z_n = x_n/\|x_n\|$. If $\lambda_1 > 0$, then $z_n \to v_1/\|v_1\|$, which is a unit eigenvector corresponding to the eigenvalue λ_1. If $\lambda_1 < 0$, then for large n, $z_n \approx \pm v_1/\|v_1\|$, with opposite signs on successive iterations. To obtain convergence in each case, let sgn(x) denote the sign of x. Then

$$(\text{sgn}(\lambda_1))^n \frac{x_n}{\|x_n\|} \to \frac{v_1}{\|v_1\|}. \tag{5}$$

Note that sgn$(\lambda_1) = $ sgn$(R(x_n))$ eventually.

As for the rate of convergence, the smaller the $|\theta_i|$ are, the faster is the convergence; but in this case the error in the normalized convergence (5) is on the order of θ^n, not θ^{2n}. One usually observes this in practice: The approximate eigenvectors converge more slowly than the approximate eigenvalues; it takes about twice as many steps to get the same accuracy in the eigenvectors.

EXAMPLE 1. THE POWER METHOD APPLIED TO A MATRIX OPERATOR $T(x) = Ax$,
WHERE

$$A = \begin{bmatrix} 5 & 4 & 3 \\ 4 & 4 & 4 \\ 3 & 4 & 6 \end{bmatrix}.$$

(Cf. Example 1 of Section 5.4.) Let $x_{n+1} = (a_n, b_n, c_n)^\mathrm{T} = Ax_n$ in the preceding iteration, with starting vector $x = x_0 = (1, 1, 1)^\mathrm{T}$. Then one obtains

n	R_n	a_n	b_n	c_n
1	12.3333	12	12	13
2	12.3545	147	148	162
3	12.3553	1813	1828	2005
4	12.3553	22,392	22,584	24,781
5	12.3553	276,639	279,028	306,198

Note that R_n converges very rapidly in this case. It is hard to see the "convergence" of y_n; i.e., it is hard to see that the y_n become "more parallel" to some fixed vector. Eventually overflow results if a computer is used.

Normalization can prevent the overflow and allow us to see the convergence of the x_n. The following revised iteration is suggested. At each step, y_n is the current estimate of the eigenvector. At each step, some normalizing weights $w_n \neq 0$ are chosen, and the next approximate eigenvector is $w_n T(y_n)$. The iteration produces a sequence y_n so that $R(y_n) \to \lambda_1$, and y_n become more and more parallel to an eigenvector. One way to write such an algorithm is as follows:

Pick $y_0 \neq 0$ and set $n = 0$.
Given n, y_n:
    Calculate $z = T(y_n)$ and $R_{n+1} = \langle z, y_n \rangle / \|y_n\|^2$.
Choose w_n and set $y_{n+1} = w_n z$.
Print $n + 1$, R_{n+1}, y_{n+1}.

There are many ways to choose the weights. For example, one may use $w_n = 1/\|z\|$, so that all the y_n are of length 1. In case $V = \mathbf{F}^n$, one could choose w_n so that a fixed entry of the column y_n is 1 (if that entry is not 0) or so that the entry of largest magnitude is 1. However the w_n are chosen, is easy to see that $y_n = c_n T^n(z_0)$, with $c_n \neq 0$. In fact, c_n is the product of $w_1, \ldots, w_n$. Since the Rayleigh quotient does not depend on the scaling, $R(y_n) = R(T^n(z_0)) = R_{n+1} \to \lambda_1$. Also, y_n "becomes more and more parallel to v_1," in the sense that a suitable normalization of y_n will converge to an eigenvector in $M(\lambda_1)$ as $n \to \infty$.

EXAMPLE 2. THE POWER METHOD WITH NORMALIZATION TO UNIT VECTORS.
Use the matrix A of Example 1. The y_n are normalized to length 1. In the table, $y_n = (a_n, b_n, c_n)^\mathrm{T}$. The same starting vector was used as in Example 1.

n	R_n	a_n	b_n	c_n
1	12.3333	0.561336	0.561336	0.608114
2	12.3545	0.556575	0.560361	0.613368
3	12.3552	0.555587	0.560184	0.614425
4	12.3553	0.555385	0.560148	0.614639
5	12.3553	0.555344	0.560140	0.614683
6	12.3553	0.555336	0.560139	0.614692
7	12.3553	0.555334	0.560138	0.614694
8	12.3553	0.555334	0.560138	0.614694

Observe that the components of y_n converge about half as fast as the Rayleigh quotients. This is exactly what Eqs. (2) and (3) predict.

If $v_1 = 0$ and the computations are done in exact arithmetic, R_n will converge to a different eigenvalue; this is possible but rare. It is not easy to pick x_0 so that $v_1 = 0$, and even if this happened, roundoff error will usually produce an x_1 with small component v_1 in the eigenspace $M(\lambda_1)$. After a slow start, the convergence $R_n \to \lambda_1$ will be observed.

Approximation of the Smallest Eigenvalue in Magnitude

Assume T is nonsingular and $|\lambda_r| < |\lambda_i|$, $1 \geq i > r$. Since T^{-1} has eigenvalues λ_i^{-1} with the same eigenvectors (and eigenspaces) as T, λ_r^{-1} is the dominant eigenvalue of T^{-1}. The power method may be applied to T^{-1} to recover λ_r^{-1} and a corresponding eigenvector e_r.

The iteration without normalization becomes

Choose x_0, and set $n = 0$.
Given n, x_n:
  Compute $x_{n+1} = T^{-1}(x_n)$ and $R_{n+1} = \langle x_{n+1}, x_n \rangle / \|x_n\|^2$.
Print $n + 1$, R_{n+1}, x_{n+1}.

The result is: $R_n \leq \lambda_r^{-1}$ and $R_n \to 1/\lambda_r$, while x_n becomes "more parallel to" some $e_r \in M(\lambda_r)$.

For obvious reasons, this is called the *inverse power method.*

EXAMPLE 3. Let $T(x) = Ax$, where A is the matrix of the previous examples. Apply the inverse power method to find $\lambda_r = \lambda_3$ and a corresponding eigenvector. Here

$$A^{-1} = \begin{bmatrix} 2 & -3 & 1 \\ -3 & 5.25 & -2 \\ 1 & -2 & 1 \end{bmatrix}.$$

Using the starting vector $(1, -2, 1)^T$, after four iterations we have about five digits of accuracy in the eigenvectors, and $R_n = 7.771614$, $e_3 \approx (0.48016, -0.81946, 0.31294)^T$. $\lambda_3 \approx 1/R_n = 0.128673$.

Remark. When $T(x) = Ax$, where A is a large matrix, and a computer is used, the inverse A^{-1} should not be computed. Rather, the system $Ax_{n+1} = x_n$ should be solved for x_{n+1}, because there is less error in appropriately done Gauss elimination. This is not as bad as it seems, since much of the Gauss elimination need not be repeated after the computation of x_1. Many linear algebra packages save an *LU factorization* of the matrix, which makes solving $Ax = y$ for subsequent y much faster. Consult the manual to see how this is done in your package.

Sharpening Good Estimates of Eigenvalues and Eigenvectors

Suppose μ is a "good estimate" of one of the eigenvalues λ_p of T. Such an estimate may come from a variety of sources, including another numerical method to get eigenvalues. Consider the operator $S = T - \mu I$. This operator has the same eigenvectors as T but has the eigenvalues $\lambda_i - \mu$. Thus if the estimate μ is good enough so that $0 < |\mu - \lambda_p| \ll |\mu - \lambda_i|$ when $\lambda_i \neq \lambda_p$, then the inverse power method may be applied to S to recover the smallest eigenvalue $\lambda_p - \mu$ of S and a corresponding eigenvector e. In that algorithm, $R_n \to \rho = 1/(\lambda_p - \mu)$ Therefore, $\lambda_p = \mu + 1/\rho$, and e is the corresponding eigenvector. This occurs in *very* few iterations if the estimate is good. This is called the *Shifted inverse power method.*

EXAMPLE 4. Let

$$A = \begin{bmatrix} 1 & -1 & 1 \\ -1 & 6 & 4 \\ 1 & 4 & 7 \end{bmatrix}.$$

Estimate the dominant eigenvalue, and find it and a corresponding eigenvector using the shifted inverse power method.

Gershgorin's first theorem gives us $|\lambda| \leq 12$. Trying $x = (1, 1, 1)^T$, $Ax = (1, 9, 12)^T$ and $R(x) = x^T Ax/\|x\|^2 = 22/3$. Try again with $x = (0, 1, 1)^T$ (which resembles Ax as just computed), and get $Ax = (0, 10, 11)^T$ and $R(x) = 10.5$. Thus $10.5 \leq \lambda_1 \leq 12$. Now try inverse power iteration with $\mu = 10.5$. Take the starting vector to be $(0, 1, 1)^T$. The second and third iterations have Rayleigh quotients that agree to seven places and $R_3 = 31.28856$. Thus $\lambda_1 \approx 10.5 + 1/R_3 = 10.53196$. The approximate unit eigenvector is $x_3/\|x_3\| = (0.00946, 0.66053, 0.75074)^T$. This agrees with $x_2/\|x_2\|$ to five places.

EXAMPLE 5. Let A be the matrix in Example 4. Suppose a graphing calculator was used to plot the characteristic polynomial of A, and it was discovered that the middle eigenvalue λ_2 is about 3.33. Improve the accuracy of λ_2 and find the corresponding eigenvector.

Use the shifted inverse power method with $\mu = 3.33$ and starting vector $(0, -1, 1)^T$, which is approximately perpendicular to the eigenvector found in Example 4. The second and third iterations have Rayleigh quotients that agree to six places: $R_3 = -211.438$. Thus $\lambda_2 \approx 3.33 + 1/R_3 = 3.32527$. The approximate unit eigenvectors for the second and third iterations agree to five places. An approximate unit eigenvector is $x_3/\|x_3\| = (0.51871, -0.64510, 0.56105)^T$.

Approximation of Secondary Eigenvalues and Eigenvectors

Having found e_1 and λ_1 approximately, one can proceed to approximate the next dominant eigenvalue λ_2 and corresponding eigenvector e_2 using the power method, as follows. Since $e_2 \perp e_1$, the iteration ought to be started with $x_0 \perp e_1$. The following proposition shows what happens.

PROPOSITION 1. If T is Hermitian and e_1 is an eigenvector of T, then $T : \{e_1\}^\perp \to \{e_1\}^\perp$. Hence if $x_0 \perp e_1$, then all the iterates $x_{n+1} = T(x_n)$ are orthogonal to e_1.

Proof. If x is orthogonal to e_1, then $\langle T(x), e_1 \rangle = \langle x, T(e_1) \rangle = \langle x, \lambda_1 e_1 \rangle = \lambda_1 \langle x, e_1 \rangle$. Hence $T(x)$ is orthogonal to e_1, which proves $T : \{e_1\}^\perp \to \{e_1{}^\perp\}$. Since $x_0 \perp e_1$, by what was just proved, $x_1 = T(x_0) \perp e_1$. Next, $x_2 = T(x_1) \perp e_1$, and so on: If $x_n \perp e_1$, then $x_{n+1} = T(x_n) \perp e_1$.

Now suppose $x_0 \perp e_1$, and the iteration $x_{n+1} = T(x_n)$ is performed. By the proposition, all $x_n \perp e_1$. Since $T : V_1 \to V_1$, where $V_1 = \{e_1\}^\perp$, the original theory applies to T restricted to V_1, and again it may be concluded that x_n becomes more parallel to an eigenvector and $R(x_n)$ tends to the dominant eigenvalue in V_1, if one exists. Next, the cases when λ_1 is simple or degenerate are more carefully examined. The method adduced here is called *suppression of the dominant eigenvector*.

Case 1: λ_1 is a simple eigenvalue: $M(\lambda_1) = \text{span}\{e_1\}$.

Assume there is a second dominant eigenvalue, i.e., $|\lambda_2| > |\lambda_i|$, for $r \geq i > 2$; then the maximum eigenvalue of T on $V_1 = M(\lambda_1)^\perp$ is λ_2, and if x_0 contains a component in $M(\lambda_2)$, then $R(x_n) \to \lambda_2$, and x_n becomes more parallel to $v_2 \in M(\lambda_2)$ as $n \to \infty$. Two problems usually arise when a computer is used: The computed e_1, call it $\tilde{e}_1$, is not exactly e_1, and the computed x_n are not exactly orthogonal to e_1. These problems may be remedied by the following devices. First, $\tilde{e}_1$ should be computed accurately. Next, on each step, the projection (component) of $T(x_n)$ along the $\tilde{e}_1$ axis should be removed to keep x_{n+1} orthogonal to e_1 as nearly as possible:

$$x_{n+1} = T(x_n) - (\langle Tx_n, \tilde{e}_1 \rangle / \|\tilde{e}_1\|^2) \tilde{e}_1.$$

Note that the component along e_1, the actual eigenvector, is not actually being removed but merely suppressed. These devices will work if the eigenvalues are not too closely spaced near λ_1 and $\tilde{e}_1$ is a good approximation to e_1.

Case 2: λ_1 is a degenerate eigenvalue: $\dim M(\lambda_1) > 1$.

In this case, the dominant eigenvalue of T on V_1 is still λ_1. Thus the Rayleigh quotients $R(x_n)$ approach λ_1, and the x_n become more parallel to some $e_2 \in M(\lambda_1)$, with e_2 orthogonal to e_1. Thus a second independent eigenvector in the same eigenspace is obtained. Subtracting out the approximate $\tilde{e}_1$ component at each step does no harm in this case, so the same algorithm works.

Here is an algorithm for λ_2, e_2 with normalization at each step. It does not require the starting vector to be chosen orthogonal to e_1 beforehand.

Given: an approximate eigenvector $\tilde{e}_1$.
Choose a starting vector z_0, and set $n = 0$.
Given n, z_n:
  Compute $\alpha = \langle z_n, \tilde{e}_1 \rangle / \|\tilde{e}_1\|^2$.
Set $z_{n+1} = (z_n - \alpha \tilde{e}_1) / \|z_n - \alpha \tilde{e}_1\|$.
  Compute $w = T(z_{n+1})$ and $R_{n+1} = R(z_{n+1}) = \langle w, z_{n+1} \rangle$.
Print $n + 1, R_{n+1}, z_{n+1}$.
Set $z_{n+1} = w$.

EXAMPLE 6. COMPUTATION OF THE SECOND DOMINANT EIGENVALUE AND EIGENVECTOR. The matrix of Examples 1 and 2 is used. It was found that $\tilde{e}_1 = (0.555334, 0.560138, 0.614694)^T$, which is a unit vector. The starting vector $z_0 = (1, -1, 1)^T$ is used. In the first step, $\alpha = 0.60989$, $z_1 = (0.407932, -0.827589, 0.385600)^T$, $w = (-0.113897, -0.136229, 0.227039)^T$, and $R_1 = 0.153826$. Set $z_1 = w$ and repeat. In the next step, $\alpha = 1.516 \times 10^{-6}$, confirming Proposition 1 (approximately), and the correction is only in the trailing digits. This is typical of the subsequent steps. In the following table, $z_n = (a_n, b_n, c_n)^T$ is printed.

n	R_n	a_n	b_n	c_n
1	0.1538	0.407932	−0.827589	0.385600
2	2.0453	−0.395160	−0.472641	0.787692
3	2.5145	−0.666621	−0.142094	0.731728
4	2.5160	−0.678382	−0.122432	0.724437
5	2.5160	−0.678973	−0.121424	0.724052
6	2.5160	−0.679003	−0.121372	0.724033
7	2.5160	−0.679005	−0.121370	0.724032
8	2.5160	−0.679005	−0.121369	0.724032
9	2.5160	−0.679005	−0.121369	0.724032

The Fourier coefficients α of the correction are on the order of roundoff error after the first correction. One could possibly correct less often, but if no correction were made after the one at $n = 0$, at $n = 8$ we would see the beginning of a very rapid change back toward the dominant eigenvalue.

The same trick can be applied after e_1 and e_2 are found approximately, and so on. However, at some stage, approximation error in the $\tilde{e}_i$ and roundoff error will make it extremely difficult to keep the x_n from having a component along the dominant eigenaxis, and the strong (relative) magnification along this axis will contaminate the calculations to a great extent and prevent convergence to another eigenvalue. This even happens for relatively small matrices, e.g., 15×15 matrices, after just a few eigenvectors have been found. Use of higher precision helps, but it only delays the inevitable. Another method with which the next few dominant eigenvalues may be computed is presented in the exercises.

A Subspace Method for Several Eigenvectors and Eigenvalues

There is a whole family of subspace methods for several eigenvalues and eigenvectors. Only one will be presented. This method is similar to the power method, but is applied to a set of vectors simultaneously. At every step, the new set of approximate eigenvectors is orthogonalized to keep the eigenvectors from collapsing into the dominant eigenspace. One may also think of the process as "suppression on the fly." The method presented here is also called the orthogonal iteration method. See Golub and Van Loan (1989).

The simultaneous iteration method

Let T be a Hermitian operator with eigenvalues $\lambda_1 \geq \lambda_2 \geq \cdots \geq \lambda_n > 0$.
Initialization: Select p orthonormal vectors $e_1^{(0)}, e_2^{(0)}, \ldots, e_p^{(0)}$ as starting vectors.
Step k: Suppose $e_1^{(k)}, e_2^{(k)}, \ldots, e_p^{(k)}$ have been computed.
Compute $f_i = Te_i^{(k)}, i = 1, 2, \ldots, p$.
Find an orthonormal basis $\beta^{(k+1)} = \{e_1^{(k+1)}, e_2^{(k+1)}, \ldots, e_p^{(k+1)}\}$ for span($f_1, f_2,$
$\ldots, f_p$) using the Gram–Schmidt method as described in (4) of Section 3.2.

This iteration produces

$$e_i^{(k)} \to v_i \in M(\lambda_i) \quad i = 1, 2, \ldots, p,$$

$$r_i^{(k+1)} = \langle e_i^{(k+1)}, f_i \rangle \to \lambda_i.$$

Notes

1. Any method of construction of $\beta^{(k+1)}$ could be used, as long as

   $$e_i^{(k+1)} \in \text{span}(f_1, f_2, \ldots, f_i) \quad \text{and} \quad \langle e_i^{(k+1)}, f_i \rangle > 0 \quad \text{for } i = 1, 2, \ldots, p.$$

   This means that the change of basis matrix is upper triangular with positive entries on the diagonal. When T is a matrix operator, this is equivalent to making the QR factorization $[f_1\ f_2\ \cdots\ f_p] = QR$, where the ith column of Q is $e_i^{(k+1)}$ and $r_{kk} > 0$, i.e., the diagonal entries of R are positive.
2. Since the Gram–Schmidt method is time-consuming, the re-orthonormalization could be done every few steps, rather than every step.

For further discussions, see Rutishauser (1969) and Stewart (1973).

EXAMPLE 7. For matrix A of the previous examples, apply the simultaneous iteration method to obtain the two largest eigenvalues and corresponding eigenvectors.

One step will be done in detail and the rest of the steps reported. Take $e_1^{(0)} = \delta_1$, $e_2^{(0)} = \delta_2$. Then $f_1 = Ae_1^{(0)} = (5, 4, 3)^T$, $f_2 = Ae_2^{(0)} = (4, 4, 4)^T$. The first approximations to the eigenvalues are $r_1 = \langle f_1, e_1^{(0)} \rangle = 5$ and $r_2 = \langle f_2, e_2^{(0)} \rangle = 4$. The Gram–Schmidt method applied to $\{f_1, f_2\}$ yields $e_1^{(1)} = f_1/\|f_1\| = (0.7071, 0.5657, 0.4243)^T$, $g_2 = f_2 - \langle f_2, e_1^{(1)} \rangle e_1^{(1)}$, and $e_2^{(1)} = g_2/\|g_2\| = (-0.5773, -0.5773, 0.8083)^T$. $e_1^{(1)}$ and $e_2^{(1)}$ are the first approximations to the eigenvectors. In the next step, $e_1^{(0)}$ and $e_2^{(0)}$ are replaced by $e_1^{(1)}$ and $e_2^{(1)}$, and the process is repeated. The following table summarizes the results. The superscripts are suppressed. Results are accurate to the digits shown.

k	r	e
1	$r_1 = 5$	$e_1 = (0.707, 0.566, 0.424)^T$
	$r_2 = 4$	$e_2 = (-0.577, -0.577, 0.808)^T$
2	$r_1 = 11.78$	$e_1 = (0.589, 0.565, 0.577)^T$
	$r_2 = 3.05$	$e_2 = (-0.639, -0.639, 0.761)^T$
3	$r_1 = 12.33$	$e_1 = (0.562, 0.561, 0.577)^T$
	$r_2 = 3.61$	$e_2 = (-0.662, -0.662, 0.737)^T$
4	$r_1 = 12.35$	$e_1 = (0.557, 0.560, 0.613)^T$
	$r_2 = 3.88$	$e_2 = (-0.667, -0.667, 0.732)^T$
5	$r_1 = 12.36$	$e_1 = (0.556, 0.560, 0.614)^T$
	$r_2 = 3.94$	$e_2 = (-0.668, -0.668, 0.731)^T$
6	$r_1 = 12.36$	$e_1 = (0.555, 0.560, 0.615)^T$
	$r_2 = 3.96$	$e_2 = (-0.668, -0.668, 0.731)^T$

Exercises 5.5

Exercises 1 through 6 should be done by hand. Many of the other exercises can be done with only small computational aids. Use them if they are available.

1.
$$A = \begin{bmatrix} 2 & 2 & 3 \\ 2 & 5 & 5 \\ 3 & 5 & 6 \end{bmatrix}.$$

has three positive eigenvalues. Use the power method and inverse power method to estimate the largest and smallest eigenvalues and corresponding eigenvectors of A.

2.
$$A = \begin{bmatrix} -3 & 4 & 5 \\ 4 & -4 & -4 \\ 5 & -4 & -3 \end{bmatrix}.$$

Find (approximately) the dominant eigenvalue and corresponding eigenvector.

3.
$$A = \begin{bmatrix} -2 & 5 & 2 \\ 5 & -2 & 2 \\ 2 & 2 & 5 \end{bmatrix}.$$

(a) Apply the power method exactly (i.e., without normalization) for several steps, using the following starting vectors: (i) $x_0 = (1, 0, 0)^T$, (ii) $x_0 = (1, 1, 0)^T$.
(b) Find an orthogonal eigenbasis for A, and for each of (i) and (ii) expand x_0 in terms of the eigenbasis and find $x_n = A^n x_0$ explicitly. Then explain what you saw in (a).

4. Let
$$A = \begin{bmatrix} 1 & -2 & 0 \\ -2 & 8 & -2 \\ 0 & -2 & 1 \end{bmatrix}, \qquad x_0 = \begin{bmatrix} 3 \\ 1 \\ 1 \end{bmatrix}.$$

(a) Do the power method exactly (i.e., without normalization) for four steps.
(b) Normalize x_0 to length 1, and round to three digits. Apply the following normalized power method for six steps, which rounds at each step to three digits: Given y with $\|y\| = 1$ (approximately), compute $z = Ay$, compute $R = \langle z, y \rangle$, find $w = z/\|z\|$, set $y = w$ (rounded to three digits), and repeat.
(c) Find the eigenvalues and eigenvectors of A exactly, and explain what happened in parts (a) and (b).

5. Let
$$A = \begin{bmatrix} 6 & 2 & -4 \\ 2 & 9 & 2 \\ -4 & 2 & 6 \end{bmatrix}, \qquad x_0 = \begin{bmatrix} 1 \\ 1 \\ 1 \end{bmatrix}.$$

(a) Apply the power method exactly, starting with x_0 to estimate the dominant eigenvalue and its eigenvector.
(b) Verify that $(2, -1, 2)^T$, $(1, 0, -1)^T$, $(1, 4, 1)^T$ are eigenvectors of A. Find x_n exactly, using the exact (unnormalized) power method. Check this against your results in (a).

6. Let

$$A = \begin{bmatrix} 5 & 0 & -4 \\ 0 & 9 & 0 \\ -4 & 0 & 5 \end{bmatrix}.$$

A has a dominant eigenvalue $\lambda_1 = 9$ and corresponding eigenvector $e_1 = \delta_2$. Use the method of suppression exactly (i.e., without normalization) to find the next most dominant eigenvalue and corresponding eigenvector for the starting vectors (i) $x_0 = (1, 1, 1)^T$ and (ii) $x_0 = (1, 1, 0)^T$. Solve the eigenvalue problem exactly for A and explain what happens in the iteration in both cases.

7. Consider the following matrices. For each of them, do one or more of parts a through d.

$$(i) \begin{bmatrix} 1 & -1 & 1 \\ -1 & 10 & 9 \\ 1 & 9 & 11 \end{bmatrix} \quad (ii) \begin{bmatrix} 3 & -5 & 2 \\ -5 & 9 & -4 \\ 2 & -4 & 3 \end{bmatrix} \quad (iii) \begin{bmatrix} 1 & 1 & 0 \\ 1 & 1 & 1 \\ 1 & 1 & 10 \end{bmatrix} \quad (iv) \begin{bmatrix} 9 & 3 & 7 \\ 3 & 2 & 4 \\ 7 & 4 & 8 \end{bmatrix}$$

$$(v) \begin{bmatrix} 2.5 & 3.5 & 1 \\ 3.5 & 5.5 & 1 \\ 1 & 1 & 1 \end{bmatrix} \quad (vi) \begin{bmatrix} 3 & 3 & 4 \\ 3 & 5 & 5 \\ 4 & 5 & 6 \end{bmatrix} \quad (vii) \begin{bmatrix} 10 & 3 & -3 \\ 3 & 3 & -2 \\ -3 & -2 & 1 \end{bmatrix}$$

$$(viii) \begin{bmatrix} 7 & -3 & 3 \\ -3 & 7 & 3 \\ 3 & 3 & 7 \end{bmatrix} \quad (ix) \begin{bmatrix} 6 & 2 & -4 \\ 2 & 9 & 2 \\ -4 & 2 & 6 \end{bmatrix} \quad (x) \begin{bmatrix} 1 & 1/2 & 1/3 \\ 1/2 & 1/3 & 1/4 \\ 1/3 & 1/4 & 1/5 \end{bmatrix}.$$

(a) Use the power method to find the dominant eigenvalue and eigenvector.
(b) Use the inverse power method to find the smallest eigenvalue and the corresponding eigenvector.
(c) Use the method of suppression to find the second dominant eigenvalue and eigenvector.
(d) Use the simultaneous iteration method to obtain the two most dominant eigenvalues and the corresponding eigenvectors.

8. Consider the generalized eigenvalue problem $Ae = \lambda Ge$, $e \neq 0$, for the following pairs:

$$(i)\ A = \begin{bmatrix} 1 & -1 & 0 \\ -1 & 10 & 20 \\ 0 & 20 & 45 \end{bmatrix} \quad G = \text{diag}(1, 2, 5)$$

$$(ii)\ A = \begin{bmatrix} 10 & -8 & 10 \\ -8 & 10 & -10 \\ 10 & -10 & 15 \end{bmatrix} \quad G = \text{diag}(1, 2, 5)$$

$$(iii)\ A = \begin{bmatrix} 1 & 0 & 3 \\ 0 & 12 & 18 \\ 3 & 18 & 48 \end{bmatrix} \quad G = \text{diag}(1, 2, 5)$$

Here $T(x) = G^{-1}Ax$ is the Hermitian operator in the inner product $\langle x, y \rangle = y^T Gx$ that should be used in the power method.

(a) Use the power method to find the dominant eigenvalue and eigenvector.
(b) Use the inverse power method to find the smallest eigenvalue and the corresponding eigenvector.

(c) Use the method of suppression to find the second most dominant eigenvalue and eigenvector.

(d) Use the simultaneous iteration method to find the two most dominant eigenvalues and their corresponding eigenvectors.

In these calculations, find the eigenvalues to four significant digits and the eigenvectors to two.

9. For

$$(i) \quad \begin{bmatrix} 3 & 2 & 1 & -5 \\ 2 & 2 & 1 & -5 \\ 1 & 1 & 3 & -5 \\ -5 & -5 & -5 & 15 \end{bmatrix}$$

$$(ii) \quad \begin{bmatrix} 82 & 0 & -36 & 0 \\ 0 & 28 & 0 & -36 \\ -36 & 0 & 28 & 0 \\ 0 & -36 & 0 & 82 \end{bmatrix}$$

$$(iii) \quad \begin{bmatrix} 1 & 1/2 & 1/3 & 1/4 \\ 1/2 & 1/3 & 1/4 & 1/5 \\ 1/3 & 1/4 & 1/5 & 1/6 \\ 1/4 & 1/5 & 1/6 & 1/7 \end{bmatrix}$$

estimate the two dominant eigenvalues of the matrix and the corresponding eigenvectors, accurate to three significant digits in the eigenvectors.

10. Let

$$A = \begin{bmatrix} 15 & 0 & 5 & 5 \\ 0 & 10 & 5 & 5 \\ 5 & 5 & 6 & 5 \\ 5 & 5 & 5 & 5 \end{bmatrix}, \quad e_1 = \begin{bmatrix} 1.000 \\ 0.5807 \\ 0.7087 \\ 0.6763 \end{bmatrix}.$$

(a) Check that e_1 is approximately an eigenvector of A, and find the corresponding eigenvalue.

(b) Find the second most dominant eigenvalue of A and its corresponding eigenvector (to three significant digits for the eigenvector).

11. Let

$$A = \begin{bmatrix} 2 & 3 & 3 & 3 \\ 3 & 5 & 6 & 6 \\ 3 & 6 & 8 & 9 \\ 3 & 6 & 9 & 11 \end{bmatrix}, \quad e_1 = \begin{bmatrix} 0.3473 \\ 0.6527 \\ 0.8794 \\ 1.000 \end{bmatrix}.$$

(a) Check that e_1 is approximately an eigenvector of A, and find the corresponding eigenvalue.

(b) Find the second most dominant eigenvalue of A and its corresponding eigenvector (to three significant digits for the eigenvector).

12. For each of

$$(i) \quad \begin{bmatrix} 1 & 1 & 0 \\ 1 & 10 & 2 \\ 0 & 2 & 50 \end{bmatrix}$$

$$(ii) \quad \begin{bmatrix} 1 & 0.5 & 0 & 0 \\ 0.5 & 5 & 0.5 & 0 \\ 0 & 0.5 & 30 & 0.5 \\ 0 & 0 & 0.5 & 150 \end{bmatrix}$$

(a) Use Gershgorin's theorem and the Rayleigh quotient to estimate the eigenvalues.

(b) Use the shifted inverse power method to find the eigenvalues and eigenvectors.

13. Let

$$A = \begin{bmatrix} 11 & 1 & 0 \\ 1 & 8 & 8 \\ 0 & 8 & 10 \end{bmatrix}.$$

Given that two eigenvalues are known approximately, $\lambda_1 \approx 17.1$ and $\lambda_2 \approx 0.763$, compute $\det(A)$ and estimate the third eigenvalue. Can you think of another way to estimate λ_3? Use the shifted inverse power method to get the third eigenvalue to high accuracy.

14. A *deflation method* **to find secondary eigenvalues and eigenvectors of a Hermitian matrix.** Suppose A is an $n \times n$ Hermitian matrix with eigenvalues λ_i and corresponding eigenvectors E_i, and $|\lambda_1| > |\lambda_2| > \cdots > |\lambda_n| > 0$.

(a) Suppose the dominant eigenvalue λ_1 and corresponding eigenvector E_1, $\|E_1\| = 1$, have been found. Put $A_1 = A - \lambda_1 E_1 E_1^H$. Show that A_1 has the eigenvalues $0, \lambda_2, \lambda_3, \ldots, \lambda_n$, and the same eigenvectors. Thus λ_2 is a dominant eigenvector for A_1, and the power method on A_1 will produce an λ_2 and E_2. We can take $\|E_2\| = 1$.

(b) Now put $A_2 = A_1 - \lambda_2 E_2 E_2^H$. Show that A_2 is Hermitian with eigenvalues $0, 0, \lambda_3, \ldots, \lambda_n$. Now λ_3 is dominant, and the power method on A_2 can be used.

(c) Continue the process. This method is clearly too good to be true, because if it worked, then every eigenvalue problem for Hermitian matrices would be routinely solved, thwarting much research in numerical analysis. What goes wrong?

(d) Try this method on some of the preceding matrices.

(e) Prove $A = \sum_{i=1}^{n} \lambda_i E_i E_i^H$.

15. Suppose A is a real diagonalizable matrix with a dominant eigenvalue λ_1 and corresponding eigenvector e_1. Show that the power method still works, with $R(x_n) \to \lambda_1$, but in general the order of convergence is only of the form $(\lambda_2/\lambda_1)^n$, where $|\lambda_2| \geq |\lambda_i|$ for $i \geq 2$. That is, the Rayleigh quotients converge no faster than the iterates for the eigenvectors. Try it on

$$A = \begin{bmatrix} 6 & 1 \\ -5 & 0 \end{bmatrix},$$

for example.

5.6
THE RAYLEIGH–RITZ METHOD

The Rayleigh–Ritz method is extremely important in the sciences. For example, it is used to determine the natural frequencies of vibration of large and small structures as well as to estimate the shapes of the natural modes of oscillation of these structures. It may be used in a similar way in electric circuits. It is also used to estimate the buckling loads and modes of structures, and it is incorporated in many finite element codes for these purposes. In fact, many people base their careers on variants of this method.

The idea of the method runs as follows. Given a Hermitian operator $T : V \rightarrow W$, we wish to find an approximation of the smallest several eigenvalues and eigenvectors of T. These are usually the most important ones physically. T is usually a differential or integral operator. Choose a finite-dimensional subspace M of V. Since the Rayleigh quotient characterizes eigenvalues and eigenvectors through the extended minimum principle, look at the Rayleigh quotient of T restricted to M. Solve the extended minimization problem in this setting, and use the results to approximate the eigenvalues and eigenvectors of T on V. Bounds for the eigenvalues can be obtained by using an analog of the Courant–Fischer theorem.

It is crucial to point out that $T : V \rightarrow V$ is *not* included in the following list of assumptions and that V may be infinite-dimensional. In many crucial applications, T is often a differential or integral operator that acts on an infinite-dimensional space V of functions.

Another related method, the Ritz method, will be presented in Chapter 7 using an ad hoc approach. More closely tied with the Rayleigh–Ritz method is the Galerkin method, also illustrated in Chapter 7. These methods can be tied together theoretically, but this is not the appropriate book in which to pursue the connections.

Assumptions. Let $T : V \rightarrow W$, with V a subspace of W and with $\langle \, , \rangle$ an inner product on W. Suppose T is linear and Hermitian on V with respect to $\langle \, , \rangle$. Assume there exists a sequence of vectors $\{e_i\}_{i=1}^{N} \subset V$ ($N \leq \infty$), that are orthogonal with respect to $\langle \, , \rangle$ and that satisfy

$$T e_i = \lambda_i e_i \tag{1}$$

$$\lambda_1 \leq \lambda_2 \leq \cdots \leq \lambda_n \leq \cdots \tag{2}$$

The minimum principle and the extended minimum principle hold:

$$\lambda_1 = \min\{R(x) : x \neq 0, x \in V\} \tag{3}$$

and $\qquad \lambda_{i+1} = \min\{R(x) : x_0, x \perp e_1, e_2, \ldots, e_i, x \in V\}, \quad i = 1, 2, \ldots \tag{4}$

where $\qquad\qquad\qquad R(x) = \dfrac{\langle T x, x \rangle}{\|x\|^2}$

is the Rayleigh quotient of T.

These assumptions hold if $V = W$ and V is finite-dimensional, as was shown in Section 5.4. For a great many integral and differential operators commonly occurring in the sciences and engineering, there is an infinite orthogonal set $\{e_i : i = 1, 2, \ldots\}$ satisfying (1)–(4).

The task is to find estimates on the first few λ_i and e_i. The Rayleigh–Ritz method obtains numbers Λ_i that estimate the eigenvalues on the high side: $\lambda_i \leq \Lambda_i$. This kind of estimate is called a *one-sided estimate*. Thus, when these estimates are used in vibration analysis, the natural frequencies of vibration $\omega_k = \sqrt{\lambda_k}$ are estimated on the high side. Finding an estimate μ_i from the other side, $\mu_i \leq \lambda_i$, is harder and will not be discussed. The interested reader is invited to consult the monograph by H. F. Weinberger (1974).

Now let us turn to a derivation of the method. Finding estimates of λ_1 and e_1 is fairly easy. For example, if a single vector $f \neq 0$ is chosen, $\lambda_1 \leq R(f)$, by the minimum principle (3). To extend this idea and make computations more systematic, let $\{f_k\}_{k=1}^n$ be a fixed linearly independent set in V, and set $V_n = \mathrm{span}\{f_k\}_{k=1}^n$. The f_i are called *basis functions*. By (3), for each $x = \sum_{i=1}^n \xi_i f_i \in V_n$, we have $\lambda_1 \leq R(x)$. The idea is to minimize $R(x)$ with respect to the variables ξ_i. A minimizing $\tilde{x} = \sum_{i=1}^n \xi_i f_i$ will be a member of V_n that gives the best approximation of λ_1 among all contestants in V_n:

$$\lambda_1 \leq R(\tilde{x}) \leq R(x) \quad \forall x \in V_n. \tag{*}$$

Now take $\Lambda_1 \equiv R(\tilde{x})$ as an approximation of λ_1, and $e_1 \equiv \tilde{x}/\|\tilde{x}\|$ as an approximation of e_1. In the following computation of Λ_1 and $\tilde{e}_1$, other information will be presented that turns out to be relevant and very important.

The first step is to convert the problem of minimizing $R(x)$ on V_n to a problem of minimizing a function of $\xi = [x] = (\xi_1, \xi_2, \ldots, \xi_n)^T \in \mathbf{F}^n$. Substituting $x = \sum_{i=1}^n \xi_i f_i$ into $\langle T(x), x \rangle$ and $\langle x, x \rangle$, we obtain

$$\langle T(x), x \rangle = \xi^H A \xi, \quad \text{where} \quad a_{ij} = \langle T f_j, f_i \rangle, \tag{5}$$

$$\langle x, x \rangle = \xi^H G \xi, \quad \text{where} \quad g_{ij} = \langle f_j, f_i \rangle. \tag{6}$$

Note that A and G are Hermitian, and G is nonsingular since it is the Gram matrix of a linearly independent set. Recall also that the coordinate map [ ] is an isomorphism (i.e., a linear one-to-one correspondence) between $x \in V_n$ and $y \in \mathbf{F}^n$.

Introducing the inner product $\langle \xi, \eta \rangle_G = \eta^H G \xi$ on $\mathbf{F}^n$ yields

$$\langle x, y \rangle = \langle \xi, \eta \rangle_G, \quad \text{if } [x] = \xi \text{ and } [y] = \eta, \quad x, y \in V_n, \tag{7}$$

so that the two inner products yield the same scalar when x and y correspond to the columns ξ and η. Now introduce the operator $L(\xi) = G^{-1} A \xi$ on $\mathbf{F}^n$, just as was done in the generalized eigenvalue theorem. Recall from that theorem that L is Hermitian with respect to $\langle \, , \, \rangle_G$. The Rayleigh quotient for L is $R_L(\xi) = \langle L(\xi), \xi \rangle_G / \langle \xi, \xi \rangle_G = \xi^H G(G^{-1} A)\xi / \xi^H G \xi = \xi^H A \xi / \xi^H G \xi$, so that the two Rayleigh quotients agree:

$$R(x) = R_L(\xi) = \xi^H A \xi / \xi^H G \xi, \quad \forall x \in V_n. \tag{8}$$

By Theorem 2 of Section 5.4, R_L minimizes precisely at the smallest eigenvalue Λ_1 of L in $\mathbf{F}^n$, and a minimizing ξ is $\xi = E_1$, an eigenvector of L associated with the eigenvalue Λ_1. Let $\tilde{e}_1 = \sum_{i=1}^n \xi_i f_i$ (where $\xi = E_1 = [\tilde{e}_1]$). Then

$\Lambda_1 = R_1(E_1) = R(\tilde{e}_1) \geq \lambda_1$ by (8) and (*). Λ_1 is taken as the approximation of λ_1, and $\tilde{e}_1 = \sum_{i=1}^{n} \xi_i f_i = [\,]^{-1}(E_1)$ as the approximation of e_1.

However, in solving the eigenvalue problem for L, one obtains eigenvalues $\Lambda_1 \leq \Lambda_2 \leq \cdots \leq \Lambda_n$ and corresponding eigenvectors E_i that are orthogonal with respect to $\langle\,,\,\rangle_G$. These turn out to be useful. In fact, it will be shown that

$$\lambda_i \leq \Lambda_i, \quad i = 2, 3, \ldots, n. \tag{9}$$

These Λ_i will be taken as the approximations to the λ_i and the vectors e_i in V_n corresponding to the E_i via the coordinate map as the approximations for the eigenvectors

$$e_i \approx \tilde{e}_i, \quad \text{where} \quad [\tilde{e}_i] = E_i. \tag{10}$$

Now (9) will be proved. Assume E_i and Λ_i have been found such that

$$L(E_i) = \Lambda_i E_i, \quad E_i \neq 0, \tag{11a}$$

$$\langle E_i, E_j \rangle_G = 0 \quad \text{if } i \neq j, \quad i, j = 1, 2, \ldots, n, \tag{11b}$$

$$\Lambda_1 \leq \Lambda_2 \leq \cdots \leq \Lambda_n. \tag{11c}$$

$\{\tilde{e}_i : i = 1, 2, \ldots, n\}$ is an orthogonal set in V_n, since by Eqs. (7), (10), and (11c),

$$\langle \tilde{e}_i, \tilde{e}_j \rangle = \langle E_i, E_j \rangle_G = 0 \quad \text{if} \quad i \neq j.$$

Let $k > 1$ be given. Following the proof of the Courant–Fischer theorem, it will be shown that there exist $u_1, u_2, \ldots, u_k$ not all 0 so that for $z = \sum_{i=1}^{k} u_i \tilde{e}_i$ we have $z \perp e_1, \ldots, e_{k-1}$, i.e.,

$$\langle z, e_j \rangle = \sum_{i=1}^{k} u_i \langle \tilde{e}_i, e_j \rangle = 0, \quad \text{for} \quad j = 1, 2, \ldots, k - 1. \tag{**}$$

In fact, since there are fewer equations in (**) than unknowns, there is a nontrivial solution for $u_1, u_2, \ldots, u_k$. Further, $\sum_{i=1}^{k} u_i \tilde{e}_i \neq 0$ since $\{\tilde{e}_1, \ldots, \tilde{e}_k\}$ is linearly independent. Note that $z \in V_n$ since each $\tilde{e}_i \in V_n$. By the extended minimality property (4) for T,

$$\lambda_k \leq R(z) = R_L([z]).$$

But $[z] = \sum_{i=1}^{k} u_i[\tilde{e}_i] = \sum_{i=1}^{k} u_i E_i$ because the coordinate map $[\,]$ is linear. By (11), from Parseval's identity and the fact that $\Lambda_1 \leq \Lambda_2 \leq \cdots \leq \Lambda_k$,

$$\langle L[z], [z] \rangle_G = \left\langle \sum_{i=1}^{k} u_i \Lambda_i E_i, \sum_{i=1}^{k} u_i E_i \right\rangle = \sum_{i=1}^{k} \Lambda_i |u_i|^2 \|E_i\|_G^2$$

$$\leq \Lambda_k \sum_{i=1}^{k} |u_i|^2 \|E_i\|_G^2 = \Lambda_k \langle [z], [z] \rangle_G.$$

(cf. Lemma 1 of Section 5.4). Thus

$$\lambda_k \leq R_L([z]) = \langle L[z], [z] \rangle_G / \langle [z], [z] \rangle_G \leq \Lambda_k,$$

which establishes (9).

A summary of the method is given next.

Rayleigh–Ritz Method

Suppose that $T : V \to W$ is a linear Hermitian operator with respect to $\langle\,,\,\rangle$; $\{e_i\}_{i=1}^{N}$ is an orthogonal sequence of eigenvectors of T, $T(e_i) = \lambda_i e_i$; and the assumptions (1) through (4) hold.

1. Select a linearly independent set $\{f_k\}_{k=1}^{n} \subset V$ of basis functions.
2. Construct the matrices $A = [a_{ij}]$, $G = [g_{ij}]$ by

$$a_{ij} = \langle T(f_j), f_i \rangle, \qquad g_{ij} = \langle f_j, f_i \rangle \quad i, j = 1, 2, \ldots, n.$$

A and G will be Hermitian.
3. Solve the generalized eigenvalue problem for A and G: Find $E_i \in \mathbf{F}^n$ and Λ_i real such that (11) holds; i.e.,

$$AE_i = \Lambda_i GE_i, \quad E_i \neq 0, \quad i = 1, 2, \ldots, n, \tag{12}$$

$$E_j^{\mathrm{H}} GE_i = 0 \quad \text{if} \quad i \neq j,$$

$$\Lambda_1 \leq \Lambda_2 \leq \cdots \leq \Lambda_n.$$

4. For each $E_j = (e_{1j}, e_{2j}, \ldots, e_{nj})^{\mathrm{T}}$, construct the vector

$$\tilde{e}_j = \sum_{k=1}^{n} e_{kj} f_k. \tag{13}$$

Then obtain the approximations

$$\lambda_j \approx \Lambda_j \quad \text{and} \quad e_j \approx \tilde{e}_j, \quad j = 1, 2, \ldots, n, \tag{14}$$

and rigorously,

$$\lambda_j \leq \Lambda_j, \quad j = 1, 2, \ldots, n. \tag{15}$$

Remarks. Note that the Λ_j and $\tilde{e}_j$ depend on $V_n = \mathrm{span}\{f_k\}_{k=1}^{n}$. If $\{g_k\}_{k=1}^{n}$ were another basis for V_n, then the same Λ_i and $\tilde{e}_i$ would result. It is possible to show that $\Lambda_j^{(n)} \to \lambda_j$, and $\tilde{e}_j \to e_j$ as $n \to \infty$ if the f_j are chosen correctly. Often, in practice, the f_j are chosen to resemble the eigenfunctions being sought, to enable a better approximation with a smaller number of f_j. This artful choice is based on good physical intuition or trial-and-error experience. Mathematically, the f_j should be chosen so that the subspaces $V_n = \mathrm{span}\{f_k\}_{k=1}^{n}$ increase, and approximate arbitrary vectors in V arbitrarily well as $n \to \infty$. More rigorously, this means that $V_n \subset V_{n+1}$, and given an $x \in V$ and an $\varepsilon > 0$, there is an f in some V_n such that $\|f - x\| < \varepsilon$.

A simple differential operator $T(f) = -f''$ will be used as an example. This operator arises in considering the (normalized) heat equation $u_t = u_{xx}$ or the wave equation $u_{tt} = u_{xx}$ with boundary conditions $u(0, t) = u(1, t) = 0$. In both of these partial differential equations, the general solution is of the form $u(x, t) = \sum_{k=1}^{\infty} T_k(t) e_k(x)$, where e_k is an eigenfunction of T. In the case of the heat equation, $T_k(t) = c_k e^{-\lambda_k t}$, and for the wave equation $T_k(t) = a_k \cos(\omega_k t) + b_k \sin(\omega_k t)$, where $T(e_k) = \lambda_k e_k$, $\lambda_k = \sqrt{\omega_k}$, and a_k, b_k, c_k depend on the initial conditions.

In the first part of the following example, polynomial approximations are made. In the second part, hat functions are used to approximate the eigenfunctions, and it is possible to find the approximate eigenfunctions for arbitrarily large numbers of basis functions. A full analysis is also possible.

EXAMPLE 1

$$Ty = -y'', \quad 0 \le t \le 1$$

$$V : y, y', y'' \text{ are continuous and } y(0) = y(1) = 0$$

$$\langle y, x \rangle = \int_0^1 yx\, dt \quad \text{(real scalars)}$$

Approximate the eigenvalues and eigenvectors of T on V.

T is Hermitian on V, by Example 1 of Section 5.1. The eigenfunctions of T are $e_k(t) = \sin(k\pi t)$ and $\lambda_i = (k\pi)^2, k = 1, 2, \ldots$, and they are known to satisfy assumptions (1) through (4).

Many different $\{f_k\}_{k=1}^n$ can be used. Here,

$$g_{ij} = g_{ji} = \int_0^1 f_i f_j\, dt.$$

An advantage can be gained if integration by parts is used in the determination of the a_{ij}, as will be explained more fully in the second part of this example. Thus

$$a_{ij} = a_{ji} = \langle Tf_j, f_i \rangle = -\int_0^1 (f_j')' f_i\, dt = -f_j' f_i \Big|_0^1 + \int_0^1 f_j' f_i'\, dt = \int_0^1 f_j' f_i'\, dt,$$

since f_i, f_j both satisfy $f(0) = f(1) = 0$.

Polynomial approximation. Approximate the eigenvalues and eigenvectors using the polynomial basis functions $f_i(x) = x^i(1 - x), i = 1, 2, \ldots, n$. Compute the results for $n = 2, 4$.

Integration and simplification yield

$$g_{ij} = \int_0^1 x^{i+j}(1 - x)^2\, dx = \frac{2}{(i + j + 1)(i + j + 2)(i + j + 3)},$$

$$a_{ij} = \int_0^1 x^{i+j-2}\{i + (i + 1)x\}\{j + (j + 1)x\}\, dx = \frac{2ij}{(i + j - 1)(i + j)(i + j + 1)}.$$

For $n = 2$,

$$G = \begin{bmatrix} 1/30 & 1/60 \\ 1/60 & 1/105 \end{bmatrix}, \qquad A = \begin{bmatrix} 1/3 & 1/6 \\ 1/6 & 2/15 \end{bmatrix}.$$

Solving the eigenvalue problem $AE = \Lambda GE, E \ne 0$, yields

$$\Lambda_1 = 10, \qquad E_1 = (1, 0)^\mathrm{T}, \qquad \Lambda_2 = 42, \qquad E_2 = (0.5, -1)^\mathrm{T}.$$

The actual eigenvalues are $\lambda_1 = \pi^2 = 9.8696$ and $\lambda_2 = 4\pi^2 = 39.478$. Not too bad. The approximate eigenfunctions are $\tilde{e}_1(x) = f_1(x) = x(1 - x)$ and $\tilde{e}_2(x) = 0.5 f_1(x) - f_2(x) = x(1 - x)(0.5 - x)$. The following graphs compare these to the actual eigenfunctions. $\tilde{e}_1$ and $\tilde{e}_2$ have been multiplied by appropriate constants so that the resulting functions have a maximum value of 1 on $[0, 1]$. The dotted lines are the actual eigenfunctions and the solid lines the approximate eigenfunctions. Not bad either.

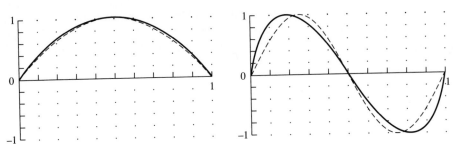

$n = 2$: Polynomial approximations.

For $n = 4$, the matrices are

$$
G = \begin{bmatrix} 1/30 & 1/60 & 1/105 & 1/168 \\ 1/60 & 1/105 & 1/168 & 1/252 \\ 1/105 & 1/168 & 1/252 & 1/36 \\ 1/168 & 1/252 & 1/360 & 1/495 \end{bmatrix}, \quad A = \begin{bmatrix} 1/3 & 1/6 & 1/10 & 1/15 \\ 1/6 & 2/15 & 1/10 & 8/105 \\ 1/10 & 1/10 & 3/35 & 1/14 \\ 1/15 & 8/105 & 1/14 & 4/63 \end{bmatrix}.
$$

Solving the eigenvalue problem, $AE = \lambda GE$, $E \neq 0$, yields

Actual	Approximate	Eigenvector
$\lambda_1 = 9.8696$	$\Lambda_1 = 9.8697$	$E_1 = (0.885, 1, -1, 0)^{\mathrm{T}}$
$\lambda_2 = 39.478$	$\Lambda_2 = 39.502$	$E_2 = (-0.0779, -0.1776, 1, -.667)^{\mathrm{T}}$
$\lambda_3 = 88.826$	$\Lambda_3 = 102.13$	$E_3 = (0.2158, -1, 1, 0)^{\mathrm{T}}$
$\lambda_3 = 157.9$	$\Lambda_4 = 200.50$	$E_4 = (0.0577, -0.4487, 1, -0.6667)^{\mathrm{T}}$

Observe that the approximations to the first two λ_i are very good, and the accuracy decreases progressively. This is typical. The corresponding approximate eigenfunctions are $\tilde{e}_1 = 0.885 f_1 + f_2 - f_3$, $\tilde{e}_2 = -0.0779 f_1 - 0.1776 f_2 + f_3 - 0.667 f_4$, $\tilde{e}_3 = 0.2158 f_1 - f_2 + f_3$, $\tilde{e}_4 = 0.0577 f_1 - 0.4487 f_2 + f_3 - 0.6667 f_4$. The following graphs compare these to the actual eigenfunctions. In each case the $\tilde{e}_i$ are multiplied by a suitable constant to make an approximate eigenfunction with a maximum value of 1 on $[0, 1]$. Solid lines are the approximate eigenfunctions, dotted ones the exact eigenfunctions. The next two graphs show the approximations for $k = 1, 2$, which are excellent. For $k = 1$, the graphs are indistinguishable.

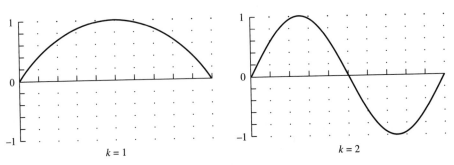

$k = 1$ $k = 2$

$n = 4$: Polynomial approximations.

For $k = 3, 4$ the approximations worsen, as the next graph shows.

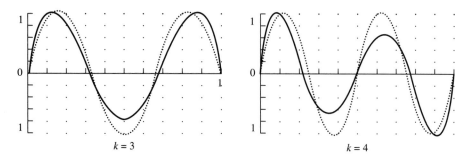

$n = 4$: Polynomial approximations.

To obtain a large number of eigenvalues and eigenfunctions would require large matrices. The matrices G and A are full matrices and may well become ill conditioned and generate unreliable values in the solution of the generalized eigenvalue problem. The next approach can help with this problem. In this example, G is tridiagonal and computationally well behaved. Furthermore, the generalized eigenvalue problem for A and G can be solved exactly, so that the behavior for large n may be examined closely.

Hat functions. Approximate the eigenvalues and eigenfunctions using the "hat" functions $\{f_1, \ldots, f_m\}$ with equispaced nodes. Let $n = m + 1$ and $t_j = j/n$. f_j is linear in each segment $[t_{k-1}, t_k]$, $k = 1, 2, \ldots, n$ and has the interpolation property

$$f_j(t_k) = \delta_{jk}, \quad \text{for } j = 1, 2, \ldots, n - 1 \text{ and } k = 0, 1, \ldots, n.$$

Specifically,

$$f_j(t) = \begin{cases} nt - (j - 1) & t_{j-1} \le t \le t_j \\ -nt + (j + 1) & t_j \le t \le t_{j+1} \\ 0 & \text{otherwise} \end{cases} \quad j = 1, 2, \ldots, n - 1.$$

See Example 6 of Section 1.3.

One advantage of the formula

$$a_{ij} = \int_0^1 f_j' f_i' \, dt \tag{16}$$

is that less continuity is required of $\{f_1, \ldots, f_{n-1}\}$ in the sense that Eq. (16) requires only one integrable derivative, while $a_{ij} = -\int_0^1 f_j'' f_i \, dt$ requires two. Notice that $a_{ij} = g_{ij} = 0$ if $|i - j| > 1$, since the intervals where f_i and f_j are nonzero do not overlap. Computation gives (cf. Section 3.6)

$$a_{ij} = a_{ji} = \begin{cases} -n & |i \pm j| = 1 \\ 2n & i = j \\ 0 & \text{otherwise} \end{cases} \tag{17}$$

$$g_{ij} = g_{ji} = \begin{cases} 1/(6n) & |i \pm j| = 1 \\ 2/(3n) & i = j \\ 0 & \text{otherwise} \end{cases} \tag{18}$$

Let $K \equiv \mathrm{Trid}(a, b, c)$ denote the tridiagonal matrix with b on the main diagonal, a just below the main diagonal, and c just above the main diagonal: $k_{ii} = b$, $k_{i,i+1} = c$, $k_{i,i-1} = a$. In this notation,

$$G = \frac{1}{6n}\,\mathrm{Trid}(1, 4, 1), \qquad A = n\,\mathrm{Trid}(-1, 2, -1).$$

If a numerical package has to be used to solve the generalized eigenvalue problem, tridiagonal matrices, in general, are much preferred to full matrices. However, in this problem the generalized eigenvalue problem can be solved exactly, as follows. The matrix $T = \mathrm{Trid}(-1, 2, -1)$ has known eigenvalues and eigenvectors (see Example 9 of Section 4.2). Clearly A is a function of T, and since

$$G = \frac{1}{6n}(6I - T),$$

G is a function of T as well. Thus the generalized eigenvalue problem $AE = \Lambda GE$, $E \neq 0$ may be solved using the eigenvalues and eigenvectors of T (cf. Theorem 8 of Section 5.3). Note that $AE = \Lambda GE$, $E \neq 0$, is equivalent to $LE = G^{-1}AE = \Lambda E$, $E \neq 0$. But

$$L = G^{-1}A = 6n(6I - T)^{-1}nT = 6n^2(6I - T)^{-1}T.$$

Hence,

$$L = f(T), \quad \text{with } f(x) = 6n^2 x(6 - x)^{-1}.$$

T is a Hermitian matrix with eigenvalues $\mu_k = 4\sin^2(\frac{1}{2}\pi k/n)$ and associated eigenvector $E_k = (\sin(k\pi/n), \sin(2k\pi/n), \dots, \sin((n-1)k\pi/n))^\mathrm{T}$, so that L has the same eigenbasis as T and the eigenvalues of L are $\Lambda_k = f(\mu_k) = 6n^2\mu_k(6 - \mu_k)^{-1}$ by Theorem 1 of Section 4.3. The columns E_k are orthogonal both with respect to the standard inner product and with respect to the inner product $\langle\,,\,\rangle_G$ generated by G.

At this point, the following have been determined:

1. The approximate eigenvalues:

$$\Lambda_k = 6n^2\mu_k(6 - \mu_k)^{-1}, \quad \mu_k = 4\sin^2(\tfrac{1}{2}\pi kh) \tag{19}$$

2. The approximate eigenvectors, $\tilde{e}_k$, where $[\tilde{e}_k] = E_k$:

$$e_k(t) = \sum_{j=1}^{n-1} \sin(k\pi t_j) f_j(t), \quad \text{where } t_j = j/n. \tag{20}$$

By the preceding theory, $\Lambda_k \geq \lambda_k = (k\pi)^2$. If the approximations are to be good, it should be the case that, for large n,

$$\Lambda_k = \Lambda_k^{(n)} \approx \lambda_k = (k\pi)^2 \tag{21}$$

$$\tilde{e}_k(t) = \tilde{e}_k^{(n)}(t) \approx e_k(t) = \sin(k\pi t). \tag{22}$$

First, let us see how well Λ_k approximates λ_k. By l'Hospital's rule, with k fixed, $n^2\mu_k = n^2 \cdot 4\sin^2(\frac{1}{2}k\pi/n) \to (k\pi)^2$ as $n \to \infty$. Since $\mu_k \to 0$ as $n \to \infty$,

$$\Lambda_k = 6n^2\mu_k(6 - \mu_k)^{-1} \to \lambda_k \quad \text{as } n \to \infty, \text{ for each fixed } k. \tag{23}$$

Let us also check some specific numerical values. These are presented in the following table.

k	$\Lambda_k = \Lambda_k^{(n)}$					λ_k
	$n = 5$	$n = 10$	$n = 15$	$n = 20$	$n = 25$	"$n = \infty$"
1	10.198	9.951	9.906	9.890	9.883	9.870
2	44.888	40.794	40.059	39.804	39.687	39.478
3	116.117	95.575	91.786	90.482	89.883	88.826
4	227.839	179.553	167.348	163.174	161.266	157.914
5	None	300.000	270.000	259.666	254.960	246.740
6	None	464.470	403.993	382.302	372.437	355.306

Note that, for fixed k, as n increases, the approximation gets better, in accordance with the limit (23). Also, for a fixed n, as k increases, the approximation $\Lambda_k \approx \lambda_k$ worsens. This is typical.

Next, let us investigate how well the approximate eigenfunctions $\tilde{e}_k$ match the actual eigenfunctions e_k. From the interpolation property $f_k(t_i) = \delta_{ik}$ (cf. Examples 6 and 7 of Section 1.3), it follows that a linear combination $f(t) = \sum_{k=1}^{n-1} c_k f_k(t)$ takes on the values c_i at the points t_i

$$f(t_i) = \sum_{k=1}^{n-1} c_k f_k(t_i) = \sum_{k=1}^{n-1} c_k \delta_{ik} = c_i.$$

Applying this to Eq. (20) yields

$$\tilde{e}_k(t_i) = \sin(k\pi t_i), \quad \text{where } t_i = i/n, \quad i = 0, 1, \ldots, n. \tag{24}$$

Hence the approximate eigenfunction $\tilde{e}_k(t)$ is the piecewise linear continuous function interpolating eigenfunction $e_k(t)$ at the $n + 1$ equispaced points $t_i = i/n, i = 0, 1, \ldots, n$.

For fixed k, it can be shown that $\tilde{e}_k \to e_k$ uniformly as $n \to \infty$. The next group of graphs shows what happens for $n = 10$ as k varies. Again, notice that as k increases, for fixed n, the approximations get worse. For $k = 1$, it is just barely possible to see a difference between $\tilde{e}_1$ and e_1.

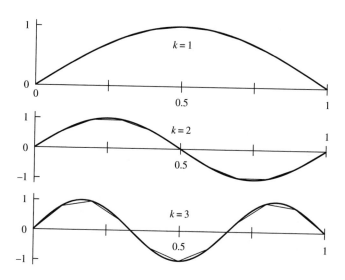

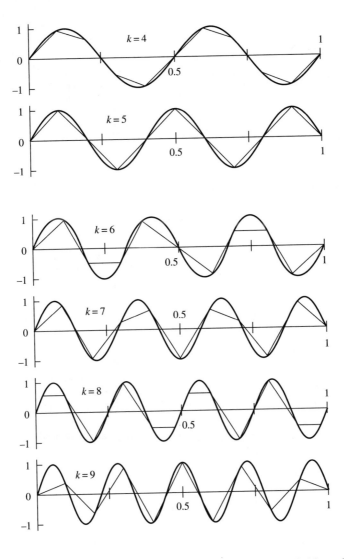

A final word on these computations. The astute reader has probably noticed that the integration by parts in (16) is not valid for linear combinations of the hat functions. This problem is circumvented by the following argument. Recall that

$$\langle Tx, y \rangle = -\int_0^1 x'' y \, dt,$$

and set

$$\tau(x, y) = \int_0^1 x' y' \, dt.$$

By (16), $\langle Tx, y \rangle = \tau(x, y)$ if x and y lie in the vector space V of functions that have two continuous derivatives and vanish at 0 and 1. The idea is that, given an $x \in V_n$ and a small $\varepsilon > 0$, one may find a function $z \in V$ such that $|\langle Tz, z \rangle - \tau(x, x)| < \varepsilon$, by "rounding the corners of x," for example. The preceding inequalities hold for these rounded functions since ε was arbitrary. Next, let $\varepsilon \to 0$ and argue that the final inequalities (i.e., $\lambda_k \le \Lambda_k$) for the span of the hat functions is valid, using the continuity of the forms $\langle Tz, z \rangle$ and $\tau(z, z)$.

This same kind of reduction of the continuity requirement is common to many finite element procedures.

Exercises 5.6

The small versions of the following problems are set up for hand calculations, some of which might be tedious. Yet they reflect the power of the method. A calculator is advised for most of these problems. Much more interesting results can be obtained if more (or different) basis functions are used and the required integrals and eigenvalue problems are calculated numerically.

The domains of the operators are composed of functions with at least two continuous derivatives on the set in question.

1. Let $T(f) = -f''$ act on the vector space V of functions satisfying $f'(0) = f(1) = 0$. Let $\langle f, g \rangle = \int_0^1 f(t)g(t)\,dt$. T has eigenfunctions $e_k(x) = \cos((k - \frac{1}{2})\pi x)$.
   (a) Show that T is Hermitian on V in the given $\langle \, , \rangle$.
   (b) Use polynomials to approximate the first two eigenvalues and eigenvectors. Normalize your approximate eigenvectors so that $\tilde{e}_k(0) = 1$. Compare your results with the exact solution.
   (c) Use polynomials to approximate the first four eigenvalues and eigenvectors. Compare your results with the exact solution.
   (d) Use a computer and polynomials to approximate the first n eigenvalues and eigenvectors, for larger n. What problems do you encounter?

2. Let $T(f) = -(f'' + 2f') = -e^{-2x}(e^{2x}f')'$ act on the vector space V of functions satisfying $f(0) = f(1) = 0$. Let $\langle f, g \rangle = \int_0^1 f(x)\overline{g(x)}e^{2x}\,dt$. The eigenfunctions are $e_k(x) = e^{-x}\sin(k\pi x)$ with corresponding eigenvalues $(k\pi)^2 + 1$.
   (a) Show that T is Hermitian on V in the given $\langle \, , \rangle$.
   (b) Show that, if f,g are real and satisfy the boundary conditions, then

$$\langle T(f), g \rangle = \int_0^1 f'(x)g'(x)e^{2x}\,dx.$$

   Express a_{ij} in terms of integrals of this type.
   (c) Use the basis functions $f_k = e^{-x}x^k(1 - x)$, $k = 1, 2$, to approximate the first two eigenvalues and eigenvectors. Compare your results with the exact solutions.

3. Let $T(f) = -(x^2 f'' + xf') = -x(xf')'$ act on the vector space V of functions satisfying $f(1) = f(2) = 0$. Let $\langle f, g \rangle = \int_1^2 f(x)\overline{g(x)}(1/x)\,dx$. The eigenfunctions are $e_k(x) = \sin((k\pi/\log 2)\log(x))$ with corresponding eigenvalues $\lambda_k = (k\pi/\log 2)^2$.
   (a) Show that T is Hermitian on V in the given $\langle \, , \rangle$.

(b) Show that if $f, g \in V$ and are real, then

$$\langle T(f), g \rangle = \int_1^2 f'(x)g'(x)x\,dx.$$

Express a_{ij} in terms of integrals of this type.

(c) Approximate the first two eigenvalues and eigenvectors using the following basis functions (which must satisfy the boundary conditions):
   (i) $f_1(x) = (x - 1)(x - 2)$, $f_2(x) = (x - 1)(x - 2)^2$
   (ii) $f_i(x) = (\log(x))^i \log(x/2)$, $i = 1, 2$

(d) Compare your results with the exact solutions.

4. Let $T(f) = -(f'' + (1/x)f') = -(1/x)(xf')'$ act on the vector space V of functions satisfying the following conditions: $f(x)$ and $f'(x)$ are bounded as $x \to 0^+$, $f(1) = 0$. Let $\langle f, g \rangle = \int_0^1 f(t)\overline{g(t)}t\,dt$. The eigenfunctions are $e_k(x) = J_0(\alpha_k x)$ with corresponding eigenvalues $-\alpha_k^2$, where $J_0(z)$ is the Bessel function of order 0 of the first kind and α_k is the kth positive zero of J_0. Note that $e_k(0) = 1$ and each eigenfunction is even in x.

(a) Show that T is Hermitian on V in the given $\langle\, ,\, \rangle$.

(b) Show that if f, g are real and satisfy the boundary conditions,

$$\langle T(f), g \rangle = \int_0^1 f'(x)g'(x)x\,dx.$$

Express a_{ij} in terms of integrals of this type.

(c) Use the following basis functions to approximate the first two eigenvalues and eigenvectors. Normalize $\tilde{e}_i$ so that $\tilde{e}_i(0) = 1$. Compare your results with the exact solutions. If you cannot evaluate $J_0(z)$, then just compare the eigenvalues.
   (i) $f_k(x) = (1 - x)^k$, $k = 1, 2$
   (ii) $f_k(x) = 1 - x^{k+1}$, $k = 1, 2$

(d) Do part (c) with $k = 1, 2, 3, 4$.

(e) The basis functions in (ii) should give better results. Can you explain why?

5. Let $T(f) = -(f'' + (2/x)f') = -(1/x^2)(x^2 f')'$ act on the vector space V of functions satisfying the following conditions: $f(x)$ and $f'(x)$ are bounded as $x \to 0^+$, $f(1) = 0$. Let $\langle f, g \rangle = \int_0^1 f(t)\overline{g(t)}t^2\,dt$. The eigenfunctions are $e_k(x) = (\sin(k\pi x)/kx)$ with corresponding eigenvalues $(k\pi)^2$. Note that $e_k(0) = 1$.

(a) Show that T is Hermitian on V in the given $\langle\, ,\, \rangle$. (Hint: Use $\langle f, g \rangle = \lim_{c \to 0^+} \int_c^1 f(t) \times \overline{g(t)}t^2\,dt$ and integration by parts.)

(b) Show that if f, g are real and satisfy the boundary conditions,

$$\langle T(f), g \rangle = \int_0^1 f'(x)g'(x)x^2\,dx.$$

Express a_{ij} in terms of integrals of this type.

(c) Use the following basis functions to approximate the first two eigenvalues and eigenvectors. Normalize the $\tilde{e}_k$ so that $\tilde{e}_k(0) = 1$. Compare your answers with the exact solution.
   (i) $f_k(x) = (1 - x)^k$, $k = 1, 2$
   (ii) $f_k(x) = 1 - x^{k+1}$, $k = 1, 2$

(d) Do (c) with $k = 1, 2, 3, 4$.

(e) The basis functions in (ii) should give better results. Can you explain why?

6. Let $T(u) = -(u_{xx} + u_{yy})$ (the negative of the Laplacian) act on the vector space of functions smooth in the unit square Ω and vanishing on the boundary. In the inner product

$$\langle f, g \rangle = \int_{\Omega} f(x, y)\bar{g}(x, y)\, dA,$$

T can be shown to be self-adjoint on the vector space in question by using one of Green's identities (cf. the exercises in Section 5.1). The eigenfunctions and eigenvalues are $e_{nm}(x, y) = \sin(n\pi x)\sin(m\pi y)$, $\lambda_{nm} = \pi(n^2 + m^2)$. Invent some basis functions and obtain approximations of the eigenvalues and eigenvectors of T. Perhaps you might try some of the form $f_i(x)g_j(y)$.

CHAPTER 6

Bilinear and Quadratic Forms

Quadratic and bilinear forms arise in a variety of ways in mathematics and the sciences. The simplest quadratic forms in the plane are $ax^2 + bxy + cy^2$. Conic sections can be described as level curves of these quadratic forms. An inner product is an example of a bilinear form, and the potential and kinetic energy of a linear mechanical system are examples of quadratic forms. Surfaces can be described by the first and second fundamental quadratic forms of differential geometry. There are relationships between quadratic forms, bilinear forms, and the Rayleigh quotient. The fact that the potential energy of many (linearized) physical systems is a quadratic form that is minimized at equilibrium may be used to formulate finite element and other schemes to solve for the equilibrium.

In this chapter quadratic and bilinear forms will be considered in a general setting in order to unify the ideas and to develop a general theory. Many examples and applications will be given. The goals of the chapter are to introduce the language of quadratic forms, to study some of the essential theory, and to indicate how computations proceed. The primary tool for computations will be the reduction of a general bilinear form to one on $\mathbf{F}^n$.

6.1
PRELIMINARIES

The usual burst of technical definitions and elementary relationships will initiate the chapter.

DEFINITIONS. A *bilinear form* b on the vector space V is a scalar-valued function of two variables $b(x, y)$, $x, y \in V$, satisfying the following rules for all $x, y \in V$ and for all scalars α, β:

(a) $$b(\alpha x + \beta y, z) = \alpha b(x, z) + \beta b(y, z) \quad \text{(Linear in the first variable)}$$

282

(b) $b(z, \alpha x + \beta y) = \overline{\alpha} b(z, x) + \overline{\beta} b(z, y)$ (Conjugate linear in the second variable)

A bilinear form b is said to be *Hermitian on V* if and only if

$$b(x, y) = \overline{b(y, x)} \quad \forall x, y \in V.$$

If the scalars are real, then $\overline{\alpha} = \alpha$ and $\overline{\beta} = \beta$, so the conjugate signs disappear in the definitions. When the scalars are complex, many authors use the term *sesquilinear form* in place of *bilinear form*, since b is not linear in the second variable. *Bilinear* is used here to simplify the exposition. Expressions will usually be written as if the scalars were complex. All that must be done to recover the real case is to drop the conjugation where it appears. The following remark indicates the strong difference between the real and complex cases.

Remark. Let b be a nonzero bilinear form, and set $b^*(x, y) \equiv \overline{b(y, x)}$ and $\tilde{b}(x, y) \equiv b(y, x)$.

(a) Complex scalars: b^* is a bilinear form, but $\tilde{b}$ is not.
(b) Real scalars: $b^* = \tilde{b}$ is linear in both variables, and so is a bilinear form. Further, b is Hermitian iff b is *symmetric:* $b(x, y) = b(y, x)$.

The following examples are important ones from mathematics. Examples from the sciences will be given shortly.

EXAMPLE 1. THE BILINEAR FORM GENERATED BY A LINEAR OPERATOR AND INNER PRODUCT. Let $T : V \to W$ be a linear operator with V a subspace of W and $\langle , \rangle$ an inner product on W. Define b by

$$b(x, y) = \langle T(x), y \rangle \quad x, y \in V. \tag{1}$$

Then

b is bilinear.
b is Hermitian iff T is Hermitian.

In fact, $b(\alpha x + \beta y, z) = \langle T(\alpha x + \beta y), z \rangle = \langle \alpha T(x) + \beta T(y), z \rangle = \alpha \langle T(x), z \rangle + \beta \langle T(y), z \rangle$
$= \alpha b(x, z) + \beta b(y, z)$. Likewise, $b(z, \alpha x + \beta y) = \langle z, T(\alpha x + \beta y) \rangle = \langle z, \alpha T x + \beta T y \rangle$
$= \overline{\alpha} \langle z, T x \rangle + \overline{\beta} \langle z, T y \rangle = \overline{\alpha} b(z, x) + \overline{\beta} b(z, y)$. To verify the second statement (assuming T^* exists),

$$\overline{b(y, x)} = \overline{\langle T(y), x \rangle} = \langle x, T(y) \rangle = \langle T^*(x), y \rangle = \langle T(x), y \rangle = b(x, y) \quad \forall x, y \in V$$

iff $T = T^*$.

EXAMPLE 2. THE BILINEAR FORM ON F^n GENERATED BY A MATRIX A. Let $A = [a_{ij}]$ be an $n \times n$ matrix. Define b by

$$b(x, y) = y^H A x = \sum_{i,j=1}^{n} a_{ij} \overline{y}_i x_j. \tag{2}$$

Then

(i) b is a bilinear form on F^n.
(ii) $b^*(x, y) = \overline{b(y, x)}$ is the bilinear form generated by A^H.
(iii) b is Hermitian iff A is Hermitian.

In fact, $b(\alpha x + \beta y, z) = z^H A(\alpha x + \beta y) = z^H(\alpha A x + \beta A y) = \alpha z^H A x + \beta z^H A y = \alpha b(x, z) + \beta b(y, z)$. Conjugate linearity is similar. The simple but crucial identity

$$\overline{x^H A y} = (\overline{x^H A y})^T = [x^H A y]^H = y^H A^H x \tag{3}$$

implies that $\overline{b(y, x)} = y^H A x$, so that b^* is the bilinear form generated by A^H. For (iii), $b(y, x) = y^H A^H x = y^H A x = b(x, y) \; \forall x, y \in \mathbf{F}^n \Leftrightarrow A^H = A$.

Of course, this example may be derived from the first one by writing $y^H A x = \langle A x, y \rangle$ and noting that the adjoint of the matrix operator is $T^*(x) = A^H x$ relative to the standard inner product. Equation (3) may also be derived by $x^H A y = \langle A y, x \rangle = \langle x, A y \rangle = \langle A^H x, y \rangle = y^H A^H x$.

DEFINITION. A *quadratic form* on V is a function q of the form $q(x) = b(x, x)$, where b is a bilinear form on V. b is said to *generate* q. q is said to be *Hermitian* iff it has a Hermitian generator b.

With real scalars, a quadratic form is always real-valued. The next proposition asserts that the same is true for complex scalars when the bilinear form is Hermitian.

PROPOSITION 1. A Hermitian quadratic form is real-valued.

Proof

$$q(x) = b(x, x) = \overline{b(x, x)} = \overline{q(x)},$$

interchanging x and x.

EXAMPLE 3. An inner product on V is a Hermitian bilinear form. Its quadratic form is $\|x\|^2 = \langle x, x \rangle$.

EXAMPLE 4. THE QUADRATIC FORM GENERATED BY A MATRIX A. If $b(x, y) = y^H A x$, then

$$q(x) = x^H A x = \sum_{i,j=1}^{n} a_{ij} \overline{x}_i x_j.$$

EXAMPLE 5. THE POTENTIAL AND KINETIC ENERGIES OF A VIBRATING STRING ARE REAL HERMITIAN QUADRATIC FORMS. Let $u(x, t)$ denote the vertical displacement at x of an ideal vibrating string at time t. Assume that the string has fixed ends at $x = 0, L$. Let ρ be the linear density of the string and τ the tension in the string. If the displacements and slopes $u_x(x, t)$ are small and there are no external forces acting on the string, then the potential energy V and kinetic energy T of the string are given (approximately) by

$$U = q(u) = \frac{1}{2}\tau \int_0^L u_x^2(x, t)\,dx \quad u_x \equiv \frac{\partial u}{\partial x}$$

$$T = r(u) = \frac{1}{2}\rho \int_0^L u_t^2(x, t)\,dx \quad u_t \equiv \frac{\partial u}{\partial t}$$

For a fixed t, U and T are real quadratic forms in u. U and T are generated, respectively, by the symmetric bilinear forms

$$b(u, v) = \frac{\tau}{2}\int_0^L u_x v_x\,dx \quad \text{and} \quad \psi(u, v) = \frac{\rho}{2}\int_0^L u_t v_t\,dx.$$

b and ψ are clearly linear in the functions u and v, due to the linearity of the derivatives and integrals. If the scalars were complex, then b and ψ would not be bilinear forms since they are not conjugate linear in the second variable v. To make them bilinear with complex scalars, replace v by $\bar{v}$.

Following are additional examples of quadratic forms on function spaces occurring in the sciences. The scalars are real. It is left to the reader to find Hermitian bilinear forms generating these quadratic forms.

EXAMPLE 6

(a) The potential energy of a static electric field in a dielectric occupying a region Ω, with dielectric constant ε, is

$$q(\phi) = \frac{1}{2} \int_{\Omega} \varepsilon \|\nabla \phi\|^2 \, dV,$$

where ϕ is the potential for the field: $-\nabla \phi = \mathbf{E}$.

(b) The potential energy of an unloaded static string fixed at $x = 0$ and elastically restrained at $x = L$ is

$$q(u) = \frac{\tau}{2} \int_0^L u_x^2 \, dx + \frac{1}{2} ku(L)^2,$$

where $u(x)$ is the vertical displacement of the string at x, τ is the tension in the string, and k is the spring constant of the elastic restraint.

(c) The potential energy in an unloaded beam with clamped or simply supported ends is

$$q(y) = \frac{1}{2} \int_0^L EI(y'')^2 \, dx,$$

where $y(x)$ is the vertical displacement of the beam at x, E is Young's modulus, and I is the cross-sectional moment of inertia about the neutral axis of the beam.

The next example comes from the theory of surfaces.

EXAMPLE 7. Let S be a surface in $\mathbf{R}^3$ parametrized by $\mathbf{r} = f(u_1, u_2)$ with f having two continuous derivatives. Put $\mathbf{r}_i = \partial \mathbf{r}/\partial u_i$, $\mathbf{r}_{ij} = \partial^2 \mathbf{r}/\partial u_i \partial u_j$, and let $\mathbf{n} = \mathbf{r}_1 \times \mathbf{r}_2 / \|\mathbf{r}_1 \times \mathbf{r}_2\|$. Note that each $\mathbf{r}_i$ is tangent to S and $\mathbf{n}$ is a unit normal to S. It is assumed that $\{\mathbf{r}_1, \mathbf{r}_2\}$ is linearly independent. Define $a_{ij} = \mathbf{r}_i \circ \mathbf{r}_j$, $b_{ij} = \mathbf{n} \circ \mathbf{r}_{ij}$, where $\circ$ is the usual dot product. The first and second fundamental forms of S are the quadratic forms

(i) $q(v) = v^T A v$ (First fundamental form)
(ii) $Q(v) = v^T B v$ (Second fundamental form)   where $v = (v_1, v_2)^T$.

q is used to measure arc length along the surface, and Q is used to measure various curvatures of the surface. Notice that A is simply the Gram matrix of the tangent vectors $\mathbf{r}_1, \mathbf{r}_2$ and is therefore nonsingular.

Recovery of a Bilinear Form from a Quadratic Form

Often the quadratic form is given without reference to a generating bilinear form. In order to perform an analysis of the quadratic form, a generating bilinear form must be found. Frequently this is easy, but there are cases where it is not. It turns out that if the quadratic form is generated by a Hermitian bilinear form, many desirable results occur. It is therefore of great usefulness to be able to construct a Hermitian bilinear form for the given quadratic form when possible. The real and complex cases are quite different in this respect.

PROPOSITION 2. Let V be a real vector space and q be a quadratic form generated by a bilinear form b. Then $\tilde{b}$ also generates q. Define $b_s = \frac{1}{2}(b + \tilde{b})$. Then b_s is a symmetric generator of q. Thus every quadratic form on a real vector space is Hermitian.

Proof. b_s is a bilinear form, since both b and $\tilde{b}$ are linear in each variable. Clearly, $q(x) = b(x, x) = \tilde{b}(x, x) = b_s(x, x)$.

EXAMPLE 8. On $\mathbf{R}^n$, let $q(x) = x^T A x$. Then q is generated by A^T and by the symmetric matrix $\frac{1}{2}(A + A^T)$.

Clearly, q is generated by $b(x, y) = y^T A x$. From (3), $\tilde{b}(x, y) = b(y, x) = x^T A y = y^T A^T x$, so that A^T also generates q. Thus

$$q(x) = b_s(x, x) = x^T \left[\frac{1}{2}(A + A^T) \right] x. \tag{4}$$

EXAMPLE 9

$$q(x) = 2x_1^2 + 3x_2^2 - x_3^2 + 4x_1 x_2 + 5x_1 x_3 - 6x_2 x_3$$

Show that q is a quadratic form, and find a symmetric matrix generating q.

We have

$$q(x) = x_1(2x_1 + 4x_2 + 5x_3) + x_2(3x_2 - 6x_3) - x_3^2$$

$$= (x_1, x_2, x_3) \begin{bmatrix} 2x_1 + 4x_2 + 5x_3 \\ 3x_2 - 6x_3 \\ -x_3 \end{bmatrix} = (x_1, x_2, x_3) \begin{bmatrix} 2 & 4 & 5 \\ 0 & 3 & -6 \\ 0 & 0 & -1 \end{bmatrix} \begin{bmatrix} x_1 \\ x_2 \\ x_3 \end{bmatrix} = x^T A x,$$

so that q is the quadratic form generated by $b(x, y) = y^T A x$.

The symmetric matrix generating q is

$$A_s = \frac{1}{2} \left\{ \begin{bmatrix} 2 & 4 & 5 \\ 0 & 3 & -6 \\ 0 & 0 & -1 \end{bmatrix} + \begin{bmatrix} 2 & 0 & 0 \\ 4 & 3 & 0 \\ 5 & -6 & -1 \end{bmatrix} \right\} = \begin{bmatrix} 2 & 2 & 5/2 \\ 2 & 3 & -3 \\ 5/2 & -3 & -1 \end{bmatrix}.$$

Note that $(A_s)_{ii}$ is the coefficient of x_i^2, while $(A_s)_{ij}$ is half the coefficient of $x_i x_j$. This will be proved in general later.

EXAMPLE 10. Let $q(f) = 2 \int_0^1 f'(x) f(x) \, dx$, real scalars. Show that q is a quadratic form and find the symmetric bilinear b_s form generating q.

Put $b(f, g) = 2 \int_0^1 f'(x) g(x) \, dx$. Clearly, b is linear in f and g and so is a real bilinear form. Also $b(f, f) = q(f)$, so b generates q. Then

$$b_s(f, g) = \frac{1}{2} \{ b(f, g) + b(g, f) \} = \int_0^1 \{ f'(x) g(x) + g'(x) f(x) \} \, dx$$

also generates q.

The following theorem addresses the question of the number of generators a quadratic form may have, and explains how to recover a generating bilinear form in the general case. It again demonstrates the differences between real and complex vector spaces. Examples 8, 9, and 10 show that quadratic forms on real vector spaces may have many generators, one of which is Hermitian. The uniqueness of this generator is addressed.

THEOREM 3. POLAR IDENTITIES

(a) If the complex numbers are the scalar field, then there is a one-to-one correspondence between bilinear forms and quadratic forms. The unique bilinear form generating a quadratic form q may be recovered from the *complex polar identity*

$$b(x, y) = \tfrac{1}{4}\{q(x + y) - q(x - y) + i[q(x + iy) - q(x - iy)]\}. \tag{5}$$

(b) If the real numbers are the scalar field, there is a one-to-one correspondence between quadratic forms and symmetric bilinear forms. The symmetric bilinear form b_s generating q is recovered from q by the *real polar identity*

$$b_s(x, y) = \tfrac{1}{4}\{q(x + y) - q(x - y)\}. \tag{6}$$

Proof

(a) *Complex scalars.* If $q(x) = b(x, x)$, the identity (5) may be verified by expanding $q(x \pm y) = b(x \pm y, x \pm y)$ and $q(x \pm iy) = b(x \pm iy, x \pm iy)$ using the bilinearity of b. Thus (5) recovers b from q, so that b and q uniquely determine each other, which establishes the one-to-one correspondence.

(b) *Real scalars.* The identity (6) can be verified just by expanding $q(x \pm y) = b_s(x \pm y, x \pm y)$ using bilinearity and the symmetry of b_s. This establishes the one-to-one correspondence between symmetric bilinear forms and quadratic forms in the real case.

Remark. In $\mathbf{C}^n$, if the quadratic form is given by

$$q(x) = \sum_{i=1}^{n} b_{ii}|x_i|^2 + \sum_{i \neq j} b_{ij} x_i \bar{x}_j,$$

there is only one way to write $q(x) = x^H A x$, namely, $a_{ij} = b_{ij}$.

Matrix Representations of Bilinear and Quadratic Forms

Given a basis β for V, a matrix A is sought such that the following matrix representation holds:

$$b(x, y) = [y]^H A[x], \quad \forall x, y \in V.$$

When this is possible, bilinear forms on finite-dimensional spaces are connected to bilinear forms on $\mathbf{F}^n$. This matrix representation enables matrix theory and computer algorithms to be used in the analysis of bilinear forms. Theoretical results about general bilinear forms can be applied to bilinear forms on $\mathbf{F}^n$ and vice versa. Many finite element algorithms are based on this correspondence between bilinear forms in function spaces and their matrix representation.

THEOREM 4. MATRIX REPRESENTATION OF BILINEAR FORMS. Let $\beta = \{e_1, e_2, \ldots, e_n\}$ be a basis for V and b a bilinear form on V. If

$$A = [a_{ij}], \quad a_{ij} = b(e_j, e_i), \tag{7}$$

then

$$b(x, y) = [y]^H A[x], \tag{8}$$

where $[x]$, $[y]$ are the coordinate columns of x and y in the basis β. Furthermore, b is Hermitian iff A is a Hermitian matrix.

A is called the *matrix of the bilinear form b* (and of the quadratic form $q(x) = b(x, x)$) in the basis $\beta = \{e_1, e_2, \ldots, e_n\}$, and one writes $A = [b]_\beta = [q]_\beta$.

Proof. Let $x = \sum_{j=1}^n x_j e_j$. $y = \sum_{i=1}^n y_i e_i$, so that $[x] = (x_1, x_2, \ldots, x_n)^T$ and $[y] = (y_1, y_2, \ldots, y_n)^T$. Bilinearity yields

$$b(x, y) = b\left(\sum_{j=1}^n x_j e_j, \sum_{i=1}^n y_i e_i\right) = \sum_{j=1}^n x_j b\left(e_j, \sum_{i=1}^n y_i e_i\right)$$

$$= \sum_{i,j=1}^n x_j \bar{y}_i b(e_j, e_i).$$

Comparing this with Example 2 gives $b(x, y) = [y]^H A[x]$. If b is Hermitian, then $a_{ij} = b(e_j, e_i) = \overline{b(e_i, e_j)} = \bar{a}_{ji}$, so that A is Hermitian. Now suppose A is Hermitian. Then the argument following Eq. (3) (with x replaced by $[x]$ and y replaced by $[y]$) shows that b is Hermitian.

Remark. For a fixed basis β, we have $[\alpha_1 b_1 + \alpha_2 b_2]_\beta = \alpha_1[b_1] + \alpha_2[b_2]_\beta$ where the α_i are scalars and the b_i are bilinear forms.

EXAMPLE 11. *Consistency:* If $b(x, y) \equiv y^H A x$ is the bilinear form generated by the $n \times n$ matrix A, then the matrix of b with respect to the standard basis is A.

To see this, just compute

$$b(\delta_j, \delta_i) = \delta_i^T A \delta_j = \delta_i^T A_j = a_{ij}.$$

EXAMPLE 12. THE MATRIX OF AN INNER PRODUCT IS THE GRAM MATRIX. Let $\langle , \rangle$ be an inner product for V and $\{e_k\}_{k=1}^n$ a basis for V. Then

$$\langle x, y \rangle = \left\langle \sum_{j=1}^n x_j e_j, \sum_{i=1}^n y_i e_i \right\rangle = \sum_{i,j} x_j \bar{y}_i \langle e_j, e_i \rangle = [y]^T G[x],$$

where G is the Gram matrix with respect to this basis: $g_{ij} = \langle e_j, e_i \rangle$.

Often quadratic forms on $\mathbf{R}^n$ are given by expressions such as $q(x) = \sum_{i=1}^n b_{ii} x_i^2 + \sum_{1 \le i < j \le n} b_{ij} x_i x_j$ (cf. Example 9). The next example shows how to express them as $x^T A x$, with A symmetric.

EXAMPLE 13. Let $q(x) = \sum_{1 \le i \le j \le n} b_{ij} x_i x_j$ with b_{ij} real. Show that q is a real quadratic form on $\mathbf{R}^n$. Find a symmetric matrix A so that $q(x) = x^T A x$.

Put $b(x, y) = \sum_{1 \le i \le j \le n} b_{ij} y_i x_j$. Then b is linear in y for fixed x, and linear in x for fixed y. Thus b is a bilinear form generating q. The matrix C of b in the standard basis is

$$c_{pq} = b(\delta_q, \delta_p) = \sum_{1 \le i \le j \le n} b_{ij} \delta_{pi} \delta_{qj} = \begin{cases} 0 & p > q \\ b_{pq} & p \le q \end{cases},$$

which is upper triangular. A symmetric matrix generating q is $A = \frac{1}{2}(C + C^T)$. That is, $a_{ij} = \frac{1}{2}(c_{ij} + c_{ji})$, so that

$$a_{ii} = b_{ii} \quad \text{and} \quad a_{ij} = a_{ji} = \tfrac{1}{2} b_{ij} \quad \text{if } i < j. \tag{9}$$

EXAMPLE 14. Let $V = \mathbf{P}_2$, with real scalars. Consider $q(f) = 2\int_0^1 f'(t)f(t)\,dt$. Find a symmetric matrix $S = [s_{ij}]$ such that, if $f(t) = x_0 + x_1 t + x_2 t^2$, then $q(f) = x^{\mathrm{T}} S x$, where $x = (x_0, x_1, x_2)^{\mathrm{T}}$.

A generator of q is $b(f, g) = 2\int_0^1 f'(t)g(t)\,dt$, and the symmetric generator of q is $b_s(f, g) = \int_0^1 \{f'(x)g(x) + g'(x)f(x)\}\,dx$ (cf. Example 10). There are several approaches to finding A_s. *First method:* Find the matrix A of b: $a_{ij} = b(t^j, t^i)$, $i, j = 0, 1, 2$, and put $S = (A + A^{\mathrm{T}})/2$. *Second method:* Obtain the matrix S directly as the matrix of b_s: $s_{ij} = b_s(t^j, t^i)$. *Third method:* Substitute $f(t) = x_0 + x_1 t + x_2 t^2$ into the definition of $q(x)$, multiply out the product $f(t)f'(t)$, and integrate. An expression of the form $q(f) = \sum_i b_{ii} x_i^2 + \sum_{i<j} b_{ij} x_i x_j$ will be obtained, from which we can write down the symmetric matrix of a real bilinear form on $\mathbf{R}^3$ according to the rule in Example 8 or 13. All these procedures yield

$$S = \begin{bmatrix} 0 & 1 & 1 \\ 1 & 1 & 1 \\ 1 & 1 & 1 \end{bmatrix}.$$

The next example is an applied problem in electrical engineering. It shows a good application of the polar identities and the matrix representation of bilinear forms.

EXAMPLE 15. COMPUTATION OF THE CAPACITANCE MATRIX OF A SYSTEM OF CONDUCTORS. A conducting surface S_0 encloses a dielectric and n conductors S_i. Assume the dielectric is homogeneous and isotropic with dielectric constant ε. Let v_i and q_i denote the voltage and charge on the ith conductor, and define the columns v and q by $v = (v_0, v_1, \ldots, v_n)^{\mathrm{T}}$, $q = (q_0, q_1, \ldots, q_n)^{\mathrm{T}}$. The voltages and charges are known to be related by $q = Cv$, where $C = [c_{ij}]$ is called the capacitance matrix and is symmetric. An important practical problem is to compute the capacitance matrix of a system of conductors. For example, if they form a cable, then these capacitances determine the amount of cross-talk experienced in the transmission of data or other signals.

The energy U in the electric field in the dielectric is given two ways:

$$U = \frac{1}{2}\int_\Omega \varepsilon \|\nabla\phi\|^2\,dV, \tag{*}$$

where Ω is the region occupied by the dielectric and ϕ is the potential for the electric field in the dielectric, and

$$U = \tfrac{1}{2} v^{\mathrm{T}} C v. \tag{**}$$

Problem: Given a numerical package that computes ϕ, find a method by which to recover the capacitance matrix C.

Since ϕ depends linearly on the boundary voltages v, both expressions are quadratic forms in v, and (**) is the (symmetric) matrix representation of (*). The symmetric bilinear form (in terms of the v_i) generating (*) cannot be explicitly written out, except in a very few special cases. However, for each v, ϕ and then $U = q(v)$ can be computed from (*). From the real polar identity (6), one can then recover the symmetric bilinear form $b_s(u, v) = \frac{1}{4}\{q(u + v) - q(u - v)\}$. Then, from the symmetric bilinear form, the matrix C may be obtained by $c_{ij} = b_s(\delta_j, \delta_i)$. For example, to find $c_{12} = \{q(\delta_1 + \delta_2) - q(\delta_1 - \delta_2)\}$ the numerical package must be run twice.

It is possible to show that every bilinear form b on a finite-dimensional space may be represented in the form $b(x, y) = \langle T(x), y \rangle \; \forall x, y \in V$, for some T, where

$\langle , \rangle$ is a given inner product. This is an aesthetically pleasing result, and it is useful in theoretical arguments. In many problems, the operator T is easy to find. However, in many practical problems T is often difficult to find, and it has no clear physical meaning when found. In such cases, when V is finite-dimensional, it is more advantageous to use the matrix representation of b for theoretical and computational considerations. See the exercises for one approach to the construction of T.

Change of Basis and the Effect on the Matrix of a Bilinear Form

Suppose the change of basis is given by $f_i = \sum_{k=1}^n p_{ki} e_k$, or, equivalently, the change of coordinates is given by $P\tilde{c} = c$, where $x = \sum_{i=1}^n c_i e_i = \sum_{i=1}^n \tilde{c}_i f_i$. Let A be the matrix of b with respect to $\beta = \{e_k\}_{k=1}^n$ and B be the matrix of b with respect to $\gamma = \{f_k\}_{k=1}^n$. Then with $y = \sum_{i=1}^n d_i e_i = \sum_{i=1}^n \tilde{d}_i f_i$,

$$b(x, y) = d^H[b]_\beta c = (P\tilde{d})^H A(P\tilde{c}) = \tilde{d}^H(P^H AP)\tilde{c} = \tilde{d}^H[b]_\gamma \tilde{c} \quad \text{for all } \tilde{c}, \tilde{d}.$$

Hence

$$B = P^H AP. \tag{10}$$

An alternative proof may be made by substituting the expressions for f_i and f_j into $b_{ij} = b(f_j, f_i)$ and using bilinearity. It is easy to remember the rule (10): Substitute $c = P\tilde{c}$ in the quadratic form $c^H Ac$ or the bilinear form $d^H Ac$.

Two matrices A and B that are connected by Eq. (10) are said to be *congruent*. Thus congruent matrices represent the same bilinear form, but in different bases.

Exercises 6.1

1. (a) Find the bilinear and quadratic forms on $\mathbf{C}^2$ generated by

$$A = \begin{bmatrix} 1 & 3i \\ -3i & 2 \end{bmatrix}.$$

   (b) Find the bilinear and quadratic forms on $\mathbf{C}^2$ generated by

$$A = \begin{bmatrix} 1 & 3 \\ 1 & 2 \end{bmatrix}.$$

   Find a symmetric matrix generating the quadratic form generated by A.

2. For each of the following quadratic forms, find a matrix A (symmetric in the real case) generating q.
   (a) $q(x) = 2x_1 x_2$ $V = \mathbf{R}^2$
   (b) $q(x) = x_1^2 + 6x_1 x_2 - x_2^2$ $V = \mathbf{R}^2$
   (c) $q(x) = |x_1|^2 + 6x_1 \bar{x}_2 - |x_2|^2$ $V = \mathbf{C}^2$
   (d) $q(x) = |x_1|^2 + 2ix_1 \bar{x}_2 + 4i\bar{x}_1 x_2 + |x_2|^2$ $V = \mathbf{C}^2$
   (e) $q(x) = 2x_1^2 + 2x_2^2 + 3x_3^2 + 2x_1 x_2 + 4x_2 x_3$ $V = \mathbf{R}^3$
   (f) $q(x) = 2x_1^2 + x_2^2 + 2x_3^2 + 2x_1 x_2 + 2x_1 x_3$ $V = \mathbf{R}^3$
   (g) $q(x) = x_1^2 + 3x^2 + 10x_3^2 + 10x_4^2 - 2x_1 x_2 + 2x_1 x_3 + 4x_1 x_4 + 6x_2 x_3 - 8x_2 x_4$ $V = \mathbf{R}^4$

3. Define q on P_2 (real scalars) by $q(x) = \int_0^1 3x'(t)^2\, dt + x(0)^2 + x(1)x'(1)$.
  (a) Find a symmetric bilinear form b such that $b(x, x) = q(x)$.
  (b) Find a nonsymmetric matrix B so that $q(x) = [x]^T B[x]$ in the basis $\{1, t, t^2\}$.
  (c) Find a symmetric matrix A so that $q(x) = [x]^T A[x]$ in the basis $\{1, t, t^2\}$.

4. Define q on P_2 (real scalars) by $q(x) = \int_0^1 x'(t)^2\, dt + x(0)^2$.
  (a) Find a symmetric bilinear form b such that $b(x, x) = q(x)$.
  (b) Find a symmetric matrix A so that $q(x) = [x]^T A[x]$ in the basis $\{1, t, t^2\}$.

5. Define q on P_2 (complex scalars) by

$$q(x) = \int_0^1 3|x'(t)|^2\, dt + |x(0)|^2 + x(1)\overline{x'(1)}.$$

  (a) Find a bilinear form b such that $b(x, x) = q(x)$.
  (b) Find a matrix A so that $q(x) = [x]^H A[x]$ in the basis $\{1, t, t^2\}$.
  (c) Is q Hermitian?

6. Let A and B be real symmetric matrices. Suppose $x^T A x = x^T B x\ \forall x$. Show that $A = B$. Give an example to show this is false if A or B is not symmetric.

7. Let $B_k = (b_{1k}, b_{2k}, \ldots, b_{nk})^T \in \mathbf{R}^n$ and α_i be real numbers. Define q by $q(x) = \sum_{k=1}^n d_k[B_k^T x]^2$. Show that q is a quadratic form, and find a symmetric matrix A so that $q(x) = x^T A x$.

8. Prove the following theorem:
  Let $\dim(V) = n < \infty$, and let $\langle\,,\,\rangle$ be an inner product on V and b be a bilinear form on V. Then there exists a linear $T : V \to V$ such that $b(x, y) = \langle Tx, y\rangle\ \forall x, y$ in V. b is Hermitian iff T is Hermitian with respect to $\langle\,,\,\rangle$. Further, if β is a basis for V, then $[b]_\beta = G_\beta[T]_\beta$, where G_β is the Gram matrix of the inner product with respect to the basis β. (*Hint:* Essentially, the idea is to define T by the relation $[b]_\beta = G_\beta[T]_\beta$.)

9. Suppose L is linear on V and $b(x, y) = \langle Lx, Ly\rangle$. Show that b is a bilinear form, and determine conditions on L in order for b to be Hermitian.

10. Let ϕ_i be linear functionals on the real vector space V and c_i be real scalars. Show that $q(x) = \sum_{i=1}^n c_i \phi_i(x)^2$ is a quadratic form.

6.2
CLASSIFICATION OF HERMITIAN QUADRATIC FORMS

The classification of Hermitian quadratic forms is important for many reasons. For example, if $q(x) = \langle Tx, x\rangle$, from properties of q certain properties of T can be deduced. Conic sections and quadric surfaces are represented by equations of the type $Q(x) = c$, and classification of Q indicates the type of curve or surface described. The classification of quadratic forms can also be used to describe the behavior of functions of several variables near critical points and to give a qualitative description of a surface in terms of the classification of its second fundamental form.

DEFINITIONS. Let q be a Hermitian quadratic form generated by the Hermitian bilinear form b on V. Then

(a) q and b are *positive definite on V* if and only if

$$q(x) > 0 \quad \text{for all } x \neq 0, \ x \in V.$$

(b) q and b are *positive on V* if and only if

$$q(x) \geq 0 \quad \text{for all } x \in V.$$

(c) q and b are *indefinite on V* if and only if

$$q(x_1) > 0 > q(x_2) \quad \text{for some } x_1, x_2 \in V.$$

(d) q and b are *negative definite on V* if and only if

$$q(x) < 0 \qquad \text{for all } x \neq 0, \ x \in V.$$

(e) q and b are *negative on V* if and only if

$$q(x) \leq 0 \quad \text{for all } x \in V.$$

Classification of Hermitian matrices and operators

The preceding language can be extended to matrices and operators in the following way. A Hermitian operator T on a vector space V is classified relative to an inner product $\langle\,,\,\rangle$ according to the classification of the quadratic form $q(x) = \langle Tx, x\rangle$ on V : T inherits the classification of q. For example, the linear operator T is *positive definite with respect to* $\langle\,,\,\rangle$ *on* V iff $\langle T(x), x\rangle > 0$ for all $x \neq 0$, $x \in V$. Similarly, an $n \times n$ Hermitian matrix A inherits the classification of the quadratic form generated by A : $q(x) = x^H A x = \langle Ax, x\rangle$. For example, A is positive definite iff $x^H A x > 0$ for all $x \neq 0$.

In (a) some authors replace the words *positive definite* by *positive*, and in (b) they replace *positive* by *nonnegative* or *positive semidefinite*, with similar language used for (d) and (e). Among authors, there seems to be an entrenched agreement to disagree.

PROPOSITION 1. A bilinear form b on V defines an inner product by the rule $\langle x, y\rangle = b(x, y)$ iff b is Hermitian and positive definite.

Proof. The statement follows immediately from the definitions relating to bilinear forms and inner products.

EXAMPLE 1. Let A be an $m \times n$ matrix. Then $A^H A$ is positive, and positive definite when the columns of A are linearly independent.

In fact, $q(x) = x^H A^H A x = \|Ax\|^2 \geq 0$. Note that $q(x) = 0 \Leftrightarrow x \in N(A)$. Thus q is definite exactly when the columns of A are independent.

EXAMPLE 2. Trid$(-1, 2, -1)$ is positive definite.

Verify that $q(x) = x^T \text{Trid}(-1, 2, -1)x = x_1^2 + \sum_{i=1}^{n-1}(x_i - x_{i+1})^2 + x_n^2 \geq 0$. If $q(x) = 0$, it is easy to see that x must be 0.

EXAMPLE 3. Let V be the set of real functions that vanish at 0 and 1 and that have two continuous derivatives. Let T be defined by $T(f) = -f''$, where $' = d/dt$, and let $\langle f, g\rangle = \int_0^1 f(t)g(t)\,dt$. Then T is positive definite on V with respect to this inner product.

Integrating by parts, we have for $f \neq 0$, $f \in V$,

$$\langle T(f), f \rangle = -\int_0^1 f''(t)f(t)\,dt = -f'f|_0^1 + \int_0^1 f'(t)f'(t)\,dt = \int_0^1 f'(t)f'(t)\,dt > 0,$$

since, if $\int_0^1 f'(t)^2\,dt = 0$, then $f'(t) = 0 \ \forall t$, whence $f(t) = \text{constant} = f(0) = 0$.

EXAMPLE 4. The potential and kinetic energies of an unloaded string with fixed ends are positive (real) quadratic forms.

According to Example 5 of Section 6.1,

$$U = \frac{1}{2}\tau\int_0^L u_x^2(x, t)\,dx \geq 0, \qquad T = \frac{1}{2}\rho\int_0^L u_t^2(x, t)\,dx \geq 0.$$

An application of the idea of positive definiteness occurs in the study of minimization of functions of several variables.

EXAMPLE 5. MINIMA OF FUNCTIONS OF SEVERAL VARIABLES. Let $a = (a_1, a_2, \ldots, a_n)^T$ be a critical point of $f(x)$, $x = (x_1, x_2, \ldots, x_n)^T$. Suppose f has continuous third-order partials. If the matrix $H(a) = [\partial^2 f(a)/\partial x_i \partial x_j]$ is positive definite, then f has a local minimum at a. H is called the *Hessian* of f at a.

This can be proved from the Taylor expansion

$$f(a + h) = f(a) + \nabla f(a)^T h + \tfrac{1}{2}h^T H(a)h + O(\|h\|^3),$$

where $h = (h_1, \ldots, h_n)^T$, and $O(\|h\|^3)$ denotes a remainder term R such that $|R| \leq C\|h\|^3$ for all small h. At a critical point a, the gradient of f at a vanishes:

$$\nabla f(a) = \left(\frac{\partial f(a)}{\partial x_1}, \frac{\partial f(a)}{\partial x_2}, \ldots, \frac{\partial f(a)}{\partial x_n}\right)^T = 0.$$

The behavior of f near a is therefore determined by the quadratic form $h^T H(a)h$, since it is larger than $O(\|h\|^3)$. Since the Hessian is positive definite, $f(x + h) > f(a)$ for all small $h \neq 0$, and thus f has a local minimum at a. In general, at a critical point of f, the behavior of $h^T H(a)h$ will indicate the behavior of f (provided $H(a)$ is nonsingular).

PROPOSITION 2. Let G be the Gram matrix of a set of vectors $\sigma = \{v_k\}_{k=1}^n$ with respect to an inner product $\langle, \rangle$. Then G is positive definite iff σ is linearly independent, and positive in any case.

Proof. From Example 12 of Section 6.1, we have $\|\sum_{i=1}^n x_i v_i\|^2 = x^H G x$, where $x = (x_1, x_2, \ldots, x_n)^T$. Thus G is positive, since norms can never be negative. If σ is linearly independent, then $x \neq 0$ implies $\sum_{i=1}^n x_i v_i \neq 0$ and therefore $\|\sum_{i=1}^n x_i v_i\|^2 > 0$. If σ is linearly dependent, then some $x \neq 0$ permits $\sum_{i=1}^n x_i v_i = 0$, so $\|\sum_{i=1}^n x_i v_i\|^2 = x^H G x = 0$, and G is not definite.

Diagonalization of quadratic forms

One method of classification of quadratic forms on a finite-dimensional vector space V is to reduce the quadratic form to a simple expression from which its classification may easily be determined. This method consists of finding a basis $\gamma = \{f_k\}_{k=1}^n$ and some scalars w_i for which $x = \sum_{i=1}^n x_i f_i$ implies $q(x) = \sum_{i=1}^n w_i |x_i|^2$. Such a basis γ is said to *diagonalize* q. The numbers w_i are called *weights*. For Hermitian quadratic forms, the w_i must be real by Proposition 1 of Section 6.1. The matrix of q in γ is thus $[q]_\gamma = \text{diag}(w_1, w_2, \ldots, w_n)$.

EXAMPLE 6. Let $q(x) = 2x_1^2 + 4(x_2 - x_1)^2 + 3(x_2 - x_3)^2$, where $x = (x_1 x_2, x_3) \in \mathbf{R}^3$. Classify q.

Make the change of coordinates $y_1 = x_1$, $y_2 = x_2 - x_3$, $y_3 = x_3 - y_3$. Since this transformation is nonsingular, it corresponds to a change of basis. In these coordinates, $q = 2y_1^2 + 4y_2^2 + y_3^2$. Clearly, $q > 0$ unless all the $y_i = 0$, which means that $x = 0$.

PROPOSITION 3. Suppose there is a basis $\{f_k\}_{k=1}^n$ that diagonalizes the Hermitian quadratic form q:

$$q\left(\sum_{i=1}^n x_i f_i\right) = \sum_{i=1}^n w_i |x_i|^2.$$

Then

(a) q is positive definite iff all $w_i > 0$.
(b) q is positive iff all $w_i \geq 0$.
(c) q is indefinite iff for some i and j, $w_i > 0 > w_j$.

Similar statements hold for the negative and negative definite cases.

Proof. Observe that $q(f_k) = w_k$. Suppose $x = \sum_{i=1}^n x_i f_i$.

(a) If all $w_i > 0$, then $x \neq 0 \Rightarrow$ some $x_k \neq 0 \Rightarrow q(\sum_{i=1}^n x_i f_i) = \sum_{i=1}^n w_i |x_i|^2 \geq w_k |x_k|^2 > 0$, and so q is positive definite. If q is positive definite, then $w_i = q(f_i) > 0$ for each i.
(b) The proof is similar.
(c) This follows from $q(f_k) = w_k$, $k = 1, 2, \ldots, n$ and exhausting the other cases.

In the following sections, several methods for diagonalizing quadratic forms will be developed. In general, each of these will produce a different set of weights w_i and a different basis $\{f_k\}_{k=1}^n$. The problem of uniqueness must therefore be addressed. Suppose $\{f_k\}_{k=1}^n$ is a basis diagonalizing the Hermitian quadratic form q: $q(\sum_{i=1}^n x_i f_i) = \sum_{i=1}^n w_i |x_i|^2$. Clearly, the vectors f_i (or, equivalently, the coordinates x_i) may be scaled so that the magnitudes of the nonzero w_i can be arbitrarily set. Thus it makes sense to ask only about the distribution of the signs of the w_i and the number of zero w_i. The next theorem resolves the uniqueness in this sense.

SYLVESTER'S LAW OF INERTIA. Let q be a Hermitian quadratic form on a finite-dimensional nontrivial vector space V. Then the number of positive, negative, and zero weights are the same in each diagonalization of q.

Proof. Suppose there are two diagonalizations,

$$q(x) = \sum_{i=1}^p u_i |y_i|^2 - \sum_{i=p+1}^r u_i |y_i|^2 \quad \text{with respect to the basis } \{e_k\}_{k=1}^n \qquad (\alpha)$$

$$q(x) = \sum_{i=1}^{p'} v_i |z_i|^2 - \sum_{i=p'+1}^{r'} v_i |z_i|^2 \quad \text{with respect to the basis } \{f_k\}_{k=1}^n, \qquad (\beta)$$

with all the coefficients u_i, v_i positive. Let $n = \dim(V)$ and suppose $p > p'$. The set $\{e_1, e_2, \ldots, e_p, f_{p'+1}, f_{p'+2}, \ldots, f_n\}$ is linearly dependent because it contains more than n vectors; hence there are y_i, z_i not all zero so that

$$0 = \sum_{i=1}^p y_i e_i + \sum_{i=p'+1}^n z_i f_i. \qquad (*)$$

It will be shown that some of the y_i are nonzero. If this were not so, then $\sum_{i=p'+1}^{n} z_i f_i = 0$, and so all $z_i = 0$ since the f_i are linearly independent; thus all y_i and all z_i are 0. But this contradicts the fact that the linear combination in $(*)$ is nontrivial, so some $y_i \neq 0$. Now let $x = \sum_{i=1}^{p} y_i e_i$, so that $x = -\sum_{i=p'+1}^{n} z_i f_i$, from $(*)$. Putting this x in each of (α) and (β) yields $q(x) > 0$ by the first expansion and $q(x) \leq 0$ by the second expansion. This is impossible, and thus $p > p'$ is false. Similarly, $p' > p$ is false, so $p = p'$. Analogously, $r - p = r' - p'$, and the theorem is proved.

DEFINITION. The *rank* of a quadratic form is the number of nonzero terms in a diagonalization. The *rank* of a bilinear form b is the rank of the quadratic form $b(x, x)$.

Minimization Problems

Minimization problems of the following type occur often in practice. Examples will be given involving the equilibria of physical systems.

EXAMPLE 7. Let A be a positive definite real symmetric matrix, b a given vector in $\mathbf{R}^n$, and γ a real number. Define

$$q(x) = x^{\mathrm{T}} A x - 2b^{\mathrm{T}} x + \gamma. \tag{1}$$

Show that q is minimal at $x = x_0 = A^{-1}b$, and $\min q(x) = \gamma - b^{\mathrm{T}} x_0$.

First, make a computation analogous to completing the square: $(x - x_0)^{\mathrm{T}} A(x - x_0) = x^{\mathrm{T}} A x - x_0^{\mathrm{T}} A x - x^{\mathrm{T}} A x_0 + x_0^{\mathrm{T}} A x_0 = x^{\mathrm{T}} A x - 2(A x_0)^{\mathrm{T}} x + x_0^{\mathrm{T}} A x_0$ since $x_0^{\mathrm{T}} A = x_0^{\mathrm{T}} A^{\mathrm{T}} = (A x_0)^{\mathrm{T}}$ and $x^{\mathrm{T}} A x_0 = (x^{\mathrm{T}} A x_0)^{\mathrm{T}} = (A x_0)^{\mathrm{T}} x$. Thus if $A x_0 = b$ and $\alpha = \gamma - x_0^{\mathrm{T}} A x_0 = \gamma - b^{\mathrm{T}} x_0$, then

$$Q(x) = (x - x_0)^{\mathrm{T}} A(x - x_0) + \alpha. \tag{2}$$

Since A is positive definite, it is nonsingular (see the exercises), so that x_0 is uniquely defined. If $x \neq x_0$, then $(x - x_0)^{\mathrm{T}} A(x - x_0) > 0$, whence $Q(x) > Q(x_0) = \alpha$. Thus Q has a strict minimum at x_0, and the minimum value is α.

The Ritz method

For many static physical systems under a load, the potential energy of the system is given by

$$V(u) = \tfrac{1}{2} a(u, u) - \phi(u), \tag{3}$$

where u represents the displacement of the system, $a(,\,)$ is a Hermitian positive definite bilinear form representing the internal energy of the system, and ϕ is a linear functional representing the work done on the system by the external loads.

Theorems in the sciences (often proved by mathematicians) assert that the equilibrium state u is determined by minimizing V, subject to the condition that u satisfies certain "essential" boundary conditions. Using calculus of variations, one can deduce the differential equation satisfied by u and certain other implied boundary conditions, called *natural boundary conditions*, that u must satisfy. Two examples follow:

1. For an ideal string under tension τ, fixed at $x = 0$, with the end $x = L$ allowed to slide vertically with no friction, enduring an applied body force density $f(x)$

at x, $u(x)$ = vertical displacement at x:

$$a(u, u) = \tau \int_0^L (u')^2 \, dx, \qquad \phi(u) = \int_0^L f(x)u(x) \, dx$$

Essential boundary condition: $u(0) = A$.
Natural boundary condition: $u'(L) = 0$.
Differential equation: $\tau u'' + f = 0$.

2. For a cantilever beam with constant stiffness EI, clamped at $x = 0$, with applied body force density $f(x)$, and with force P applied at $x = L$, $u(x)$ = vertical displacement at x:

$$a(u, u) = \int_0^L EI(u'')^2 \, dx, \qquad \phi(u) = \int_0^L f(x)u(x) \, dx + Pu(L)$$

Essential boundary conditions: $u(0) = A$, $u'(0) = B$.
Natural boundary conditions: $u''(L) = 0$, $EIu^{(3)}(L) + P = 0$.
Differential equation: $EIu^{(4)} = f$.

The Ritz method is extremely useful for approximating the solutions to such problems and is widely used. It goes as follows: Select a function u_0 that satisfies the inhomogeneous essential boundary conditions. Then select linearly independent functions $\{f_k\}_{k=1}^n$ satisfying the homogeneous essential boundary conditions. (These are called *basis functions* in the trade.) Minimize V over the set of all functions of the form $u = u_0 + w$, $w = \sum_{i=1}^n c_i f_i$. One beauty of the method is that the natural boundary conditions need not be enforced in the approximating functions, although they are needed to solve the differential equation (boundary value problem). A better approximation will result, however, if it can be easily arranged for the f_k to satisfy the natural boundary conditions as well. The functions f_k are chosen by physical insight, by blind guessing, or by selecting some functions that are known, through mathematical theory, to be good approximators. Upon setting $u = u_0 + v$ in $V(u)$, we obtain an expression of the form

$$V = \tfrac{1}{2}a(v, v) - \psi(v) + \gamma. \tag{4}$$

In fact, $a(u_0 + v, u_0 + v) = a(u_0, u_0) + 2a(v, u_0) + a(v, v)$, $\phi(u_0 + v) = \phi(u_0) + \phi(v)$ so that $\psi(v) = \phi(v) - a(v, u_0)$, $\gamma = \tfrac{1}{2}a(u_0, u_0) - \phi(u_0)$. Next, setting $v = \sum_{i=1}^n c_i f_i$ in (4) yields

$$V = \tfrac{1}{2}c^\mathrm{T} Ac - b^\mathrm{T} c + \gamma, \tag{5}$$

where A is the matrix of a with respect to $\{f_k\}_{k=1}^n$ and b^T is the matrix of ψ, i.e.,

$$a_{ij} = a_{ji} = a(f_i, f_j), \qquad b_i = \psi(f_i) \tag{6}$$

Equations (6) may also be obtained from (4) by directly setting $u = u_0 + \sum_{i=1}^n c_i f_i$ in $V(u)$, expanding, collecting terms, and rewriting the result in the form (5). The advantage of the longer analysis is that (6) gives simple formulae for A and b. The c minimizing (5) is found by the method of Example 7: c satisfies $Ac = b$ (minimize $2q = c^\mathrm{T} Ac - 2b^\mathrm{T} c + 2\gamma$). This is the Ritz method, and it provides an approximation to the actual minimizer u. This method is very powerful and is the basis for many approximation schemes, including some in finite elements.

EXAMPLE 8. A string is attached at height $\frac{1}{2}$ at $x = 0$. The end at $x = 1$ moves vertically without friction. The applied body force per unit length is

$$f(x) = \begin{cases} 0 & x \le 0.5 \\ 2 & x > 0.5 \end{cases}.$$

The tension of the string is 1. Estimate the position of the string, using four functions. In general, $u = u_0 + v$ with $u_0 = \frac{1}{2}$, $v(0) = 0$. We have

$$V(u) = \frac{1}{2} \int_0^1 [(u_0 + v)']^2 \, dx - \int_0^1 f(x)[u_0 + v] \, dx$$

$$= \frac{1}{2} \int_0^1 [v']^2 \, dx - \int_0^1 f(x) v \, dx - \int_0^1 f(x) u_0 \, dx$$

$$= \frac{1}{2} a(v, v) - \psi(v) - \gamma.$$

Now setting $v = \sum_{i=1}^4 c_i f_i$, we obtain

$$a(v, v) = \int_0^1 [v']^2 \, dx = c^T A c, \quad a_{ij} = a(f_j, f_i) = \int_0^1 f_j' f_i' \, dx$$

$$\phi(v) = \int_0^1 f(x) v \, dx = b^T c, \quad b_i = \phi(f_i) = \int_0^1 f(x) f_i \, dx$$

so that

$$V(u) = \frac{1}{2} c^T A c - b^T c - \gamma.$$

A is the matrix of the bilinear form $a(h, g) = \int_0^1 h' g' \, dx$ and will be positive definite whenever $\{f_k'\}_{k=1}^n$ is a linearly independent set, since A is the gram matrix of $\{f_k'\}_{k=1}^n$. According to Example 7, V is minimized by $u_0 = \frac{1}{2} + \sum_{i=1}^4 c_i f_i$, with $c = A^{-1} b$.

(a) *Solution using polynomials.* Let $f_k(x) = x^k$, $k = 1, 2, 3, 4$, and $u = \frac{1}{2} + v$, $v = \sum_{i=1}^4 c_i x^i$. Here

$$a_{ij} = \frac{ij}{i + j - 1} \quad \text{and} \quad b_i = \frac{2}{i+1}(1 - (0.5)^{i+1}).$$

Solving the 4×4 system $Ac = b$ yields $c = (0.9375, 0.4375, -0.625, 0)^T$. The approximation is $u_1 = \frac{1}{2} + 0.9375x + 0.4375x^2 - 0.625x^3$.

(b) *Solution using hat functions with nodes.* $x_0 = 0$, $x_1 = 0.25$, $x_2 = 0.5$, $x_3 = 0.75$, $x_4 = 1$. f_k is piecewise linear and continuous, $f_k(x_j) = \delta_{kj}$, $k, j = 1, 2, 3, 4$. (See Example 6 of Section 1.3.) Here A is tridiagonal with -4 above and below the diagonal, and 8s on the diagonal except for $a_{44} = 4$. Further, $b = \frac{1}{4}(0, \frac{1}{2}, 1, \frac{1}{2})^T$, so that $c = A^{-1} b = (0.25, 0.5, 0.6875, 0.75)^T$. The approximation is $u_2 = \frac{1}{2} + 0.25 f_1 + 0.5 f_2 + 0.6875 f_3 + 0.75 f_4$.

(c) *Exact solution*

$$u = \frac{1}{2} + x + \begin{cases} 0 & x \le 0.5 \\ (x - 0.5)^2 & x > 0.5 \end{cases}.$$

This is found by directly integrating $u'' = -f$ twice and applying the boundary conditions $u(0) = \frac{1}{2}$, $u'(1) = 0$.

Comparing the approximations at $x = k/4$:

x	u	u_1	u_2
0.25	0.75	0.752	0.75
0.5	1.00	1.00	1.00
0.75	1.188	1.186	1.188
1	1.25	1.25	1.25

Both are excellent approximations. For a larger number of functions, use of the hat functions have two advantages. First, the matrix A is tridiagonal and positive definite, and there are very fast special routines for solving $Ac = b$. The matrix for the polynomials is full and has the potential for numerical inaccuracies when $Ac = b$ is solved (ill conditioning). (One almost never computes $c = A^{-1}b$ for large matrices.) Second, if $u = \frac{1}{2} + \sum_{i=1}^{n} c_i f_i$, then $u(k/n) = \frac{1}{2} + c_k$ due to the interpolation property of the hat functions, so that the function values are easy to read from the coefficients. In this application, the eigenvectors of A have no particular physical meaning. They depend on the basis functions, which can be fairly arbitrary.

Exercises 6.2

1. Classify each of the following real quadratic forms by using the definition.
   (a) $q(x) = x_1 x_2$ $V = \mathbf{R}^2$
   (b) $q(x) = x_1^2 - 2x_1 x_2$ $V = \mathbf{R}^2$
   (c) $q(x) = x_1^2 + 2x_1 x_2 + 2x_2^2$ $V = \mathbf{R}^2$
   (d) $q(x) = x^T \text{Trid}(-1, 3, -1)x$ $V = \mathbf{R}^3$
   (e) $q(x) = x^2(0) + 2x^2(1) + x^2(2)$ $V = \mathbf{P}_2$
   (f) $q(x) = x^2(0) + 2x^2(1) + x^2(2)$ $V = \mathbf{P}_3$
   (g) $q(x) = \int_0^1 x^2(t)e^t\, dt$ $V = \mathbf{P}_3$

2. Show that AA^H is positive. Under what conditions is it definite?

3. Let $A = P^T D P$, where D is a diagonal matrix and P is real and nonsingular. Find a change of coordinates diagonalizing $q(x) = x^T A x$, $x \in \mathbf{R}^n$. What is the classification of q (i.e., of A)? What is the rank of q?

4. Let $A = \text{Trid}(-1, 4, -1)$. Show that A is positive definite.

5. Let $Tf(x) = -e^{-x}(d/dx)(e^x f'(x))$ act on the vector space V of functions having two continuous derivatives on $[0, a]$ and satisfying $f(0) = f(a) = 0$. Define $\langle f, g \rangle = \int_0^a f(x)\bar{g}(x)e^x\, dx$. Show that T is positive definite on V in the given $\langle , \rangle$.

6. Let $Tf = f^{(4)}$ act on the vector space V of smooth functions satisfying $f(0) = f'(0) = f(1) = f'(1) = 0$. Let $\langle f, g \rangle = \int_0^1 f(x)\bar{g}(x)\, dx$. Classify T.

7. Let ∇^2 be the Laplacian and Ω a smooth bounded domain in the plane. Let V be the vector space of functions vanishing on $\partial\Omega$ (the boundary of Ω). Give V the inner product $\langle f, g \rangle = \int_\Omega f(p)\bar{g}(p)\, dA$. Show that ∇^2 is negative definite. (Hint: Use one of Green's identities.)

8. If $q(f) = \sum_{k=0}^{3} 2^{-k}[f^{(k)}(0)]^2$ is a real quadratic form on $\mathbf{P}_4$, classify q and find the rank of q.

9. On $\mathbf{T}_n$ with real coefficients, define

$$q(f) = \sum_{k=0}^{n} \frac{1}{k+1} \left[\int_0^{2\pi} f(t)\cos(kt)\,dt \right]^2.$$

Find the rank of q.

10. On $\mathbf{T}_n$ with complex coefficients, define $q(f) = \int_{-\pi}^{\pi} |f'(t)|^2\,dt$. Show that $\{e^{ikt}\}_{k=-n}^{n}$ diagonalizes q, classify q, and find the rank of q.

11. Suppose x_0 is a critical point of $f(x)$ and the Hessian of f is negative definite. Show that x_0 is a local maximum of f.

12. Suppose L is a real lower triangular matrix with all $l_{kk} = 1$, and D is a diagonal matrix with all $d_{kk} > 0$. Show $A = LDL^T$ is Hermitian and positive definite.

13. Suppose that A is positive definite. Show that A is nonsingular.

14. Suppose $b(f, g) = \langle L(f), L(g) \rangle$, where L is linear on V. State conditions under which ϕ is an inner product on V.

15. Let $b(x, y) = \langle T(x), y \rangle$ be a Hermitian bilinear form on a finite-dimensional vector space V. Show that the rank of b is the same as the rank of T.

16. Let $Q(x) = x^T A x - 2b^T x + \gamma$, with A positive definite, $x \in \mathbf{R}^n$. Find the necessary condition that $Q(x)$ be minimal, using calculus.

17. Let $q(x) = x^T A x - 2b^T x + \gamma$, and A be positive. Prove the following.
(a) If x_0 satisfies $Ax_0 = b$, then $q(x_0)$ is minimal.
(b) Suppose $Ax_0 = b$, and let x^* be the projection of x_0 on $R(A)$. Then x^* is the vector that minimizes $q(x)$ and has minimal length.

18. A cable of length L that is fixed at both ends at the same elevation has weights attached to it at uniform intervals with spacing $h = L/(n+1)$. Let T be the tension in the cable. For small deflections, T is approximately constant, and the potential energy in the cable is (neglecting gravity)

$$U = \frac{T}{2h} \sum_{i=0}^{n} (y_{i+1} - y_i)^2, \quad y_i = \text{vertical displacement at station } i.$$

If a force f_i is applied at station i, then the potential energy of the system is

$$V = U - \sum_{i=1}^{n} f_i y_i.$$

Knowing that the equilibrium position of the string occurs at a minimizer of V, derive the equations for the equilibrium position of the cable both using calculus and without using calculus. Discuss the existence and uniqueness of the minimizer.

19. Apply the Ritz method to find the approximate position of a loaded cable fixed at both ends. Here

$$V = \frac{1}{2}\tau \int_0^L u_x^2(x)\,dx - \int_0^L f(x)u(x)\,dx, \quad u(0) = a, u(L) = b.$$

Note that the exact solution satisfies $\tau u'' + f = 0$, $u(0) = a$, $u(L) = b$.

Take $u_0 = [a(L - x) + bx]/L$.

Use $\tau = L = 1$, $f(x) = \begin{cases} -8 & x < 1/2 \\ 0 & x > 1/2 \end{cases}$, $a = 0, b = 1, n = 3$.

(a) Find the approximate solution, using the following as basis functions:
  (i) Hat functions with nodes $0, 0.25, 0.5, 0.75, 1$.
  (ii) $\sin(k\pi x)$, $k = 1, 2, 3$
  (iii) $(1 - x)x^k$, $k = 1, 2, 3$
(b) Find the exact solution and compare it with the approximations.

20. An ideal string is fixed at $x = 0$ with height 1, has an elastic restraint at $x = L$, and has a distributed applied force density $f(x)$. Let $u(x)$ be the height of the string at point x, so that $u(0) = 1$. The potential energy is given by

$$V(u) = \frac{\tau}{2}\int_0^L u_x^2(x)\,dx + \frac{1}{2}ku(L)^2 - \int_0^L f(x)u(x)\,dx.$$

The exact solution is given by the solution of

$$\tau u'' + f = 0, \qquad u(0) = 1, \qquad \tau u'(L) + ku(L) = 0.$$

Take

$$L = \tau = 1, \qquad k = 2, \qquad f = \begin{cases} 0 & x < 2/3 \\ 6 & x \geq 2/3 \end{cases}$$

for simplicity. The basis functions need only satisfy the essential boundary condition $u(0) = 0$.
(a) Find the approximate solution, using the basis functions
  (i) x^k, $k = 1, 2, 3$
  (ii) Hat functions f_1, f_2, f_3 with the nodes $x_0 = 0$, $x_1 = \frac{1}{3}$, $x_2 = \frac{2}{3}$, $x_3 = 1$.
(b) Find the exact solution and compare the approximations to it.

21. A cantilever beam with applied body force density f and applied end force P has potential energy

$$V(u) = \frac{1}{2}\int_0^L EI(u'')^2\,dx - \int_0^L f(x)u(x)\,dx - Pu(L).$$

The beam is welded to a wall at $x = 0$ with height $u(0) = 0$ and slope $u'(0) = 0$. The exact solution satisfies

$$EIu^{(4)} = f, \qquad u(0) = 0, \qquad u'(0) = 0, \qquad u''(L) = 0, \qquad EIu^{(3)}(L) + P = 0.$$

The basis functions need only satisfy the essential boundary conditions $u(0) = u'(0) = 0$.
(a) Set up the general equations for the minimizer, for an arbitrary set of basis functions $\{f_k\}_{k=1}^n$. Find $[a_{ij}]$ if $f_k(x) = x^{k+1}$, $k = 1, 2, \ldots, n$.

(b) Take $EI = 1 = L, P = -30$, and $n = 3$. Use polynomial basis functions to find the approximate equilibrium position if

(i) $f(x) = 96$

(ii) $f(x) = 96\sqrt{x}$

(iii) $f(x) = \begin{cases} 0 & x \le 0.5 \\ 96 & x > 0.5 \end{cases}$

In each case, compare the approximate solution with the exact solution.

6.3
ORTHOGONAL DIAGONALIZATION

Given a Hermitian quadratic form q on a finite-dimensional space V, there are two approaches to its study. One is to select an inner product $\langle , \rangle$ (which is presumably natural to the physical or mathematical problem being studied) and then to find a Hermitian operator so that $q(x) = \langle T(x), x \rangle$. The study of q is thus reduced to the study of T. The second is to introduce a basis $\{g_k\}_{k=1}^n$ into V and to represent the quadratic form in terms of a Hermitian matrix A: $q(x) = [x]^H A[x]$. In this case A, or, equivalently, the quadratic form on $\mathbf{F}^n$ generated by A, is studied.

The first theorem proved is called the orthogonal diagonalization theorem. It states that not only can a Hermitian quadratic form of the type $\langle T(x), x \rangle$ be diagonalized, but that we may do so by means of an orthogonal basis. The second theorem is called the principal axis theorem and is the analog of the first theorem for quadratic forms generated by a matrix. The extra geometric information beyond mere diagonalization gives some very powerful computational tools. Some applications are given in this chapter, and many more are found in Chapter 7.

THE ORTHOGONAL DIAGONALIZATION THEOREM. Let $T : V \to V$ be Hermitian with respect to $\langle , \rangle$, and $q(x) = \langle T(x), x \rangle$. Let $\beta = \{e_1, e_2, \ldots, e_n\}$ be an orthogonal eigenbasis for V and $T(e_i) = \lambda_i e_i$. Then β diagonalizes q, and the weights are the eigenvalues λ_i. Further, such a basis exists and

(a) q and T are positive definite iff all $\lambda_i > 0$.

(b) q and T are positive iff all $\lambda_i \ge 0$.

(c) q and T are indefinite iff some $\lambda_i > 0 > \lambda_j$.

Similar statements apply for the negative and negative definite cases.

(d) The rank of $q = r(T)$.

Proof. The spectral theorem guarantees the existence of the basis. Let $x = \sum_{i=1}^n y_i e_i$. Then $Tx = \sum_{i=1}^n \lambda_i y_i e_i$, and

$$q(x) = \langle T(x), x \rangle = \left\langle \sum_{i=1}^n \lambda_i y_i e_i, \sum_{i=1}^n y_i e_i \right\rangle = \sum_{i=1}^n \lambda_i |y_i|^2 \|e_i\|^2 \tag{1}$$

by Parseval's identity. Statements (a)–(c) follow from Proposition 3 of Section 6.2. (d) follows from the definition of $r(q)$ and Theorem 2 of Section 4.2, which asserts that the rank of T is the number of nonzero λ_i.

The next theorem is a special version of the orthogonal diagonalization theorem. A matrix theory proof will be given to illustrate matrix methods.

THE PRINCIPAL AXIS THEOREM. Let $Q(x) = x^H A x$ be a quadratic form on $\mathbf{F}^n$ with A Hermitian. Then there is a unitary matrix U and real numbers λ_i such that the change of coordinates $Uy = x$ implies $Q(x) = \sum_{i=1}^{n} \lambda_i |y_i|^2$. Specifically, the columns of U are an orthonormal set of eigenvectors of A and the λ_i are the corresponding eigenvalues $U = [E_1 \ E_2 \ \cdots \ E_n]$, where $AE_i = \lambda_i E_i$, $E_j^H E_i = \delta_{ij}$. These columns are called *principal axes*. Further,

(a) Q and A are positive definite iff all $\lambda_i > 0$.
(b) Q and A are positive iff all $\lambda_i \geq 0$.
(c) Q and A are indefinite iff for some $i, j, \lambda_i > 0 > \lambda_j$.

Similar statements apply for the negative and negative definite cases.

(d) The rank of $Q = r(A)$.

Proof. This theorem follows from the orthogonal diagonalization theorem by setting $T(x) = Ax$ and $\langle x, y \rangle = y^H x$. A matrix theory proof goes as follows. From Theorem 1 of Section 5.3, there is a unitary matrix U and a real diagonal matrix Λ such that $U^H A U = \Lambda$. Thus, if $x = Uy$, it follows that

$$Q(x) = (Uy)^H A(Uy) = y^H(U^H AU)y = y^H \Lambda y = \sum_{i=1}^{n} \lambda_i |y_i|^2. \tag{2}$$

In this theorem, $U = [E_1 \ E_2 \ \cdots \ E_n]$ and $AE_i = \lambda_i E_i$, $E_j^H E_i = \delta_{ij}$. The λ_i are real because A is Hermitian.

Remark. If the eigenvectors E_i in the matrix U of the principal axis theorem are orthogonal but not normalized, a useful diagonalization still results. If $Uy = x$, then

$$Q(x) = (Uy)^H A(Uy) = y^H(U^H AU)y = y^H(U^H[AE_1 \ AE_2 \ \cdots \ AE_n])y$$
$$= y^H(U^H[\lambda_1 E_1 \ \lambda_2 E_2 \ \cdots \ \lambda_n E_n])y = y^H([\lambda_j E_i^H E_j])y = y^H \text{diag}(\lambda_j \|E_j\|^2)y$$
$$= \sum_{i=1}^{n} \lambda_i |y_i|^2 \|E_i\|^2$$

which is the same as Eq. (1) but derived by matrix methods.

The following examples apply the principal axis theorem and some extremum principles from Section 5.4 to quadratic forms on $\mathbf{R}^2$ and $\mathbf{R}^3$. A conic section and a quadric surface are analyzed in the process.

EXAMPLE 1. Let $Q(x) = -2x_1^2 + 4x_1 x_2 + x_2^2$, $V = \mathbf{R}^2$.

(a) Classify Q.
(b) Sketch the level curve $Q(x) = 4$.
(c) Find the max and min of $Q(x)$ subject to $x_1^2 + x_2^2 = 1$ and the vectors at which the max and min are attained.

$$Q(x) = x^T A x, \quad \text{where } A = \begin{bmatrix} -2 & 2 \\ 2 & 1 \end{bmatrix}.$$

Solving the orthonormal eigenvalue problem for A yields $AE_i = \lambda_i E_i$, where $\lambda_1 = -3$, $E_1 = (2, -1)^T/\sqrt{5}$, $\lambda_2 = 2$, $E_2 = (1, 2)^T/\sqrt{5}$.

(a) Q is indefinite, since $\lambda_1 < 0$, $\lambda_2 > 0$.

(b) $x = y_1 E_1 + y_2 E_2$ implies $Q(x) = -3y_1^2 + 2y_2^2$, by the principal axis theorem. In the y_1, y_2 coordinate system, which is defined by the vectors E_1, E_2 and the origin, the nature of the curve $Q(x) = -3y_1^2 + 2y_2^2 = 4$ is clear: $Q(x) = 4$ is the hyperbola shown in the figure.

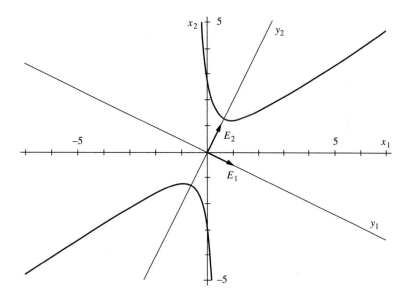

(c) Observe that $Q(x) = \langle Ax, x \rangle$ and $\|x\|^2 = x_1^2 + x_2^2$, using the standard inner product on $\mathbf{R}^2$. Thus (c) may be phrased as: Minimize and maximize $\langle Ax, x \rangle$ subject to $\|x\|^2 = 1$. This is a constrained Rayleigh quotient min-max problem, and Theorem 3 of Section 5.4 asserts that the solution is as follows: The minimum of Q is $\lambda_1 = -3$ and is attained at $\pm E_1$ (i.e., at $y_1 = \pm 1$, $y_2 = 0$), and the maximum of Q is $\lambda_2 = 2$ and is attained at $\pm E_2$ (i.e., at $y_1 = 0$ and $y_2 = \pm 1$).

Remark. The constrained minimization problem could also be solved by Lagrange multipliers. The necessary condition for a constrained minimum or maximum is $\nabla Q(x) = \mu \nabla \|x\|^2$, which turns out to be $Ax = \lambda x$, with $\lambda = 2\mu$. However, the Lagrange multiplier theorem does not supply a test for a minimum or maximum.

EXAMPLE 2. Let $Q(x) = 2x_1^2 + 2x_2^2 + 2x_3^2 - 2x_1 x_2 - 2x_1 x_3 + 2x_2 x_3$.

(a) Find an orthogonal matrix M and real numbers α_i such that the transformation $y = Mx$ reduces Q to $Q(x) = \alpha_1 y_1^2 + \alpha_2 y_2^2 + \alpha_3 y_3^2$.

(b) Classify Q.

(c) Sketch the level surface $Q(X) = 4$.

(d) Find the max and min of $Q(X)$ subject to $\sum x_i^2 = 1$ and the vectors at which the max and min are attained.

The symmetric matrix of Q is

$$A = \begin{bmatrix} 2 & -1 & -1 \\ -1 & 2 & 1 \\ -1 & 1 & 2 \end{bmatrix}.$$

The eigenvalues and an orthonormal basis of eigenvectors of A are $\lambda_1 = \lambda_2 = 1$, $E_1 = (1, 0, 1)^T/\sqrt{2}$, $E_2 = (1, 2, -1)^T/\sqrt{6}$, $\lambda_3 = 4$, $E_3 = (-1, 1, 1)^T/\sqrt{3}$. Let $U = [E_1 \ E_2 \ E_3]$. Then, by the principal axis theorem,

$$x = Uy = y_1 E_1 + y_2 E_2 + y_3 E_3 \ \Rightarrow \ Q(x) = y_1^2 + y_2^2 + 4y_3^2. \qquad (*)$$

(a) If $M = U^T = U^{-1}$, then $y = Mx$ and $\alpha_i = \lambda_i$ give the desired result.

(b) Q is positive definite since all the eigenvalues are positive.

(c) If the axes defined by the basis vectors E_i and the corresponding coordinates y_i are used, then (*) holds. The level surface $Q(x) = y_1^2 + y_2^2 + 4y_2^2 = 4$ is easy to sketch in the y_1, y_2, y_3 coordinate system: It is an ellipsoid that intersects the plane $y_3 = 0$ in a circle of unit radius. Notice that the eigenspace $M(1)$ is the E_1, E_2 plane and does not have a unique orthonormal basis. This corresponds to the fact that the ellipsoid is invariant under rotations about the E_3 axis.

(d) Observe that $Q(x) = x^T A x = \langle Ax, x \rangle$, where $\langle,\rangle$ is the standard inner product on $\mathbf{R}^3$, and that the constraint is $\|x\|^2 = 1$. By Theorem 3 of Section 5.4 on minimizing and maximizing constrained Rayleigh quotients, the maximum is $\lambda_3 = 4$ and the minimum is $\lambda_1 = 1$. The vectors maximizing Q subject to the constraint are unit vectors in $M(4)$, and these are $\pm E_3$. The vectors minimizing Q subject to the constraint are the unit vectors in $M(1)$, and these are of the form $\alpha E_1 + \beta E_2$, where $\|\alpha E_1 + \beta E_2\|^2 = \alpha^2 \|E_1\|^2 + \beta^2 \|E_2\|^2 = \alpha^2 + \beta^2 = 1$. That is, these vectors lie on the unit circle in the E_1, E_2 plane.

EXAMPLE 3. PRINCIPAL AXES OF STRAIN AND THE STRAIN ELLIPSOID. The following is a very simplified description. Consider a deformed material. x denotes the position of a particle before deformation, and y denotes the position of the same particle after deformation. At a particular point in a material, the deformed shape may be approximated by an affine transformation $y = Lx + b$, where L is a nonsingular 3×3 matrix and b is in $\mathbf{R}^3$; L and b are independent of x. Two nearby particles at x and x' before deformation have a displacement $v = x' - x$, and after deformation the displacement is $y' - y = Lv$. To describe the deformation, it is helpful to know the shape of the surface $S = \{Lv : \|v\| = 1\}$ and those unit vectors in which Lv has maximal or minimal length. Note that $\|Lv\|^2 = v^T L^T Lv$ is simply a Hermitian quadratic form with the symmetric matrix $L^T L$. Let $\{E_1, E_2, E_3\}$ be orthonormal eigenvectors of $L^T L$, and let λ_i be the associated eigenvalues, i.e., $L^T L E_i = \lambda_i E_i$. Assume $\lambda_1 < \lambda_2 < \lambda_3$. Note that the quadratic form is clearly positive definite, so the eigenvalues must be positive. The set S turns out to be an ellipsoid with principal axes $F_i = LE_i/\|LE_i\|$ and semi-axes $\sqrt{\lambda_i}$. To see this, let $w = \sum_{i=1}^3 d_i F_i = Lv$, with $\|v\| = 1$. It must be shown that $\gamma = \{F_1, F_2, F_3\}$ is orthonormal and $\sum_{i=1}^3 d_i^2/\lambda_i = 1$. But $\langle LE_i, LE_j \rangle = (LE_j)^T LE_i = E_j^T L^T LE_i = \lambda_i E_j^T E_i = 0$ if $i \neq j$ and $\langle LE_i, LE_i \rangle = \|LE_i\|^2 = \lambda_i$, from which it follows that γ is orthonormal. Note that $F_i = LE_i/\sqrt{\lambda_i}$. Now suppose $v = \sum_{i=1}^3 c_i E_i$, with $\|v\|^2 = \sum_{i=1}^3 c_i^2 = 1$. Then $w = Lv = \sum_{i=1}^3 c_i LE_i = \sum_{i=1}^3 c_i \sqrt{\lambda_i} F_i$, so that $d_i = c_i \sqrt{\lambda_i}$. Thus $\sum_{i=1}^3 d_i^2/\lambda_i = 1$. The meaning of this result is this: If the displacement in the undeformed body varies over the unit sphere, then the displacement in the deformed body traces out the strain ellipsoid.

In which directions must the displacements in the undeformed body be in order to maximize or minimize the displacements in the deformed body? To maximize or minimize $\|Lv\|^2$ subject to $\|v\|^2 = 1$, use Theorem 3 of Section 5.4, which says that the extreme values are the largest and smallest eigenvalues and occur along the corresponding eigenaxes. For example, if the displacement in the undeformed body is along E_1, then the ratio $\|Lv\|/\|v\|$ is minimal (and equal to $\sqrt{\lambda_1}$), while if the displacement in the undeformed body is along E_3, then the ratio $\|Lv\|/\|v\|$ is maximal (and equal to $\sqrt{\lambda_3}$). Thus the strain ellipsoid gives a good geometric feel for the geometry of the deformation.

The following theorem characterizes positive definite matrices in terms of certain determinants. Let $A^{(i)}$ denote the submatrix of A obtained by deleting rows and columns $i + 1, \ldots, n$ of A. $A^{(i)}$ is called a *principal submatrix* of A.

SYLVESTER'S TEST. Let A be an $n \times n$ Hermitian matrix. Then A is positive definite iff $|A^{(i)}| > 0$, $i = 1, 2, \ldots, n$.

Proof. Suppose A is positive definite. Let $x = (x_1, x_2, \ldots, x_i, 0, \ldots, 0)^T$ and $y = (x_1, x_2, \ldots, x_i)$. Then $x^T A x = y^T A^{(i)} y > 0$. Now suppose all $|A^{(i)}| > 0$, and it must be shown that A is positive definite. A proof by induction is made. For $n = 1$, $A = [a_{11}]$ with $|A| = a_{11} > 0$, and A is clearly positive definite. Suppose the assertion is true for $n \le k$, and let $n = k + 1$. The eigenvalues of $A^{(k)}$ are all positive. By the interlacing theorem in Section 5.4, eigenvalues of A are all positive, except possibly for the smallest one, λ_1. But $|A| = \prod_{i=1}^{n} \lambda_i > 0$, so $\lambda_1 > 0$ as well. Thus A is positive definite.

Geometric Minimization Problems

In Example 7 of Section 6.2, it was seen that $Q(x) = x^T A x - 2b^T x + \gamma$ is minimal when $x = x_0 \equiv A^{-1}b$, assuming that A is positive definite. The problem at hand is to describe the surface $Q(x) = \beta$. The surface $Q(x) = \beta$ is also given by $y^T A y = \phi$, with $\phi = \beta - \gamma - b^T x_0$ and $y = x - x_0$. If $\phi > 0$, the surface is an ellipsoid with center at x_0 whose principal axes are the eigenvectors of A.

EXAMPLE 4. Let $Q(x, y) = 2x^2 + 4xy + 5y^2 - 6y - 1$. Minimize Q, and describe the curve $Q(x, y) = 5$.

In terms of Example 7 of Section 6.2,

$$A = \begin{bmatrix} 2 & 2 \\ 2 & 5 \end{bmatrix}, \qquad b = \begin{bmatrix} 0 \\ 3 \end{bmatrix}, \quad \text{and} \quad \gamma = -1.$$

Thus $(x_0, y_0)^T = (-1, 1)^T$, and $\alpha = -1 - 3 = -4$. Equation (2) of Section 6.2 becomes $Q = 2(x + 1)^2 + 4(x + 1)(y - 1) + 5(y - 1)^2 - 4$, which checks. The eigenvalues of A are $\lambda_1 = 1$ and $\lambda_2 = 2$, with corresponding unit eigenvectors $E_1 = (1/\sqrt{3})(2, -1)^T$, $E_2 = (1/\sqrt{3})(1, 2)^T$. Thus A is positive definite and Q has a minimum value of -4 at $x = x_0 = -1$, $y = y_0 = 1$. The translation and change of coordinates (rotation)

$$\begin{bmatrix} x + 1 \\ y - 1 \end{bmatrix} = \xi E_1 + \eta E_2 = [E_1 \; E_2] \begin{bmatrix} \xi \\ \eta \end{bmatrix} = U \begin{bmatrix} \xi \\ \eta \end{bmatrix}$$

yields

$$Q = [\xi, \eta] U^T A U \begin{bmatrix} \xi \\ \eta \end{bmatrix} - 4 = \xi^2 + 6\eta^2 - 4.$$

Thus the curve $Q = 5$ is the same as $\xi^2 + 6\eta^2 = 9$; it is an ellipse with center $(-1, 1)$ with axes parallel to E_1 and E_2 and semi-axes 3 and $\sqrt{1.5}$ along the ξ and η axes, respectively. See the following figure.

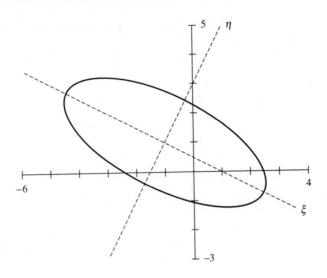

Exercises 6.3

1. For each of the following quadratic forms Q, find
   (i) An orthonormal basis E_i and real α_i such that $x = \sum y_i E_i \Rightarrow Q = \sum \alpha_i y_i^2$.
   (ii) An orthogonal matrix M and real α_i such that $y = Mx \Rightarrow Q = \sum \alpha_i y_i^2$.
   (iii) The classification of Q.
   (iv) The max and min of Q on the set of x such that $\sum x_i^2 = 1$, and the vectors at which the minimum is attained.
   (v) Describe the surface $Q(x) = 1$.
       (a) $Q(x) = 4x_1^2 + 2x_2^2 + 2x_3^2 + 2x_1 x_2 - 2x_1 x_3 - 2x_2 x_3$
       (b) $Q(x) = x_1^2 - 2x_2^2 + x_3^2 + 4x_1 x_2 + 8x_1 x_3 + 4x_2 x_3$
       (c) $Q(x) = x_1^2 + 3x_2^2 + x_3^2 + 2x_1 x_3$
       (d) $Q(x) = x_2^2 + 2x_1 x_3$
       (e) $Q(x) = -x_1^2 + 2x_2^2 + 2x_3^2 - 2x_1 x_2 + 2x_1 x_3 + 4x_2 x_3$
       (f) $Q(x) = x_1^2 + x_2^2 + x_3^2 + 2x_1 x_2 - 2x_1 x_3 + 2x_2 x_3$
       (g) $Q(x) = 3x_1^2 + 2x_2^2 + 3x_3^2 + 2x_1 x_2 + 2x_2 x_3$

2. Test the following matrices for positive definiteness.
   (a) $\begin{bmatrix} 2 & -1 & -1 \\ -1 & 2 & 1 \\ -1 & 1 & 2 \end{bmatrix}$
   (b) $\begin{bmatrix} 4 & 1 & -1 \\ 1 & 2 & -1 \\ -1 & -1 & 2 \end{bmatrix}$
   (c) $\begin{bmatrix} 1 & 1 & 2 \\ 1 & 2 & 1 \\ 2 & 1 & 1 \end{bmatrix}$

3. Find a matrix A for which $|A^{(i)}| \geq 0$, $i = 1, 2, \ldots, n$, yet A is not positive.

4. **Reciprocal strain ellipsoid.** Let L be a nonsingular 3×3 matrix. Following the discussion of the strain ellipsoid, show that the set $S_r = \{v : \|Lv\| = 1\}$ is an ellipsoid, and find its axes and semi-axes. S_r is called the *reciprocal strain ellipsoid*. Give a physical interpretation of S_r. What significance do the vectors have, for which $\|v\|$ is maximal or minimal, subject to $\|Lv\| = 1$?

5. Minimize (without calculus) and describe the surface $q = c$.
   (a) $q = 2x^2 + 2y^2 + 2z^2 - 2xy + 2xz - 2yz + 8x - 16y + 1, c = 21$
   (b) $q = 4x^2 + 2y^2 + 2z^2 + 2xy - 2xz - 2yz - 2x + 6y - 8z, c = 24$
   (c) $q = 2x^2 + 4y^2 + 4z^2 + 2yz - 4x + 16y + 4z + 20, c = 7$

6. Describe and sketch the following plane curves by first reducing the function on the left to the form $(x - x_0)^{\mathrm{T}} A(x - x_0) + \alpha$.
   (a) $3y^2 + 4xy - 8x - 16y + 10 = 6$
   (b) $2x^2 + 5y^2 + 4xy - 8x - 14y = 1$
   (c) $3x^2 + 3y^2 + 2xy - 10x + 2y - 1 = 3$
   (d) $x^2 + y^2 - 4xy - 2x + 4y + 1 = 9$

7. For $V = \mathbf{P}_2$ (real scalars), classify and diagonalize

$$q(p) = \frac{1}{4} \int_{-1}^{1} \int_{-1}^{1} (3 + 12xt - 15x^2t^2) p(t) p(x) \, dt \, dx.$$

   (*Hint:* Let $K(x, t) = \frac{1}{4}(3 + 12xt - 15x^2t^2)$, define T by $Tp(x) = \int_{-1}^{1} K(x, t)p(t) \, dt$, and define an $\langle , \rangle$ by $\langle f, g \rangle = \int_{-1}^{1} f(x)g(x) \, dx$. Then $q(p) = \langle Tp, p \rangle$.)

8. Let G be an $n \times n$ matrix and define $b(x, y) = y^{\mathrm{H}} Gx$. Give necessary and sufficient conditions for $b(,)$ to be an inner product on $\mathbf{C}^n$ in terms of G and its eigenvalues.

9. Let A be a Hermitian positive definite matrix.
   (a) Show that there are matrices B and C with C (Hermitian) positive definite such that $A = B^{\mathrm{H}} B = C^2$.
   (b) **Cholesky factorization.** Show that there is a lower triangular matrix L with positive diagonal entries such that $A = LL^{\mathrm{H}}$. (*Hint:* Apply QR factorization to B in part a.)

10. Show that G is positive definite iff there is a matrix P such that $P^{\mathrm{H}} G P = I$.

11. Let $A = LDL^{\mathrm{T}}$, with L a real lower triangular $n \times n$ matrix, D diagonal, and $l_{kk} = 1$, $d_{kk} > 0$ for $k = 1, \ldots, n$. Show that A is diagonalizable and that all the eigenvalues of A are positive.

12. Let $T : V \to V$ be linear and V be finite-dimensional with inner product $\langle , \rangle$. Give necessary and sufficient conditions on T in terms of T and its eigenvalues for $b(x, y) = \langle T(x), y \rangle$ to be an inner product on V.

6.4
OTHER METHODS OF DIAGONALIZATION

Orthogonal diagonalization of Hermitian quadratic forms is computationally demanding: An eigenvalue problem with orthogonal eigenvectors must be solved. In this section, three other methods to accomplish a diagonalization are presented. Each will produce a different basis that diagonalizes the quadratic form in question and a different set of weights. If A is the matrix of the Hermitian quadratic form, diagonalizing the quadratic form is equivalent to finding a matrix P such that $P^H A P = D$ is diagonal. There are many more ways to do this than there are to find a matrix P for which $P^{-1} A P = \Lambda$ is diagonal. The latter is the eigenvalue problem, in which the columns of P are eigenvectors of A. The eigenspaces are unique, and if the eigenvalues of A are distinct, the columns of P are unique up to scalar multiples and order. That is, congruence transformations to a diagonal matrix have much more latitude than similarity transformations to a diagonal matrix.

The goal of this section is to quickly describe alternative methods so that they may be used on small problems; there is no attempt to develop a complete theory. The first two methods presented are computationally much simpler than the orthogonal diagonalization method. A user of a sophisticated package who is doing calculations on small matrices would probably not detect much computation time in solving the eigenvalue problem, and therefore would not be impressed.

Diagonalization by the Completion of Squares

Real scalars will be considered in the following discussion. The complex case is similar and slightly more complicated. Again, the strategy is to reduce the computations to a quadratic form on $\mathbf{R}^n$ and to do the work there. Therefore, given a Hermitian quadratic form, the first step is to choose a basis $\{e_1, e_2, \ldots, e_n\}$ for the original vector space V and write $q(\sum_{i=1}^{n} x_i e_i) = x^T A x = Q(x)$, where $x = (x_1, x_2, \ldots, x_n)^T$ and A is a real, symmetric matrix. The next step is to construct a coordinate transformation $Py = x$ such that

$$Q(x) = y^T P^T A P y = y^T \Lambda y = \sum_{i=1}^{n} \lambda_i |y_i|^2,$$

from which the form may be classified. In the following procedure, the diagonalization is done first, and P must be recovered later.

The first step is to group all terms involving x_1, then group all the remaining terms involving $x_2, \ldots, x_n$ in the form

$$
\begin{aligned}
Q(x) = \sum_{i \le j} b_{ij} x_i x_j = {}& b_{11} x_1^2 + b_{12} x_1 x_2 + b_{13} x_1 x_3 + \cdots + b_{1n} x_1 x_n \\
& + b_{22} x_2^2 + b_{23} x_2 x_3 + \cdots + b_{2n} x_2 x_n \\
& + b_{33} x_3^2 + \cdots + b_{3n} x_3 x_n \\
& + \cdots + b_{nn} x_n^2.
\end{aligned}
\tag{1}
$$

The method is based on the identity

$$\left(\sum u_i \right)^2 = \sum_{i,j} u_i u_j = \sum u_i^2 + 2 \sum_{i<j} u_i u_j \tag{2}$$

and proceeds inductively.

If the sum is "square free," i.e., all $b_{kk} = 0$, then perform a preparatory step to introduce squared terms. By reindexing the x_i if necessary, one may assume that $b_{12} \neq 0$. Then make the substitution (change of coordinates)

$$x_1 = y_1 + y_2, \qquad x_2 = y_2 - y_2, \qquad y_k = x_k \quad \text{for } k \geq 3.$$

Then

$$Q(x) = b_{12}[y_1^2 - y_2^2] + \text{terms not involving } y_1^2, y_2^2,$$

that is, Q is now of the form (1) with $b_{11} \neq 0$.

When the sum contains squared terms, reindex the x_i if necessary to ensure that $b_{11} \neq 0$. Now use the identity (2) to complete the square in the group of terms involving x_1:

$$b_{11}x_1^2 + b_{12}x_1x_2 + b_{13}x_1x_3 + \cdots + b_{1n}x_1x_n$$

$$= b_{11}(x_1 + c_2x_2 + \cdots + c_nx_n)^2 + \text{terms involving only } x_2, \ldots, x_n, \quad (3)$$

where $c_i = \frac{1}{2}b_{1i}/b_{11}$. Now make the substitution $y_1 = x_1 + c_2x_2 + \cdots + c_nx_n$, which gives us

$$Q = b_{11}y_1^2 + \text{quadratic form in } x_2, x_3, \ldots, x_n.$$

Continue the reduction on the quadratic form in fewer variables. It is easy to see that each substitution (change of coordinates) is reversible (i.e., nonsingular).

When the process is finished, the quadratic form appears as $q(x) = \sum_{i=1}^{n} d_i z_i^2$, where z_i are the labels for the last set of variables used. From this one can classify the quadratic form. If the change of coordinate transformation is desired, the sequence of substitutions generated at each step is written as $z = Mx$. M is nonsingular since each of the substitutions was reversible. Thus $q(x) = x^T A x = z^T D z$, where $P^T A P = D = \text{diag}(d_1, \ldots, d_n)$, and $P = M^{-1}$ is the change of basis matrix.

EXAMPLE 1

$$Q(x) = 2x_1^2 - \tfrac{1}{2}x_2^2 + 2x_1x_2 - 4x_1x_3$$

Diagonalize Q by completion of squares, and find the change of basis matrix and the new basis in which Q is diagonalized.

For the x_1 terms, by Eqs. (2) and (3),

$$2(x_1^2 + x_1x_2 - 2x_1x_3) = 2[(x_1 + \tfrac{1}{2}x_2 - x_3)^2 - \tfrac{1}{4}x_2^2 - x_3^2 + x_2x_3],$$

so that setting $y_1 = x_1 + \tfrac{1}{2}x_2 - x_3$ and replacing the x_1 terms gives

$$Q = 2y_1^2 - x_2^2 - 2x_3^2 + 2x_2x_3.$$

For the x_2 terms,

$$-[x_2^2 - 2x_2x_3] = -[(x_2 - x_3)^2 - x_3^2],$$

whence, after putting $y_2 = x_2 - x_3$ and $y_3 = x_3$,

$$Q = 2y_1^2 - y_2^2 - y_3^2.$$

Clearly, the form is indefinite (see Proposition 3 of Section 6.2). To recover the coordinate transformation, our substitutions are

$$y_1 = x_1 + \tfrac{1}{2}x_2 - x_3$$
$$y_2 = x_2 - x_3 \qquad \text{or} \qquad y = \begin{bmatrix} 1 & 0.5 & -1 \\ 0 & 1 & -1 \\ 0 & 0 & 1 \end{bmatrix} x = Mx.$$
$$y_3 = x_3$$

The change of basis matrix is $P = M^{-1} = [E_1 \ E_2 \ E_3]$, with $E_1 = (1, 0, 0)^T$, $E_2 = (-\tfrac{1}{2}, 1, 0)^T$, $E_3 = (\tfrac{1}{2}, 1, 1)^T$. Thus, if $x = \sum_{i=1}^{3} y_i E_i = Py$, then

$$Q(x) = 2y_1^2 - y_2^2 - y_3^2.$$

Note that one can complete the square on any squared term desired in this process; it is not necessary to slavishly proceed in order.

EXAMPLE 2. Let $q(p) = 2\int_0^1 tp(t)^2 \, dt$ be a real quadratic form on $\mathbf{P}_2$. Diagonalize q by completion of squares, and find the change of basis matrix and the new basis in which q is diagonalized.

Choose the usual basis $\{1, t, t^2\}$. By computing the matrix of q with respect to this basis, or by simply substituting $p(t) = x_0 + x_1 t + x_2 t^2$ into q, expanding and doing the integration, we obtain

$$q(p) = Q(x) = 20x_0 x_1 + 12x_1 x_2, \qquad \text{where } x = (x_0, x_1, x_2)^T.$$

This is square free, so make the substitution

$$x_0 = y_0 + y_1, \qquad x_1 = y_0 - y_1, \qquad x_2 = y_2$$

and find

$$q(p) = 20y_0^2 - 20y_1^2 + 12y_0 y_2 - 12y_1 y_2$$
$$= 20(y_0^2 + 0.6y_0 y_2) - 20y_1^2 - 12y_1 y_2$$
$$= 20[(y_0 + 0.3y_2)^2 - 0.09y_2^2] - 20y_1^2 - 12y_1 y_2$$
$$= 20(y_0 + 0.3y_2)^2 - 20(y_1^2 + 0.6y_1 y_2 - 0.09y_2^2)$$
$$= 20(y_0 + 0.3y_2)^2 - 20(y_1 + 0.3y_2)^2.$$

Setting $z_0 = y_0 + 0.3y_2$, $z_1 = y_1 + 0.3y_2$ yields

$$q(x) = 20z_0^2 - 20z_1^2,$$

and clearly the form is indefinite by Proposition 3 of Section 6.2. If an overall change of coordinates must be found, first observe that $y_0 = (x_0 + x_1)/2$, $y_1 = (x_0 - x_1)/2$; then, putting $z_2 = x_2$ (this substitution can be chosen freely as long as the resulting transform is nonsingular),

$$z_0 = 0.5(x_0 + x_1) + 0.3x_2$$
$$z_1 = 0.5(x_0 - x_1) + 0.3x_2$$
$$z_2 = x_2;$$

i.e.,
$$z = \begin{bmatrix} 0.5 & 0.5 & 0.3 \\ 0.5 & -0.5 & 0.3 \\ 0 & 0 & 1 \end{bmatrix} x = Mx.$$

The change of basis matrix is

$$P = \begin{bmatrix} 1 & 1 & -0.6 \\ 1 & -1 & 0 \\ 0 & 0 & 1 \end{bmatrix}.$$

The corresponding basis is $e_0(t) = 1 + t$, $e_1(t) = 1 - t$, and $e_2(t) = t^2 - 0.3$. Thus, if $p(t) = \sum_{i=0}^{2} z_i e_i(t)$, then $q(p) = 20z_0^2 - 20z_1^2$; hence $\{e_0, e_1, e_2\}$ is a basis diagonalizing q.

Simultaneous Row and Column Operations

Let q be a Hermitian quadratic on V, let the scalars be real, and assume that a basis has been introduced so that $q(v) = x^T A x = Q(x)$, where $x = [v]$ is the coordinate vector of v, and A is symmetric. Recall that if E is an elementary matrix, then EA is the result of doing the corresponding row operation on A, and note that AE^T is the result of doing the corresponding column operation on A. Hence EAE^T is the result of doing the same row and column operation on A and is symmetric. Thus, if $S = E_m \cdots E_2 E_1$, then SAS^T is the result of doing the sequence of corresponding operations on the rows and columns of A, and $S = SI = E_m \cdots E_2 E_1 I$ is the result of doing these same row operations on I. The calculation of PAP^T is referred to as "simultaneous row and column operations." With the preceding language, the first part of the next theorem has been proved.

THEOREM 1. If a sequence of simultaneous elementary row and column operations reduces the symmetric matrix A to a diagonal matrix D, then the same sequence of row operations on the identity matrix yields a nonsingular matrix S such that $SAS^T = D$. In addition, such a sequence of operations exists, and every method of diagonalization of $q(x) = x^T A x$ may be obtained in this way.

Proof. If one uses the change of coordinates $Py = x$, where $P = S^T$, the quadratic form is thereby diagonalized: $q(x) = (Py)^T A (Py) = y^T P^T A P y = y^T S A S^T y = y^T D y$. If q is diagonalized by $Py = x$, then write $S = P^T$ as a product of elementary matrices. The corresponding simultaneous row and column operations on A must produce a diagonal matrix.

EXAMPLE 3

$$Q(x) = x_1^2 + 3x_2^2 + 5x_3^2 + 4x_1 x_2 + 2x_1 x_3 + 8x_2 x_3$$

Diagonalize Q by simultaneous row and column transformations. Find the new basis in which Q is diagonalized.

Recall our notation: $R_3 \leftarrow R_3 - 2R_1$, for example, means "subtract 2 times row 1 from row 3 and replace row 3 by the result." Reduce the matrix A of Q as follows:

$$A = \begin{bmatrix} 1 & 2 & 1 \\ 2 & 3 & 4 \\ 1 & 4 & 5 \end{bmatrix} \rightarrow \begin{bmatrix} 1 & 0 & 1 \\ 0 & -1 & 2 \\ 1 & 2 & 5 \end{bmatrix} \rightarrow \begin{bmatrix} 1 & 0 & 0 \\ 0 & -1 & 2 \\ 0 & 2 & 4 \end{bmatrix} \rightarrow \begin{bmatrix} 1 & 0 & 0 \\ 0 & -1 & 0 \\ 0 & 0 & 8 \end{bmatrix} = D$$

$$R_2 \leftarrow R_2 - 2R_1 \quad R_3 \leftarrow R_3 - R_1 \quad R_3 \leftarrow R_3 + 2R_2$$
$$C_2 \leftarrow C_2 - 2C_1 \quad C_3 \leftarrow C_3 - C_1 \quad C_3 \leftarrow C_3 + 2C_2$$

The same sequence of row operations on I yields

$$S = \begin{bmatrix} 1 & 0 & 0 \\ -1 & 1 & 0 \\ -5 & 2 & 1 \end{bmatrix}.$$

Hence $SAS^T = D$, which is easy to check. The change of coordinates $X = S^T y$ gives $Q(x) = y^T SAS^T y = y^T Dy = y_1^2 - y_2^2 + 8y_3^2$. Thus the change of basis matrix is $P = [E_1 \, E_2 \, E_3] = S^T$, and the basis in which Q is diagonalized is $\{E_1, E_2, E_3\}$.

Jacobi's Diagonalization

Jacobi's method of diagonalizing a Hermitian quadratic form has some similarities to the Gram–Schmidt process. It is described next. Let b be a Hermitian bilinear form on the n-dimensional vector space V, $\beta = \{e_k\}_{k=1}^n$ be a basis for V, and $A = [a_{ij}] = [b(e_j, e_i)]$ be the matrix of b with respect to β. Assume that the determinants of the principal submatrices $A^{(k)}$ (which are called the *principal minors* of A) of A are all nonzero:

$$\Delta_k \equiv \det \begin{bmatrix} a_{11} & \cdots & a_{1k} \\ & \cdots & \\ a_{k1} & \cdots & a_{kk} \end{bmatrix} = \det(A^{(k)}) \neq 0 \qquad \text{for } k = 1, 2, \ldots, n.$$

Construct a new basis $\{f_k\}_{k=1}^n$ with the following properties:

1. The change of basis matrix is upper triangular: $f_i \in \text{span}\{e_1, e_2, \ldots, e_i\}$ for $i = 1, 2, \ldots, n$.
2. $b(f_i, f_j) = 0$ if $i \neq j$ (Orthogonality).
3. $b(f_i, e_i) = 1$ (Normalization).

The construction proceeds as follows. Let $f_1 = p_{11}e_1$, and set $p_{11} = 1/\Delta_1$ so that $b(f_1, e_1) = 1$. Now properties 1 and 2 hold for $i = 1$. Put $f_k = \sum_{j=1}^k p_{jk}e_j$, and require that

$$b(f_k, e_i) = \sum_{j=1}^k p_{jk}b(e_j, e_i) = \sum_{j=1}^k p_{jk}a_{ij} = 0 \quad \text{for } i = 1, 2, \ldots, k-1, \quad (4a)$$

$$b(f_k, e_k) = \sum_{j=1}^k p_{jk}b(e_j, e_k) = \sum_{j=1}^k p_{jk}a_{kj} = 1, \qquad (4b)$$

i.e.,

$$\begin{bmatrix} a_{11} & a_{12} & \cdots & a_{1k} \\ \vdots & \vdots & \vdots & \vdots \\ a_{k1} & a_{k2} & \cdots & a_{kk} \end{bmatrix} \begin{bmatrix} p_{1k} \\ p_{2k} \\ \vdots \\ p_{kk} \end{bmatrix} = \begin{bmatrix} 0 \\ 0 \\ \vdots \\ 0 \\ 1 \end{bmatrix}. \qquad (5)$$

The matrix of this system is precisely $A^{(k)}$, which has the nonzero determinant Δ_k, and so the system has a unique solution for $p_{1k}, p_{2k}, \ldots, p_{kk}$. This defines the f_k for

$k = 2, \ldots, n$. By construction, the change of basis matrix is upper triangular, and property 3 holds. Let us show that property 2 holds. If $i < k$, then

$$b(f_k, f_i) = b\left(f_k, \sum_{j=1}^{i} p_{ji}e_i\right) = \sum_{j=1}^{i} \overline{p_{ji}}b(f_k, e_i) = 0$$

by (4a). If $i > k$, then $b(f_k, f_i) = \overline{b(f_i, f_k)} = 0$ by the previous case. By property 2, the matrix of b with respect to basis $\{f_k\}_{k=1}^{n}$ is diagonal, and the last step is to calculate the diagonal entries w_k. By (4a),

$$w_k = b(f_k, f_k) = b\left(f_k, \sum_{j=1}^{k} p_{jk}e_j\right) = b(f_k, p_{kk}e_k) = \overline{p_{kk}}.$$

Since b is Hermitian, $\overline{p_{kk}} = p_{kk}$. But p_{kk} can be found from the system (5) by using Cramer's rule:

$$p_{kk} = \det \begin{bmatrix} A^{(k-1)} & \begin{matrix} 0 \\ 0 \\ 0 \end{matrix} \\ * & \cdots & 1 \end{bmatrix} \Delta_k^{-1} = \Delta_{k-1}/\Delta_k \quad \text{for } k \geq 2,$$

where the cofactor expansion by the last column was used. Hence, if $x = \sum_{i=1}^{n} y_i f_i$, then $q(x) = \sum_{i=1}^{n} w_i |y_i|^2$, where

$$w_1 = 1/\Delta_1, \qquad w_i = \Delta_{i-1}/\Delta_i, \quad i = 2, 3, \ldots, n. \tag{6}$$

In the following two examples, the procedure in Jacobi's method is closely followed.

EXAMPLE 4

$$V = \mathbf{R}^3, \qquad Q(x) = x_1^2 + x_3^2 - 2x_1 x_2 + 2x_2 x_3$$

(a) Use Jacobi's method to classify Q without finding the new basis.
(b) Find the basis in which Q is diagonalized, by Jacobi's method.

Solution

(a) Q is generated by the symmetric bilinear form

$$b(x, y) = y^T Ax = x_1 y_1 - x_2 y_1 + x_3 y_2 - x_1 y_2 + x_2 y_3 + x_3 y_3.$$

A is the matrix of b with respect to the basis $\{\delta_1, \delta_2, \delta_3\}$ and is given by $a_{ij} = b(\delta_j, \delta_i)$, so that

$$A = \begin{bmatrix} 1 & -1 & 0 \\ -1 & 0 & 1 \\ 0 & 1 & 1 \end{bmatrix}.$$

Here

$$\Delta_1 = 1, \qquad \Delta_2 = \begin{vmatrix} 1 & -1 \\ -1 & 0 \end{vmatrix} = -1, \qquad \Delta_3 = -2.$$

Thus the weights of Jacobi's method are $w_1 = 1/\Delta_1 = 1$, $w_2 = \Delta_1/\Delta_2 = -1$, and $w_3 = \Delta_2/\Delta_3 = 2$. Thus Q is indefinite.

(b) p_{ij} must be found so that f_1, f_2, f_3, defined by

$$f_1 = p_{11}\delta_1$$
$$f_2 = p_{12}\delta_1 + p_{22}\delta_2$$
$$f_3 = p_{13}\delta_1 + p_{23}\delta_2 + p_{33}\delta_3,$$

satisfy the conditions of the construction. Take $p_{11} = 1/\Delta_1 = 1$. The system of equations (5) for f_2 is

$$\begin{bmatrix} b(f_2, \delta_1) \\ b(f_2, \delta_2) \end{bmatrix} = A^{(2)} \begin{bmatrix} p_{12} \\ p_{22} \end{bmatrix} = \begin{bmatrix} 1 & -1 \\ -1 & 0 \end{bmatrix} \begin{bmatrix} p_{12} \\ p_{22} \end{bmatrix} = \begin{bmatrix} 0 \\ 1 \end{bmatrix},$$

which yields $p_{12} = -1$, $p_{22} = -1$. The system (5) for f_3 is

$$\begin{bmatrix} b(f_3, \delta_1) \\ b(f_3, \delta_2) \\ b(f_3, \delta_3) \end{bmatrix} = A^{(2)} \begin{bmatrix} p_{13} \\ p_{23} \\ p_{33} \end{bmatrix} = \begin{bmatrix} 1 & -1 & 0 \\ -1 & 0 & 1 \\ 0 & 1 & 1 \end{bmatrix} \begin{bmatrix} p_{13} \\ p_{23} \\ p_{33} \end{bmatrix} = \begin{bmatrix} 0 \\ 0 \\ 1 \end{bmatrix},$$

which gives $p_{13} = p_{23} = p_{33} = \frac{1}{2}$.

Thus the basis $\{f_1, f_2, f_3\}$ in which Q is diagonalized and the change of basis matrix P are

$$f_1 = \begin{bmatrix} 1 \\ 0 \\ 0 \end{bmatrix}, \qquad f_2 = \begin{bmatrix} -1 \\ -1 \\ 0 \end{bmatrix}, \qquad f_3 = \frac{1}{2}\begin{bmatrix} 1 \\ 1 \\ 1 \end{bmatrix}, \qquad P = \begin{bmatrix} 1 & -1 & 1/2 \\ 0 & -1 & 1/2 \\ 0 & 0 & 1/2 \end{bmatrix}.$$

Finally, if $x = y_1 f_1 + y_2 f_2 + y_3 f_3$, then $Q(x) = y_1^2 - y_2^2 + 2y_3^2$. As a check, it must be that $w_i = b(f_i, f_i)$, and this is so.

As an immediate corollary, the following test for positive definiteness emerges. The scalars may be real or complex.

SYLVESTER'S TEST. Let q be a Hermitian quadratic form on a vector space V with matrix A in some basis. Then q is positive definite if and only if all the principal minors of A are positive.

Proof. If all the $\Delta_k > 0$, then from Jacobi's method, $q(x) = \sum_{i=1}^n w_i |y_i|^2 > 0$ if $x \neq 0$. On the other hand, suppose q is positive definite. Restrict q to the subspace $M_k = \text{span}\{e_1, e_2, \ldots, e_k\}$. Here the matrix of q is $A^{(k)}$, and the product of eigenvalues of $A^{(k)}$ is Δ_k by Proposition 2 of Section 4.1. Since the restricted quadratic form is still positive definite and $q(x) = [x]^H A^{(k)} [x]$ when $x \in M_k$, $A^{(k)}$ is then a positive definite matrix, so by the principal axis theorem, the eigenvalues of $A^{(k)}$ are all positive, and therefore $\Delta_k > 0$.

Remark. From $\Delta_k \geq 0$ for all k, it does not follow that q is positive. For example, if the first row and column of A are zero, then all the $\Delta_k = 0$, while the rest of A is arbitrary.

EXAMPLE 5. Consider the following quadratic form on $\mathbf{T}_1$:

$$q(x) = \int_0^{2\pi} |x'(t)|^2 \, dt + |x(0)|^2.$$

(a) Use Sylvester's test to classify q.
(b) Use Jacobi's method of diagonalization to find a basis in which q is diagonalized.

Using the basis $1, e^{it}, e^{-it}$ for $\mathbf{T}_1$, the matrix of the Hermitian bilinear form generating q is

$$A = \begin{bmatrix} 1 & 1 & 1 \\ 1 & 2\pi + 1 & 1 \\ 1 & 1 & 2\pi + 1 \end{bmatrix}.$$

The principal minors are $\Delta_1 = 1, \Delta_2 = 2\pi, \Delta_3 = 4\pi^2$; hence q is positive definite by Sylvester's test. Pursuing the construction of the Jacobi method further, the weights in the diagonalization are $w_1 = 1/\Delta_1 = 1$, $w_2 = \Delta_1/\Delta_2 = 1/(2\pi)$, and $w_3 = \Delta_2/\Delta_3 = 1/(2\pi)$. To find the change of coordinates, put

$$f_1 = p_{11}$$
$$f_2 = p_{12} + p_{22}e^{it}$$
$$f_3 = p_{13} + p_{23}e^{it} + p_{33}e^{-it}.$$

Solving the system in (5) with $k = 1, 2, 3$ yields $p_{11} = 1, p_{22} = -p_{12} = 1/(2\pi), p_{33} = -p_{13} = 1/(2\pi), p_{23} = 0$. Thus a new basis in which q is diagonalized is $f_1(t) = 1$, $f_2(t) = (1 - e^{it})/(2\pi)$, $f_3(t) = (1 - e^{-it})/(2\pi)$.

Remark. There is now an indirect way of finding the number of positive, negative, and zero eigenvalues of a Hermitian matrix A. The quadratic form $x^H A x$ can be diagonalized in some way other than calculating the eigenvalues and eigenvectors of A. Since the eigenvalues appear in the w_i resulting from the principal axis reduction to diagonal form, by the law of inertia the number of positive, negative, and zero eigenvalues must be the same as the number of positive, negative, and zero w_i obtained in every other diagonalization.

EXAMPLE 6. Show several ways to determine that the following real quadratic form is positive definite:

$$Q(x) = x_1^2 + 17x_2^2 + 3x_3^2 + 4x_1x_2 - 2x_1x_3 - 14x_2x_3$$

$$= x^T \begin{bmatrix} 1 & 2 & -1 \\ 2 & 17 & -7 \\ -1 & -7 & 3 \end{bmatrix} x = x^T A x.$$

(a) $|A - \lambda I| = p(\lambda) = -\lambda^3 + 21\lambda^2 - 17\lambda + 1$. For $\lambda \geq 0$, $p(-\lambda) < 0$; hence no eigenvalue is negative or zero, and Q is positive definite, using the principal axis theorem.

(b) Computing the determinants of the principal minors gives $\Delta_1 = 1$, $\Delta_2 = 13$, and $\Delta_3 = 1$, whence Q is positive definite by Sylvester's test.

(c) Diagonalization by completion of squares yields $Q(X) = z_1^2 + 2z_2^2 + 0.5z_3^2$, where $z_1 = x_1 + 2x_2 - x_3$, $z_2 = x_3 + 2.5x_2$, $z_3 = x_2$. Q is positive definite by Proposition 2 of Section 6.2

(d) Diagonalization by simultaneous row and column operations yields $A \rightarrow D = \text{diag}(1, 13, 1/13)$, so that $Q(x) = y_1^2 + 13y_2^2 + (1/13)y_3^2$ for appropriate coordinates y_i in $\mathbf{R}^3$. Again, Q is positive definite.

From the fact that Q is positive definite, we can infer that for $c > 0$, the surface $Q(x) = c$ is an ellipsoid. It would be hard to sketch or describe this ellipsoid, even if the new basis corresponding to the diagonalization were computed for each of the preceding diagonalizations. Clearly, the eigenvectors and eigenvalues from the principal

axis theorem give us the information needed to satisfactorily describe the ellipsoid geometrically.

Exercises 6.4

1. Diagonalize the following real quadratic forms by completion of the square. In addition,
   (i) Find a symmetric matrix A such that $Q(x) = x^T A x$.
   (ii) Find a matrix P and real numbers w_k such that $x = Py \Rightarrow Q(x) = \sum w_k y_k^2$.
   (iii) Find a basis $E_2, E_2, \ldots, E_n$ and real numbers w_k such that

$$x = \sum y_k E_k \Rightarrow Q(x) = \sum w_k y_k^2.$$

   (iv) Classify Q.
   (v) Determine the number of positive eigenvalues, the number of negative eigenvalues, and the number of zero eigenvalues of A.
   (a) $Q(x) = x_1^2 + 3x_2^2 + 10x_3^2 + 10x_4^2 - 2x_1 x_2 + 2x_1 x_3 + 4x_1 x_4 + 6x_2 x_3 - 8x_2 x_4$
   (b) $Q(x) = x_1^2 + 3x_2^2 + 9x_3^2 + 9x_4^2 + 2x_1 x_2 - 2x_1 x_3 + 4x_1 x_4 - 10x_2 x_3 + 8x_2 x_4 - 18x_3 x_4$
   (c) $Q(x) = 2x_1 x_2 - 3x_1 x_3 + x_2 x_3$
   (d) $Q(x) = x_1^2 + 2x_2^2 + 10x_3^2 + 2x_1 x_2 + 2x_2 x_3$
   (e) $Q(x) = x_1^2 + 6x_2^2 + 6x_3^2 - 4x_1 x_2 + 2x_1 x_3 - 8x_2 x_3$
   (f) $Q(x) = -2x_1^2 - 5x_2^2 + 12x_3^2 - 8x_1 x_2 + 4x_1 x_3 - 4x_2 x_3$

2. Repeat Exercise 1, but use simultaneous row and column operations.

3. For the following real quadratic forms on P_2,
   (i) Find a symmetric matrix A such that

$$x(t) = \sum c_k t^k \Rightarrow q(x) = x^T A x, \quad \text{where } c = (c_0, c_1, c_2)^T = [x].$$

   (ii) Find a basis $\{e_0, e_1, e_2\}$ and real numbers w_k such that

$$x(t) = \sum d_k e_k(t) \Rightarrow Q(x) = \sum w_k d_k^2.$$

   (iii) Classify q.
   (a) $q(x) = 3 \int_0^1 x'(t)^2 \, dt + x(0)^2 + x(1)x'(1)$
   (b) $q(x) = \int_0^1 x'(t)^2 \, dt + x(0)x(1)$
   (c) $q(x) = x(2)^2 + x'(0)x''(0)$
   (d) $q(x) = \int_0^1 x'(t)^2 \, dt + x(0)^2 + x'(1)^2$

4. Define $q(x) = \int_0^1 x''(t)^2 \, dt + x(0)x(1)$, $x \in P_3$, real scalars.
   (a) Find a symmetric bilinear form ϕ such that $q(x) = \phi(x, x)$.
   (b) Find a symmetric matrix A such that $q(x) = c^T A c$ if $x = c_0 + c_1 t + c_2 t^2 + c_3 t^3$.
   (c) Use completion of squares to find a basis $\{e_0, e_1, e_2, e_3\}$ for P_3 and real numbers w_i such that $x = \sum_{i=0}^{3} d_i e_i \Rightarrow q(x) = \sum_{i=0}^{3} w_i d_i^2$.
   (d) Classify q.

5. Diagonalize the quadratic forms in the preceding exercises using Jacobi's method, if possible.

6. Without computing the characteristic polynomial, find the signs of the eigenvalues of

$$A = \begin{bmatrix} 1 & 3 & -3 \\ 3 & 1 & -1 \\ -3 & -1 & 7 \end{bmatrix}.$$

6.5
SIMULTANEOUS DIAGONALIZATION OF QUADRATIC FORMS

Simultaneous diagonalization of quadratic forms arises frequently in mechanical and electrical systems when fundamental modes of oscillation are to be found, and in mechanics when the buckling modes are to be found. It also arises in the theory of surfaces when the curvatures and principal directions need to be computed for the description of surfaces. Simultaneous diagonalization of quadratic forms is essential to the Rayleigh–Ritz method, which provides an approximation of eigenvalues and eigenvectors for operators on infinite-dimensional spaces.

First, the problem will be reduced to one for matrices. The theory will be developed in the context of matrices.

Let two Hermitian quadratic forms q and w on an n-dimensional vector space V, with w positive definite, be given. This means that w is generated by a bilinear form that is an inner product on V. The first step is to pick a basis $\gamma = \{f_k\}_{k=1}^n$ for V and to compute the matrices of the bilinear forms, obtaining, if $x = \sum_{i=1}^n c_i f_i$, and $[x] = (c_1, c_2, \ldots, c_n)^T$,

$$q(x) = [x]^H A[x] = Q([x]) \qquad A \text{ Hermitian}$$

$$w(x) = [x]^H G[x] = W([x]) \qquad G \text{ Hermitian, positive definite.}$$

G is positive definite since it is the Gram matrix of β with respect to the inner product generating w. Now Q, W can be diagonalized in $\mathbf{C}^n$. Observe that $\langle c, d \rangle \equiv d^H Gc$ is an inner product on $\mathbf{C}^n$ since G is positive definite. The simultaneous diagonalization theorem will be stated and proved in $\mathbf{C}^n$. To get back to V, one has only to reattach coefficients. To get the correct theorem for $\mathbf{R}^n$, simply drop the conjugate signs where they appear.

THEOREM 1. Let A be a Hermitian matrix and G a positive definite Hermitian matrix, both $n \times n$, and $W(x) = x^H Gx$, $Q(x) = x^H Ax$. There exists a basis $\beta = \{E_k\}_{k=1}^n$ for $\mathbf{C}^n$ that is orthonormal with respect to the inner product $\langle x, y \rangle_G = y^H Gx$ and that diagonalizes both quadratic forms W, Q with real weights λ_i. Further, if $P = [E_1 \; E_2 \; \cdots \; E_n]$ and $\Lambda = \text{diag}(\lambda_1, \lambda_2, \ldots, \lambda_n)$, then

$$P^H GP = I \quad \text{and} \quad P^H AP = \Lambda. \tag{1}$$

If A and G are real, the E_i may be taken to be real.

Proof. By Theorem 4 of Section 5.3, there is a basis $\{E_k\}_{k=1}^n$ orthonormal with respect to $\langle, \rangle_G$ for which $AE_i = \lambda_i GE_i$ with real λ_i. The rest of the proof follows from Parseval's identity: If $x = \sum_{i=1}^n y_i E_i$, then $G^{-1}Ax = \sum_{i=1}^n \lambda_i y_i E_i$.

$$W(x) = \|x\|_G^2 = \left\langle \sum_{i=1}^{n} y_i E_i, \sum_{i=1}^{n} y_i E_i \right\rangle_G = \sum_{i=1}^{n} |y_i|^2 \|E_i\|_G^2 = \sum_{i=1}^{n} |y_i|^2 \qquad (*)$$

and $Q(x) = x^H A x = x^H G(G^{-1} A x) = \langle G^{-1} A x, x \rangle_G = \left\langle \sum_{i=1}^{n} \lambda_i y_i E_i, \sum_{i=1}^{n} y_i E_i \right\rangle_G$

$$= \sum_{i=1}^{n} \lambda_i |y_i|^2 \|E_i\|_G^2 = \sum_{i=1}^{n} \lambda_i |y_i|^2. \qquad (**)$$

Let $N = \mathrm{diag}(\|E_1\|_G, \ldots, \|E_n\|_G)$. Equations (1) follow from the calculations

$$P^H G P = [E_i^H G E_j] = [\langle E_j, E_i \rangle] = N^2$$
$$P^H A P = [E_i^H A E_j] = [E_i^H \lambda_j G E_j] = [\lambda_j \langle E_j E_i \rangle] = \Lambda N^2 \qquad (2)$$

and the fact that $N = I$ when the basis is orthonormal. Alternatively, using the change of basis formula for bilinear and quadratic forms (Eq. (10) of Section 6.1) along with $(*)$ and $(**)$, it follows that

$$I = [W]_\beta = P^H G P \quad \text{and} \quad \Lambda = [Q]_\beta = P^H A P.$$

Remark. In some problems it may be desirable not to normalize the E_i. In this case (2) is used in place of (1).

Remark. The problem of finding the vectors $E \neq 0$ and scalars λ satisfying $AE = \lambda GE$ is called the *generalized eigenvalue problem*. The procedure for solving this was discussed in Section 5.3. At the risk of being repetitious, some of the discussion is briefly repeated here. The following procedure is primarily for hand calculations or theoretical discussions. Note that

$$AE = \lambda GE, \quad E \neq 0 \Leftrightarrow (A - \lambda G)E = 0, \quad E \neq 0$$

$$\Leftrightarrow |A - \lambda G| = 0 \quad \text{and} \quad 0 \neq E \in N(A - \lambda G).$$

Thus the eigenvalues are exactly the zeros of the characteristic polynomial $p(\lambda) \equiv |A - \lambda G|$, and for each such root, the associated eigenspace is $M(\lambda) = N(A - \lambda G)$. First, determine the roots λ_k of the characteristic equation $|A - \lambda G| = 0$. For each of the distinct eigenvalues λ_i, find the eigenspace $M(\lambda_i) = N(A - \lambda_i G)$. Recall that the dimension of the eigenspace is the multiplicity of the eigenvalue as a root of the characteristic equation, and that eigenvectors from distinct eigenspaces are automatically orthogonal with respect to $\langle,\rangle_G$. Thus, for each eigenspace of dimension 1, select a nonzero vector; for each eigenspace of dimension greater than 1, select an orthogonal basis, e.g., by the Gram–Schmidt process. The collection of all the eigenvectors so found will do the job.

EXAMPLE 1. Consider the quadratic forms on $\mathbf{R}^3$,

$$f(x) = -4x_1^2 + 3x_2^2 + 3x_3^2 - 4x_1 x_2 - 6x_2 x_3$$
$$g(x) = 4x_1^2 + 3x_2^2 + x_3^2 + 4x_1 x_2 - 2x_2 x_3.$$

Find a change of coordinates $x = Py$ that simultaneously diagonalizes the given quadratic forms.

First write $f(x) = x^{\mathrm{T}}Ax$, $g(x) = x^{\mathrm{T}}Bx$, where

$$A = \begin{bmatrix} -4 & -2 & 0 \\ -2 & 3 & -3 \\ 0 & -3 & 3 \end{bmatrix}, \qquad B = \begin{bmatrix} 4 & 2 & 0 \\ 2 & 3 & -1 \\ 0 & -1 & 1 \end{bmatrix}.$$

f is clearly not positive definite, since $f(\delta_1) = -4$. One can verify that g is positive definite by Sylvester's test, for example. Thus $g(x) = \|x\|_B^2 = \|x\|^2$ in this problem. Computing $|A - \lambda B| = -4(\lambda + 1)(\lambda - 1)(\lambda - 3)$ yields the eigenvalues $\lambda_1 = -1$, $\lambda_2 = 1$, $\lambda_3 = 3$, so that the eigenspaces are one-dimensional and the eigenvectors are automatically orthogonal with respect to the inner product generated by B. $(A + B)x = 0$ has the solution $x = F_1 = (1, 0, 0)^{\mathrm{T}}$, and $\|F_1\|^2 = g(x) = 4$, so that a normalized eigenvector is $E_1 = 0.5F_1$. $(A - B)x = 0$ has the solution $x = F_2 = (-1, 2, 2)^{\mathrm{T}}$, and $g(F_2) = 4$, so a normalized eigenvector is $E_2 = 0.5F_2$. $(A - 3B)x = 0$ has the solution $x = F_3 = (0, 0, 1)^{\mathrm{T}}$ with $g(F_3) = 1$, and $E_3 = F_3$ is a suitable choice for a normalized eigenvector. Thus $P = [E_1\ E_2\ E_3]$ will simultaneously diagonalize both matrices via $P^{\mathrm{H}}AP = \mathrm{diag}(-1, 1, 3)$, $P^{\mathrm{H}}BP = I$, and both quadratic forms will be diagonalized via the substitution $x = Py$. If the nonnormalized vectors F_i were used, i.e., $P = [F_1\ F_2\ F_3]$, then $P^{\mathrm{H}}AP = \mathrm{diag}(\lambda_i\|F_i\|^2) = \mathrm{diag}(-4, 4, 3)$ and $P^{\mathrm{H}}BP = \mathrm{diag}(\|F_i\|^2) = \mathrm{diag}(4, 4, 1)$.

EXAMPLE 2. Consider the quadratic forms on $\mathbf{R}^3$

$$f(x) = 2x_1^2 + 3x_2^2 + 6x_3^2 + 6x_1x_3$$

$$g(x) = 2x_1^2 + 2x_2^2 + 5x_3^2 + 6x_1x_3 + 2x_2x_3.$$

Find a transformation $y = Qx$ changing one quadratic form to $\sum_{i=1}^{3} y_i^2$ and the other to $\sum_{i=1}^{3} w_i y_i^2$.

Write $f(x) = x^{\mathrm{T}}Ax$, $g(x) = x^{\mathrm{T}}Bx$, where

$$A = \begin{bmatrix} 2 & 0 & 3 \\ 0 & 3 & 0 \\ 3 & 0 & 6 \end{bmatrix}, \qquad B = \begin{bmatrix} 2 & 0 & 3 \\ 0 & 2 & 1 \\ 3 & 1 & 5 \end{bmatrix}.$$

A is positive definite by Sylvester's test, so A plays the role of the Gram matrix G. Thus $\langle x, y \rangle = y^{\mathrm{T}}Ax$, and $f(x) = \|x\|^2$ in this problem, and the new basis should be orthonormal with respect to the inner product generated by A since $f(x) = \sum_{i=1}^{3} y_i^2$ is required. Calculating $|B - \lambda A| = -9\lambda(\lambda - 1)^2$, the distinct eigenvalues are 0 and 1, and $M(1)$ is two-dimensional. Let $\lambda_1 = 0$ and $\lambda_2 = \lambda_3 = 1$. Solving $(B - 0A)x = 0$ gives $x = G_1 = (3, 1, -2)^{\mathrm{T}}$. Solving $(B - A)x = 0$ gives $F_1 = (0, 1, 1)^{\mathrm{T}}$ and $F_2 = (1, 0, 0)^{\mathrm{T}}$ as a basis for $M(1)$. The Gram–Schmidt procedure gives $F_3 = F_2 - (F_1^{\mathrm{T}}AF_2/F_1^{\mathrm{T}}AF_1)F_1 = F_2 - \frac{1}{3}F_1 = \frac{1}{3}(3, -1, -1)^{\mathrm{T}}$, so F_1 and F_3 form an orthogonal basis for $M(1)$. Thus an orthogonal basis of eigenvectors for L corresponding to the λ_i is $\{G_1, G_2, G_3\}$ where $G_2 = F_1$, $G_3 = 3F_3 = (3, -1, -1)^{\mathrm{T}}$. All $f(G_i) = \|G_i\|^2 = 9$. Thus an orthornormal basis is $E_i = \frac{1}{3}G_i$. If $P = [E_1\ E_2\ E_3]$, then

$$P^{\mathrm{H}}AP = [E_i^{\mathrm{H}}AE_j] = \mathrm{diag}(\|E_i\|^2) = I \qquad\qquad (\#)$$

$$P^{\mathrm{H}}BP = [E_i^{\mathrm{H}}BE_j] = [E_i^{\mathrm{H}}(\lambda_j AE_j)] = \mathrm{diag}(\lambda_i\|E_i\|^2) = \mathrm{diag}(0, 1, 1).$$

The substitution $x = Py$ gives $f(x) = y^{\mathrm{H}}P^{\mathrm{H}}APy = y^{\mathrm{H}}Iy = \sum_{i=1}^{3} y_i^2$, and $g(x) = y^{\mathrm{H}}P^{\mathrm{H}}BPy = y^{\mathrm{H}}\mathrm{diag}(0, 1, 1)y = y_2^2 + y_3^2$. Thus $Q = P^{-1} = P^{\mathrm{H}}A$, using $(\#)$.

EXAMPLE 3. Let $q(v) = v^T A v$, $Q(v) = v^T B v$ be the first and second fundamental forms for the smooth surface S given by $\mathbf{r} = f(u_1, u_2)$. Let $\kappa_1 < \kappa_2$ be the generalized eigenvalues for the generalized eigenvalue problem $Bv = \kappa A v$, $v \neq 0$. These κ_i are called the *principal curvatures* of the surface, at the point P having $\mathbf{r}$ as a position vector. The following property holds: If a plane passes through P and contains the normal $\mathbf{n}$ to S at P, then this plane cuts the surface in a curve whose curvature κ at P satisfies $\kappa_1 \leq \kappa \leq \kappa_2$. Thus the generalized eigenvalues are extreme values of such curvatures. For the computation of A and B, see Example 7 of Section 6.1. For a complete classical treatment of the geometry of surfaces (and much more), see Sokolnikoff (1964).

An alternative proof and method for simultaneous diagonalization

The procedure is as follows: First choose a basis $\{e_i\}_1^n$ and represent the quadratic forms by Hermitian matrices A and G, with G positive definite. Next, find in any way a change of coordinates $x = Py$ such that the positive definite quadratic form reduces to $w(x) = \sum_{i=1}^n w_i|y_i|^2$, with $w_i > 0$. This is possible because G is Hermitian and positive definite. Next make the substitution $z_i = \sqrt{w_i} y_i$, so that $w(x) = \sum_{i=1}^n |z_i|^2$. This is the same as the change of coordinates $z = Ny$, where $N = \text{diag}(\sqrt{w_1}, \sqrt{w_2}, \ldots, \sqrt{w_n})$. Now, using the fact that N is Hermitian, it follows that $q(x) = x^H A x = z^H (N^{-1} P^H A P N^{-1}) z = z^T B z$, where $B = N^{-1} P^H A P N^{-1}$. Next find a unitary matrix $U = [U_1 \, U_2 \, \cdots \, U_n]$ so that the change of coordinates $z = U\zeta$ reduces the second quadratic form to a weighted sum of squares $q(x) = \sum_{i=1}^n \mu_i|\zeta_i|^2$. This is done by solving the Hermitian eigenvalue problem for B; i.e., find U_i such that $U_i^H U_j = \delta_{ij}$ and $BU_i = \mu_i U_i$, as indicated by the spectral theorem. Note that $w(x) = \sum_{i=1}^n \mu_i|\zeta_i|^2$, since a unitary change of basis does not change the computation of inner products and norms. This is also evident from $z^H z = (U\zeta)^H U\zeta = \zeta^H U^H U\zeta = \zeta^H \zeta$. Thus the overall change of coordinates $x = PN^{-1}U\zeta$ will simultaneously diagonalize both quadratic forms.

This method is apparently much more complex than solving the generalized eigenvalue problem. However, it is attractive for computer use (if no algorithm for the generalized eigenvalue problem is available) because there are some extremely good computer programs that solve the orthogonal eigenvalue problem for Hermitian matrices. Using these algorithms, the first diagonalization would produce a unitary matrix $P = [E_1 \, E_2 \, \cdots \, E_n]$ of eigenvectors of G, and $w_i = \lambda_i > 0$, where $GE_i = \lambda_i E_i$. The next diagonalization solves the orthogonal eigenvalue problem for the Hermitian matrix $B = N^{-1} P^H A P N^{-1}$ to get the matrix U mentioned previously. The columns of U satisfy $BU_i = \mu_i U_i$, which gives the numbers μ_i.

Simultaneous Diagonalization of Three Quadratic Forms

The problem of simultaneously diagonalizing three quadratic forms arises in solving the differential equations for a damped oscillating system. Examples are found in Chapter 7.

Suppose it is desired to diagonalize the quadratic forms $x^H G x$, $x^H A x$, and $x^H B x$ by the same substitution $x = Py$. Assume all are Hermitian and G is positive definite. Specifically, it is desired to find a basis $\{E_k\}_{k=1}^n$ of $\mathbf{F}^n$ such that

$$E_i^H GE_j = 0 \quad \text{if } i \neq j$$

$$BE_i = \gamma_i GE_i \tag{3}$$

$$AE_i = \lambda_i GE_i.$$

These relations are equivalent to (with $P = [E_1 \ E_2 \ \cdots \ E_n]$)

$$P^H GP = N^2$$

$$P^H BP = \Gamma N^2 \tag{4}$$

$$P^H AP = \Lambda N^2,$$

where

$$\Lambda = \text{diag}(\lambda_1, \lambda_2, \ldots, \lambda_n)$$

$$\Gamma = \text{diag}(\gamma_1, \gamma_2, \ldots, \gamma_n) \tag{5}$$

$$N^2 = \text{diag}(\|E_1\|_G^2, \|E_2\|_G^2, \ldots, \|E_n\|_G^2), \quad \|E\|_G^2 \equiv E^H GE.$$

Just as for the simultaneous diagonalization of two quadratic forms, the possibility of three-way diagonalization is equivalent to the simultaneous diagonalization of $G^{-1}A$ and $G^{-1}B$ with an eigenbasis orthogonal with respect to the inner product generated by G.

THEOREM 2. Let G, A, and B be $n \times n$ Hermitian matrices and G be positive definite. Then there is a basis $\{E_k\}_{k=1}^n$ for $\mathbf{C}^n$ such that (3) holds if and only if

$$BG^{-1}A \text{ is Hermitian.} \tag{6}$$

Proof. Take $V = \mathbf{C}^n$, and define $\langle x, y \rangle = y^H Gx$, $S = G^{-1}B$, $T = G^{-1}A$. Note that T and S are Hermitian with respect to this $\langle , \rangle$ (cf. Theorem 1). Observe that $ST = G^{-1}BG^{-1}A$ and $TS = G^{-1}AG^{-1}B$. Hence $TS = ST \Leftrightarrow BG^{-1}A = AG^{-1}B = A^H(G^{-1})^H B^H = (BG^{-1}A)^H$. Now apply Theorem 6 of Section 5.2.

In certain situations it can be determined in advance that the multiple simultaneous diagonalization may be done. These follow from Propositions 7 and 8 of Section 5.3. The proofs are left as exercises for now. Examples of the use of these theorems are given in Chapter 7.

THEOREM 3. Let A and G be $n \times n$ Hermitian matrices and G be positive definite. Suppose $\beta = \{E_k\}_{k=1}^n$ simultaneously diagonalizes $x^H Ax$ and $x^H Gx$; i.e.,

$$AE_i = \lambda_i GE_i, \qquad E_i^H GE_j = 0 \quad \text{if } i \neq j, \quad i, j = 1, 2, \ldots, n.$$

If $B = Gf(G^{-1}A)$, then $\{E_k\}_{k=1}^n$ diagonalizes $x^H Bx$ as well; specifically,

$$BE_i = \gamma_i GE_i \quad \text{with } \gamma_i = f(\lambda_i), \quad i = 1, 2, \ldots, n. \tag{7}$$

THEOREM 4. If G, A, and B are all-real valued functions of a Hermitian matrix L, and G is positive definite, then an orthogonal eigenbasis for L simultaneously diagonalizes all three quadratic forms $x^H Gx$, $x^H Ax$, and $x^H Bx$ in the sense of (3).

Exercises 6.5

1. For the following pairs of real quadratic forms,
   (i) Find one that is positive definite and call it $n(x)$ and the other $q(x)$.
   (ii) Find a basis $\{E_1, E_2, E_3\}$ of $\mathbf{R}^3$ and real λ_i such that $x = \sum c_k E_k$ implies $n(x) = \sum c_k^2$
   and $q(x) = \sum \lambda_k c_k^2$.
   (iii) Find a matrix M and real λ_i so that $y = Mx$ implies $n(x) = \sum c_k^2$ and $q(x) = \sum \lambda_k c_k^2$.
   (iv) Find the max and min of $q(x)$ subject to $n(x) = 1$. Also find those vectors where the max and min are attained.
   (a) $f(x) = 2x_1^2 + 3x_2^2 + x_3^2 + 4x_1 x_2 + 2x_2 x_3$
       $g(x) = 5x_1^2 + 3x_2^2 + x_3^2 - 2x_1 x_2 + 2x_2 x_3$
   (b) $f(x) = 5x_1^2 + 3x_2^2 + 4x_3^2 + 2x_1 x_2 - 4x_2 x_3$
       $g(x) = 3x_1^2 + x_2^2 + 4x_3^2 + 6x_1 x_2 - 4x_2 x_3$
   (c) $f(x) = 4x_1^2 + 6x_2^2 + x_3^2 - 2x_1 x_3 - 6x_2 x_3$
       $g(x) = 4x_1^2 + 2x_2^2 + 3x_3^2 - 6x_1 x_3 - 2x_2 x_3$
   (d) $f(x) = 2x_1^2 + 4x_2^2 + 10x_3^2 + 4x_2 x_3$
       $g(x) = 2x_1^2 + 4x_2^2 + 10x_3^2 + 4x_1 x_2 - 4x_1 x_3 + 4x_2 x_3$
   (e) $f(x) = 5x_1^2 + 3x_2^2 + 4x_3^2 + 2x_1 x_2 - 4x_2 x_3$
       $g(x) = 2x_1^2 + 4x_3^2 + 8x_1 x_2 - 4x_2 x_3$
   (f) $f(x) = x_1^2 + 3x_2^2 + 2x_3^2 + 2x_1 x_2 + 4x_2 x_3$
       $g(x) = 5x_1^2 + 3x_2^2 + x_3^2 - 2x_1 x_2 + 2x_2 x_3$
   (g) $f(x) = 2x_1^2 + 2x_2^2 + 3x_3^2 + 2x_1 x_2 + 4x_2 x_3$
       $g(x) = 2x_1^2 + x_2^2 + 2x_3^2 + 2x_1 x_2 + 2x_2 x_3$

2. Suppose G is positive definite and $P^H G P = I$, $P = [E_1 \ E_2 \ \cdots \ E_n]$. Show that
   (a) $P^{-1} = P^H G$
   (b) $GPP^H = I = PP^H G$
   (c) $G^{-1} = \sum_{i=1}^{n} E_i E_i^H$

3. Suppose A and G are Hermitian with G positive definite, $\beta = \{E_1, E_2, \ldots, E_n\}$ is orthogonal with respect to $\langle,\rangle_G$, and $AE_i = \lambda_i G E_i$. Show that
   (a) If $\lambda_k \neq \alpha$ for all k, then $(A - \alpha G)^{-1} = \sum_{i=1}^{n} \dfrac{E_i E_i^H}{(\lambda_i - \alpha)\|E_i\|^2}$.
   (b) β also diagonalizes $x^H B x$, where $B = \alpha A + \gamma G$.
   (c) β also diagonalizes $x^H B x$, where $B = G\left\{\sum_{i=1}^{n} c_i \dfrac{E_i E_i^H}{\|E_i\|^2}\right\} G$.

4. **Quadratic forms generated by Hermitian tridiagonal matrices.** Let $L = \text{Trid}(-1, 2, -1)$, and suppose $\{e_k\}_{k=1}^{n}$ is an orthogonal eigenbasis for L with $Le_k = \mu_k e_k$. In fact, $\mu_k = 4\sin^2(k\pi/2N)$, $e_k = (\sin(k\pi x_1), \sin(k\pi x_2), \ldots, \sin(k\pi x_n))^T$, where $N = n + 1$, $x_i = i/N$.
   (a) Show that $\{e_k\}_{k=1}^{n}$ is orthogonal in the standard inner product. It can be shown that $\|e_k\|^2 = N/2$.
   (b) Show that $A = \text{Trid}(a, b, a)$ is a linear combination of L and I, and find the eigenvalues of A in terms of the μ_i and an eigenbasis for A.

(c) Find sufficient conditions in terms of a and b, which are independent of n, in order for Trid(a, b, a) to be positive definite.

(d) Suppose $A = $ Trid(a, b, a), $G = $ Trid(u, v, u), and G is positive definite. Show that $\{e_k\}_{k=1}^n$ simultaneously diagonalizes $x^H A x$ and $x^H G x$. Find the generalized eigenvalues λ_k in $A e_k = \lambda_k G e_k$.

(Note: In Example 9 of Section 4.2, the eigenvalues and vectors of Trid(a, b, c) were derived. The point of this exercise is to deduce (a)–(d) from only the information on L.)

5. Prove Theorem 3.

6. Prove Theorem 4.

CHAPTER 7

Small Oscillations

The purpose of this chapter is to describe one of the major areas of application of linear algebra: the solution of the differential equations of small oscillations. These equations will be derived in several ways: from Newton's and Kirchhoff's laws, Lagrange's equations, the Ritz method, and Galerkin's method. The language of small oscillations and interpretation of solutions will be discussed. Examples from mechanical and electrical systems will be given. Free undamped oscillations, damped oscillations, and driven oscillations will be treated. The ideas presented here are fundamental to the understanding of small oscillations and to many finite element codes involved in the solution of small-oscillation and buckling problems.

7.1
DIFFERENTIAL EQUATIONS OF SMALL OSCILLATIONS

The matrix differential equation for undamped small oscillations is

$$M\ddot{x} + Kx = f, \quad \text{where } \dot{} = \frac{d}{dt}. \tag{1}$$

This equation arises in a variety of contexts and can be deduced by a number of methods. This section illustrates a few of the methods by which this differential equation is derived, and is perhaps more of a cultural exposure than an attempt to teach these methods in detail. In the next section more detailed examples will be given. In the last section, Galerkin 's method is presented and, again, the differential equation of small oscillations will arise.

Newton's and Kirchhoff's Laws

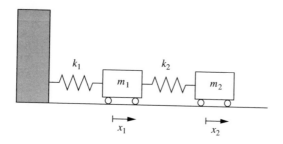

EXAMPLE 1. A MECHANICAL SYSTEM. Newton's second law asserts that the mass times the acceleration of a body equals the net force acting on the body. In the figure, x_i denotes the displacement of mass i from its equilibrium position. Thus x_1 is the stretch in spring 1 while $x_2 - x_1$ is the stretch in spring 2. Assuming linear springs, force $= k \times$ stretch, Newton's law gives

$$m_1 \ddot{x}_1 = -k_1 x_1 + k_2(x_2 - x_1)$$

$$m_2 \ddot{x}_2 = -k_2(x_2 - x_1)$$

or

$$M \ddot{x} + K x = 0,$$

where

$$M = \begin{bmatrix} m_1 & 0 \\ 0 & m_2 \end{bmatrix}, \qquad K = \begin{bmatrix} k_1 + k_2 & -k_2 \\ -k_2 & k_2 \end{bmatrix}, \qquad x = \begin{bmatrix} x_1 \\ x_2 \end{bmatrix}.$$

EXAMPLE 2. AN ELECTRICAL CIRCUIT. One of Kirchhoff's laws asserts that the sum of the voltage drops around a loop is zero. Another says that the net current into a junction is zero. In the accompanying figure, I_i is the ith loop current and $I_i = \dot{q}_i$, where q_i is the charge on the ith capacitor due to the current I_i. The voltage drop across C_1 is q_1/C_1, while the drop across C_2 is $(q_2 - q_1)/C_2$. The drop across inductor k is $L_k \dot{I}_k$. Kirchhoff's voltage law then yields

$$L_1 \dot{I}_1 + q_1/C_1 - (q_2 - q_1)/C_2 = 0$$

$$L_2 \dot{I}_2 + (q_2 - q_1)/C_2 = 0.$$

Using $\dot{q}_k = I_k$, we obtain

$$\begin{bmatrix} L_1 & 0 \\ 0 & L_2 \end{bmatrix} \ddot{q} + \begin{bmatrix} 1/C_1 + 1/C_2 & -1/C_2 \\ -1/C_2 & 1/C_2 \end{bmatrix} q = 0, \qquad \text{where } q = \begin{bmatrix} q_1 \\ q_2 \end{bmatrix}.$$

If one identifies $L = m$, $k = 1/C$, $q = x$, this is the same matrix differential equation as in the last example, and it has the form (1) with $f = 0$.

More examples using Newton's and Kirchhoff's laws will be found in the following sections.

Lagrange's Equations of Motion for Discrete Systems

Consider a discrete system of N particles with time-independent constraints subjected to conservative forces. Assume there are n degrees of freedom, which means that there are n generalized coordinates q_i describing the location of the particles. Lagrange's theory asserts that, whatever generalized coordinates are used, the motion of the system obeys the following differential equations, called *Lagrange's equations of motion:*

$$\frac{d}{dt}\left[\frac{\partial L}{\partial \dot{q}_k}\right] = \frac{\partial L}{\partial q_k}, \quad \dot{} = \frac{d}{dt}, \quad k = 1, 2, \ldots, n, \tag{2}$$

where $L = T - V$, T is the kinetic energy of the system, and V is the potential energy of the system. L is called the *Lagrangian* of the system. In mechanics, these differential equations are shown to be equivalent to Newton's laws.

Small oscillations about an equilibrium

Let $\eta_i = q_i - q_i^{(\text{eq})}$ denote the displacement from equilibrium in the generalized coordinates. From Newton's laws and Taylor expansions, it can be shown that the potential energy V and kinetic energy T are expressible as

$$V = \frac{1}{2}\sum_{i,j=1}^{n} k_{ij}\eta_i\eta_j + O(\|\eta\|^3), \qquad T = \frac{1}{2}\sum_{i,j=1}^{n} m_{ij}\dot{\eta}_i\dot{\eta}_j + O(\|\dot{\eta}\|^3),$$

where $m_{ij} = m_{ji}$, $k_{ij} = k_{ji}$. $O(\|\eta\|^3)$ denotes a remainder term that is $\leq C\|\eta\|^3$ for all small η. The terms in the sums are of order 2, so the remainder terms are smaller and may be neglected to obtain a first approximation of the energies. This yields

$$V = \frac{1}{2}\eta^{\mathsf{T}}K\eta, \qquad T = \frac{1}{2}\dot{\eta}^{\mathsf{T}}M\dot{\eta}, \tag{3}$$

where $K = [k_{ij}]$, $M = [m_{ij}]$ are symmetric matrices of constants, $\eta = (\eta_1, \eta_2, \ldots, \eta_n)^{\mathsf{T}}$. M is called the *inertial* or *mass* matrix, and K is called the *stiffness* matrix. Thus the potential and kinetic energies are quadratic forms in the generalized displacements from equilibrium and velocities, respectively. The kinetic energy is known to be positive, unless all the velocities are 0, which means that T is a positive definite quadratic form in the generalized velocities $\dot{\eta}_i$. From a theorem in differential equations, the potential energy quadratic form will be positive definite if and only if the equilibrium is stable.

The next step is to apply the Lagrange equations of motion:

$$-\frac{\partial L}{\partial \eta_k} = \frac{\partial V}{\partial \eta_k} = \frac{1}{2} \sum_{i,j=1}^{n} k_{ij}(\eta_j \delta_{ik} + \eta_i \delta_{kj}) = \frac{1}{2}\left(\sum_{j=1}^{n} k_{kj}\eta_j + \sum_{i=1}^{n} k_{ik}\eta_i\right)$$

$$= \frac{1}{2}\left(\sum_{j=1}^{n} k_{kj}\eta_j + \sum_{i=1}^{n} k_{ki}\eta_i\right) = \sum_{i=1}^{n} k_{ki}\eta_i \tag{4}$$

using $k_{ik} = k_{ki}$. Similarly,

$$\frac{\partial L}{\partial \dot{\eta}_k} = \frac{\partial T}{\partial \dot{\eta}_k} = \sum_{i=1}^{n} m_{ki}\dot{\eta}_i. \tag{5}$$

And the system of differential equations (4) and (5) can be written in matrix form

$$M\ddot{\eta} + K\eta = 0, \tag{6}$$

which is the same form as Eq. (1). This same differential equation can be derived directly from Newton's laws as well, if the system is linear or has been linearized.

In the following example, the process of (1) finding the potential and kinetic energies, (2) approximating them by quadratic forms near an equilibrium, and (3) deriving the equations of motion is illustrated.

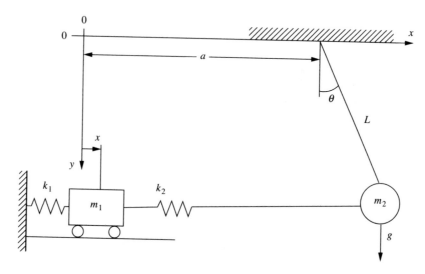

EXAMPLE 3. A mass m_2 is connected by a massless rod of length L to a frictionless pivot in a rigid support, and swings only in the xy plane. A cart of mass m_1 travels on linear frictionless rails parallel to the x axis and cannot leave the rails. m_1 and m_2 are linked by a massless spring, and m_2 is connected to a rigid support through a massless spring. At equilibrium, both springs are unstretched; the centers of mass of m_1 and m_2 are on the line $y = L$; the center of mass of m_1 is at $x = 0$, $y = L$; and m_2 and the pivot are on the line $x = a$. In the figure, x and θ denote the displacement from equilibrium and are the generalized displacements.

The potential energy is the sum of the gravitational potential energy and the energy stored in the springs:

$$V = m_2 g L(1 - \cos\theta) + \tfrac{1}{2}k_1 x^2 + \tfrac{1}{2}k_2(d - a)^2,$$

where d is the distance between the centers of mass of m_1 and m_2:

$$d = [(a + L\sin\theta - x)^2 + L^2(1 - \cos\theta)^2]^{1/2}.$$

Using the Taylor expansions of $\sqrt{1 + z}$, $\sin\theta$, and $\cos\theta$,

$$\sqrt{1 + z} = 1 + z/2 - z^2/8 + O(z^3)$$

$$\sin\theta = \theta - \theta^3/6 + O(\theta^5)$$

$$\cos\theta = 1 - \theta^2/2 + O(\theta^4),$$

and assuming that $1 \gg |\theta|$, $a \gg |x|$, $L|\theta|$ so that the Taylor expansions are valid and the higher-order terms are small, it follows (with some work) that

$$d = a - x + L\theta + e_0 \quad \text{and} \quad 1 - \cos(\theta) = \tfrac{1}{2}\theta^2 + f_0,$$

where the errors e_0 and f_0 satisfy $|e_0| = O(\theta^4)$, $|f_0| = O(h^2)$, with $h = \max(|x|, |\theta|)$. Hence

$$V = \tfrac{1}{2}[m_2 g L\theta^2 + k_1 x^2 + k_2(x - L\theta)^2] + e_1, \quad |e_1| = O(|h|^3).$$

The kinetic energy is

$$T = \tfrac{1}{2}(m_2 L^2\dot{\theta}^2 + m_1\dot{x}^2).$$

Dropping e_1 from the potential energy expression gives

$$V = \tfrac{1}{2}[m_2 g L\theta^2 + k_1 x^2 + k_2(x - L\theta)^2]$$

$$T = \tfrac{1}{2}[m_2 L^2\dot{\theta}^2 + m_1\dot{x}^2].$$

Lagrange's equations are

$$\frac{d}{dt}\frac{\partial(T - V)}{\partial\dot{x}} - \frac{\partial(T - V)}{\partial x} = m_1\ddot{x} + k_1 x + k_2(x - L\theta) = 0$$

$$\frac{d}{dt}\frac{\partial(T - V)}{\partial\dot{\theta}} - \frac{\partial(T - V)}{\partial\theta} = m_2 L^2\ddot{\theta} + m_2 g L\theta + Lk_2(L\theta - x) = 0.$$

In the study of electrical circuits containing only inductances and capacitances, Lagrange's equations can be applied by analogy to get Eq. (6). In this case the η_i are certain charges and the $\dot{\eta}_i$ the corresponding currents, T is the energy stored in magnetic fields associated with inductances, and V is the energy stored in electric fields in capacitors.

Ritz Method for the Oscillations of a Continuous System

The Ritz method is a means of converting a continuous system to a discrete one, so that Lagrange's equations of motion may be applied. It is quite general, extremely powerful, and extensively used. It is the method of choice for many problems. Its rigorous foundation will not be discussed here, since the mathematical machinery needed is not available. Part of the rigorous foundation was given in Section 5.5, and the rest of it involves calculus of variations and elementary Hilbert space theory. Some references are provided in the last section of this chapter. The method will be introduced by example here, and an "abstract" version will be proposed in the exercises. In the next section, another example will be given and specific computations using both examples will be made. The method presented here is also called the mixed Ritz method. There is a close relation between this method and Galerkin's method. The Ritz method is somewhat more general than Galerkin's method in that the so-called natural boundary conditions need not be satisfied by the basis functions (cf. the discussion of the Ritz method for static systems in Section 6.2).

EXAMPLE 4. VIBRATIONS OF A DRUM. Consider an unloaded ideal membrane stretched over a region Ω and clamped on the boundary. $u(p, t)$ denotes the vertical displacement at point p and time t. For small displacements and slopes, the potential energy V and kinetic energy T are (approximately) given by the quadratic forms

$$V = \frac{\tau}{2} \int_\Omega \|\nabla u\|^2 \, dA, \qquad T = \frac{1}{2} \int_\Omega \rho u_t^2 \, dA,$$

where τ is the tension (force per unit length) in the membrane and ρ is its density (mass per unit area). The first step of the Ritz method is to choose linearly independent functions $f_1(p), f_2(p), \ldots, f_n(p)$ called *basis functions*. An approximate solution of the form

$$U(p, t) = \sum_{i=1}^n c_i(t) f_i(p), \qquad c_i(t) \text{ unknown,}$$

is then sought. u is replaced by U in T and V to get

$$V = \frac{\tau}{2} \int_\Omega \nabla U \circ \nabla U \, dA = \frac{\tau}{2} \int_\Omega \left[\sum_{i=1}^n c_i(t) \nabla f_i(p)\right] \circ \left[\sum_{j=1}^n c_j(t) \nabla f_j(p)\right] dA$$

$$= \frac{\tau}{2} \sum_{i,j=1}^n c_i(t) c_j(t) \int_\Omega \nabla f_i(p) \circ \nabla f_j(p) \, dA = \frac{1}{2} c^{\mathrm{T}} K c,$$

where $c = (c_1, c_2, \ldots, c_n)^{\mathrm{T}}$, $\circ$ is the usual dot product in $\mathbf{R}^3$, and

$$k_{ij} = \tau \int_\Omega \nabla f_i(p) \circ \nabla f_j(p) \, dA = k_{ji}.$$

Similarly,

$$T = \frac{1}{2} \dot{c}^{\mathrm{T}} M \dot{c}, \qquad \text{where } m_{ij} = m_{ji} = \int_\Omega \rho f_i(p) f_j(p) \rho \, dA.$$

The next step is to apply Lagrange's equations as if this were a discrete system. This yields

$$M\ddot{c} + Kc = 0. \tag{7}$$

M is positive definite, since it is the Gram matrix of a linearly independent set with respect to the (real) weighted inner product $\langle f, g \rangle = \int_\Omega f(p) g(p) \rho \, dA$. K is positive because it is the Gram matrix of vector functions with inner product $\langle \mathbf{f}, \mathbf{g} \rangle = \tau \int_\Omega \mathbf{f} \circ \mathbf{g} \, dA$. It is positive definite if $\{\nabla f_k\}_{k=1}^n$ is linearly independent. It can be shown that the frequencies of oscillation are overestimated, and that the approximation is good if the basis functions are well chosen. Part of the theory behind this was discussed in Section 5.6.

Exercises 7.1

1. **A beaded string.** Suppose n beads of equal mass m are equally spaced on a taut string of length L. The ends of the string are fixed on the line $y = 0$. The spacing between the beads is thus $h = L/(n+1)$. Assume the beads experience only small vertical displacements y_i, and neglect the mass of the string. The tension τ may be assumed to be constant.
   (a) Show that the potential energy and kinetic energy of the system are (approximately)

$$V = \frac{\tau}{2h} \sum_{i=0}^{n} (y_{i+1} - y_i)^2, \quad \text{(taking } y_0 = y_{n+1} = 0\text{)}.$$

$$T = \frac{m}{2} \sum_{i=1}^{n} \dot{y}_i^2.$$

   (b) Derive the differential equations of motion from part a and from Newton's second law.

2. For the systems of pendulums illustrated,
   (a) Find the approximate potential and kinetic energies of the system as quadratic forms.
   (b) Find the differential equations of small oscillations.

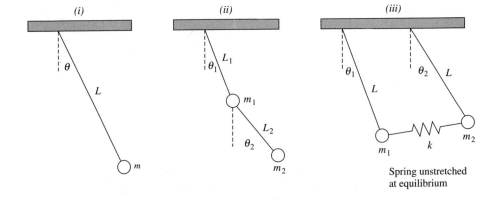

3. A system consists of three pendulums suspended by rods of length L from a horizontal ceiling and having the same mass m. Two springs of the same spring constant k connect masses 1 to 2 and 2 to 3, and the springs are unstretched at equilibrium, which occurs when the rods are vertical. θ_i denotes the counterclockwise angular displacement of mass i from equilibrium. For small displacements, show that the kinetic and potential energies are approximately

$$T = \tfrac{1}{2} mL^2 (\dot{\theta}_1^2 + \dot{\theta}_2^2 + \dot{\theta}_3^2)$$
$$V = \tfrac{1}{2} [mgL(\theta_1^2 + \theta_2^2 + \theta_3^2) + kL^2(\theta_2 - \theta_1)^2 + kL(\theta_3 - \theta_2)^2].$$

4. Apply the Ritz method to each of the following continuous systems. In each case assume a set of basis functions $\{f_1, \ldots, f_n\}$, and set $v = \sum_{i=1}^{n} c_i(t) f_i(x)$. Replace u by v in the given kinetic and potential energy quadratic forms to obtain expressions of the form $V = \tfrac{1}{2} c^T K c$ and $T = \tfrac{1}{2} \dot{c}^T M \dot{c}$. Identify M and K. Then find the differential equations for c using

Lagrange's equations of motion. Identify the boundary conditions that the basis functions must satisfy.

(a) *End-loaded vibrating string.* Each particle of the string moves vertically. The end at $x = 0$ is fixed at 0 displacement, and the end at $x = L$ is free to slide vertically with no friction, but with a mass m attached. $u(x, t)$ is the vertical displacement of the particle over x at time t. τ is the constant tension and ρ the variable mass per unit length. Subscripts denote partial derivatives.

$$V = \frac{\tau}{2} \int_0^L u_x^2 \, dx, \qquad T = \frac{1}{2} \int_0^L \rho u_t^2 \, dx + \frac{1}{2} m u_t^2(L, t)$$

(b) *Cantilever with elastic restraint.* A beam is welded to a wall at $x = 0$ with 0 height and with 0 slope. At $x = L$, the beam is attached to an elastic restraint with spring constant k. $u(x, t)$ is the vertical displacement of the particle over x at time t. E is Young's modulus, I is the cross-sectional moment of inertia about the neutral axis of the beam, and ρ is the mass per unit length. Subscripts denote partial derivatives. If the cantilever is made of uniform material but is tapered, I and ρ will be variable.

$$V = \frac{1}{2} \int_0^L E I u_{xx}^2 \, dx + \frac{1}{2} k u^2(L, t), \qquad T = \frac{1}{2} \int_0^L \rho u_t^2 \, dx$$

(c) *Longitudinal vibrations of a rod with elastic end load.* A rod is welded to a wall at $x = 0$, and the other end, at $x = L$, is attached through a spring with spring constant k to a wall. $u(x, t)$ denotes the horizontal displacement at time t of the section originally at x. E is Young's modulus, ρ is the density of the rod, and A is the cross-sectional area at x. For a tapered uniform rod, A will be variable. For small u and u_x

$$V = \frac{1}{2} \int_0^L E A u_x^2 \, dx + \frac{1}{2} k u^2(L, t), \qquad T = \frac{1}{2} \int_0^L \rho A u_t^2 \, dx.$$

5. **An abstract version of the Ritz method**

(a) Suppose the potential and kinetic energies are given by

$$V = \tfrac{1}{2} a(L_x u, L_x u) - \psi(u), \qquad T = \tfrac{1}{2} b(u_t, u_t),$$

where $a(\,,\,)$ and $b(\,,\,)$ are symmetric real quadratic forms, ψ is a linear functional, and L_x is a linear differential operator involving differentiation with respect to x. $\psi(u)$ represents the work done by external loads. Apply the Ritz method to find the differential equations of motion.

(b) Identify a, b, ψ, and L_x in each part of Exercise 4 and in Example 4.

7.2
UNDAMPED SMALL OSCILLATIONS

In this section undamped systems with and without externally applied forcing functions are treated. Methods for solving the matrix differential equation $M\ddot{x} + Kx = f$ are presented. The language of small oscillations is described and applied to the interpretation of the solution of problems. Examples are given from mechanics, from electrical circuits, and from applying the Ritz method.

Solution of the Homogeneous Differential Equations

Consider

$$M\ddot{x} + Kx = 0, \quad \text{where } \dot{} = \frac{d}{dt} \qquad (1)$$

with initial conditions

$$x(0) = x_0, \qquad \dot{x}(0) = v_0. \qquad (2)$$

M and K are $n \times n$ Hermitian matrices and M is positive definite; i.e., $\langle x, y \rangle = y^H M x$ is an inner product. Two approaches to solving these equations are discussed here. They appear to be different, but both methods are logically equivalent and simply represent different viewpoints or tastes. Each has its own advantages. The connection will be given later.

Vector space approach

A change of basis

$$x(t) = \sum_{i=1}^{n} y_i(t) E_i \qquad (3)$$

will be made in such a way that (1) becomes easy to solve. Observe that (3) does not restrict generality, since every $x(t)$ must be expressible in terms of the basis $\{E_k\}_{k=1}^{n}$. Substituting this expression for x into the differential equation yields

$$\sum_{i=1}^{n} (\ddot{y}_i M E_i + y_i K E_i) = 0. \qquad (*)$$

From Theorem 4 of Section 5.3, it is known that the generalized eigenvalue problem $KE = \lambda ME$ can be solved; that is, vectors E_i and real numbers λ_i may be found such that $KE_i = \lambda_i ME_i$ and $\{E_k\}_{k=1}^{n}$ is orthogonal with respect to $\langle x, y \rangle = y^T M x$. Inserting this into (*) yields

$$\sum_{i=1}^{n} (\ddot{y}_i + \lambda_i y_i) M E_i = 0.$$

Since M is nonsingular and $\{E_k\}_{k=1}^{n}$ is linearly independent, $\{ME_k\}_{k=1}^{n}$ is linearly independent, and so

$$\ddot{y}_i + \lambda_i y_i = 0, \quad i = 1, 2, \dots, n. \qquad (4)$$

These equations are no longer coupled, and they have the complete solution $y_i = a_i \phi_i(t) + b_i \psi_i(t)$, where $\phi_i(t), \psi_i(t)$ are a basis of solutions for (4). Let us choose them so that

$$\phi_i(0) = 1, \qquad \psi_i(0) = 0, \qquad \dot{\phi}_i(0) = 0, \qquad \dot{\psi}_i(0) = 1. \qquad (5)$$

Thus

$$\phi_i(t) = \cosh(r_i t), \qquad \psi_i(t) = r_i^{-1} \sinh(r_i t) \quad \text{where } r_i = \sqrt{-\lambda_i} \text{ when } \lambda_i < 0$$

$$\phi_i(t) = 1, \qquad \psi_i(t) = t \quad \text{when } \lambda_i = 0 \tag{6}$$

$$\phi_i(t) = \cos(\omega_i t), \qquad \psi_i(t) = \omega_i^{-1} \sin(\omega_i t) \quad \text{where } \omega_i = \sqrt{\lambda_i} \text{ when } \lambda_i > 0.$$

Hence the complete solution of the original differential equation is

$$x = \sum_{i=1}^{n} y_i E_i = \sum_{i=1}^{n} [a_i \phi_i(t) + b_i \psi_i(t)] E_i. \tag{7}$$

The initial conditions are used to find the constants a_i, b_i. In fact,

$$x_0 = x(0) = \sum_{i=1}^{n} y_i(0) E_i = \sum_{i=1}^{n} a_i E_i$$

$$v_0 = \dot{x}(0) = \sum_{i=1}^{n} \dot{y}_i(0) E_i = \sum_{i=1}^{n} b_i E_i. \tag{8}$$

But these are just Fourier expansions, since the E_i are orthogonal with respect to $\langle x, y \rangle = y^H M x$. Thus

$$a_i = \langle x_0, E_i \rangle / \|E_i\|^2 = E_i^T M x_0 / (E_i^T M E_i)$$

$$b_i = \langle v_0, E_i \rangle / \|E_i\|^2 = E_i^T M x_0 / (E_i^T M E_i). \tag{9}$$

Matrix theory approach

Make the substitution (change of coordinates)

$$x = Py \tag{10}$$

in the differential equation (again without loss of generality), and multiply the result on the left by P^H to obtain

$$P^H(MP\ddot{y} + KPy) = P^H MP\ddot{y} + P^H KPy = 0. \tag{**}$$

By Corollary 5 of Section 5.3, a matrix P can be found such that $P^H MP = N^2$ and $P^H KP = \Lambda N^2$, where $\Lambda = \text{diag}(\lambda_1, \ldots, \lambda_n)$, $N = \text{diag}(\|E_1\|, \ldots, \|E_n\|)$, $\|E_i\|^2 = E_i^H M E_i$. For such a P, (**) becomes $\ddot{y} N^2 + \Lambda N^2 y = 0$; thus

$$\ddot{y} + \Lambda y = 0;$$

i.e.,

$$\ddot{y}_i + \lambda_i y_i = 0, \quad i = 1, 2, \ldots, n.$$

After these decoupled equations are solved, the general solution is $x = Py$.

To satisfy the initial conditions, let us use $y_i = a_i \phi_i(t) + b_i \psi_i(t)$, where the ϕ_i and ψ_i are defined by (6), so that

$$y(0) = a = (a_1, a_2, \ldots, a_n)^T$$

$$\dot{y}(0) = b = (b_1, b_2, \ldots, b_n)^T.$$

Then the initial conditions demand that

$$x_0 = Pa \quad \text{and} \quad v_0 = Pb.$$

Observe that $P^H M P = N^2$ implies $P^{-1} = N^{-2} P^H M$, so that

$$a = N^{-2} P^H M x_0 \quad \text{and} \quad b = N^{-2} P^H M v_0. \tag{11}$$

Remark. The matrix theory and vector space solutions are connected by the identity $x = Py = \sum_{i=1}^{n} y_i E_i$, where $P = [E_1 \; E_2 \; \cdots \; E_n]$. Furthermore, the requirement that $P^H M P = N^2$ and $P^H A P = \Lambda N^2$ just means that the E_i are orthogonal with respect to $\langle x, y \rangle = y^H M x$ and that $K E_i = \lambda_i M E_i$ (cf. Corollary 5 in Section 5.3). To see that (9) and (11) agree, note that $a = N^{-2} P^H M x_0$ has the ith entry $a_i = E i^H M x_0 / \|E_i\|^2 = \langle x_0, E_i \rangle / \|E_i\|^2$, which agrees with (9).

Remark. In many modern applications, the matrices M and K are very large. Such matrices arise in finite element analysis, for example. The computation of P^{-1} could therefore be onerous and contaminated with error. In such a case the relation $P^{-1} = N^{-2} P^H M$ is very valuable. The calculations are slightly more convenient if the eigenvectors E_i have been normalized so that $\|E_i\|^2 = 1$. In this case $N = I$, and (11) becomes

$$a = P^H M x_0 \quad \text{and} \quad b = P^H M v_0. \tag{11'}$$

Remark. If all $\lambda_i = \omega_i^2 > 0$, then a compact form for the general solution is

$$x = P \cos(\omega t) a + P \sin(\omega t) b,$$

where $\omega = \text{diag}(\omega_1, \ldots, \omega_n)$ and $P = [E_1 \; E_2 \; \cdots \; E_n]$ is the modal matrix. The proof is left as an exercise.

Terminology Used in Discussing Small Oscillations

When $\lambda_j > 0$, the solution for y_j can be written $y_j = c_j \sin(\omega_j t + \theta_j)$, where $\omega_j = \sqrt{\lambda_j}$. These ω_j are called *natural frequencies*. The eigenvectors E_i are called *normal modes* or *modal vectors*, and $P = [E_1 \; E_2 \; \cdots \; E_n]$ is called the *modal matrix*. When $\lambda_j > 0$, each vector function $x^{(j)} = \sin(\omega_j t + \theta_i) E_j$ is a solution of $M\ddot{x} + Kx = 0$ and is called a *fundamental solution*, a *principal mode*, or a *natural mode*. When all $\lambda_j > 0$, the general solution is a linear combination of the principal modes: $x = \sum_{i=1}^{n} c_i x^{(i)}$. Thus the general solution is made up of a superposition of solutions involving only a finite number of frequencies $\omega_1, \omega_2, \ldots, \omega_n$. This is a fundamental and perhaps surprising fact: Only n frequencies and normal modes are needed to describe the general motion of the system.

If the system giving rise to these differential equations is a system of coupled particles, then the meaning of the fundamental modes may be interpreted as follows. When only the kth mode is active, a solution is $x = x^{(k)} = \sin(\omega_k t + \theta_k) E_k$, and all particles in the system oscillate with the same frequency ω_k. Let $E_k =$

$(p_{1_k}, p_{2_k}, \ldots, p_{n_k})^T$. Evidently the jth particle has displacement $x_j^k = p_{jk} \sin(\omega_k t + \theta_k)$. This implies that the particles $j = 1, 2, \ldots, n$ either have the same phase or are $180°$ out of phase, depending on whether p_{jk} is positive or negative. The relative magnitudes of the coordinates p_{jk} of E_k give the relative amplitudes of the displacements from equilibrium. The next example illustrates this.

Applications in Mechanics and Electrical Circuit Theory

EXAMPLE 1. MECHANICS. Three masses are constrained to move along a line, and they are coupled by linear elastic forces. One may think of a linear triatomic molecule or of three cars on a track coupled by springs. If l_i is the unstretched length of spring i, and z_i is the distance of mass i from a fixed origin on the line of motion, then $x_1 = z_1$, $x_2 = z_2 - l_1$, $x_3 = z_3 - (l_1 + l_2)$.

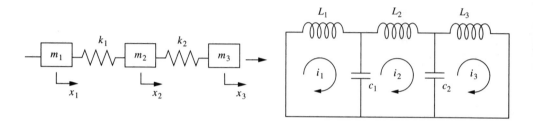

Electrical circuit. The circuit is shown. To make the correspondence, set

$$m_j = L_j, \qquad k_j = 1/c_j, \qquad \dot{x}_j = i_j.$$

The kinetic and potential energies for the mechanical system are

$$T = \tfrac{1}{2}[m_1 \dot{x}_1^2 + m_2 \dot{x}_2^2 + m_3 \dot{x}_3^2].$$

$$V = \tfrac{1}{2}[k_1(x_1 - x_2)^2 + k_2(x_2 - x_3)^2]$$

Newton's laws are expressed by

$$m_1 \ddot{x}_1 = k_1(x_2 - x_1)$$

$$m_2 \ddot{x}_2 = k_1(x_1 - x_2) + k_2(x_3 - x_2)$$

$$m_3 \ddot{x}_3 = k_2(x_2 - x_3).$$

To illustrate the derivation of the equations expressing Newton's laws, consider the second equation. When $x_1 > x_2$, spring 1 is compressed and exerts a force on m_2 in the positive x_2 direction of $k_1(x_1 - x_2)$ units; when $x_3 > x_2$, spring 2 is stretched, which causes a force $k_2(x_3 - x_2)$ acting in the positive x_2 direction on m_2. Thus the net force on m_2 is $k_1(x_1 - x_2) + k_2(x_3 - x_2)$, which has the correct sign and must equal the mass times the acceleration. These equations may also be derived from T and V by Lagrange's equations of motion.

For the electrical circuit, each of these three equations asserts that the net voltage drop around each loop is zero. The voltage across a capacitor is q/C, where q is the charge on the capacitor and C is its capacitance, while the voltage drop across an inductor is $L(di/dt)$, where L is the inductance and $i = \dot{q}$ is the current through the inductor.

Thus the differential equations are

$$M\ddot{x} + Kx = 0, \quad \text{where} \quad K = \begin{bmatrix} k_1 & -k_1 & 0 \\ -k_1 & k_1 + k_2 & -k_2 \\ 0 & -k_2 & k_2 \end{bmatrix} \quad \text{and} \quad M = \text{diag}(m_1, m_2, m_3).$$

Notice that the potential energy V is diagonalized by the substitution $y_1 = x_1$, $y_2 = x_1 - x_2$, $y_3 = x_2 - x_3$, and that there are two positive weights and one zero weight. Since V is also diagonalized by solving the generalized eigenvalue problem, there will be one zero and two positive generalized eigenvalues, by Sylvester's law of inertia. $E_1 = (1, 1, 1)^T$ is an eigenvector associated with the eigenvalue $\lambda_1 = 0$, since $KE_1 = 0 \cdot ME_1 = 0$. Such a system is called a *semidefinite system*. Thus the form of the general solution will be $x = (a_1 + b_2t)E_1 + [a_2 \cos(\omega_2 t) + b_2 \sin(\omega_2 t)]E_2 + a_3 \cos(\omega_3 t) + b_3 \sin(\omega_3 t)]E_3$, where $K_iE_i = \lambda_i ME_i$, and $\omega_i = \sqrt{\lambda_i}$. The principal mode $(a_1 + b_1 t)E_1$ is interpreted as a pure (rigid body) translation of all masses simultaneously. Thus the general motion is a superposition of a uniform translation with two oscillations.

For the sake of illustration, take $m_1 = m_3 = 1$, $m_2 = 3$, $k_1 = k_2 = 6$. Then

$$k = \begin{bmatrix} 6 & -6 & 0 \\ -6 & 12 & -6 \\ 0 & -6 & 6 \end{bmatrix}, \quad M = \text{diag}(1, 3, 1).$$

Thus $p(\lambda) = |K - \lambda M| = 3\lambda(6 - \lambda)(\lambda - 10)$. The eigenvalues are $\lambda_1 = 0$, $\lambda_2 = 6$, $\lambda_3 = 10$, and solving $(K - \lambda_i M)E_i = 0$ gives the (non-normalized) associated eigenvectors $E_1 = (1, 1, 1)^T$, $E_2 = (1, 0, -1)^T$, $E_3 = (3, -2, 3)^T$. The natural frequencies are $\omega_2 = \sqrt{6}$, $\omega_3 = \sqrt{10}$. Note that the generalized eigenvectors E_i must be orthogonal with respect to the inner product generated by the mass matrix: $\langle E_i, E_j \rangle = E_i^T M E_j = 0$ if $i \neq j$. This is a good computational check.

Mode 2: $\omega_2 = \sqrt{6}$

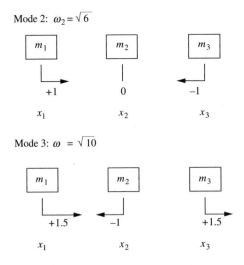

Mode 3: $\omega = \sqrt{10}$

When the system oscillates in the second mode, all masses oscillate with frequency $\sqrt{6}$, the second mass is stationary, and the first and third oscillate with the same amplitude but are 180° out of phase; i.e., their directions of motions oppose. In the third principal mode, all masses oscillate with frequency $\sqrt{10}$, the two outer masses are in phase, and the inner mass is $180°$ out of phase with the outer two, with amplitude of oscillation two-thirds that of the outer two.

If the initial conditions are $x(0) = x_0 = (1, 1, 0)^T$ and $\dot{x}(0) = v_0 = (1, 2, 1)^T$, then we must find a_i and b_i so that

$$x_0 = a_1 E_1 + a_2 E_2 + a_3 E_3$$
$$v_0 = b_1 E_1 + \sqrt{6} b_2 E_2 + \sqrt{10} b_3 E_3.$$

Computing the Fourier coefficients yields

$$a_1 = E_1^H M x_0 / \|E_1\|^2 = \tfrac{4}{5}, \quad a_2 = E_2^H M x_0 / \|E_2\|^2 = \tfrac{1}{2}, \quad a_3 = E_3^H M x_0 / \|E_3\|^2 = -\tfrac{1}{10}$$

and

$$b_1 = E_1^H M v_0 / \|E_1\|^2 = \tfrac{8}{5}, \quad \sqrt{6} b_2 = E_2^H M v_0 / \|E_2\|^2 = 0, \quad \sqrt{10} b_3 = E_3^H M v_0 / \|E_3\|^2 = -\tfrac{1}{5}$$

so that $b_2 = 0, b_3 = -\frac{\sqrt{10}}{50}$.

Examples of the Ritz Method

The first example will concern a vibrating string, and the second the radially symmetric vibrations of a circular drum. In these examples, the interest is only in finding approximations to the fundamental frequencies and normal modes of oscillation. When the Ritz method is used this way, it is sometimes called the Rayleigh–Ritz method (cf. Section 5.6). However, in the Ritz method, the so-called natural boundary conditions need not be satisfied by the basis functions.

EXAMPLE 2. Consider a vibrating string of length L that is fixed at $x = 0$ and has an elastic restraint at $x = L$. Use the Ritz method to derive the differential equations of motion. Specialize the physical parameters and approximate fundamental frequencies and mode shapes using two and four basis functions.

The potential and kinetic energies of the system are

$$T(u) = \frac{1}{2} \int_0^L \rho u_t^2 \, dt, \qquad V(u) = \frac{\tau}{2} \int_0^L u_x^2 \, dx + \frac{1}{2} k u^2(L, t)$$

where u is the vertical displacement, ρ is the mass per unit length, τ is the tension, and k is the spring constant of the elastic restraint at $x = L$. The boundary condition is $u(0, t) = 0$. Variational methods imply that at $x = L$, a natural boundary condition for the exact solution holds. It is $\tau u_x(L, t) + k u(L, t) = 0$.

Let $\{f_k\}_{k=1}^n$ be linearly independent and satisfy $f_k(0) = 0$. Put $U = \sum_{i=1}^n c_i(t) f_i(x)$. Put U into T and V, which yields

$$T(U) = \frac{1}{2} \int_0^L \rho \left[\sum_{i=1}^n \dot{c}_i(t) f_i(x) \right]^2 dx = \frac{1}{2} \sum_{i,j=1}^n \dot{c}_i(t) \dot{c}_j(t) \int_0^L \rho f_i(x) f_j(x) \, dx$$

$$V(U) = \frac{1}{2} \int_0^L \rho \left[\sum_{i=1}^n c_i(t) f_i'(x) \right]^2 dx + \frac{1}{2} k \left[\sum_{i=1}^n c_i(t) f_i(L) \right]^2$$

$$= \frac{1}{2} \sum_{i,j=1}^n c_i(t) c_j(t) \left\{ \int_0^L \rho f_i'(x) f_j'(x) \, dx + k f_i(L) f_j(L) \right\}.$$

Thus $T(U) = \frac{1}{2}\dot{c}^T M \dot{c}$ and $V(U) = \frac{1}{2}c^T K c$, where

$$m_{ij} = m_{ji} = \int_0^L \rho f_i(x) f_j(x)\, dx$$

and

$$k_{ij} = k_{ji} = \int_0^L \rho f_i'(x) f_j'(x)\, dx + k f_i(L) f_j(L).$$

Applying Lagrange's equations of motion yields $M\ddot{c} + Kc = 0$. As we have seen, solving this involves solving the generalized eigenvalue problem $KE_i = \lambda_i ME_i$, with $\langle E_i, E_j \rangle_M = 0$ if $i \neq j$. The general solution of Lagrange's equations is a superposition (linear combination) of the fundamental solutions $c^{(j)}(t) = \sin(\tilde{\omega}_j t + \psi_j)E_j$, where $\tilde{\omega}_j = \sqrt{\lambda_j}$. Each of these solutions $c^{(j)}(t)$ corresponds to an approximate solution $U^{(j)}(t)$ for the motion of the string, which is found by attaching the entries of $c^{(j)}$ to the corresponding f_i. Specifically, if $E_j = (e_{1j}, e_{2j}, \ldots, e_{nj})^T$, then

$$U^{(j)}(x, t) = \sum_{i=1}^n \sin(\tilde{\omega}_j t + \psi_j) e_{ij} f_i(x) = \sin(\tilde{\omega}_j t + \psi_j) \sum_{i=1}^n e_{ij} f_i(x)$$

$$= \sin(\tilde{\omega}_j t + \psi_j)\tilde{e}_j(x)$$

where

$$\tilde{e}_j(x) = \sum_{i=1}^n e_{ij} f_i(x), \qquad E_j = (e_{1j}, e_{2j}, \ldots, e_{nj})^T. \tag{12}$$

In the language of coordinate maps, $[U^{(j)}] = c^{(j)}$ and $[\tilde{e}_j] = E_j$. The approximate frequencies of oscillation are $\tilde{\omega}_j = \sqrt{\lambda_j}$, and the approximate normal modes are the $\tilde{e}_j$. In the motion $U^{(j)}$, every point on the string oscillates with angular frequency $\tilde{\omega}_j$, and the amplitude of a point at x is $\tilde{e}_j(x)$. The function $\tilde{e}_j(x)$ is an approximation to an actual normal mode $e_j(x)$, and $\tilde{\omega}_j$ is an (over) approximation to an actual frequency of vibration.

It can be shown that the exact solutions for the normal modes are $e_k(x) = \sin(\omega_k x)$, where ω_k is the kth positive root of $k\tan(\omega) = -\tau\omega$.

For the computational example, take $k = L = \rho = \tau = 1$. The exact solutions are: $\omega_1 = 2.0288$, $\omega_2 = 4.9132$, $\omega_3 = 7.9787$, $\omega_4 = 11.086$ (to five significant digits). Choose the basis functions $f_j(x) = x^j$, $1 \le j \le n$. Then

$$m_{ij} = \int_0^1 x^i x^j\, dx = \frac{1}{i + j + 1}$$

$$k_{ij} = \int_0^1 ijx^{i-1}x^{j-1}\, dx + 1 = \frac{ij}{i + j - 1} + 1.$$

We will consider $n = 2, 4$ and find the approximate normal modes and the approximate frequencies. These will be compared to the exact solutions.

For $n = 2$,

$$M = \begin{bmatrix} 1/3 & 1/4 \\ 1/4 & 1/5 \end{bmatrix}, \qquad K = \begin{bmatrix} 2 & 2 \\ 2 & 7/3 \end{bmatrix},$$

and solving $Ke = \lambda Me$, $e \neq 0$, yields the approximate frequencies $\tilde{\omega}_i = \sqrt{\lambda_i} = 2.038, 6.206$, which is excellent for the first frequency. The generalized eigenvectors are $E_1 = (1, -0.640)^T$ and $E_2 = (1, -1.421)^T$, so that the approximate normal modes (to a scalar multiple) are $\tilde{e}_1(x) = x - 0.640x^2$ and $\tilde{e}_2(x) = x - 1.421x^2$. If the $\tilde{e}_i$ are normalized so that their maximum value on [0, 1] is 1, then the graph of $\tilde{e}_1$ shows it is a fairly good approximation to $e_1(x)$ while the $\tilde{e}_2$ approximates e_2 only poorly. This is shown in the following graphs, where the $\tilde{e}_k$ are plotted with solid lines and e_k with dotted lines.

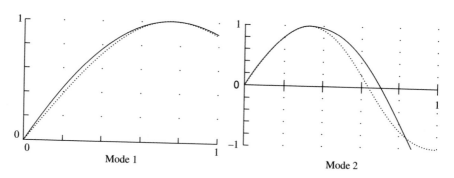

Mode 1   Mode 2

For $n = 4$ (with computational aids), $\tilde{\omega}_i = \sqrt{\lambda_i} = 2.028, 4.932, 8.45, 16.86$. The lowest two frequencies are approximated very well. For example, the generalized eigenvector corresponding to λ_1 is $E_1 = (1, 0.0635, -0.8993, 0.2805)^{\mathrm{T}}$, and the corresponding approximate eigenfunction is $\tilde{e}_1(x) = x + 0.0635x^2 - 0.8993x^3 + 0.2805x^4$. The other approximate eigenfunctions (mode shapes) are $\tilde{e}_2(x) = x - 0.9260x^2 - 2.2314x^3 + 1.9783x^4$, $\tilde{e}_3(x) = x - 4.852x^2 - 6.910x^3 - 3.005x^4$, $\tilde{e}_4(x) = x - 6.5964x^2 + 12.493x^3 - 7.106x^4$. The following graphs show the approximate normal modes $\tilde{e}_k$ (solid lines) and the exact normal modes e_k (dotted lines). The $\tilde{e}_k$ are scaled to have a maximum of 1 in order to compare the graphs.

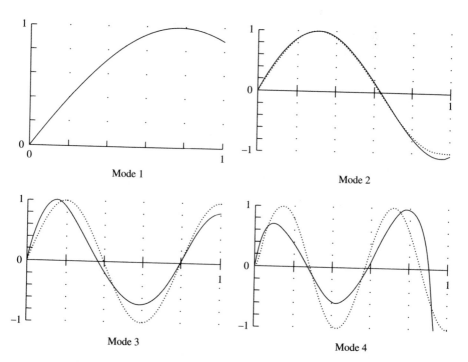

Mode 1   Mode 2

Mode 3   Mode 4

There is no visual difference between the two for the first mode. The second mode is good, and the last two are progressively worse. Nevertheless, the mode shapes are well captured for such a few approximating functions.

EXAMPLE 3. Use the Ritz method to approximate the first two normal modes and the natural frequencies of a radially symmetric drum whose membrane is stretched over the unit disk and clamped at the edges.

The potential and kinetic energies are given by

$$V = \frac{\tau}{2} \int_\Omega \|\nabla u\|^2 \, dA, \qquad T = \frac{1}{2} \int_\Omega u_t^2 \rho \, dA,$$

where Ω is the unit disk. Due to radial symmetry, $\nabla u = u_r \mathbf{i}_r$, where $\mathbf{i}_r$ is a unit vector pointing radially outward. Since $dA = r \, dr \, d\phi$,

$$V = \pi\tau \int_0^1 u_r^2 r \, dr, \qquad T = \pi \int_0^1 u_t^2 \rho r \, dr. \tag{*}$$

Selecting linearly independent basis functions $f_1(r), \ldots, f_n(r)$ and replacing u by $U = \sum_{i=1}^n c_i(t) f_i(r)$ in (*) gives (as in Section 7.1)

$$V = \tfrac{1}{2} c^T K c, \qquad T = \tfrac{1}{2} \dot{c}^T M \dot{c}$$

with $\qquad k_{ij} = 2\pi\tau \int_0^1 f_i'(r) f_j'(r) r \, dr, \qquad m_{ij} = 2\pi \int_0^1 f_i(r) f_j(r) \rho r \, dr. \tag{**}$

Note that M is positive definite since it is the Gram matrix of linearly independent functions. K is positive and will be definite if the derivatives f_i' are linearly independent.

Lagrange's equations are $M\ddot{c} + Kc = 0$. Suppose the generalized eigenvalue problem $KE_i = \lambda_i M E_i$ has been solved, with the E_i being a basis orthogonal with respect to the inner product generated by M. As in Example 2, the approximate frequencies are $\omega_k = \sqrt{\lambda_k}$, and the approximate normal modes are

$$\tilde{e}_j(r) = \sum_{i=1}^n e_{ij} f_i(r), \qquad E_j = (e_{1_j}, e_{2_j}, \ldots, e_{nj})^T.$$

In the language of coordinate maps, $[U^{(j)}] = c^{(j)}$ and $[\tilde{e}_j] = E_j$.

To give a simple illustration of the method, take $2\pi\tau = 2\pi\rho = 1$ and $f_1(r) = (1 - r)$, $f_2 = (1 - r)r$. From (**),

$$K = \begin{bmatrix} 1/2 & 1/6 \\ 1/6 & 1/6 \end{bmatrix} \quad \text{and} \quad M = \begin{bmatrix} 1/12 & 1/30 \\ 1/30 & 1/60 \end{bmatrix}.$$

Solving the generalized eigenvalue problem (approximately) yields

$$\lambda_1 = 5.858, \qquad E_1 = \begin{bmatrix} 1 \\ 0.1414 \end{bmatrix}, \qquad \text{so} \quad \tilde{\omega}_1 = 2.420, \qquad \tilde{e}_1(r) = f_1(r) + 0.414 f_2(r)$$

and

$$\lambda_2 = 34.14, \qquad E_2 = \begin{bmatrix} 1 \\ -2.1414 \end{bmatrix}, \qquad \text{so} \quad \tilde{\omega}_2 = 5.834, \qquad \tilde{e}_2(r) = f_1(r) - 2.414 f_2(r).$$

The approximate mode shapes have been normalized so that $\tilde{e}_j(0) = 1$.

The actual normal modes of oscillation are $e_j(r) = J_0(\omega_j r)$, where J_0 is the 0th-order Bessel function of the first kind,

$$J_0(z) = \sum_{k=0}^{\infty} \frac{(-\frac{1}{4}z^2)^k}{k!^2},$$

and ω_j is the jth positive zero of $J_0(z)$. For this example, the actual frequencies of oscillation are the ω_j. The first three are $\omega_1 = 2.4048256$, $\omega_2 = 5.52008$, $\omega_3 = 8.65372$. The approximation of the frequencies is remarkable, given that just two (uninspired) basis functions were used. The next two graphs illustrate the agreement of the normal modes. The solid lines are the approximate eigenfunctions, and the dotted lines are the exact eigenfunctions.

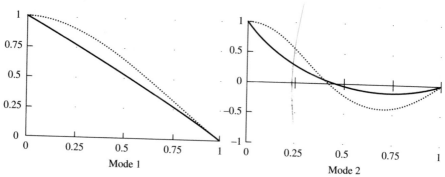

Mode 1   Mode 2

Notice that $\tilde{e}_1$ is a better approximation than $\tilde{e}_2$ to the corresponding normal mode. Neither approximation is superlative, but both are surprisingly good considering the fact that a set $\{f_1, f_2\}$ of just two simple functions was used. The results are typical: The lower frequencies and modes are better approximated. Looking at the graphs, it becomes apparent that the choice of basis functions is not good on physical grounds. These graphs show half of a cross section of the drum. The whole cross section extends from $r = -1$ to $r = 1$ and is symmetric across $r = 0$. For a smooth drum head, one needs $e_i'(0) = 0$. A better choice of basis functions would satisfy $f_i'(0) = 0$ as well as $f_i(1) = 0$, e.g., $f_1(r) = 1 - r^2$, $f_2(r) = 1 - r^3$. The results are better.

The Power Method for the Extreme Eigenvalues and Eigenvectors

The generalized eigenvalue problem $KE = \lambda ME$ is equivalent to the ordinary eigenvalue problem for the operator $L = M^{-1}K : M^{-1}KE = \lambda E$. It has been shown that L is Hermitian with respect to the inner product $\langle x, y \rangle = y^H M x$ generated by M. Thus the power method of Section 5.5 applies. With normalization by $w_n = \|z_n\|_M^{-1}$, the method of Section 5.5 in this context is

Select a starting vector $u \neq 0$, and set $v_0 = u/\|u\|_M$, $n = 0$.
Given n, v_n, with $\|v_n\|_M = 1$:
    Set $w = Lv_n$.
    Compute $R_{n+1} = \langle w, v_n \rangle / \|v_n\|_M$.
    Set $v_{n+1} = w/\|w\|_M$.
    Print $n + 1$, R_{n+1}, v_{n+1}.

When M is positive definite and K is positive, a dominant eigenvalue λ_1 exists. If u has a component in the eigenspace of λ_1, then $R_n \to \lambda_1$, and $v_n \to E_1$ lying in the eigenspace $M(\lambda_1)$.

In the inverse power method, simply replace L by L^{-1} (if it exists) in the preceding algorithm, and the Rayleigh quotients will converge to the reciprocal of the smallest eigenvalue and v_n will converge to a corresponding eigenvector. These two simple iterations will recover the highest and lowest frequencies of oscillation, which are sometimes the most important. In large structures, often the lowest frequencies are the most important. In this case, we can use the inverse power method, and then the inverse power method with suppression to get the first few lowest frequencies. Alternatively, the subspace iteration with the inverse power method can be used to obtain several lowest frequencies all at once.

EXAMPLE 4. MECHANICAL PROBLEM. Three masses are constrained to move in a line. The first mass is connected to a rigid support through a spring. The others are connected as shown. It is desired to find the motion in the highest and lowest modes.

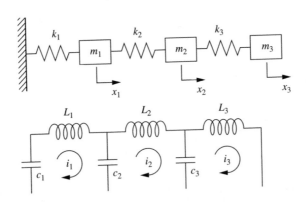

Electric Circuit Problem. Find the loop currents in the circuit shown in the highest and lowest modes. The differential equations will be the same, with the correspondence

$$m_j = L_j, \qquad k_j = 1/c_j, \qquad \dot{x}_j = i_j.$$

The kinetic and potential energies are

$$T = \tfrac{1}{2}[m_1 \dot{x}_1^2 + m_2 \dot{x}_2^2 + m_3 \dot{x}_3^2]$$
$$V = \tfrac{1}{2}[k_1 x_1^2 + k_2(x_1 - x_2)^2 + k_3(x_2 - x_3)^2].$$

V is diagonalized by the substitution $y_1 = x_1, y_2 = x_1 - x_2, y_3 = x_2 - x_3$ and has positive weights in this diagonalization. Thus V is positive definite. By Sylvester's law of inertia, there must be three positive generalized eigenvalues.

The system of differential equations derived from Newton's laws or by the Lagrange equations of motion is

$$m_1 \ddot{x}_1 + k_1 x_1 + k_2(x_1 - x_2) = 0$$
$$m_2 \ddot{x}_2 + k_2(x_2 - x_1) + k_3(x_2 - x_3) = 0$$
$$m_3 \ddot{x}_3 + k_3(x_3 - x_2) = 0.$$

In matrix form, this is

$$M\ddot{x} + Kx = 0,$$

where
$$K = \begin{bmatrix} k_1 + k_2 & -k_2 & 0 \\ -k_2 & k_2 + k_3 & -k_3 \\ 0 & -k_3 & k_3 \end{bmatrix}, \qquad M = \text{diag}(m_1, m_2, m_3).$$

Take $k_1 = k_2 = k_3 = 4$, $m_1 = m_2 = 1$, $m_3 = 2$. Then $M = \text{diag}(1, 1, 2)$ and

$$L = M^{-1}K = \begin{bmatrix} 8 & -4 & 0 \\ -4 & 8 & -2 \\ 0 & -2 & 2 \end{bmatrix}, \qquad L^{-1} = \frac{1}{4}\begin{bmatrix} 1 & 1 & 2 \\ 1 & 2 & 4 \\ 1 & 2 & 6 \end{bmatrix}.$$

To determine the highest frequency, the starting vector $u = (2, -2, 1)^T$ was chosen because it was suspected that the lighter two masses would oscillate out of phase with one another, and the heavier mass would have smaller amplitude. The results are given next, with $v = v_n = (x_n, y_n, z_n)^T$ normalized to be of unit length in $\| \ \|_M$.

n	R_n	x_n	y_n	z_n
1	11.6000	0.63422	-0.73992	0.15855
2	12.3911	0.64813	-0.73341	0.14498
3	12.4012	0.65464	-0.72891	0.14166
4	12.4026	0.65734	-0.72697	0.14038
5	12.4029	0.65844	-0.72617	0.13986
6	12.4029	0.65890	-0.72585	0.13965
7	12.4029	0.65908	-0.72571	0.13956
8	12.4029	0.65916	-0.72566	0.13953
9	12.4029	0.65919	-0.72563	0.13951
10	12.4029	0.65920	-0.72562	0.13951

Observe that R_n converges more rapidly to the maximum eigenvalue than v_n converges to approximate eigenvectors, as predicted in Section 5.5. To a good approximation, $\lambda_1 = \lambda_{max} = 12.4094$, so $\omega_{max} = \sqrt{\lambda_{max}} \approx 3.52178$. A corresponding eigenvector (normal mode) is $E_1 = (0.6592, -0.7256, 0.1395)^T$.

Next, the iteration for the smallest eigenvalue is exhibited. The notation is the same. The starting vector is $(1, 2, 2)^T$.

n	R_n	x_n	y_n	z_n
1	1.9423	0.24811	0.46077	0.60255
2	1.9727	0.24255	0.45365	0.60637
3	1.9729	0.24189	0.45305	0.60673
4	1.9729	0.24182	0.45300	0.60676
5	1.9729	0.24182	0.45299	0.60676
6	1.9729	0.24182	0.45299	0.60676

Thus the smallest eigenvalue is $\lambda_3 = 1/1.972918$, and the corresponding eigenvector is $E_3 = (0.2418, 0.4530, 0.6067)^T$. The lowest frequency of oscillation is about $\sqrt{\lambda_3} \approx 0.71194$. The last eigenvector E_2 is orthogonal to E_1 and E_3 relative to $\langle \ , \ \rangle_M$, and thus it may be found easily. Once E_2 is known, λ_2 may be found from $KE_2 = \lambda_2 ME_2$. Since there are small errors, λ_2 is better calculated by the Rayleigh quotient: $R(E_2) = E_2^T KE_2/E_2^T ME_2$, since it is a weighted average.

Solution of Inhomogeneous Systems

The equation to be solved is

$$M\ddot{x} + Kx = f(t) \tag{13}$$

with the initial conditions (2): $x(0) = x_0$, $\dot{x}(0) = v_0$.
    The same two equivalent solution schemes are discussed.

Vector space approach
Begin again with a basis $\{E_k\}_{k=1}^n$ of $\mathbf{F}^n$ that is orthogonal with respect to $\langle x, y \rangle = y^H M x$ and consists of generalized eigenvectors: $KE_i = \lambda_i ME_i$. Putting $x = \sum_{i=1}^n y_i E_i$ into (13) yields $\sum_{i=1}^n (\ddot{y}_i + \lambda_i y_i)ME_i = f(t)$. $\{ME_k\}_{k=1}^n$ is a basis, since it is a linearly independent set of n vectors in $\mathbf{F}^n$. Thus $f(t)$ can be expanded as

$$f(t) = \sum_{i=1}^n g_i(t)ME_i, \tag{*}$$

so that $\sum_{i=1}^n g_i(t)ME_i = \sum_{i=1}^n (\ddot{y}_i + \lambda_i y_i)ME_i$. Thus $\ddot{y}_i + \lambda_i y_i = g_i(t)$, by the independence of the ME_i. The coefficients $g_i(t)$ may be recovered as follows: Since $\langle E_i, E_j \rangle = E_j^H ME_i = 0$ if $i \neq j$, multiplying (*) on the left by E_j^H yields $E_j^H f(t) = E_j^H \sum_{i=1}^n g_i(t)ME_i = g_j(t)\|E_j\|^2$. Hence

$$\ddot{y}_j + \lambda_j y_j = g_j(t), \tag{14a}$$

where
$$g_j(t) = E_j^H f(t)/\|E_j\|^2, \quad j = 1, 2 \ldots, n. \tag{14b}$$

These decoupled equations are solvable by elementary means. The solutions of (14) are of the form $y_j = b_j\phi_j(t) + c_j\psi_j(t) + h_j(t)$, where ϕ_i and ψ_i are a basis of solutions of $\ddot{y}_j + \lambda_j y_j = 0$, and h_i is a particular solution of (14a). If the $\phi_i(t)$ and $\psi_i(t)$ as defined in (6) are used, the initial conditions give

$$x_0 = x(0) = \sum_{i=1}^n y_i(0) = \sum_{i=1}^n (a_i + h_i(0))E_i,$$

and
$$v_0 = \dot{x}(0) = \sum_{i=1}^n \dot{y}_i(0) = \sum_{i=1}^n (b_i + \dot{h}(0))E_i.$$

The orthogonality of the E_i yields

$$a_i + h(0) = \langle x_0, E_i \rangle/\|E_i\|^2 \quad \text{and} \quad b_i + \dot{h}(0) = \langle v_0, E_i \rangle/\|E_i\|^2,$$

where $h(t) = (h_1(t), h_2(t), \ldots, h_n(t))^T$.

Matrix approach

Make the substitution $x = Py$, and multiply the differential equation on the left by P^H to obtain

$$P^H M P y + P^H K P y = P^H f. \tag{*}$$

According to Corollary 5 of Section 5.3, P can be chosen so that $P^H M P = N^2$, $P^H K P = \Lambda N^2$, where $\Lambda = \text{diag}(\lambda_1, \lambda_2, \ldots, \lambda_n)$, $N = \text{diag}(\|E_i\|)$, and $P = [E_1 \ E_2 \ \cdots \ E_n]$. Hence (*) becomes

$$\ddot{y} + \Lambda y = N^{-2} P^H f(t).$$

Writing this out for each component yields

$$\ddot{y} + \lambda_i y_i = E_i^H f(t)/\|E_i\|^2, \quad i = 1, 2, \ldots, n,$$

as before. The formulae are a little cleaner if the columns of P are normalized, i.e., $N = I$.

The fact that the methods are equivalent comes from the identity $Px = \sum_{i=1}^n x_i E_i$, if $P = [E_1 \ E_2 \ \cdots \ E_n]$, and from the fact that the conditions $P^H M P = N^2$, $P^H K P = \Lambda N^2$ are equivalent to the fact that the columns E_i of P must be solutions of the generalized eigenvalue problem: $K E_i = \lambda_i M E_i$, $E_j^H M E_i = 0$ if $k \neq j$.

EXAMPLE 5. MECHANICAL PROBLEM. Three masses are constrained to move in a line. The first is connected to a rigid support through a spring. The other two masses are connected as shown. A force is applied to the third mass along the line.

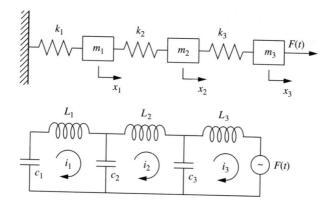

Electric circuit problem. Find the loop currents in the circuit shown. $F(t)$ is a voltage generator. The differential equations will be the same, with the correspondence

$$m_j = L_j, \quad k_j = 1/c_j, \quad \dot{x}_j = i_j.$$

The system of differential equations derived from Newton's laws is

$$m_1 \ddot{x}_1 + k_1 x_1 + k_2(x_1 - x_2) = 0$$
$$m_2 \ddot{x}_2 + k_2(x_2 - x_1) + k_3(x_2 - x_3) = 0$$
$$m_3 \ddot{x}_3 + k_3(x_3 - x_2) = F(t).$$

In matrix form, this is

$$M\ddot{x} + Kx = \delta_3 F(t),$$

where

$$K = \begin{bmatrix} k_1 + k_2 & -k_2 & 0 \\ -k_2 & k_2 + k_3 & -k_3 \\ 0 & -k_3 & k_3 \end{bmatrix}, \quad M = \text{diag}(m_1, m_2, m_3), \quad \delta_3 = \begin{bmatrix} 0 \\ 0 \\ 1 \end{bmatrix}.$$

If there were no external forces, the potential energy would be

$$V = \tfrac{1}{2}[k_1 x_1^2 + k_2(x_1 - x_2)^2 + k_3(x_2 - x_3)^2].$$

Thus V is diagonalized by the substitution $y_1 = x_1$, $y_2 = x_1 - x_2$, $y_3 = x_2 - x_3$ and is positive definite. Since the generalized eigenvectors also diagonalize V, by Sylvester's law of inertia, there must be three positive generalized eigenvalues.

For example, let us choose $m_1 = 25$, $m_2 = 4$, $m_3 = 1$, $k_1 = 40$, $k_2 = 10$, $k_3 = 2$, and $F(t) = e^{-t}$. Suppose $x(0) = (1, 0, -1)^T$, $\dot{x}(0) = 0$. Now

$$M = \text{diag}(25, 4, 1) \quad \text{and} \quad K = \begin{bmatrix} 50 & -10 & 0 \\ -10 & 12 & -2 \\ 0 & -2 & 2 \end{bmatrix}.$$

In this problem, the matrix method of solution will be used. First the generalized eigenvalue problem, $KE = \lambda ME$, $E \neq 0$, must be solved. Computing $|K - \lambda M| = 100(2 - \lambda)(\lambda - 4)(\lambda - 1)$, the eigenvalues are $\lambda_1 = 1$, $\lambda_2 = 2$, $\lambda_3 = 4$. Solving $(K - \lambda M)x = 0$ for each of the eigenvalues yields the associated eigenvectors $E_1 = (0.2, 0.5, 1)^T$, $E_2 = (-0.2, 0, 1)^T$, $E_3 = (0.2, -1, 1)^T$ with norms $\|E_1\|^2 = 3$, $\|E_2\|^2 = 2$, $\|E_3\|^2 = 6$, where $\|E\|^2 = E^T M E$. Setting $P = [E_1 \, E_2 \, E_3]$ yields

$$P^T M P = [E_i^T M E_j] = \text{diag}(\|E_1\|^2, \|E_2\|^2, \|E_3\|^2) \equiv N^2$$

$$P^T K P = [E_i^T K E_j] = [E_i^T M (M^{-1}K) E_i] = \text{diag}(\lambda_i \|E_i\|^2) = \Lambda N^2,$$

where $\Lambda = \text{diag}(\lambda_1, \ldots, \lambda_n)$. Now make the substitution $x = Py$ in the differential equation, and multiply the result on the left by P^T, which yields

$$N\ddot{y} + \Lambda N y = P^T \delta_3 e^{-t} = (1, 1, 1)^T e^{-t}.$$

Writing this out for each component and dividing by the $\|E_i\|^2$ yields

$$\ddot{y}_1 + y_1 = e^{-t}/3$$

$$\ddot{y}_2 + 2y_2 = e^{-t}/2$$

$$\ddot{y}_3 + 4y_3 = e^{-t}/6.$$

Each equation has a solution of the form $y = y_h + y_p$, where y_h is a solution of the homogeneous equation and y_p is a particular solution. A particular solution can be obtained by trying $y_p = Ce^{-t}$, and determining C. This gives

$$y_1 = a_1 \cos(t) + b_1 \sin(t) + e^{-t}/6$$

$$y_2 = a_2 \cos(\sqrt{2}t) + b_2 \sin(\sqrt{2}t) + e^{-t}/6$$

$$y_3 = a_3 \cos(2t) + b_3 \sin(2t) + e^{-t}/30.$$

To determine the constants of integration, use the initial conditions. First, from $x(0) = Py(0) = (1, 0 - 1)^T$ and $P^{-1} = N^{-2}P^T$, it follows that $y(0) = (a_1 + \frac{1}{6}, a_2 + \frac{1}{6}, a_3 + \frac{1}{30})^T = P^{-1}x(0) = N^{-2}P^T Mx(0) = (\frac{4}{3}, -3, \frac{2}{3})^T$, which determines the a_i:

$$a_1 = \tfrac{7}{6}, \qquad a_2 = -\tfrac{19}{6}, \qquad a_3 = \tfrac{19}{30}.$$

Next $\dot{x}(0) = P\dot{y}(0) = 0$, giving

$$(b_1 - \tfrac{1}{6}, \sqrt{2}b_2 - \tfrac{1}{6}, 2b_3 - \tfrac{1}{30})^T = N^{-2}P^T M0 = 0, \text{ so that}$$

$$b_1 = \tfrac{1}{6}, \qquad b_2 = \tfrac{\sqrt{2}}{12}, \qquad b_3 = \tfrac{1}{60}.$$

Response to Harmonic Excitation

If the inhomogeneous term has the special form $f(t) = ae^{i\omega t}$, a particular solution to $M\ddot{x} + Kx = f(t)$ can be found in a very easy manner: Look for particular solutions of the form $ce^{i\omega t}$, $c \in \mathbf{F}^n$. Putting $x = ce^{i\omega t}$ in the differential equation gives $M(-\omega^2 ce^{i\omega t}) + Kce^{i\omega t} = ae^{i\omega t}$; hence

$$(-\omega^2 M + K)c = a. \tag{**}$$

Thus

$$c = (K - \omega^2 M)^{-1}a, \tag{15}$$

provided the inverse exists. The inverse will exist as long as ω is not a resonant frequency of the system $|K - \omega^2 M| \neq 0$, i.e., ω^2 is not a generalized eigenvalue of K and M.

Equation (15) may not be an attractive formula for certain computations. For example, if K and M are large, the inverse may be difficult to obtain accurately. If the response c to a great many frequencies ω is desired, the frequent computation of $(K - \omega^2 M)^{-1}$ will be expensive.

If the generalized eigenvalue problem for K and M can be solved, another expression for c can be obtained. This expression can also be used to great computational advantage and to gain some insight about the solution theoretically. Assume that an orthogonal basis of generalized eigenvectors of K and M (both $n \times n$) has been constructed,

$$KE_i = \lambda_i ME_i \qquad\qquad i = 1, 2, \ldots, n \tag{16a}$$

$$E_i^H ME_j = 0 \quad \text{if } i \neq j \quad i, j = 1, 2, \ldots, n, \tag{16b}$$

or equivalently, with $P = [E_1\ E_2\ \cdots\ E_n]$,

$$P^H KP = \Lambda N^2, \qquad P^H MP = N^2, \tag{17}$$

where $\Lambda = \text{diag}(\lambda_1, \ldots, \lambda_n)$, $N = \text{diag}(\|E_1\|, \ldots, \|E_n\|)$, $\|x\|^2 = x^H Mx$.

As usual, two approaches will be demonstrated. For the vector space approach, expand $c = \sum_{i=1}^{n} b_i E_i$, put this in (**), and use (16a):

$$(K - \omega^2 M)c = (K - \omega^2 M) \sum_{i=1}^{n} b_i E_i = \sum_{i=1}^{n} b_i (K - \omega^2 M) E_i$$

$$= \sum_{i=1}^{n} b_i (\lambda_i M E_i - \omega^2 M E_i) = \sum_{i=1}^{n} b_i (\lambda_i - \omega^2) M E_i = a.$$

Multiply this on the left by E_j^H and use the orthogonality relations (16b) to get $b_j(\lambda_j - \omega^2) \|E_j\|^2 = E_j^H a$. Since $-\omega^2 \neq \lambda_i$ for all i,

$$c = \sum_{i=1}^{n} \frac{E_i^H a}{(\lambda_i - \omega^2) \|E_i\|^2} E_i. \tag{18}$$

Now let us use a matrix approach. Set $c = Pb$ in (**), multiply on the left by P^H, and use (17) to get

$$P^H(-\omega^2 M + K)Pb = (-\omega^2 P^H M P + P^H K P)b = (-\omega^2 N^2 + \Lambda N^2)c$$

$$= (-\omega^2 I + \Lambda)N^2 b = P^H a.$$

Since $-\omega^2 \neq \lambda_i$ for all i, $(-\omega^2 I + \Lambda)^{-1}$ exists and $b = N^{-2}(-\omega^2 I + \Lambda)^{-1} P^H a$ so that

$$c = PN^{-2}(\Lambda - \omega^2 I)^{-1} P^H a. \tag{19}$$

The matrix multiplication identities of Section 0.4 show that (18) and (19) are the same. Since a was arbitrary, one obtains the interesting identity

$$(K - \omega^2 M)^{-1} = PN^{-2}(\Lambda - \omega^2 I)^{-1} P^H = \sum_{i=1}^{n} \frac{E_i E_i^H}{(\lambda_i - \omega^2) \|E_i\|^2}. \tag{20}$$

Remark. It is clear from (20) that if for some k, $\omega \approx \omega_k \equiv \sqrt{\lambda_k}$, then the solution $x = ce^{i\omega t}$ has very large amplitude, and $c \approx (E_k^H a/(\lambda_k - \omega^2)\|E_k\|^2)E_k$. On the other hand, if $\omega >> \omega_i$ for all i, then $x = ce^{i\omega t}$ has very small amplitude.

Sometimes complex arithmetic may be used to find real solutions.

PROPOSITION 1. If x is a solution to $M\ddot{x} + Kx = f(t)$, and M, K are real, then $u = \mathrm{Re}(x)$ is a solution of $M\ddot{u} + Ku = \mathrm{Re}(f(t))$ and $v = \mathrm{Im}(x)$ is a solution of $M\ddot{v} + Kv = \mathrm{Im}(f(t))$.

Proof. Let $u = \mathrm{Re}(x)$. Then $\ddot{u} = \mathrm{Re}(\ddot{x})$, and $\mathrm{Re}(\ddot{x} + Kx) = M\,\mathrm{Re}(\ddot{x}) + K\,\mathrm{Re}(x) = M\ddot{u} + Ku = \mathrm{Re}(f(t))$. The proof for v is similar.

EXAMPLE 6. Consider the system of Example 5 with the same m_i and k_i. Let $F(t) = 2\cos(\omega t)$. Find the solution of

$$M\ddot{x} + Kx = 2\delta_3 \cos(\omega t) \tag{α}$$

of the form $x = A\cos(\omega t)$, for $\omega = 1.4$, $\omega = 10$.

According to Proposition 1, the solution $z = ce^{i\omega t}$ of $M\ddot{z} + Kz = 2\delta_3 e^{i\omega t}$ will yield a solution $x = \mathrm{Re}(z)$ for (α). The solution of $M\ddot{z} + Kz = 2\delta_3 e^{i\omega t}$ is found by $c = (K - \omega^2 M)^{-1}(2\delta_3)$ according to (15). But then c is real, since a, K, M, and ω are real. Thus the solution of the problem is $x = c\cos(\omega t)$.

In Example 5, it was found that $\lambda_1 = 1, \lambda_2 = 2, \lambda_3 = 4$ and the associated eigenvectors are $E_1 = (0.2, 0.5, 1)^T$, $E_2 = (-0.2, 0, 1)^T$, $E_3 = (0.2, -1, 1)^T$ with norms $\|E_1\|^2 = 3, \|E_2\|^2 = 2, \|E_3\|^2 = 6$. Thus $\omega_1 = 1, \omega_2 = \sqrt{2}, \omega_3 = 2$. If $P = [E_1\,E_2\,E_3]$, then

$$(K - \omega^2 M)^{-1} = PN^{-2}(\Lambda - \omega^2 I)^{-1} P^H.$$

For $w = 1.4$,

$$(K - \omega^2 M)^{-1} \approx \begin{bmatrix} 0.4894 & -0.05106 & -2.553 \\ -0.05106 & -0.05106 & -0.2553 \\ -2.553 & -0.2553 & 12.23 \end{bmatrix}.$$

Hence $c = 2(K - \omega^2 M)^{-1} a = (0.9788, -0.1021, -5.1062)^T$. Since $x = c\cos(\omega t)$, the third mass is most severely affected by the applied force.
For $w = 10$,

$$(K - \omega^2 M)^{-1} \approx 10^{-4} \begin{bmatrix} -4.082 & 0.1052 & -0.002147 \\ 0.1052 & -25.78 & 0.5261 \\ -0.002147 & 0.5261 & 102.1 \end{bmatrix}.$$

Hence $c = 2(K - \omega^2 M)^{-1} a = 10^{-4}(-8.164, 0.2104, 0.004295)^T$. Mass 1 is most affected, but all amplitudes are very small.

Exercises 7.2

1. A system consists of three pendulums suspended by rods of length L from a horizontal ceiling and having the same mass m. Two springs of the same spring constant k connect masses 1 to 2 and 2 to 3, and the springs are unstretched at equilibrium, at which the rods are vertical. θ_i denotes the angular displacement of mass i from equilibrium. For small displacements, the kinetic and potential energies are approximately

$$T = \tfrac{1}{2} mL^2(\dot{\theta}_1^2 + \dot{\theta}_2^2 + \dot{\theta}_3^2)$$
$$V = \tfrac{1}{2}[mgL(\theta_1^2 + \theta_2^2 + \theta_3^2) + kL^2(\theta_2 - \theta_1)^2 + kL(\theta_3 - \theta_2)^2].$$

For simplicity, take $L = g, k = 3m$.
(a) Find the normal frequencies and normal modes of oscillation.
(b) Find the general solution for $\theta(t)$.
(c) Describe the behavior of the system in each normal mode.

2. A system has potential and kinetic energies given by

$$T = \tfrac{1}{2}\{\dot{x}_1^2 + \dot{x}_2^2 + 4\dot{x}_3^2\}, \qquad V = \tfrac{1}{2}\{x_1^2 + (x_1 - x_2)^2 + 2(x_2 - x_3)^2 + 6x_4^2\}.$$

(a) Show that V is positive definite.
(b) Find and obtain the general solution of Lagrange's equations of motion:
$M\ddot{x} + Kx = 0.$
(c) Describe the motion in each of the normal modes.
(d) Solve $M\ddot{x} + Kx = (24,0,0)^T$, $x(0) = (0, 2, 2)^T$, $\dot{x} = 0.$

3. Let $T = \tfrac{1}{2}\{2\dot{x}_1^2 + \dot{x}_2^2 + \dot{x}_3^2\}$, $V = \tfrac{1}{2}\{2x_1^2 + 3x_2^2 + x_3^2 - 4x_1 x_2 - 2x_2 x_3\}$ be the kinetic and potential energies of a system.
(a) Find the equations of motion and the general solution of them.

(b) Describe the motion of the system in each normal mode.

(c) Classify V.

4. A system has potential and kinetic energies given by

$$T = \tfrac{1}{2}\{4\dot{x}_1^2 + \dot{x}_2^2 + \dot{x}_3^2\}, \qquad V = \tfrac{1}{2}\{6x_1^2 + 2(x_1 - x_2)^2 + (x_2 - x_3)^2 + x_3^2\}.$$

(a) Show that V is positive definite.

(b) Find and obtain the general solution of Lagrange's equations of motion:
$$M\ddot{x} + Kx = 0.$$

(c) Describe the motion in each of the normal modes.

(d) Solve $M\ddot{x} + Kx = 24(1,0,0)^{\mathrm{T}}$, $x(0) = 6(0, 1, 0)^{\mathrm{T}}$, $\dot{x}(0) = 0$.

5. A system has the equations of motion $M\ddot{x} + Kx = 0$, where

$$M = \begin{bmatrix} 2 & 0 & -1 \\ 0 & 3 & 0 \\ -1 & 0 & 2 \end{bmatrix}, \qquad K = \begin{bmatrix} 1 & -1 & 0 \\ -1 & 5 & 4 \\ 0 & 4 & 4 \end{bmatrix}.$$

(a) Find and obtain the general solution of the equations of motion.

(b) Solve $M\ddot{x} + Kx = 9(1, 1, -1)^{\mathrm{T}}$, $x(0) = 0$, $\dot{x}(0) = (0, 3, 0)^{\mathrm{T}}$.

6. Find the general solution of $M\ddot{x} + Kx = 0$, where

(a) $M = \begin{bmatrix} 5 & -1 & 0 \\ -1 & 3 & 1 \\ 0 & 1 & 1 \end{bmatrix}, K = \begin{bmatrix} 2 & 2 & 0 \\ 2 & 3 & 1 \\ 0 & 1 & 1 \end{bmatrix}$

(b) $M = \begin{bmatrix} 2 & -1 & 0 \\ -1 & 2 & 0 \\ 0 & 0 & 1 \end{bmatrix}, K = \begin{bmatrix} 3 & -3 & 1 \\ -3 & 6 & 2 \\ -1 & 2 & 2 \end{bmatrix}$

(c) $M = \begin{bmatrix} 2 & 1 & 0 \\ 1 & 2 & 2 \\ 0 & 2 & 3 \end{bmatrix}, K = \begin{bmatrix} 4 & 4 & 4 \\ 4 & 3 & 3 \\ 4 & 3 & 3 \end{bmatrix}.$

7. For each of the following systems with the given kinetic and potential energies, use the power method and inverse power method to find the highest and lowest natural frequencies of oscillation to four significant digits. Find the corresponding normal modes, and normalize them so that the largest entry in magnitude it $+1$.

(a) $T = \tfrac{1}{2}\{10\dot{x}_1^2 + 2\dot{x}_2^2 + 15\dot{x}_3^2\}$, $V = \tfrac{1}{2}\{10x_1^2 + 20(x_1 - x_2)^2 + 30(x_2 - x_3)^2\}$

(b) $T = \tfrac{1}{2}\{5\dot{x}_1^2 + \dot{x}_2^2 + 10\dot{x}_3^2\}$, $V = \tfrac{1}{2}\{10x_1^2 + 10(x_1 - x_2)^2 + 50(x_2 - x_3)^2\}$

(c) $T = \tfrac{1}{2}(\dot{x}_1^2 + \dot{x}_2^2 + 4\dot{x}_3^2 + 4\dot{x}_4^2)$, $V = \tfrac{1}{2}[10x_1^2 + 10(x_1 - x_2)^2 + (x_2 - x_3)^2 + (x_3 - x_4)^2]$

8. Let M and K be real and positive definite. Show that the following iteration yields the smallest generalized eigenvalue and corresponding eigenvector. x_n denotes the current approximation to the eigenvector. One cycle is given.

Given x_n:

Compute $y = K^{-1}Mx_n$ and $r_n = y^{\mathrm{T}}Ky/y^{\mathrm{T}}My$

Set $x_{n+1} = r_n y$.

Show that $r_n \to \lambda_{\min}$ and $x_n/\|x_n\|$ converges to the corresponding eigenvector.

9. The potential and kinetic energies of a certain vibrating string are given by

$$V = \frac{1}{2}\int_0^1 u_x^2\,dx, \qquad T = \frac{1}{2}\int_0^1 u_t^2\,dx.$$

$u(x, t)$ is the vertical displacement at station x and time t, and is constrained by $u(1, t) = 0$ $\forall t > 0$. The natural boundary condition for this problem is $u_x(1, t) = 0$. The first two normal modes are $e_1(x) = \cos(\pi x/2)$ and $e_2(x) = \cos(3\pi x/2)$ with natural frequencies $\omega_1 = \pi/2$, $\omega_1 = 3\pi/2$ rad/sec.

(a) For a general set $\{f_k\}_{k=1}^n$ of basis functions, apply the Ritz method and find the differential equations of motion $M\ddot{c} + Kc = 0$.

(b) Use the Rayleigh–Ritz method to approximate the solution by linear combinations of

  (i) $f_1(x) = 1 - x$, $f_2(x) = (1 - x)x$
  (ii) $f_1(x) = 1 - x^2$, $f_2(x) = 1 - x^3$ (which also satisfy $f_i'(0) = 0$).
  In each case normalize the approximate normal modes by $e(0) = 1$. Compare the results against the exact answer. Which is better? Can you explain why?

10. A certain vibrating string with an elastic constraint at the right end has potential and kinetic energies

$$V = \frac{1}{2}\int_0^1 u_x^2\,dx + \frac{1}{2}u^2(1, t), \qquad T = \frac{1}{2}\int_0^1 u_t^2\,dx$$

and is fixed at $x = 0 : u(0, t) = 0$ $\forall t$. The natural boundary condition is $u_x(1, t) + u(1, t) = 0$. Use the Ritz method to estimate the first two natural frequencies ω_i and normal modes $e_i(x)$ using the following basis functions:

(a) x, x^2, x^3

(b) $f_1(x), f_2(x), f_3(x)$, the hat functions with nodes $x_i = 0, 1/3, 2/3, 1$.

In each case, normalize the approximate normal modes by requiring each mode to have a derivative of 1 at $x = 0$. Compare your solutions with Example 2.

11. A longitudinally vibrating rod of length L with fixed ends has potential and kinetic energies

$$V = \frac{1}{2}\int_0^L EAu_x^2\,dx, \qquad T = \frac{1}{2}\int_0^L \rho Au_t^2\,dx$$

and satisfies $u(0, t) = 0$, $u(L, 0) = 0$. E is Young's modulus, ρ is the density of the rod, and A is the cross section at x. E, ρ, and A may depend on x.

(a) For a general set of basis functions $\{f_k\}_{k=1}^n$, apply the Ritz method to find the differential equation of motion $M\ddot{c} + Kc = 0$.

Now suppose $E = \rho = L = 1$. The aim is to compare strings with constant and nonconstant density (e.g., tapered and nontapered strings) in a simple situation.

(b) Use the basis functions $x(1 - x)$, $x^2(1 - x)$ to approximate the fundamental frequencies and normal modes when (i) $A = 1$, (ii) $A = 1 - x$.

(c) Use the basis functions $x^k(1 - x)$, $k = 1, \ldots, n$ to approximate the fundamental frequencies and normal modes when (i) $A = 1$, (ii) $A = 1 - x$. Try this for $n = 4$, ... Is there a point at which computations become unreliable?

(d) Use the hat functions $\{f_k\}_{k=1}^{n-1}$ with equispaced nodes to approximate the fundamental frequencies and normal modes when (i) $A = 1$, (ii) $A = 1 - x$. Try this for $n = 4$, ... Is there a point at which computations become unreliable?

(e) If in part (c) the basis functions $x(1 - x^k)$, $k = 1, \ldots, n$ were used, would the computed frequencies and modes be different? Why?
(*Note:* For $A = 1$, $e_k(x) = \sin(\omega_k x)$, where $\omega_k = k\pi$. Use this to check your calculations in parts (a)–(d).)

12. **Radially symmetric vibrating drums with variable density.** Repeat Example 3 with $2\pi\tau = 1, 2\pi\rho = 1 - r$.

13. For each of the following systems $M\ddot{x} + Kx = f(t)$, find the steady-state solution.

(a) $M = \text{diag}(25, 9, 1)$, $K = \begin{bmatrix} 75 & -15 & 0 \\ -15 & 18 & -3 \\ 0 & -3 & 3 \end{bmatrix}$

   (i) $f(t) = 2\sin(2t)\delta_1$ (ii) $f(t) = 10\cos(\sqrt{2}t)(5, 0, -1)^{\mathrm{T}}$
   (iii) $f(t) = 30e^{\sqrt{2}it}(10, 3, 2)^{\mathrm{T}}$

(b) $M = \text{diag}(1, 1, 4)$, $K = \begin{bmatrix} 2 & -1 & 0 \\ -1 & 3 & -2 \\ 0 & -2 & 8 \end{bmatrix}$

   (i) $8(1, 1, 2)^{\mathrm{T}} e^{it/\sqrt{2}}$ (ii) $4(3, -1, 2)^{\mathrm{T}} e^{i\sqrt{3}t}$ (iii) $4(1, -2, 2)^{\mathrm{T}} e^{i\sqrt{3}t}$

(c) $M = \text{diag}(4, 1)$, $K = \begin{bmatrix} 14 & -3 \\ -3 & 3.5 \end{bmatrix}$

   (i) $f(t) = 16e^{2it}(1, 1)^{\mathrm{T}}$ (ii) $f(t) = 16\cos(t)\delta_2$ (iii) $f(t) = 32\sin(\sqrt{6}t)\delta_1$

14. Find a particular solution to $M\ddot{x} + Kx = f(t)$, where $f(t) = \sum_{k=1}^{m} a_k e^{i\omega_k t}$ where $\omega_k \in \mathbf{R}$, $a_k \in \mathbf{C}^n$, and no ω_k^2 is a generalized eigenvalue.

15. In Example 1, show that the center of mass $z \equiv (x_1 m_1 + x_2 m_2 + x_3 m_3)/(m_1 + m_2 + m_3)$ moves uniformly. Can you generalize?

16. Find, if possible, a particular solution to $M\ddot{x} + Kx = f(t)$, where $f(t) = ae^{zt}$, $a \in \mathbf{C}^n$, and z is a fixed real or complex number, and where M is positive definite and K is Hermitian. Express the solution in terms of P, N, Λ. Can you generalize this to other simple functions $f(t)$?

17. Consider $M\ddot{x} + Kx = 0$. Suppose K and M are $n \times n$ and $\{E_1, \ldots, E_n\}$ is a basis for $\mathbf{F}^n$ such that $KE_i = \lambda_i ME_i$, with all $\lambda_i = \omega_i^2 > 0$. Show that a compact form for the general solution is

$$x = P\cos(\omega t)a + P\sin(\omega t)b,$$

where $\omega = \text{diag}(\omega_1, \ldots, \omega_n)$ and $P = [E_1 \ E_2 \ \cdots \ E_n]$ is the modal matrix.

18. **The beaded string.** For the beaded string of Exercise 1 in Section 7.1, determine the natural frequencies of oscillation and the corresponding normal modes. (*Hint:* See Section 4.2 for the eigenvalues and eigenvectors of the tridiagonal matrix involved.)

19. **Multistory shear buildings and earthquakes.** A building can be modeled as masses (floors) supported by cantilever beams. Each floor is subject to a wind load f_i. If the

cantilever is clamped at both ends, the equivalent spring constant is $k_i = 12EI/L_i^3$, where EI is the stiffness of the beam and L_i is the length of the beam. For a cantilever clamped at one end and pinned at the other, $k_i = 12EI/L_i^3$. During an earthquake the ground moves a distance $x_g(t)$ from the dashed reference line in the figure. Let u_i denote the displacement of mass i relative to the base of the building, so that $x_i = x_g + u_i$ is the displacement of mass i from the reference line.

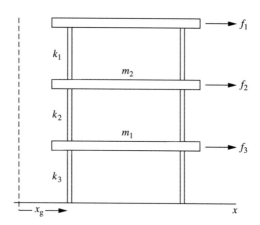

(a) Derive the differential equation of motion for a three-story building

$$M\ddot{u} + Ku = -m\ddot{x}_g + f,$$

where

$$M = \operatorname{diag}(m_1, m_2, m_3), \qquad K = \begin{bmatrix} k_1 + k_2 & -k_2 & 0 \\ -k_2 & k_2 + k_3 & -k_3 \\ 0 & -k_3 & k_3 \end{bmatrix}, \qquad m = \begin{bmatrix} m_1 \\ m_2 \\ m_3 \end{bmatrix}.$$

(b) Take each $L_i = 3$ m, $EI = 10^6$ N-m^2 (steel I-beams), and the floor masses $m_1 = 6000$ kg, $m_2 = 5000$ kg, $m_3 = 3000$ kg. Determine the fundamental frequencies of oscillation and normal modes by numerical methods when the beams are either all doubly clamped or all singly clamped. Use the method of suppression for the intermediate mode.

(c) Take each $L_i = L = 9$ m with EI the same and solve the problem as in (b). What happens?

20. **Solution of $M\ddot{x} + Kx = f$ without orthogonality.** Consider the initial value problem (IVP): Solve $M\ddot{x} + Kx = f$, $x(0) = a$, $\dot{x}(0) = b$. Let M and K be $n \times n$, not necessarily symmetric, with M nonsingular. Suppose linearly independent vectors $E_1, \ldots, E_n$ have been found such that

$$KE_i = \lambda_i ME_i. \tag{#}$$

(a) Show how to solve the IVP using the vector space approach.
(b) Rewrite (#) as a matrix multiplication involving $P = [E_1 \ E_2 \ \cdots \ E_n]$. Then show how to solve the IVP using a matrix approach.

7.3
DAMPED SMALL OSCILLATIONS

In this section linearly damped systems are considered. These are systems whose differential equations are of the form

$$M\ddot{x} + C\dot{x} + Kx = f(t), \tag{1}$$

where M, C, K are Hermitian $n \times n$ matrices and M is positive definite. For a mechanical system, a matrix differential equation such as this will occur if the damping is proportional to the generalized velocities $\dot{x}_i$ and the restoring forces are proportional to the displacements. This kind of damping is called *linear viscous damping*. In this case, the quadratic form $\frac{1}{2}\dot{x}^H C\dot{x}$ is the rate at which energy is lost due to the viscous friction.

The strategy for solution is to simultaneously diagonalize the three quadratic forms $x^H Kx$, $x^H Cx$, and $x^H Mx$ or, equivalently, to solve the simultaneous generalized eigenvalue problem for K and C with inner product generated by M; i.e., find vectors $\{E_k\}_{k=1}^n$ such that

$$KE_i = \lambda_i ME_i, \qquad CE_i = \gamma_i ME_i \quad \forall i \tag{2a}$$

and $\qquad \{E_k\}_{k=1}^n$ is orthogonal with respect to $\langle x, y \rangle = y^H Mx$. $\tag{2b}$

In matrix form, this means find P so that

$$P^H KP = \Lambda N^2, \qquad P^H CP = \Gamma N^2, \qquad P^H MP = N^2, \tag{3}$$

where

$$P = [E_1 \, E_2 \, \cdots \, E_n],$$

$$\Lambda = \text{diag}(\lambda_1, \ldots, \lambda_n), \qquad \Gamma = \text{diag}(\gamma_1, \ldots, \gamma_n),$$

$$N = \text{diag}(\|E_1\|, \ldots, \|E_n\|), \qquad \|E_i\|^2 \equiv E_i^H ME_i.$$

By Corollary 5 of Section 5.3, this can be done iff

$$KM^{-1}C \text{ is Hermitian.} \tag{4}$$

As in Section 7.2, the differential equation (1) will be decoupled by the expansion $x = \sum_{i=1}^n y_i E_i$ or, equivalently, by the change of coordinates $x = Py$. The procedure is essentially the same as for the undamped case, and it will be illustrated by an example.

EXAMPLE 1. Solve $M\ddot{x} + C\dot{x} + Kx = 0$, $x(0) = (1, 1, 0)^T$, $\dot{x} = 0$, where

$$M = \begin{bmatrix} 6 & -3 & 0 \\ -3 & 6 & 0 \\ 0 & 0 & 3 \end{bmatrix}, \qquad C = \begin{bmatrix} 5 & -1 & 1 \\ -1 & 2 & -2 \\ 1 & -2 & 2 \end{bmatrix}, \qquad K = \begin{bmatrix} 6 & 0 & 0 \\ 0 & 3 & -3 \\ 0 & -3 & 3 \end{bmatrix}.$$

A calculation shows that $KM^{-1}C$ is Hermitian. Thus the simultaneous diagonalization is possible. Solving the generalized eigenvalue problem $KE = \lambda ME$, $E \neq 0$, yields the generalized eigenvalues $\lambda_1 = 0$, $\lambda_2 = 1$, and $\lambda_3 = 2$. The corresponding eigenvectors are $E_1 = (0, 1, 1)^T$, $E_2 = (1, 0, 1)^T$, and $E_3 = (1, 1, -1)^T$ with norms (using the M

inner product) $\|E_i\|^2 = 9$. Since the eigenvalues are distinct, the eigenvectors are automatically orthogonal with respect to the inner product generated by $M : \langle x, y \rangle = y^H M x$. The eigenvectors are unique up to constant multiples and so must also diagonalize C. To find the γ_i, just compute CE_i and ME_i and find the constant of proportionality γ_i : $CE_i = \gamma_i ME_i$. This yields $\gamma_1 = 0$, $\gamma_2 = 1$, and $\gamma_3 = 1$, and is also a check on our previous work, since the theory tells us that this must happen. Making the change of basis/change of coordinates,

$$x = \sum_{j=1}^{3} y_j E_j = Py, \quad \text{where } P = [E_1 \, E_2 \, E_3],$$

we have

$$M\ddot{x} + C\dot{x} + Kx = \sum_{j=1}^{3}(\ddot{y}_j ME_j + \dot{y}_j CE_j + y_j KE_j)$$

$$= \sum_{j=1}^{3}(\ddot{y}_j ME_j + \dot{y}_j \gamma_j ME_j + y_j \lambda_j ME_j)$$

$$= \sum_{j=1}^{3}(\ddot{y}_j + \gamma_j \dot{y}_j + \lambda_j y_j)ME_j = 0.$$

Since the vectors ME_i are linearly independent, their coefficients in the last sum must be zero. Thus

$$\ddot{y}_j + \gamma_j \dot{y}_j + \lambda_j y_j = 0 \quad \forall j,$$

which is a system of differential equations with no coupling. For the current problem,

$$\ddot{y}_1 = 0, \qquad \ddot{y}_2 + \dot{y}_2 + y_2 = 0, \qquad \ddot{y}_3 + \dot{y}_3 + 2y_3 = 0. \tag{5}$$

The differential equation

$$\ddot{y} + \gamma\dot{y} + \lambda y = 0 \tag{6}$$

has the general solution

$$y = \begin{cases} e^{-\gamma t/2}[a\cosh(\rho t) + b\sinh(\rho t)] & \text{if } \rho = \frac{1}{2}\sqrt{\gamma^2 - 4\lambda} > 0 \\ e^{-\gamma t/2}[a + bt] & \text{if } \gamma^2 = 4\lambda \\ e^{-\gamma t/2}[a\cos(\omega t) + b\sin(\omega t)] & \text{if } \omega = \frac{1}{2}\sqrt{4\lambda - \gamma^2} > 0 \end{cases} \tag{7}$$

Remark. If $\gamma, \lambda > 0$, then (6) can be regarded as the differential equation of a damped spring-mass system. The three solutions in (7) correspond to overdamped, critically damped, and underdamped motion, respectively.

Thus the decoupled equations have the general solutions

$$y_1 = a_1 + b_1 t$$

$$y_2 = e^{-t/2}(a_2 \cos(\omega_2 t) + b_2 \sin(\omega_2 t)), \qquad \omega_2 = \frac{\sqrt{3}}{2}$$

$$y_3 = e^{-t/2}(a_3 \cos(\omega_3 t) + b_3 \sin(\omega_3 t)), \qquad \omega_3 = \frac{\sqrt{7}}{2}.$$

Applying the initial conditions gives

$$x(0) = (1,1,0)^T = \Sigma_{i=1}^{3} y_i(0)E_i = a_1 E_1 + a_2 E_2 + a_3 E_3$$

and $\dot{x}(0) = 0 = \sum_{i=1}^{3} \dot{y}_i(0)E_i = b_1 E_1 + (\omega_2 b_2 - \frac{1}{2} a_2)E_2 + (\omega_3 b_3 - \frac{1}{2} a_3)E_3.$

The orthogonality of the E_i with respect to the inner product generated by M may be used to compute the (Fourier) coefficients

$$a_i = \langle x(0), E_i \rangle / \|E_i\|^2,$$

and this yields $a_1 = \frac{1}{3}, a_2 = \frac{1}{3}, a_3 = \frac{2}{3}$. Since $\dot{x}(0) = 0$,

$$b_1 = 0, \qquad \omega_2 b_2 - \frac{1}{2} a_2 = 0, \qquad \text{and} \qquad \omega_3 b_3 - \frac{1}{2} a_3 = 0$$

so that $b_1 = 0, \qquad b_2 = \frac{\sqrt{3}}{9}, \qquad$ and $\qquad b_3 = \frac{2\sqrt{7}}{21}.$

The final solution is

$$x = Py = \begin{bmatrix} y_2(t) + y_3(t) \\ y_1(t) + y_3(t) \\ y_1(t) + y_2(t) - y_3(t) \end{bmatrix}.$$

A priori knowledge of diagonalizability

In some problems, the mass and stiffness matrices are known and cannot be changed, yet damping must be added to the system. An example of this is retrofitting machinery or buildings with vibration absorbers. Thus it is desirable to be able to specify certain kinds of damping matrices C for which $CE_i = \gamma_i ME_i$, where $E_1, \ldots, E_n$ are the generalized eigenvectors of K and M. The following criteria depend on the theory of functions of diagonalizable operators developed in Chapter 4. They do not depend on verifying that $KM^{-1}C$ is Hermitian. See also Sections 5.3 and 6.5.

Proportional damping. $C = \alpha M + \beta K$. From $KE_i = \lambda_i ME_i$, it follows that

$$CE_i = (\alpha M + \beta K)E_i = \alpha ME_i + \beta \lambda_i ME_i = (\alpha + \beta \lambda_i)ME_i.$$

Thus (2) holds with $\gamma_i = \alpha + \beta \lambda_i$. This construction of C is convenient, because it essentially specifies where the dampers are to be put (see Example 2, which follows).

The following theorems give substantial generalization of proportional damping.

THEOREM 1. Let K and M be $n \times n$ matrices and suppose $KE_i = \lambda_i ME_i, i = 1, 2, \ldots, n$, and $\{E_k\}_{k=1}^{n}$ is a basis for $\mathbf{F}^n$. If $C = Mf(M^{-1}K)$, then $CE_i = f(\lambda_i)ME_i \ \forall_i.$

Proof. By Theorem 1 in Section 4.3, from $(M^{-1}K)E_i = \lambda_i E_i$, it follows that $f(M^{-1}K)E_i = f(\lambda_i)E_i$. Thus $CE_i = f(\lambda_i)ME_i.$

Remark. According to Theorem 5 of Section 4.3, every function of an $n \times n$ matrix A may be written as a polynomial of degree at most $n - 1$ in A.

The next theorem gives a situation in which a simultaneous eigenvalue problem may be solved, provided the solution of a conventional orthogonal eigenvalue problem is known.

THEOREM 2. Suppose $M, K,$ and C are $n \times n$ matrices that are functions of a Hermitian matrix A:

$$M = f(A), \qquad K = g(A), \qquad C = h(A).$$

Let $\{E_k\}_{k=1}^n$ be an eigenbasis of A, orthogonal in the standard inner product, and $AE_i = \alpha_i E_i$. Suppose $f(\alpha_i) > 0$, and $g(\alpha_i)$, $h(\alpha_i)$ are real-valued $\forall i$. Then M is positive definite and K and C are Hermitian. Further,

$$KE_i = \lambda_i ME_i \quad \text{with } \lambda_i = g(\alpha_i)/f(\alpha_i) \tag{2a}$$

$$CE_i = \gamma_i ME_i \quad \text{with } \gamma_i = h(\alpha_i)/f(\alpha_i)$$

$\{E_k\}_{k=1}^n$ is orthogonal with respect to $\langle x, y \rangle = y^H M x$. \hfill (2b)

Proof. The classification of the matrices M, K, and C follows from the theory of Section 5.3. From Section 4.3,

$$ME_i = f(A)E_i = f(\alpha_i)E_i$$

$$KE_i = g(A)E_i = g(\alpha_i)E_i = \frac{g(\alpha_i)}{f(\alpha_i)} ME_i$$

$$CE_i = h(A)E_i = h(\alpha_i)E_i = \frac{h(\alpha_i)}{f(\alpha_i)} ME_i.$$

The orthogonality in the inner product generated by M follows from

$$\langle E_i, E_j \rangle \equiv E_j^H M E_i = E_j^H f(\alpha_i) E_i = \begin{cases} 0 & i \neq j \\ f(\alpha_i)\|E_i\|^2 & i = j \end{cases}$$

EXAMPLE 2. PROPORTIONAL DAMPING: $C = \alpha M + \beta K$. Two models are illustrated below. The correspondence between the quantities is the same as in the other examples, except the viscous damping coefficients c_i are related to the resistances R_i by $c_i = R_i$.

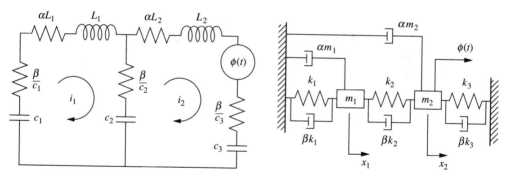

The matrix differential equation is

$$M\ddot{x} + C\dot{x} + Kx = \phi(t)\delta_2, \qquad \delta_2 = (0, 1)^T,$$

where

$$M = \begin{bmatrix} m_1 & 0 \\ 0 & m_2 \end{bmatrix}, \quad K = \begin{bmatrix} k_1 + k_2 & -k_2 \\ -k_2 & k_2 + k_3 \end{bmatrix}, \quad C = \alpha M + \beta K.$$

Take $k_1 = 1$, $k_2 = 3$, $k_3 = \frac{1}{2}$, $m_1 = 4$, $m_2 = 1$, $\alpha = \beta = \frac{2}{3}$, and $\phi(t) = 20\sin(t)$ for the calculations. Then

$$M = \begin{bmatrix} 4 & 0 \\ 0 & 1 \end{bmatrix}, \quad K = \begin{bmatrix} 14 & -3 \\ -3 & 3.5 \end{bmatrix}, \quad C = \begin{bmatrix} 12 & -2 \\ -2 & 3 \end{bmatrix}.$$

The initial conditions are: $x(0) = (4, 4)^T$, $\dot{x}(0) = 0$. Find the solution and the steady-state solution.

The first step is to solve the generalized eigenvalue problem for M and K. Solving $|K - \lambda M| = 0$ yields $\lambda_1 = 2$ and $\lambda_2 = 5$. Solving $(K - \lambda M)E = 0$ for the eigenvectors gives $E_1 = (1,2)^T$, $E_2 = (1, -2)^T$, and $\|E_j\|^2 = E_j^H M E_j = 8$, $j = 1,2$. Next

$$CE_j = \tfrac{2}{3}(M + K)E_j = \tfrac{2}{3}(1 + \lambda_j)ME_j = \gamma_j E_j$$

implies $\gamma_1 = 2$, $\gamma_2 = 4$. Computations can be checked at this point by directly calculating CE_j and ME_j to verify $CE_j = \gamma_j ME_j$. Using the matrix method, if

$$P = [E_1, E_2] = \begin{bmatrix} 1 & 1 \\ 2 & -2 \end{bmatrix},$$

then

$$P^H MP = \text{diag}(\|E_1\|^2, \|E_2\|^2) = N^2 = 8I$$
$$P^H KP = \text{diag}(\lambda_1\|E_1\|^2, \lambda_2\|E_2\|^2) = \Lambda N^2 = 8\,\text{diag}(2, 5)$$
$$P^H CP = \text{diag}(\gamma_1\|E_1\|^2, \gamma_2\|E_2\|^2) = \Gamma N^2 = 8\,\text{diag}(2, 4)$$

by Eqs. (2) and (3). The substitution/change of coordinates $x = Py$ followed by a left multiplication by P^H yields

$$P^H(M\ddot{x} + C\dot{x} + Kx) = P^H M P\ddot{y} + P^H C P\dot{y} + P^H K Py$$
$$= N^2\ddot{y} + \Gamma N^2\dot{y} + \Lambda N^2 y = 20\,\sin(t)\,P^H\delta_2.$$

Writing these out for each component gives

$$\ddot{y}_1 + 2\dot{y}_1 + 2y_1 = 5\,\sin(t)$$
$$\ddot{y}_2 + 4\dot{y}_2 + 5y_2 = -5\,\sin(t).$$

A particular solution may be found by trying $y = A\,\cos(t) + B\,\sin(t)$ in each of these equations. Once found, the general solution of the homogeneous equation is added. This gives the general solutions

$$y_1 = e^{-t}[a_1\,\cos(t) + b_1\,\sin(t)] + \sin(t) - 2\,\cos(t) \qquad (*)$$
$$y_2 = e^{-2t}[a_2\,\cos(t) + b_2\,\sin(t)] - 20\,\sin(t) + 20\,\cos(t).$$

Fitting the initial conditions requires

$$x(0) = 4(1, 1)^T = y_1(0)E_1 + y_2(0)E_2 = Py(0).$$

By (3),

$$P^{-1} = N^{-2}P^H M = \frac{1}{4}\begin{bmatrix} 2 & 1 \\ 2 & -1 \end{bmatrix}.$$

Thus, from (*),

$$y(0) = \begin{bmatrix} a_1 - 2 \\ a_2 + 20 \end{bmatrix} = N^{-2}P^H M x(0) = \begin{bmatrix} 3 \\ 1 \end{bmatrix},$$

so that $a_1 = 5$, $a_2 = -19$. Next $\dot{x}(0) = 0 = P\dot{y}(0)$, which gives $\dot{y}(0) = 0$. From (*), we find

$$\dot{y}_1(0) = -a_1 + b_1 + 1 = 0$$
$$\dot{y}_2(0) = -a_2 + b_2 - 20 = 0,$$

so that $b_1 = 44$, $b_2 = 1$. The solution is then

$$x = Py = \begin{bmatrix} y_1(t) + y_2(t) \\ 2y_1(t) - 2y_2(t) \end{bmatrix}$$

with

$$y_1 = e^{-t}[5 \cos(t) + 4 \sin(t)] + \sin(t) - 2 \cos(t)$$
$$y_2 = e^{-2t}[-19 \cos(t) + \sin(t)] - 20 \sin(t) + 20 \cos(t).$$

Response to Harmonic Excitation

Just as in the undamped case, when the external forces have the special form $f = ae^{i\omega t}$, a harmonic particular solution of the form $x = ce^{i\omega t}$ may be found quite easily. In most cases in practice there is damping, and $ce^{i\omega t}$ is the part of the solution left after the transients die out. Thus this solution is also called the "steady-state" solution even though it depends on t. Substituting $x = ce^{i\omega t}$ in the differential equation $M\ddot{x} + C\dot{x} + Kx = ae^{i\omega t}$ gives

$$(-\omega^2 M + i\omega C + K)c = a, \tag{8}$$

so that

$$c = (K + i\omega C - \omega^2 M)^{-1}a \tag{9}$$

provided the inverse exists.

Next we investigate alternative forms of the solution of (8). To this end, assume the generalized eigenvalue problem for K and C relative to M can be simultaneously solved, i.e., (2) and (3) hold: $KE_i = \lambda_i ME_i$, $CE_i = \gamma ME_i$ with $\{E_k\}_{k=1}^n$ orthogonal in $\langle\,,\,\rangle_M$. A "vector space" method will be presented first. Expand $c = \sum_{k=1}^n b_k E_k$ and put this in (8) to get

$$
\begin{aligned}
(-\omega^2 M + i\omega C + K)c &= (-\omega^2 M + i\omega C + K)\sum_{k=1}^n b_k E_k \\
&= \sum_{k=1}^n b_k(-\omega^2 M + i\omega C + K)E_k \\
&= \sum_{k=1}^n b_k(-\omega^2 ME_k + i\omega\gamma_k ME_k + \lambda_k ME_k) \\
&= \sum_{k=1}^n b_k(-\omega^2 + i\omega\gamma_k + \lambda_k)ME_k = a.
\end{aligned}
$$

Multiply this on the left by E_j^H and use the orthogonality of the E_j in $\langle\,,\,\rangle_M$ to get $b_j(-\omega^2 + i\omega\gamma_j + \lambda_j)\|E_j\|^2 = E_j^H a$. Thus,

$$c = \sum_{i=1}^n \frac{E_i^H a}{(-\omega^2 + i\omega\gamma_j + \lambda_j)\|E_i\|^2} E_i. \tag{10}$$

One advantage of this representation of the solution may be seen in a large system when it is known that ω is small and the higher frequency modes are inconsequential. In this case, only a small number of terms in (10) need to be summed. Conversely, (10) can be used to determine which terms of the solution are important and which can be safely neglected.

In the matrix method, (8) can be simplified by substituting $c = Pb$ and multiplying by P^H on the left:

$$P^H(-\omega^2 M + i\omega C + K)Pb = (-\omega^2 P^H M P + i\omega P^H C P + P^H K P)b$$
$$= (-\omega^2 N^2 + i\omega \Gamma N^2 + \Lambda N^2)b$$
$$= (-\omega^2 I + i\omega \Gamma + \Lambda)N^2 b = P^H a.$$

Hence

$$c = Pb = P(-\omega^2 I + i\omega \Gamma + \Lambda)^{-1} N^{-2} P^H a \tag{11}$$

for all F and all real ω. Thus

$$(K + i\omega C - \omega^2 M)^{-1} = P(-\omega^2 + i\omega \Gamma + \Lambda)^{-1} N^{-2} P^H \tag{12}$$

exists for all real ω. One may also write (12) as

$$(K + i\omega C - \omega^2 M)^{-1} = \sum_{k=1}^{n} \frac{E_k E_k^H}{(\lambda_k + i\omega\gamma_k - \omega^2)\|E_k\|_M^2} \tag{13}$$

using the matrix identities of Section 0.4 (cf. Eq. (20) of Section 7.2).

Remarks

1. The amplitudes no longer tend to ∞ as $\omega \to \omega_i = \sqrt{\lambda_i}$. The damping term $i\omega\Gamma$ will shift the maximum amplitudes away from the natural frequencies. Phase shifts will also occur. This is in sharp contrast to the undamped case, where the motions for the harmonic particular solution are all in phase (or 180° out of phase) when M and K are real.
2. If the response must be calculated for many frequencies, or if P, N, Λ, and Γ are known, then Eqs. (11) and (12) are more efficient than (9).
3. If M, K, and C are real, then so are P, N, Λ, and Γ. Then Eq. (11) provides an easy way to find the real and imaginary parts of $(K + i\omega C - \omega^2 M)^{-1}$:

$$\text{Re}(K + i\omega C - \omega^2 M)^{-1} = P[\text{Re}(-\omega^2 I + i\omega\Gamma + \Lambda)^{-1}]N^{-2} P^T \tag{14}$$
$$\text{Im}(K + i\omega C - \omega^2 M)^{-1} = P[\text{Im}(-\omega^2 I + i\omega\Gamma + \Lambda)^{-1}]N^{-2} P^T.$$

Thus it suffices to calculate the real and imaginary parts of the diagonal matrix $(-\omega^2 + i\omega\Gamma + \Lambda)^{-1}$ and do the required multiplications. In fact,

$$(-\omega^2 I + i\omega\Gamma + \Lambda)^{-1} = \text{diag}([\lambda_1 - \omega^2 + i\omega\gamma_1]^{-1}, \ldots, [\lambda_n - \omega^2 + i\omega\gamma_n]^{-1})$$
$$= \text{diag}([\lambda_1 - \omega^2]/d_1, \ldots, [\lambda_n - \omega^2]/d_n) \tag{15}$$
$$- i \cdot \text{diag}(\omega\gamma_1/d_1, \ldots, \omega\gamma_n/d_n)$$

where $d_j = [\lambda_j - \omega^2]^2 + \omega^2\gamma_j^2$. This also means that the calculation of $(K + i\omega C - \omega^2 M)^{-1}$ can be done in real arithmetic.
4. Just as in Proposition 1 of Section 7.2, if z is a solution of $M\ddot{z} + C\dot{z} + Kz = f(t)$, where M, K, and C are real, then $x = \text{Re}(z)$ is a solution of $M\ddot{x} + C\dot{x} + Kx = \text{Re}(f(t))$ and $y = \text{Im}(z)$ is a solution of $M\ddot{y} + C\dot{y} + Ky = \text{Im}(f(t))$.

EXAMPLE 3. Consider the system of Example 5 of Section 7.2 with $m_1 = 25$, $m_2 = 4$, $m_3 = 1$, $k_1 = 40$, $k_2 = 10$, $k_3 = 2$. Let $F(t) = 3 \sin(\omega t)$. Determine the steady-state solution when $C = 0.1K + 0.2M$, for $\omega = \sqrt{2}$, 10.

Our task is to find the solution $x = ce^{i\omega t}$ for $f = ae^{i\omega t}$, $a = 3\delta_1$, and take the imaginary part of x. To do this, compute $(K + i\omega C - \omega^2 M)^{-1} = R + iT$ with R and T real using Eqs. (14) and (15). By Eq. (9), $c = Ra + iTa$, so that our solution is $\mathrm{Im}(ce^{i\omega t}) = Ra \sin(\omega t) + Ta \cos(\omega t)$. Recall that $\lambda_1 = 1$, $\lambda_2 = 2$, $\lambda_3 = 4$, and $E_1 = (0.2, 0.5, 1)^T$, $E_2 = (-0.2, 0, 1)^T$, $E_3 = (0.2, -1, 1)^T$ with norms $\|E_1\|^2 = 3$, $\|E_2\|^2 = 2$, $\|E_3\|^2 = 6$. Thus the natural frequencies are 1, $\sqrt{2}$, and 2. Furthermore, $P = [E_1\ E_2\ E_3]$, $N^2 = \mathrm{diag}(3, 2, 6)$, $\Lambda = \mathrm{diag}(1, 2, 4)$, $\Gamma = \mathrm{diag}(0.1\lambda_j + 0.2) = \mathrm{diag}(0.3, 0.4, 0.6)$.

(a) If $\omega = \sqrt{2}$, then $(-\omega^2 I + i\omega\Gamma + \Lambda)^{-1} = \mathrm{diag}(-1/1.6, 0, 2/5.2) - i\sqrt{2}\,\mathrm{diag}(0.3/1.6, 0.4/0.8, 0.6/5.2)$, $\sqrt{2}/6)$. Computing R, T and then Ra, Ta yields

$$RF \approx -0.01(6.125, 1.977, 31.07)^T$$

$$TF \approx -0.01(10.19, 5.992, -1.978)^T,$$

and the desired solution is $x = Ra \sin(\sqrt{2}t) + Ta \cos(\sqrt{2}t)$. Let us compute the phase angle of each mass. The solution for x is

$$x_1 \approx -0.06125 \sin(\sqrt{2}t) - 0.1019 \cos(\sqrt{2}t) = 0.1193 \sin(\sqrt{2}t - 2.119)$$

$$x_2 \approx -0.01977 \sin(\sqrt{2}t) - 0.05992 \cos(\sqrt{2}t) = 0.2066 \sin(\sqrt{2}t - 2.847)$$

$$x_3 \approx -0.3107 \sin(\sqrt{2}t) + 0.01978 \cos(\sqrt{2}t) = 0.3683 \sin(\sqrt{2}t + 2.575).$$

(b) If $\omega = 10$, similar computations yield

$$x_1 \approx 8.16 \times 10^{-4} \sin(10t - 3.10)$$

$$x_2 \approx 2.97 \times 10^{-5} \sin(10t + 0.878)$$

$$x_3 \approx 8.56 \times 10^{-7} \sin(10t - 1.44)$$

Compare this with Example 5 of Section 7.2.

Exercises 7.3

1. Solve $M\ddot{x} + C\dot{x} + Kx = 0$ in each of the following cases by simultaneous diagonalization, if possible. If initial conditions $x(0) = x_0$, $\dot{x}(0) = v_0$ are given, find the specific solution; otherwise, find the general solution. In each case, verify that the simultaneous diagonalization may be done before solving the DE.

(a) $M = \begin{bmatrix} 2 & 0 \\ 0 & 2 \end{bmatrix}$, $C = \begin{bmatrix} 12 & -4 \\ -4 & 12 \end{bmatrix}$, $K = \begin{bmatrix} 16 & -8 \\ -8 & 16 \end{bmatrix}$

(b) $M = \begin{bmatrix} 2 & 0 \\ 0 & 1 \end{bmatrix}$, $C = \begin{bmatrix} 10 & -1 \\ -1 & 5.5 \end{bmatrix}$, $K = \begin{bmatrix} 12 & -2 \\ -2 & 7 \end{bmatrix}$, $x_0 = 3\delta_2$, $v_0 = 0$

(c) $M = \begin{bmatrix} 4 & 0 \\ 0 & 1 \end{bmatrix}$, $C = \begin{bmatrix} 12 & -2 \\ -2 & 3 \end{bmatrix}$, $K = \begin{bmatrix} 14 & -3 \\ -3 & 3.5 \end{bmatrix}$

(d) $M = \begin{bmatrix} 2 & 0 \\ 0 & 3 \end{bmatrix}$, $C = \begin{bmatrix} 14 & -6 \\ -6 & 6 \end{bmatrix}$, $K = \begin{bmatrix} 12 & -6 \\ -6 & 3 \end{bmatrix}$

(e) $M = \begin{bmatrix} 2 & 0 & -1 \\ 0 & 3 & 0 \\ -1 & 0 & 2 \end{bmatrix}$, $C = \begin{bmatrix} 1 & -1 & 0 \\ -1 & 2 & 1 \\ 0 & 1 & 1 \end{bmatrix}$, $K = \begin{bmatrix} 1 & -1 & 0 \\ -1 & 3 & 2 \\ 0 & 2 & 2 \end{bmatrix}$,

$x_0 = 3\delta_1, v_0 = 3\delta_2$

(f) $M = \begin{bmatrix} 2 & 0 & 0 \\ 0 & 3 & 0 \\ 0 & 0 & 6 \end{bmatrix}$, $C = \begin{bmatrix} 1 & 1 & 1 \\ 1 & 2 & 1 \\ -1 & 1 & 5 \end{bmatrix}$, $K = \begin{bmatrix} 3 & 1 & -1 \\ 1 & 4 & -1 \\ -1 & -1 & 7 \end{bmatrix}$

(g) $M = \begin{bmatrix} 2 & 0 & 0 \\ 0 & 3 & 0 \\ 0 & 0 & 6 \end{bmatrix}$, $C = \begin{bmatrix} 2 & 0 & 0 \\ 0 & 2 & -2 \\ 0 & -2 & 2 \end{bmatrix}$, $K = \begin{bmatrix} 4 & 4 & -4 \\ 4 & 5 & -2 \\ -4 & -2 & 8 \end{bmatrix}$

(h) $M = \begin{bmatrix} 3 & 0 & -2 \\ 0 & 2 & -1 \\ -2 & -1 & 2 \end{bmatrix}$, $C = \begin{bmatrix} 2 & 1 & -2 \\ 1 & 1 & -1 \\ -2 & -1 & 2 \end{bmatrix}$, $K = \begin{bmatrix} 5 & -3 & -1 \\ -3 & 5 & -1 \\ -1 & -1 & 1 \end{bmatrix}$

(i) $M = \begin{bmatrix} 3 & 0 & -2 \\ 0 & 3 & -1 \\ -2 & -1 & 2 \end{bmatrix}$, $C = \begin{bmatrix} 4 & 0 & -3 \\ 0 & 2 & -1 \\ -3 & -1 & 3 \end{bmatrix}$, $K = \begin{bmatrix} 7 & -1 & -5 \\ -1 & 3 & -1 \\ -5 & -1 & 5 \end{bmatrix}$

2. Solve $M\ddot{x} + C\dot{x} + Kx = f(t)$ in each of the following cases. If initial conditions $x(0) = x_0$, $\dot{x}(0) = v_0$ are given, find the specific solution; otherwise, find the general solution. In each case, verify that the simultaneous diagonalization may be done before solving the DE.

(a) $M = \begin{bmatrix} 2 & 0 \\ 0 & 1 \end{bmatrix}$, $C = \begin{bmatrix} 4 & -2 \\ -2 & 3 \end{bmatrix}$, $K = \begin{bmatrix} 1 & -1 \\ -1 & 1 \end{bmatrix}$, $f(t) = 3\begin{bmatrix} 0 \\ 1 \end{bmatrix}$

(b) $M = \begin{bmatrix} 2 & 0 \\ 0 & 1 \end{bmatrix}$, $C = \frac{1}{3}\begin{bmatrix} 28 & -4 \\ -4 & 16 \end{bmatrix}$, $K = \begin{bmatrix} 12 & -2 \\ -2 & 7 \end{bmatrix}$, $f(t) = \begin{bmatrix} 12 \\ 0 \end{bmatrix}$,

$x_0 = 3(0, 1)^T, v_0 = 0$

(c) $M = \begin{bmatrix} 4 & 0 \\ 0 & 1 \end{bmatrix}$, $C = \begin{bmatrix} 12 & -2 \\ -2 & 3 \end{bmatrix}$, $K = \begin{bmatrix} 14 & -3 \\ -3 & 3.5 \end{bmatrix}$, $f(t) = 8\begin{bmatrix} 2 \\ 1 \end{bmatrix}$, $x_0 = (4, 4)^T, v_0 = 0$

3. Find the steady-state solution of the following systems $M\ddot{x} + C\dot{x} + Kx = af(t)$ using M, C, K of Exercise 1(a)–(i) and
   (a) 1(a) with (i) = $f(t)16 \sin(2t)(1,1)^T$, (ii)$f(t) = 80e^{2it}(1, -1)^T$.
   (b) 1(b) with (i) = $f(t)160e^{4it}(1, -1)^T$, (ii)$f(t) = 10(2,1)^T$.
   (c) 1(c) with (i) = $f(t)16 \cos(t)(2, -1)^T$, (ii)$f(t) = 20e^{2it}(2,1)^T$.
   (d) 1(e) with (i) = $f(t)60e^{2it}(1,1, -1)^T$, (ii)$f(t) = 6 \cos(t)(1, -1,0)^T$, (iii)$f(t) = 12e^{-it}(0,1,1)^T$.
   (e) 1(g) with (i) = $f(t)24 \sin(2t)(1,1, -1)^T$, (ii)$f(t) = 27 \cos(2t)(0,1,2)^T$, (iii)$f(t) = 60e^{-2it}(1, -1,1)^T$.
   (f) 1(i) with (i) = $f(t)2 \cos(t)(1,1, -1)^T$, (ii)$f(t) = 8e^{-2it}(1, -1,0)^T$, (iii)$f(t) = 13e^{-it}(-1,0,1)^T$.

4. Find a steady-state solution to $M\ddot{x} + C\dot{x} + Kx = f(t)$, where $f(t) = \Sigma_{k=1}^{m} a_k e^{i\omega_k t}$ where $\omega_k \in \mathbf{R}, a_k \in \mathbf{C}^n$.

5. In engineering practice it is often desired to add damping to existing structures or circuits. Sometimes this is done just with pencil and paper (and computer) to see the effects of various amounts of damping. Suppose M and K are given (with the usual properties) and $E_1, \ldots, E_n$ are the normal modes generated by M and K, orthogonal with respect to the inner product generated by M. The damping is to be added in such a way that the damping coefficients are known (in the decoupled system) and the quadratic form generated by C is diagonalizable using the given normal modes E_i. To this end, consider

$$C = M \left\{ \sum_{i=1}^{n} \gamma_i \frac{E_i E_i^{H}}{\|E_i\|^2} \right\} M.$$

Show that $P^{H}CP$ is diagonal, and find the entries on the diagonal. Express C as a product of matrices (involving no summation sign).

7.4
GALERKIN'S METHOD FOR PARTIAL
DIFFERENTIAL EQUATIONS

Galerkin's method is a powerful and general method by which to solve partial differential equations. Applied to wave equations, it again yields the differential equation $M\ddot{c} + Kc = 0$, together with initial values for c. Galerkin's method will be introduced through a simple example, which is easy to generalize. There are close connections with the Ritz method of Section 7.1 and the Rayleigh–Ritz approximation of eigenvalues and eigenvectors presented in Section 5.6. Once sufficient mathematical theory is understood, these disparate fragments become unified, part of a grander whole. References are given for further reading, as the mathematical machinery is not available here to elucidate the unification. The intention here is to introduce some powerful ideas involving the use of the linear algebra developed in this book—and perhaps to whet some intellectual appetites.

The model problem is the one-dimensional wave equation

$$\rho u_{tt} = \tau u_{xx}, \quad 0 \le x \le L, t \ge 0, \tag{1}$$

with boundary conditions

$$u(0,t) = u(L,t) = 0$$

and initial conditions

$$u(x,0) = u_0(x), \qquad u_t(x,0) = v_0(x).$$

The subscripts indicate partial derivatives: $u_t = \partial/\partial t$, etc. $\rho(x)$ is the mass per unit length of the string, and τ is the constant tension in the string.

Galerkin's method proceeds as follows. Select n real-valued linearly independent functions $f_i(x)$ (called *basis* functions) that satisfy the boundary conditions

$$f_i(0) = 0 = f_i(L),$$

and set

$$U(x,t) = \sum_{i=1}^{n} c_i(t) f_i(x). \tag{2}$$

The coefficient functions $c_i(t)$ are determined so that U is a good approximation of the solution u of the partial differential equation. There are many ways to enforce a good fit, and Galerkin's method is the one that forces the error resulting from substituting the expression (2) for U into the partial differential equation to be orthogonal to each basis function f_i:

$$\int_0^L (\rho U_{tt} - \tau U_{xx}) f_j \, dx = 0, \quad j = 1, 2, \ldots, n. \tag{3}$$

Note that $U_{tt} = \sum_{i=1}^n \ddot{c}_i(t) f_i(x)$ and $U_{xx} = \sum_{i=1}^n c_i(t) f_i''(x)$, where the overdot denotes differentiation with respect to t and the prime denotes differentiation with respect to x. Putting the expression for U into the orthogonality relation (3) and rearranging yields

$$\int_0^L \rho U_{tt} f_j \, dx = \sum_{i=1}^n \ddot{c}_i(t) \int_0^L \rho f_i f_j \, dx = \tau \int_0^L U_{xx} \, dx$$

$$= \tau \sum_{i=1}^n c_i(t) \int_0^1 f_i'' f_j \, dx \tag{4}$$

$$= \tau \sum_{i=1}^n c_i(t) \left(f_i' f_j \,|_0^1 - \int_0^L f_i' f_j' \, dx \right) = -\tau \sum_{i=1}^n c_i(t) \int_0^L f_j' f_i' \, dx,$$

where the last step comes from integrating by parts and applying the boundary conditions again. Setting

$$m_{ij} = \int_0^L \rho f_i f_j \, dx \quad \text{and} \quad k_{ij} = \tau \int_0^L f_i' f_j' \, dx, \tag{5}$$

(4) can be written as the matrix differential equation

$$M\ddot{c} + Kc = 0, \qquad c = c(t) = (c_1(t), c_2(t), \ldots, c_n(t))^T, \tag{6}$$

which is the same form as Eq. (1) in Section 7.1. M is again called the mass or inertial matrix, and K is called the stiffness matrix. Note that M is symmetric and positive definite because it is the Gram matrix of the linearly independent set $\{f_k\}_{k=1}^n$ with respect to the real inner product $\langle f, g \rangle_\rho = \int_0^L f g \rho \, dx$. K is symmetric and positive, because it is the Gram matrix of $\{f_k'\}_{k=1}^n$ in the real inner product $\tau \int_0^L f g \, dx$.

Initial conditions on $c(t) = (c_1(t), c_2(t), \ldots, c_n(t))^T$ are required in order to solve the differential equations uniquely. There are various possibilities. For example, one may require that the errors from substituting U into the initial conditions be orthogonal to the basis functions f_j with weight ρ:

$$\int_0^1 [U(x, 0) - u_0(x)] f_j(x) \rho \, dx = 0 \tag{7a}$$

$$\int_0^1 [U_t(x, 0) - v_0(x)] f_j(x) \rho \, dx = 0 \quad \text{for } j = 1, 2, \ldots, n. \tag{7b}$$

Substituting $U(x, t) = \sum_{i=1}^{n} c_i(t) f_i(x)$ into (7a) yields

$$\sum_{i=1}^{n} c_i(0) \int_0^1 f_i f_j \rho \, dx = \int_0^1 u_0 f_j \rho \, dx.$$

A similar expression for (7b) holds. Thus $c(0)$ and $\dot{c}(0)$ satisfy

$$Mc(0) = a \tag{8a}$$

$$M\dot{c}(0) = b, \tag{8b}$$

where $a = (a_1, a_2, \ldots, a_n)^T$ and $b = (b_1, b_2, \ldots, b_n)^T$ with

$$a_j = \int_0^1 u_0(x) f_j(x) \rho \, dx \tag{9a}$$

$$b_j = \int_0^1 v_0(x) f_j(x) \rho \, dx. \tag{9b}$$

This just means that $U(x, 0)$ is the projection of $u_0(x)$ onto span $\{f_k\}_{k=1}^n$ in the inner product $\langle f, g \rangle_\rho$; in fact, (8a) is the normal equation for this projection. $U_t(x, 0)$ is likewise the projection of $v_0(x)$ onto span $\{f_k\}_{k=1}^n$. This seems natural for this problem. Another possibility is to require interpolation at n selected points $x_1, x_2, \ldots, x_n$:

$$U(x_i, 0) = u_0(x_i) \quad \text{and} \quad U_t(x_i, 0) = v_0(x_i) \quad \forall i. \tag{10}$$

In some cases this is highly desirable, and in others it may not be feasible.

At this point the basis functions are quite arbitrary, as long as the boundary conditions are satisfied and integration by parts is possible. Polynomials, trigonometric, hat, or other functions could be used. In general, basis functions carefully tailored to the physical problem can give remarkable results.

Recovery of the approximate solutions

$M\ddot{c} + Kc = 0$ is solved in the usual way, by solving the generalized eigenvalue problem: $KE_i = \lambda_i M E_i$, E_i orthogonal in $\langle x, y \rangle_M$. Usually $\lambda_i > 0$, so assume this and put $\tilde{\omega}_i = \sqrt{\lambda_i}$. In the basis $\{E_k\}_{k=1}^n$, with $c = \sum_{i=1}^n d_i(t) E_i$, $M\ddot{c} + Kc = 0$ decouples to $\ddot{d}_i + \lambda_i d_i = 0$, so that $d_i(t) = \alpha_i \cos(\tilde{\omega}_i t) + \beta_i \sin(\tilde{\omega}_i t)$. α_i and β_i are then found by (8) or (10).

The α_i can be found from (8a) in two ways. First, since $c(0) = \sum_{i=1}^n \alpha_i E_i$, (8a) implies $a = Mc(0) = \sum_{i=1}^n \alpha_i M E_i$. Using the orthogonality of $\{E_k\}_{k=1}^n$ in $\langle , \rangle_M$, $E_j^H a = \alpha_j E_j^H M E_j$ so that $\alpha_j = E_j^H a / \|E_j\|_M^2$. Alternatively, $c(0) = P\alpha$, where $P = [E_1 \, E_2 \cdots E_n]$. Thus (8a) becomes $a = MP\alpha$, with $\alpha = (\alpha_1, \alpha_2, \ldots, \alpha_n)^T$. Thus $\alpha = (MP)^{-1}a$. For large problems, $(MP)^{-1}$ may be computed as follows: Observe that $P^H M P = \text{diag}(\|E_1\|_M^2, \|E_2\|_M^2, \ldots, \|E_2\|_M^2) \equiv N$. Thus $(MP)^{-1} = N^{-1}P^H$. The equation $\alpha = N^{-1}P^H a$ is equivalent to the formula for the α_i derived in the first method. Similar equations hold for the β_i.

To recover U, let $\tilde{e}_j(x) = \sum_{i=1}^n e_{ij} f_i(x)$, where $E_j = (e_{1j}, \ldots, e_{nj})^T$. This means that $[\tilde{e}_j] = E_j$, where [ ] is the coordinate map with respect to $\{f_k\}_{k=1}^n$. Since $[U] = c = \sum_{i=1}^n d_i(t) E_i = [\sum_{i=1}^n d_i(t) \tilde{e}_i(x)]$, it follows that $U = \sum_{i=1}^n d_i(t) \tilde{e}_i(x)$.

Each $\tilde{e}_i(x)$ is an approximation to the normal mode of the vibration, and $\tilde{\omega}_i$ is an approximation to a fundamental frequency.

Remarks. Observe that the same differential equations as for the Ritz method of Sections 7.1 and 7.2 are produced. When ρ is constant, the same generalized eigenvalue problem will arise as in the Rayleigh–Ritz method of Section 5.6. However, for a string with an elastic restraint at $x = L$, then the boundary condition at $x = L$ is $-\tau u(L, t) = ku(L, t)$. In the Ritz method, this boundary condition does not appear in the potential and kinetic energies, and so may be ignored in the choice of the basis functions. The Rayleigh–Ritz (in Section 5.6) and Ritz (in Sections 7.1 and 7.2) methods have been used only to estimate normal modes (eigenfunctions) and fundamental frequencies. Galerkin's method provides two ways to incorporate the initial conditions in the approximate solution.

In the first example, a tapered string will be investigated. There is no easy analytic solution with which to compare the solution, so this example illustrates the utility of Galerkin's method. Modes and frequencies will be compared with results from Section 5.6 as a way to validate results. $U(x, t)$ will be calculated using (7) to determine initial values for $c(t)$.

EXAMPLE 1. Consider a vibrating string of length 1, tension 1, and linear density $\rho = \frac{1}{2}(1 + x)$. Give the string an initial displacement $u_0(x) = 4x(1 - x)$ and take $v_0 = 0$. Approximate the normal modes and fundamental frequencies. Find the approximate solution from the given initial values.

Select the basis functions $f_i(x) = x^i(1 - x), i = 1, \ldots, n$. A computation yields, from (5),

$$m_{ij} = m_{ji} = \int_0^1 f_i f_j \rho \, dx = \frac{1}{2} \int_0^1 (1 + x) x^{i+j} (1 - x)^2 \, dx = \frac{2p + 5}{(p + 1)(p + 2)(p + 3)(p + 4)}$$

$$k_{ij} = k_{ji} = \int_0^1 f_i' f_j' \, dx = \frac{2ij}{(p - 1)p(p + 1)}, \quad \text{where } p = i + j.$$

For meaningful results, take $n = 4$. To solve $M\ddot{c} + Kc = 0$, the generalized eigenvalue problem $KE_i = \lambda_i ME_i, E_i$ orthogonal in $\langle x, y \rangle_M$, must be solved. This yields $\tilde{\omega}_i = \sqrt{\lambda_i} = 3.6190, 7.2880, 11.6575, 16.8525$ $(i = 1, 2, 3, 4)$ and the eigenvectors $E_1 = (1, 1.0685, -0.2561, -0.6361)^T$, $E_2 = (1, 4.3396, -17.0344, 10.204)^T$, $E_3 = (1, -2.7609, -1.9584, 5.3547)^T$, $E_4 = (1, -7.1734, 14.995, -9.4912)^T$. The corresponding approximate normal modes are $\tilde{e}_1(x) = f_1(x) + 1.0685 f_2(x) - 0.2561 f_3(x) - 0.6361 f_4(x)$, etc. In Example 1(a) of Section 5.6, the generalized eigenvalue problem was solved for the case $\rho = 1$. The resulting approximate frequencies were $\sqrt{\Lambda_i} = 3.1416, 6.2851, 10.106, 14.160$, and the exact frequencies are $\omega_i = \pi i = 3.1416, 6.2832, 9.4248, 12.566$. Since a lighter string should have a higher pitch, it appears that the $\tilde{\omega}_i$ are reasonable. The graphs of the $\tilde{e}_k(x)$ resemble those of Example 1(a) of Section 5.6 (and so are omitted), which again confirms the reasonableness of the calculations.

In the basis $\{E_1, \ldots, E_4\}, c = Pd = \sum_{i=1}^4 d_i E_i$ decouples (6) to $\ddot{d}_i + \lambda_i d_i = 0$, so that $c = \sum_{i=1}^4 \{\alpha_i \cos(\tilde{\omega}_i t) + \beta_i \sin(\tilde{\omega}_i t)\} E_i$. Next, α_i and β_i are determined from (8) and (9). Here, $a_j = 2 \int_0^1 (1 + x) x^{j+1} (1 - x)^2 \, dx = 4m_{1j}$ and all $b_j = 0$. Clearly, all $\beta_i = 0$. From (8), $a = Mc(0) = \sum_{i=1}^4 \alpha_i ME_i$. By orthogonality in $\langle , \rangle_M$, $E_j^T a = \alpha_j E_j^T ME_j$, or

$\alpha_j = E_j^{\mathrm{T}} a / \|E_j\|_M^2$. A computation gives $a = (0.1, 0.0523810, 0.0309524, 0.0198413)^{\mathrm{T}}$, and then $\alpha_1 = 2.9463$, $\alpha_2 = 0.15567$, $\alpha_3 = 0.59336$, and $\alpha_4 = 0.30301$. $U = \sum_{i=1}^{4} d_i \tilde{e}_i = \sum_{i=1}^{4} \alpha_i \cos(\tilde{\omega}_i t) \tilde{e}_i(x)$.

In the next example, the hat functions are used as basis functions, which yields an example from elementary finite elements. It turns out that, with $\rho = 1$, $M\ddot{c} + Kc = 0$ can be solved exactly for arbitrarily many basis functions. Furthermore, if the initial conditions are determined by interpolation (10), the approximate solution can be compared with the analytic solution and a rough estimate can be made to determine how accurate the approximate solution is. All the gory details are presented. For sharper estimates and more complete discussions involving finite elements, see the references cited at the end of the section.

EXAMPLE 2. GALERKIN APPROXIMATION OF THE WAVE EQUATION USING HAT FUNCTIONS. The model problem will be solved approximately using $L = \rho = \tau = 1$ and $v_0(x) = 0$. Let us take as our basis functions $\{f_k\}_{k=1}^{n-1}$ the hat functions based on equispaced points in $[0, 1]$: $x_k = k/n$, $k = 0, 1, \ldots, n$. These hat functions were used in Example 1(b) of Section 5.6 (see also Example 6 of Section 1.3). The entries of the mass matrix M and the stiffness matrix K are

$$k_{ij} = \int_0^1 f_i' f_j' \, dx = \begin{cases} -n & |i \pm j| = 1 \\ 2n & i = j \\ 0 & \text{otherwise} \end{cases}, \qquad m_{ij} = \int_0^1 f_i f_j \, dx = \begin{cases} h/6 & |i \pm j| = 1 \\ 2h/3 & i = j \\ 0 & \text{otherwise} \end{cases}.$$

where $h = 1/n$. These are precisely the matrices $M = G$ and $K = A$ that appeared in Example 1(b) of Section 5.6 illustrating the Rayleigh–Ritz method. In that section, it was found that the generalized eigenvalues are

$$\lambda_k = \frac{24n^2 \sin^2(\frac{1}{2}k\pi/n)}{6 - 4\sin^2(\frac{1}{2}k\pi/n)}$$

and the associated generalized eigenvectors are

$$E_k = (\sin(k\pi/n), \sin(2k\pi/n), \ldots, \sin([n-1]k\pi/n))^{\mathrm{T}}$$
$$= (\sin(k\pi x_1), \sin(k\pi x_2), \ldots, \sin(k\pi x_{n-1}))^{\mathrm{T}}.$$

Also, in that section, it was seen that the E_k are eigenvectors of a certain Hermitian tridiagonal matrix, and thus they are orthogonal not only with respect to the inner product $y^{\mathrm{T}} M x$ generated by the matrix M, but also with respect to the standard inner product. The substitution $c = Py = \sum_{i=1}^{n} y_i E_i$, where $P = [E_1 E_2 \cdots E_{n-1}]$ decouples the differential equations to

$$\ddot{y}_i + \lambda_i y_i = 0,$$

and these differential equations have the solutions

$$y_i = a_i \cos(\omega_i t) + b_i \sin(\omega_i t), \quad \text{where } \omega_i = \sqrt{\lambda_i}. \tag{11}$$

Thus the general solution of the system of differential equations $M\ddot{c} + Kc = 0$ is

$$c(t) = \sum_{k=1}^{n-1} [a_k \cos(\omega_k t) + b_k \sin(\omega_k t)] E_k. \tag{12}$$

Our approximate solution is

$$U(x, t) = \sum_{j=1}^{n-1} c_j(t) f_j(x).$$

Since the basis functions have the interpolation property (namely, $f_i(x_j) = \delta_{ij}$), the interpolation initial conditions (10) will be used. Thus

$$U(x_i, 0) = \sum_{j=1}^{n-1} c_j(0) f_j(x_i) = \sum_{j=1}^{n-1} c_j(0) \delta_{ji} = c_i(0) = u_0(x_i) \tag{13}$$

$$U_t(x_i, 0) = \sum_{j=1}^{n-1} c_j(0) f_j(x_i) = \sum_{j=1}^{n-1} c_j(0) \delta_{ji} = \dot{c}_i(0) = 0.$$

Denoting $U_0 = (u_0(x_1), u_0(x_2), \ldots, u_0(x_{n-1}))^{\mathsf{T}}$, (11) and (12) yield

$$U_0 = c(0) = \sum_{i=1}^{n-1} a_i E_i \quad \text{and} \quad 0 = \sum_{i=1}^{n-1} \omega_i b_i E_i. \tag{14}$$

Thus all $b_i = 0$. It is preferable to use the standard inner product to compute the Fourier coefficients a_i in (13) because this allows a good interpretation of the a_i. It turns out that $\|E_i\|^2 = n/2$ for all i; hence

$$a_k = \frac{2}{n} E_k^{\mathsf{T}} U_0 = \frac{2}{n} \sum_{i=1}^{n-1} u_0(x_j) \sin(k\pi x_j)). \tag{15}$$

Now it remains to compare the approximate solution with the analytic solution. The analytic solution is

$$u(x, t) = \sum_{k=1}^{\infty} A_k \sin(k\pi t) \sin(k\pi x),$$

where

$$A_k = 2 \int_0^1 u_0(x) \sin(k\pi x) \, dx.$$

Writing each component of (12) using the explicit form of the E_k yields

$$c_j(t) = \sum_{k=1}^{n-1} [a_k \cos(\omega_k t) + b_k \sin(\omega_k t)] \sin(k\pi x_j). \tag{12'}$$

Substituting expression (12') for the c_j into the expression for U, noting that all $b_k = 0$, and rearranging terms yields

$$U(x, t) = \sum_{j=1}^{n-1} c_j(t) f_j(x) = \sum_{j,k=1}^{n-1} a_k \cos(\omega_k t) \sin(k\pi x_j) f_j(x) \tag{16}$$

$$= \sum_{k=1}^{k-1} a_k \cos(\omega_k t) \left\{ \sum_{j=1}^{n-1} \sin(k\pi x_j) f_j(x) \right\} = \sum_{j=1}^{n-1} a_k \cos(\omega_k t) s_k(x)$$

where

$$s_k(x) = \sum_{j=1}^{n-1} \sin(k\pi x_j) f_j(x). \tag{17}$$

Next, it will be shown that $a_k \approx A_k$, $\omega_k \approx k\pi$, $s_k(x) \approx \sin(k\pi x)$. $s_k(x)$ is exactly the piecewise linear continuous function that interpolates $\sin(k\pi x)$ at the points $x_j = j/n$, because of the interpolation property of the f_i. Let us see how well $s_k(x)$ approximates $\sin(k\pi x)$. Referring to Eq. (24) in Section 5.6, it is seen that $s_k(x) = \tilde{e}_k(x)$ for $k = 1, 2, \ldots, n - 1$. In Section 5.6, for $n = 10$, graphs of $\sin(k\pi x)$ and $s_k(x)$ were plotted together for $k = 1, 2, \ldots, 10$. Note that the approximations appear good for $k = 1, 2$ and fairly good for $k = 3, 4, 5$, and they disintegrate in quality rapidly thereafter. It can be shown that, if a function g has two continuous derivatives on $[0, 1]$ and s is the piecewise linear continuous function interpolating g at $x_j = j/n$, $j = 0, 1, \ldots, n$, then

$$|g(x) - s(x)| \le \frac{1}{8n^2} \max\{|g''(x)| : 0 \le x \le 1\}, \quad \text{for all } x \in [0, 1].$$

Hence

$$|\sin(k\pi x) - s_k(x)| \le \frac{\pi k^2}{8n^2} \quad \text{for all } x \in [0, 1].$$

These inequalities are conservative, but they show explicitly why the approximations are good when k is small compared with n.

So far it is known that $s_k(x)$ is a good approximation of $\sin(k\pi x)$, when k is small compared with n. In the Rayleigh–Ritz example of Section 5.6, it was observed that $\omega_k = \omega_k^{(n)} \to k\pi$ as $n \to \infty$, and the table there indicates that when k is small compared with n, ω_k is a good approximation of $k\pi$. The trapezoidal approximation to $\int_0^1 g(x)\,dx$ is

$$T(g) \equiv \frac{1}{n}(g_0 + 2g_1 + \cdots + 2g_{n-1} + g_n) \quad \text{where } g_j = f(j/n).$$

Thus (15) is simply the trapezoidal approximation to the integral

$$A_k = 2 \int_0^1 u_0(x)\,\sin(k\pi x)\,dx.$$

The trapezoidal approximation is fourth-order accurate if u_0 has period 1 and has four continuous derivatives on the whole real line. If the analytic solution has negligible high-frequency components after a certain point (say, $A_k \approx 0$ for $k \ge N$), n is sufficiently larger than N so that $s_k(x)$ is a good approximation of $\sin(k\pi x)$, and the trapezoidal approximation of the Fourier coefficients is good, then Galerkin's method will give good results.

For a discussion of the Ritz and Galerkin's methods in the context of finite elements and relevant error analysis, see Strang and Fix (1973) and Ciarlet (1978). For discussions of the general theory connecting many of these ideas, see Courant and Hilbert (1953).

Exercises 7.4

Some of the following problems are set up for hand calculations and may be tedious, yet they reflect the power of the method. Much more interesting results are obtained if more (or different) basis functions are used and the eigenvalue problems are calculated numerically.

Compare these with the exercises and ideas in Sections 5.6, 7.1, and 7.2. Often the number of basis functions is left undefined. Try a few, or try to get a general expression for the matrices involved.

1. $$u_t = u_{xx} \quad \text{in } 0 < x < 1, t > 0$$

Boundary conditions: $u(0, t) = u(1, t) = 0$

Initial conditions: $u(x, 0) = u_0(x), \quad \text{in } 0 < x < 1$

u_0 and exact solutions:
(i) $u_0(x) = \sin(\pi x)$, $u(x, t) = \exp(-\pi^2 t) \sin(\pi x)$
(ii) $u_0(x) = 4x(1 - x)$

$$u(x, t) = \sum_{k=1}^{\infty} c_k \exp(-(k\pi)^2 t) \sin(k\pi x), \quad c_k = 16\frac{1 - \cos(k\pi)}{(k\pi)^3}.$$

(a) Use hat functions.
(b) Use polynomial basis functions $x^k(1 - x)$, $k = 1, \ldots, n$.
Compare $U(x, t)$ with the exact solution.

2. $$u_{tt} = u_{xx} \quad \text{in } 0 < x < 1, t > 0$$

Boundary conditions: $u(0, t) = u(1, t) = 0$

Initial conditions: $u(x, 0) = u_0(x), u_t(x, 0) = 0 \quad \text{in } 0 < x < 1$

u_0 and exact solutions:
(i) $u_0(x) = \sin(\pi x)$, $u(x, t) = \cos(\pi t) \sin(\pi x)$
(ii) $u_0(x) = 4x(1 - x)$,

$$u(x, t) = \sum_{k=1}^{\infty} c_k \cos(k\pi t) \sin(k\pi x), \quad c_k = 16\frac{1 - \cos(k\pi)}{(k\pi)^3}$$

Use polynomial basis functions $\{x^k(1 - x)\}_{k=1}^{n}$ to approximate the first few modes of oscillation. Normalize your approximate eigenvectors so that their maximum is 1. Compare the approximate normal modes $\tilde{e}_k(x)$ with the exact ones $e_k(x) = \sin(k\pi x)$. Compute $U(x, t)$ and compare it with the exact solution.

3. $$u_{tt} = u_{xx} \quad \text{in } 0 < x < 1, t > 0$$

Boundary conditions: $u_x(0, t) = u(1, t) = 0$

Initial conditions: $u(x, 0) = u_0(x), u_t(x, 0) = 0 \quad \text{in } 0 < x < 1$

u_0 and exact solutions:
(i) $u_0(x) = \cos(\pi x/2)$, $u(x, t) = \cos(\pi t/2) \cos(\pi x/2)$
(ii) $u_0(x) = x$

$$u(x, t) = \sum_{k=1}^{\infty} c_k \cos(\alpha_k t) \cos(\alpha_k x), \quad c_k = \frac{2}{\alpha_k^2}, \quad \alpha_k = \left(k - \frac{1}{2}\right)\pi$$

(a) Use polynomials $f_j(x) = 1 - x^{j+1}$, $j \geq 1$, to approximate the first modes of oscillation. Normalize your approximate normal modes so that $\tilde{e}_k(0) = 1$. Compare the approximate normal modes $\tilde{e}_k(x)$ with the exact ones $e_k(x) = \cos(\alpha_k x)$. Compute $U(x, t)$ and compare it with the exact solution.

(b) Repeat part (a) using hat functions with equally spaced nodes. In this case, simply ignore the boundary condition at $x = 0$, and use the hat functions $f_0, f_1, \ldots, f_{n-1}$.

4. For a vibrating string with variable density and a free end at $x = 0$, the differential equation is (with $L = 1, \tau = 1$)

$$\rho u_{tt} = u_{xx} \quad \text{in } 0 < x < 1, t > 0$$

Boundary conditions: $u_x(0, t) = u(1, t) = 0$

Initial conditions: $u(x, 0) = 1 - x, u_t(x, 0) = 0 \quad \text{in } 0 < x < 1$

Take (i) $\rho = x$, (ii) $\rho = 1 - x$.
Use polynomials $f_j(x) = 1 - x^{j+1}, j \geq 1$, to approximate the first modes of oscillation. Normalize your approximate normal modes so that $\tilde{e}_k(0) = 1$. Compare your results with the results of Exercise 3, which is the case $\rho = 1$.

In the next two problems, a more general Galerkin procedure will be illustrated, using the concept of a Hermitian operator.

5. Longitudinal vibrations of a rod with fixed ends

$$\rho A u_{tt} = (EAu_x)_x \quad \text{in } 0 < x < L, t > 0$$

Boundary conditions: $u(0, t) = u(L, t) = 0$

Initial conditions: $u(x, 0) = u_0(x), u_t(x, 0) = 0 \quad \text{in } 0 < x < L$

A is the cross-sectional area at x, ρ is the density, and E is Young's modulus, all of which may be positive functions of x.
First, rewrite the partial differential equation as

$$u_{tt} + S(u) = 0 \tag{*}$$

where
$$S(u) \equiv \frac{1}{\rho A} \partial_x (EA\partial_x u), \quad \partial_x = \frac{\partial}{\partial x}.$$

(a) Show that S is Hermitian on the vector space V of smooth functions $f(x)$ satisfying $f(0) = f(1) = 0$ with respect to the inner product $\langle f, g \rangle = \int_0^L f(x)\overline{g}(x)\rho A \, dx$. (*Note:* This means that eigenfunctions of S in V corresponding to different eigenvalues are orthogonal with respect to this inner product.)

(b) **Galerkin's method.** Let $U(x, t) = \sum_{j=1}^n c_j(t) f_j(x)$. Put U in (*) and require that

$$U_{tt} + S(U) \perp f_i, \quad i = 1, 2, \ldots, n \tag{**}$$

using the inner product in (a). Deduce $M\ddot{c} + Kc = 0$, where $m_{ij} = \langle f_j, f_i \rangle$, $k_{ij} = \langle S(f_j), f_i \rangle$. Next show $k_{ij} = \int_0^L EAf_i' f_j' \, dx$.

(c) For the initial conditions, require that

$$U(x, 0) - u_0(x) \perp f_i \quad \text{and} \quad U_t(x, 0) \perp f_i, \quad i = 1, 2, \ldots, n$$

using the inner product in (a). Deduce the equations governing $c(0), \dot{c}(0)$, where $c = (c_1, c_2, \ldots, c_n)^T$.

(d) For simplicity, take $\rho = E = L = 1, u_0(x) = 4x(1 - x)$. Consider
  (i) $A = 1$
  (ii) $A = x$
  (iii) $A = \frac{1}{2}(1 + x)$
Use polynomial basis functions $f_k(x) = x^k(1 - x)$ to estimate the normal modes and approximate solution $U(x, t)$.

6. An advection-diffusion equation problem

$$u_t = u_{xx} + 2u_x \quad \text{in } 0 < x < 1, t > 0 \tag{\#}$$

Boundary conditions: $u(0, t) = u(1, t) = 0$

Initial conditions: $u(x, 0) = u_0(x) \quad \text{in } 0 < x < 1$

Note that $u_{xx} + 2u_x = e^{-2x}\partial_x(e^{2x}\partial_x u) \equiv -S(u)$, where $\partial_x = \partial/\partial_x$. Define an inner product $\langle f, g \rangle = \int_0^1 f(x)\bar{g}(x)e^{2x}\,dt$. Thus the partial differential equation becomes $u_{tt} + S(u) = 0$.

(a) Show that S is Hermitian on the vector space V of smooth functions $f(x)$ satisfying $f(0) = f(1) = 0$ with respect to the inner product $\langle f, g \rangle = \int_0^L f(x)\bar{g}(x)e^{2x}\,dx$. (*Note*: This means that eigenfunctions of S in V corresponding to different eigenvalues are orthogonal with respect to this inner product. The eigenfunctions of S are $e_k(x) = e^{-x}\sin(k\pi x)$ with corresponding eigenvalues $1 + (k\pi)^2$.)

(b) **Galerkin's method.** Let $U(x, t) = \sum_{j=1}^n c_j(t)f_j(x)$. Put U in (\#) and require that

$$U_{tt} + S(U) \perp f_i, \quad i = 1, 2, \ldots, n \tag{\#\#}$$

using the inner product in (a). Deduce $M\ddot{c} + Kc = 0$, where $m_{ij} = \langle f_j, f_i \rangle$, $k_{ij} = \langle S(f_j), f_i \rangle$. Next show $k_{ij} = \int_0^L e^{2x}f_i'f_j'\,dx$.

(c) For the initial conditions, require that

$$U(x, 0) - u_0(x) \perp f_i \quad \text{and} \quad U_t(x, 0) \perp f_i, \quad i = 1, 2, \ldots, n$$

using the inner product in (a). Deduce the equations governing $c(0)$, $\dot{c}(0)$, where $c = (c_1, c_2, \ldots, c_n)^T$.

(d) The exact solution of the PDE is

$$u(x, t) = \sum_{k=1}^{\infty} c_k \exp(-(k\pi)^2 t)e_k(x), \quad c_k = \frac{\langle u_0, e_k \rangle}{\|e_k\|^2},$$

where $\langle , \rangle$ is the inner product defined in part (a). Take $u_0(x) = 4x(1 - x)$.

(i) Use the basis functions $f_k = e^{-x}x^k(1 - x)$, $k = 1, 2$ to approximate the first two eigenvalues and eigenvectors. Compare your results with the exact solutions.

(ii) If sufficient numerical aids are available, repeat part (i) for a larger number of basis functions.

(iii) What are appropriate piecewise continuous basis functions to use?

CHAPTER 8

Factorizations and Canonical Forms

In this chapter some factorizations and canonical forms for arbitrary matrices and operators are discussed. Some theorems of this type were discussed in previous chapters, for special kinds of matrices. Examples of these factorizations are the QR factorization and the representation of normal matrices as $A = U \Lambda U^H$ with Λ diagonal and U unitary. The point of such representations is to gain insight into the structure of the operator in question, just as was done in the spectral theorem.

8.1
THE SINGULAR VALUE AND POLAR DECOMPOSITIONS

A general linear transformation is highly anisotropic: The length of the image $\|T(x)\|$ can depend strongly on the direction of x. Both the singular and polar decompositions provide insight about the anisotropy and geometry of an operator through special kinds of factorizations. The singular value decomposition gives an enormous amount of insight into the structure of the linear operator $T(x) = Ax$ in terms of its null space and range, where A is of arbitrary size. It is very useful in numerical analysis, statistics, and many other applications. Since excellent algorithms for computing the singular value decomposition exist, it it highly useful in computational situations. The polar decomposition (or form) is closely related theoretically and is essential in continuum mechanics.

The Singular Value Decomposition

Let A be an $m \times n$ matrix. There exists an integer r, $\sigma_1 \geq \sigma_2 \geq \cdots \geq \sigma_r > 0$, an $m \times m$ unitary matrix U, an $n \times n$ unitary matrix V, and an $m \times n$ matrix S, all of whose entries are 0 except $S_{ii} = \sigma_i$, $i = 1, 2, \ldots, r$ such that

$$A = USV^H. \tag{1}$$

More graphically,

$$A_{m,n} = U_{m,m} \begin{bmatrix} \sigma & 0 \\ 0 & 0 \end{bmatrix}_{m,n} V^H_{n,n}, \quad \text{where } \sigma = \text{diag}(\sigma_1, \sigma_2, \ldots, \sigma_r).$$

If A is real, then U and V may be taken to be real. Equation (1) is called a *singular value decomposition (SVD) of A*, and the numbers σ_i are called the *singular values of A*. The zero blocks in

$$S = \begin{bmatrix} \sigma & 0 \\ 0 & 0 \end{bmatrix}$$

may or may not be present, depending on whether or not $r = m$ or $r = n$.

Before we present the proof that this factorization is possible, some consequences of (1) will be deduced. These will allow a fuller appreciation for the geometry and structure behind the theorem, and will suggest the proof. Some of the facts have been deduced before, but the SVD gives a new proof of them. To this end, let

$$U = [U_1 \ U_2 \ \cdots \ U_m] \quad \text{and} \quad V = [V_1 \ V_2 \ \cdots \ V_n]. \tag{2}$$

COROLLARY 1. $\{V_1, V_2, \ldots, V_n\}$ is an orthonormal eigenbasis for $A^H A$. σ_i^2 $i = 1, 2, \ldots, r$, are the nonzero eigenvalues of $A^H A$, and $V_1, \ldots, V_r$ are the corresponding eigenvectors. $V_{r+1}, \ldots, V_n$ are eigenvectors corresponding to the eigenvalue 0.

Proof

$$A^H A = VS^H U^H USV^H = VSS^T V^H = V \begin{bmatrix} \sigma^2 & 0 \\ 0 & 0 \end{bmatrix}_{n,n} V^H$$

From Theorem 1 of Section 5.3, this corollary follows:

COROLLARY 2. $\{U_1, U_2, \ldots, U_m\}$ is an orthonormal eigenbasis for AA^H. σ_i^2 $i = 1, 2, \ldots, r$ are the nonzero eigenvalues of AA^H, and $U_1, \ldots, U_r$ are the corresponding eigenvectors. $U_{r+1}, \ldots, U_m$ are eigenvectors corresponding to the eigenvalue 0.

Proof

$$AA^H = USV^H VS^H U^H = U \begin{bmatrix} \sigma^2 & 0 \\ 0 & 0 \end{bmatrix}_{m,m} U^H$$

Apply the same theorem.

COROLLARY 3. The vectors U_i, V_i, $i = 1, 2, \ldots, r$, are related by

$$AV_i = \sigma_i U_i, \quad i = 1, 2, \ldots, r$$

$$AV_i = 0, \quad \text{if } i > r \text{ (should such indices exist)} \tag{3}$$

Proof. Equation (1) is equivalent to $AV = USV^H V = US$. But $AV = [AV_1 \ \cdots \ AV_r \ \cdots \ AV_n]$ and $US = [\sigma_1 U_1 \ \sigma_2 U_2 \ \cdots \ \sigma_r U_r 0 \ \cdots \ 0]$. Equating columns gives Eq. (3).

COROLLARY 4. The vectors U_i, V_i, $i = 1, 2, \ldots, r$, are related by

$$A^H U_i = \sigma_i V_i, \quad i = 1, 2, \ldots, r \tag{4}$$

$$A^H U_i = 0 \quad \text{if } i > r \text{ (should such indices exist)}$$

Proof. Left as an exercise.

The following results show that the SVD gives valuable information about the null space and range of A and A^H. The diadic expansion (5) will be extremely useful in understanding the SVD in various ways.

COROLLARY 5. Equation (1) is equivalent to the diadic expansion

$$A = \sum_{i=1}^{r} \sigma_i U_i V_i^H. \tag{5}$$

Further,

(a) $\{U_1, \ldots, U_r\}$ is an orthonormal basis for the range $R(A)$ of A.
  $\{U_{r+1}, \ldots, U_m\}$ is an orthonormal basis for $R(A)^\perp$.
(b) $\{V_1, \ldots, V_r\}$ is an orthonormal basis for $N(A)^\perp$.
  $\{V_{r+1}, \ldots, V_n\}$ is an orthonormal basis for the null space $N(A)$ of A.

Proof. Equation (5) follows from (cf. Section 0.4)

$$A = USV^H = [\sigma_1 U_1 \ \cdots \ \sigma_r U_r \ 0 \ \cdots \ 0] \begin{bmatrix} V_1^H \\ V_2^H \\ \cdots \\ V_n^H \end{bmatrix} = A = \sum_{i=1}^{r} \sigma_i U_i V_i^H.$$

Applying A to x yields

$$Ax = \sum_{i=1}^{r} \sigma_i U_i V_i^H x = \sum_{i=1}^{r} \sigma_i \langle x, V_i \rangle U_i, \tag{6}$$

where $\langle , \rangle$ is the standard inner product in $\mathbf{F}^n$. First (a) will be proved. Equation (6) implies that the range of A is a subset of $\text{span}\{U_1, \ldots, U_r\}$. If $y = \sum_{i=1}^{r} \alpha_i U_i$, let $x = \sum_{i=1}^{r} \beta_i V_i$. Then from (3), $Ax = \sum_{i=1}^{r} \beta_i A U_i = \sum_{i=1}^{r} \beta_i \sigma_i U_i = y$, provided that $\beta_i = \alpha_i / \sigma_i$. Hence $R(A) = \text{span}\{U_1, \ldots, U_r\}$. $R(A)^\perp = \text{span}\{U_{r+1}, \ldots, U_m\}$ by Theorem 1 of Section 3.5.

Next let us turn to (b). From (6), if $x \perp V_1, \ldots, V_r$, then $Ax = 0$. On the other hand, if $Ax = 0$, then all $\langle x, V_i \rangle = 0$ since $U_1, \ldots, U_r$ are linearly independent so that $x \perp V_1, \ldots, V_r$. Hence $N(A) = [\text{span}\{V_1, \ldots, V_r\}]^\perp$. Finally, $N(A) = \text{span}\{V_{r+1}, \ldots, V_n\}$, according to Theorem 1 of Section 3.5.

COROLLARY 6. Equation (1) is equivalent to the diadic expansion

$$A^H = \sum_{i=1}^{r} \sigma_i V_i U_i^H. \tag{7}$$

Further,

(a) $\{V_1, \ldots, V_r\}$ is an orthonormal basis for $R(A^H)$.
  $\{V_{r+1}, \ldots, V_n\}$ is an orthonormal basis for $R(A^H)^\perp$.
(b) $\{U_1, \ldots, U_r\}$ is an orthonormal basis for $N(A^H)^\perp$.
  $\{U_{r+1}, \ldots, U_m\}$ is an orthonormal basis for $N(A^H)$.

Proof. Left as an exercise.

COROLLARY 7. A, A^H, $A^H A$, and $A A^H$ have the same rank, which is r.

Proof. By Corollaries 5 and 6, A and A^H have rank r, by the definition of rank. The diagonalizable matrices $A^H A$ and $A A^H$ have rank r, since that is the number of nonzero eigenvalues they have, by Theorem 2 of Section 4.2.

After this information has been inferred, an informative diagram of the SVD can be drawn.

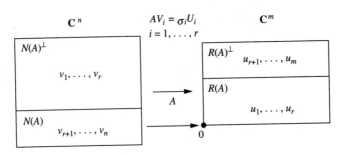

Next we turn to the proof of the SVD.

Proof of the singular value decomposition theorem

Let $\{V_k\}_{k=1}^n$ be an orthonormal eigenbasis for $A^H A$, with $A^H A V_i = \lambda_i V_i$ and $\lambda_1 \geq \lambda_2 \geq \cdots \geq \lambda_r > 0 = \lambda_{r+1} = \cdots = \lambda_n$. Set $\sigma_i = \sqrt{\lambda_i}, i = 1, 2, \ldots, r$, and $G_i = A V_i, i = 1, 2, \ldots, n$. Then

$$\langle G_i, G_j \rangle = \langle A V_i, A V_j \rangle = \langle V_i, A^H A V_j \rangle = \lambda_j \langle V_i, V_j \rangle = \lambda_j \delta_{ij}. \qquad (*)$$

Hence $\|G_i\| = \sigma_i$, for $i = 1, 2, \ldots, r$, while $\|G_i\| = 0$ if $i > r$. Defining $U_i \equiv G_i/\|G_i\| = A V_i/\sigma_i, i = 1, 2, \ldots, r$, yields an orthonormal set $\{U_1, \ldots, U_r\}$ with $A V_i = \sigma_i U_i$. Moreover $G_i = A V_i = 0$ if $i > r$. Now select $\{U_{r+1}, \ldots, U_m\}$ so that $\{U_1, U_2, \ldots, U_m\}$ is an orthonormal basis for $\mathbf{F}^m$. Define S as described in the statement of the SVD theorem. Put $U = [U_1 \ U_2 \ \cdots \ U_m]$ and $V = [V_1 \ V_2 \ \cdots \ V_n]$. Then

$$AV = [AV_1 \ AV_2 \ \cdots \ AV_n] = [\sigma_1 U_1 \ \sigma_2 U_2 \ \cdots \ \sigma_r U_r \ 0 \ \cdots \ 0] = US,$$

which is equivalent to Eq. (1).

Uniqueness in the SVD

r and S are unique, but practically nothing else is. In the construction of V, the eigenvalues and eigenspaces are uniquely determined by $A^H A$, but not the particular bases for these eigenspaces if their dimension is greater than 1. Further, the first r columns of U depend on V and are related to V as in Corollaries 3 and 4. A reversed construction could have been carried out by obtaining the U_i first and then finding the V_i in terms of the U_i, using (4). The point is that the matrices U and V are not independent. An extreme example of nonuniqueness occurs when A is unitary. In this case, $S = I$, and $V = U$ is an arbitrary unitary matrix.

If the σ_i are distinct, then $V_1, \ldots, V_r$ are unique up to scalar multiples since they lie in one-dimensional eigenspaces, and $U_i = \sigma_i^{-1} A V_i \ (i = 1, 2, \ldots, r)$ are determined from them. $\{U_{r+1}, \ldots, U_m\}$ is an arbitrary basis of $R(A)^\perp$, and $\{V_{r+1}, \ldots, V_n\}$ is an arbitrary basis for $N(A)$. Hence if $m = n = r$ and the σ_i are distinct, then the factorization is essentially unique.

EXAMPLE 1. Find the singular value decomposition of

$$A = \begin{bmatrix} 1 & -1 & -1 \\ 0 & 0 & 0 \\ 1 & 0 & 1 \\ 1 & -1 & -1 \end{bmatrix}.$$

$$A^H A = \begin{bmatrix} 3 & -2 & -1 \\ -2 & 2 & 2 \\ -1 & 2 & 3 \end{bmatrix},$$

which has eigenvalues $\lambda_i = 6, 2, 0$ and corresponding eigenvectors $E_i = (-1, 1, 1)^T$, $(1, 0, 1)^T, (1, 2, -1)^T$. The columns of V are $V_i = E_i/\|E_i\|$, the singular values are $\sigma_i = \sqrt{6}, \sqrt{2}$, and $r = 2$. To get the U_i, use (3) and compute $U_1 = AV_1/\sigma_1 = AE_1/\|AE_1\| = (0, 0, 1, 0)^T$, $U_2 = AV_2/\sigma_2 = AE_2/\|E_2\| = 2^{-1/2}(1, 0, 0, 1)^T$. Next find orthogonal unit vectors $U_3, U_4 \perp U_1, U_2$. One is $U_3 = (0, 1, 0, 0)^T$, and another is $U_4 = 2^{-1/2}(1, 0, 0, -1)^T$. The factorization is

$$A = [U_1\ U_2\ U_3\ U_4] \begin{bmatrix} \sqrt{6} & 0 & 0 \\ 0 & \sqrt{2} & 0 \\ 0 & 0 & 0 \\ 0 & 0 & 0 \end{bmatrix} \begin{bmatrix} V_1^H \\ V_2^H \\ V_3^H \end{bmatrix}.$$

Note that V_1, V_2 are unique up to a scalar multiple of ± 1, and they determine U_1 and U_2. $\{U_3, U_4\}$ can be any orthonormal basis for $\mathrm{span}\{U_1, U_2\}^\perp$.

EXAMPLE 2. Find a singular value decomposition for

$$A = \begin{bmatrix} 1 & 1 & 1 & 1 \\ 1 & -1 & -1 & 1 \end{bmatrix}.$$

AA^H is 4×4 and has two singular values, since the rank of A is 2. It is therefore more efficient to consider

$$A^H A = \begin{bmatrix} 4 & 0 \\ 0 & 4 \end{bmatrix}.$$

Here the reverse construction is used: The U_i are determined first. The eigenvalues are $\lambda_i = 4, 4$ and any orthonormal pair is an eigenbasis. Take

$$U_1 = \begin{bmatrix} 1 \\ 0 \end{bmatrix}, \qquad U_2 = \begin{bmatrix} 0 \\ 1 \end{bmatrix}.$$

From (4), $V_1 = A^H U_1/\sigma_1 = A^H U_1/\|AU_1\| = \frac{1}{2}(1, 1, 1, 1)^T$, $V_2 = A^H U_2/\sigma_2 = A^H U_2/\|A^H U_2\| = \frac{1}{2}(1, -1, -1, 1)^T$. Choose any two unit columns orthogonal to V_1, V_2 for the other V_i, e.g., $V_3 = \frac{1}{2}(1, 1, -1, -1)^T$, $V_4 = \frac{1}{2}(1, -1, 1, -1)^T$.

Remark. Good numerical methods do *not* construct the SVD as was done in the proof of the theorem and in these examples. There are problems with loss of small singular values, for example, since if $\lambda_k \approx 0$, it may be lost or contaminated with roundoff error, thereby giving unreliable σ_k. In such cases, $\sigma_k = \sqrt{\lambda_k}$ is much larger and is in less danger of being lost in the roundoff.

The Polar Decomposition

For a square matrix, the polar decomposition theorem asserts the existence of a factorization $A = MR$ with M positive and R unitary. This may be regarded as saying that the linear transformation A is a "rotation" R followed by a (anisotropic) "stretch" M. This can be considered a multidimensional analogue of the fact that every complex nonzero number may be written as $z = re^{i\theta}, r > 0$, since multiplication by $e^{i\theta}$ is a rotation and multiplication by r is a (isotropic) dilation.

POLAR DECOMPOSITION 1. Let A be an $m \times n$ matrix with $m \leq n$. There exists an $m \times m$ matrix M and an $m \times n$ matrix R such that

$$A = MR, \tag{8}$$

where M is Hermitian and positive, $RR^H = I$. $M = [AA^H]^{1/2}$ is unique. If A is real, then M is real, and R may be taken to be real.

Proof. First assume Eq. (8) holds. Then $AA^H = MRR^H M^H = M^2$, so $M = [AA^H]^{1/2}$ is the only possibility for M. Since AA^H is Hermitian and positive, it has nonnegative eigenvalues. Therefore, M is Hermitian and positive by Theorem 5 of Section 5.2 and the theory in Section 6.3. (The uniqueness of the positive square root is left to the exercises.) Next R is constructed so that the factorization in (8) holds. Let the vectors $V_i \in \mathbb{C}^n, U_i \in \mathbb{C}^m$ be as constructed in the proof of the singular value decomposition. Thus $\{U_k\}_{k=1}^m$ and $\{V_k\}_{k=1}^n$ are orthonormal bases in their respective spaces. From their construction, it follows that

$$AV_i = \sigma_i U_i, \quad i = 1, 2, \ldots, r \quad \text{and} \quad AV_i = 0 \quad \text{for } i > r. \tag{9a}$$

$$AA^H U_i = \lambda_i U_i, \quad i = 1, 2, \ldots, r \quad \text{and} \quad AA^H U_i = \lambda_i U_i = 0 \quad \text{for } i > r. \tag{9b}$$

Since $\sigma_i = \sqrt{\lambda_i}$ and $M = [AA^H]^{1/2}$, by Eq. (6a) or (2) of Section 4.3 and by (9b),

$$MU_i = \sigma_i U_i, \quad i = 1, 2, \ldots, m \quad \text{and} \quad MU_i = 0 \text{ otherwise.} \tag{9c}$$

Now define an $m \times n$ matrix R by

$$RV_i = U_i \quad \text{for } i = 1, 2, \ldots, m \quad \text{and} \quad RV_i = 0 \quad \text{for } i > m. \tag{9d}$$

But then from (9c) and (9d), $MRV_i = MU_i = \sigma_i U_i, i = 1, 2, \ldots, r$, while $MRV_i = 0$ otherwise. Comparing this with (9a), $AV_i = MRV_i$ for $i = 1, 2, \ldots, n$, and it follows that $A = MR$ since (as linear transformations) they agree on a basis. To form R specifically, let $U = [U_1 \ U_2 \cdots U_n]$ and $V = [V_1 \ V_2 \cdots V_m]$. Then (9d) means that $RV = [U \ 0]$ and so $R = [U \ 0]V^H$. Finally,

$$RR^H = [U \ 0]V^H V[U \ 0]^H = [V \ 0]\begin{bmatrix} V^H \\ 0 \end{bmatrix} = I.$$

COROLLARY 8. If A is square, then there are Hermitian positive matrices M, N and unitary matrices R, S such that

$$A = MR = SN. \tag{10}$$

Furthermore, $M = [AA^H]^{1/2}$ and $N = [A^H A]^{1/2}$ (uniquely). If A is real, so are M, N, R, S.

Proof. For the second factorization, consider A^H.

In many physical applications, the matrix A is nonsingular with positive determinant. Such is the case in elasticity, when A is the deformation gradient. Then more can be said about the polar decomposition. A quite different proof is possible.

POLAR DECOMPOSITION 2. Let A be square and nonsingular. Then there are unique positive definite Hermitian matrices M, N and a unique unitary matrix R such that

$$A = MR = RN. \tag{11}$$

$M = [AA^H]^{1/2}$ and $N = [A^H A]^{1/2}$ (uniquely). If A is real, so are M, N, R. If A is real and $|A| > 0$, then $|R| = 1$.

Proof. As before, Eq. (11) implies the formulae for M and N. Put $R = M^{-1}A$. Then $RR^H = M^{-1}AA^H M^{-H} = M^{-1}M^2 M^{-1} = I$, so R is unitary and $A = MR$. Similarly, $S = AN^{-1}$ is unitary and $A = VN$. To show $R = S$, consider the equivalent statements $R = S \Leftrightarrow M^{-1}A = AN^{-1} \Leftrightarrow AN = MA \Leftrightarrow M = ANA^{-1}$. But M is the unique positive definite matrix such that $M^2 = AA^H$, so if $(ANA^{-1})^2 = AA^H$, then $M = ANA^{-1}$ and $R = S$. But $(ANA^{-1})^2 = ANA^{-1}ANA^{-1} = AN^2 A^{-1} = AA^H AA^{-1} = AA^H$, whence $R = S$. The last statements are left to the reader.

Exercises 8.1

1. Find the SVD of the following matrices.

(a) $\begin{bmatrix} 10 & 2 & 10 & 2 \\ 5 & 11 & 5 & 11 \end{bmatrix}$
(b) $\begin{bmatrix} 1 & 1 \\ 1 & 1 \\ -2 & 1 \end{bmatrix}$
(c) $\begin{bmatrix} 1 & 1 & 1 \\ 1 & -1 & 1 \\ 1 & 0 & -2 \\ 1 & 0 & 0 \end{bmatrix}$
(d) $\begin{bmatrix} 1 & 1 \\ 2 & 2 \end{bmatrix}$

(e) $\begin{bmatrix} 1 & 1 & 1 \\ -1 & 0 & -2 \\ 1 & 2 & 0 \end{bmatrix}$
(f) $\begin{bmatrix} 1 & 0 & 1 \\ 0 & 1 & -1 \\ -1 & 1 & 0 \end{bmatrix}$
(g) $\begin{bmatrix} 2 & 1 & 0 & 0 \\ 0 & 0 & 2 & 0 \\ 2 & 0 & 0 & 1 \end{bmatrix}$

(h) $\begin{bmatrix} 2 & 1 & 0 \\ 2 & 0 & 1 \end{bmatrix}$
(i) $\begin{bmatrix} 1 & 0 & 2 \\ 0 & 4 & 0 \\ 2 & 0 & 1 \end{bmatrix}$
(j) $\begin{bmatrix} 1 & 2 & 2 \\ 2 & 1 & -2 \\ 2 & -2 & 1 \end{bmatrix}$

2. Let $A = [A_1 \ A_2 \ \cdots \ A_n]$ have orthogonal columns. Find the SVD of A.

3. Let $A = EF^H$, where $E \in \mathbf{C}^m, F \in \mathbf{C}^n, E, F \neq 0$. Find the SVD of A.

4. Suppose A is Hermitian. Describe the SVD of A. What if A is positive?

5. Prove Corollary 4.

6. Prove Corollary 6.

7. Use the SVD and its corollaries to prove
   (a) $N(A) = N(A^H)$
   (b) $R(A) = R(AA^H)$
   (c) $N(A^H) = R(A)^\perp$
   (d) $R(A^H) = N(A)^\perp$
   (*Note:* These were established by other means in Section 3.5.)

8. Show that

$$\sigma_1 = \max\left\{\frac{\|Ax\|}{\|x\|}, x \neq 0\right\} = \max\left\{\frac{\|A^H x\|}{\|x\|}, x \neq 0\right\}.$$

9. Let $A = USV^H$ be the SVD of A. Show that $b \in R(A) \Leftrightarrow U^H b$ has the last $m - r$ entries equal to 0.

10. Suppose A has singular values $\sigma_1, \ldots, \sigma_r$. Find the singular values of $\overline{A}$, A^T, A^H, αA.

11. Let A be a nonsingular $n \times n$ matrix.
  (*a*) Show that the following procedure produces a singular value decomposition.
    (*i*) Find V unitary and Λ diagonal with $\lambda_1 \geq \lambda_2 \geq \cdots \geq \lambda_n$ and $A^H A = V\Lambda V^H$.
    (*ii*) Put $S = \Lambda^{1/2}$ and $U = AV\Lambda^{-1/2}$.
  (*b*) Find a similar construction using $AA^H = U\Lambda U^H$, U unitary.

12. Suppose that A is an $n \times n$ normal matrix with eigenvalues λ_i indexed such that $|\lambda_1| \geq |\lambda_2| \geq \cdots \geq |\lambda_n|$. Let $Ae_i = \lambda_i e_i$, with $P = [e_1 \ e_2 \cdots e_n]$ unitary.
  (*a*) Find the singular values of A.
  (*b*) Show that $U = P$ may be used to form an SVD of A. What must S and V be?
  (*c*) Show that $V = P$ may be used to form an SVD of A. What must S and U be?

13. (*a*) Suppose $r(A) = m$ and the singular values of A are distinct. Exhibit the form of S, and state a uniqueness theorem.
  (*b*) Suppose $r(A) = n$ and the singular values of A are distinct. Exhibit the form of S, and state a uniqueness theorem.

14. (*a*) Generalize the SVD to a linear map $T : V \rightarrow W$ where $\dim(V) = n$, $\dim(W) = m$, $\langle,\rangle$ is an inner product on V, and $(,)$ is an inner product on W.
  (*b*) Restate your results from part (*a*) in terms of matrices, when $T(x) = Ax$, $V = \mathbf{C}^n$, $W = \mathbf{C}^m$, $\langle x, y \rangle = y^H Gx$, and $(x, y) = y^H \Gamma x$.

15. Suppose A is a positive matrix. Show that A has a unique positive square root. (*Note:* Positive matrices are always required to be Hermitian.)

16. Suppose A is a square matrix with $A = MR$ with M positive and R unitary. Show that A is normal iff $MR = RM$.

17. Let A be a square matrix. Show that there is a positive matrix M and a Hermitian matrix H such that $A = Me^{iH}$. When are M and H unique?

18. Let A be an $m \times n$ matrix. Show that there is a positive matrix N and a matrix P such that $A = PN$ and $P^H P = I$.

8.2
APPLICATIONS OF THE SVD

In this section, several applications of the singular value decomposition are given.

Solutions of $Ax = b$ and the Pseudo-Inverse

The SVD gives good insight into the nature of the solutions of $Ax = b$ when the system is compatible and provides a very nice approximate solution when the system is inconsistent. The approximate solution of inconsistent systems was studied in Chapter 3, but the SVD instantly gives the results sought and does so for more general problems. The analysis begins with the diadic expansion of Corollary 5 in Section 8.1,

$$A = \sum_{i=1}^{r} \sigma_i U_i V_i^{\mathrm{H}} \tag{1}$$

and its application to a vector x:

$$Ax = \sum_{i=1}^{r} \sigma_i U_i V_i^{\mathrm{H}} x = \sum_{i=1}^{r} \sigma_i \langle x, V_i \rangle U_i, \tag{2}$$

where $\langle, \rangle$ is the standard inner product in $\mathbf{F}^n$.

Suppose first that A is square and invertible. Then $r = n$ and all $\sigma_i > 0$. Hence

$$A^{-1} = (U\sigma V^{\mathrm{H}})^{-1} = (V^{\mathrm{H}})^{-1} \sigma^{-1} U^{-1} = V\sigma^{-1} U^{\mathrm{H}} = \sum_{i=1}^{n} \sigma_i^{-1} V_i U_i^{\mathrm{H}}. \tag{3}$$

Hence

$$x = A^{-1} b = \sum_{i=1}^{n} \sigma_i^{-1} \langle b, U_i \rangle V_i, \tag{4}$$

which gives an easily computed solution of $Ax = b$.

Now consider the case when A is singular and of arbitrary dimension. The idea is to attempt a solution of $Ax = b$ using an analogue of (4), namely,

$$x^{+} = \sum_{i=1}^{r} \sigma_i^{-1} \langle b, U_i \rangle V_i. \tag{5a}$$

This can be written as

$$x^{+} = [V_1 \ V_2 \ \cdots \ V_r] \operatorname{diag}\left(\frac{1}{\sigma_1}, \ldots, \frac{1}{\sigma_r}\right)[U_1 \ U_2 \ \cdots \ U_r]^{\mathrm{H}} b. \tag{5b}$$

Alternatively, x^{+} can be written as

$$x^{+} = A^{+} b, \tag{5c}$$

where

$$A^{+} = VS^{+} U^{\mathrm{H}} \quad \text{with} \quad S^{+} = \begin{bmatrix} \sigma^{-1} & 0 \\ 0 & 0 \end{bmatrix}_{n,m}, \tag{6}$$

i.e., S^+ is $n \times m$ and $S_{ij}^+ = 0$, except for $S_{ii}^+ = \sigma_i^{-1}, i = 1, 2, \ldots, r$. A^+ is called the *pseudo-inverse* of A. The diadic expansion of A^+ is

$$A^+ = \sum_{i=1}^{r} \frac{1}{\sigma_i} V_i U_i^H. \tag{7}$$

For computational purposes, (5b) and (7) are more economical expressions for x^+ in terms of matrix multiplications since only the columns from U and V that are actually used are included. A further discussion of the pseudo-inverse, also called the *Moore–Penrose generalized inverse*, is found in the exercises.

A variety of events may occur for a system $Ax = b$. If it is consistent, either the solution will be unique or there will be infinitely many solutions. If it is inconsistent, then minimizing $\|Ax - b\|$ (to get the best candidate for a solution) can produce one or infinitely many minimizers.

Two very nice things occur relating to x^+: $x = x^+$ minimizes $\|Ax - b\|$, and among all minimizers of $\|Ax - b\|$, x^+ has minimal norm. These are proved in the following propositions.

PROPOSITION 1. Let x^+ be defined by (5). Then Ax^+ is the projection of b on $R(A)$. In other words,

$$\min\{\|Ax - b\| : x \in F^n\} = \|Ax^+ - b\|. \tag{8}$$

Proof. Since $AV_i = \sigma_i U_i$,

$$Ax^+ = \sum_{i=1}^{r} \sigma_i^{-1} \langle b, U_i \rangle AV_i = \sum_{i=1}^{r} \langle b, U_i \rangle U_i.$$

Since $U_1, \ldots, U_r$ is an orthonormal basis for the range of A, this Fourier sum $A^+ x$ is the projection of b on $R(A)$.

Let us examine more completely the possibilities surrounding the system $Ax = b$ when A has no inverse. $s \equiv Ax^+$ is the projection of b in $R(A)$, by the last proposition. If $b \in R(A)$, then $s = b$, and there are infinitely many solutions of $Ax = b$. All of these solutions are of the form $x = x^+ + \Delta, \Delta \in N(A)$. On the other hand, if $b \notin R(A)$, then $s \neq b$. If $x = z$ also minimizes $\|Ax - b\|$, then $Az = Ax^+$ because the projection s is unique. If $r = n$, then $z = x^+$ since the columns of A are linearly independent. If $r < n$, then $z = x^+ + \Delta$ for some $\Delta \in N(A)$. Therefore, when $N(A) \neq 0$, the solution (or approximate solution) $x^+ = A^+b$ is one of infinitely many, all of the form $z = x^+ + \Delta, \Delta \in N(A)$. The next proposition tells us which of these solutions zx^+ is: It is the solution of minimal norm.

PROPOSITION 2. Let $x^+ = A^+b$. If Az is also a projection of b on the range of A, then $\|z\| \geq \|x^+\|$. In other words,

$$\min\{\|x^+ + \Delta\| : \Delta \in N(A)\} = \|x^+\|. \tag{9}$$

Proof. From (5), $x^+ = \sum_{i=1}^{r} \sigma_i^{-1} \langle b, U_i \rangle V_i$. Thus x^+ is orthogonal to $\{V_{r+1}, \ldots, V_n\}$. But these vectors span $N(A)^\perp$ by Corollary 5 of Section 8.1; hence $z \perp N(A)$. Therefore, $\|x^+ + \Delta\|^2 = \|x^+\|^2 + \|\Delta\|^2 \geq \|x^+\|^2$, by the Pythagorean law. The conclusion also follows from the fact that 0 is the best approximation to x^+ in $N(A) \Leftrightarrow x^+ - 0 \perp N(A)$. See Theorem 1 of Section 3.3.

Applications of the SVD in Numerical Analysis

Determining the rank of a matrix

One of the simplest applications of the SVD is the determination of the rank of a matrix. Theoretically, there are many ways to find the rank, and finding it should be simple, since the rank is an integer. We may row-reduce A or A^T to a row echelon form and count the number of nonzero rows remaining. Numerically, however, it is not so easy to judge what is zero and what is merely small. We can use an extended QR algorithm (or Gram–Schmidt, casting out dependent vectors) and try to detect which columns of the original matrix are dependent. We may also try to compute the eigenvalues of a matrix (if it is diagonalizable) and count the number of nonzero eigenvalues. This is perilous for the same reason. It turns out that the SVD is the best tool to use. Count the number of nonzero singular values; look at the singular values and make an educated guess as to which ones ought to be zero. The size of the matrix and of its entries should be taken into account. There are some rules of thumb; the manuals for some software packages will tell you what they are. Even for relatively small matrices, different methods can give different answers for the rank.

Solving sensitive systems of equations

Consider solving $Ax = b$ when A is square and invertible. Suppose an approximate solution $\tilde{x}$ has been found by Gauss elimination. The amazing fact is that, nearly always, the residual $r \equiv A\tilde{x} - b$ has small norm. Equally amazing is the fact that, even though r has small norm, $\tilde{x}$ may be far from the actual solution x, both in an absolute and a relative sense. Let's see why. Using an SVD of A, Eq. (1) gives

$$A\tilde{x} = \sum_{i=1}^{r} \sigma_i U_i V_i^H \tilde{x} = \sum_{i=1}^{r} \sigma_i \langle \tilde{x}, V_i \rangle U_i \quad \text{with } r = n.$$

Suppose $\sigma_1 \gg \sigma_n > 0$, and $\tilde{x} = x + \Delta$. If $\Delta = V_n$, then $A\tilde{x} = Ax + \sigma_n U_n$. If also $\langle x, V_1 \rangle$ is not small, then $\sigma_n U_n$ is a negligible term, and $A\tilde{x}$ and Ax are nearly the same but Δ is not small. The quantity σ_1/σ_n is called the *condition number of the matrix,* and most reliable numerical packages for solving $Ax = b$ give an estimate of it. The larger the ratio, the greater the possibility that the estimated solution is in error. About one digit of accuracy may be lost for every power of 10 in σ_1/σ_n.

Just as the SVD can indicate where the error is, it can disclose a remedy. First compute and examine the singular values of the matrix. The singular values (if any) that are judged to be at the roundoff level are set to 0. In this way the rank of the matrix is determined, and the "solution" $x = x^+$ to $Ax = b$ is found by (5):

$$x^+ = \sum_{i=1}^{r} \sigma_i^{-1} \langle b, U_i \rangle V_i = [V_1 \ V_2 \ \cdots \ V_r] \operatorname{diag}\left(\frac{1}{\sigma_1}, \ldots, \frac{1}{\sigma_r}\right)[U_1 \ U_2 \ \cdots \ U_r]^H.$$

Propositions 1 and 2 tell us that no great harm has been done: $\|Ax^+ - b\|$ is minimal, and among the x's achieving this minimum, the one of minimal norm has been found. From another point of view, consider (5): $x^+ = \sum_{i=1}^{r} \sigma_i^{-1} \langle b, U_i \rangle V_i$. If some σ_i are very small relative to all the others, they are best deleted from the sum since they can cause a great error in x^+. One may have to experiment a bit.

Let us study an example where this kind of error can be particularly troublesome. Suppose the task is to approximate some data by a polynomial of degree $n - 1$ on m data points, and the discrete inner product of Chapter 2 with uniform weights is being used. This amounts to minimizing $\|Ax - b\|$, where x is the vector of coefficients of the polynomial, or to solving the normal equations $A^H Ax = A^H b$. Solving these equations numerically to obtain $x = \tilde{x}$ and checking the residual $A^H A\tilde{x} - A^H b$ will probably give a small answer. However, if the data points were poorly chosen when a graph of the polynomial $p(t)$ so obtained is plotted, the graph may agree well at the data points but be very poor at points away from the data. The method of the previous paragraph can sometimes save the situation. Look at the singular values of A. Select the r' "significant" ones. Use as the coefficients $x = x^+ = \sum_{i=1}^{r'} \sigma_i^{-1}\langle b, U_i \rangle V_i$. Again, some experimentation may have to be done.

Applications in Pattern Recognition

Some of the ideas to be described are used in pattern recognition, statistical data analysis, cluster analysis, artificial intelligence, and data compression. We will only touch on some of the ideas and give a couple of examples. For a fuller discussion, see Oja (1983).

Suppose N data vectors $X_1, X_2, \ldots, X_N$ in $\mathbf{C}^m$ are given. We want to fit the best directions $e_1, e_2, \ldots$ to the data in the following sense. Observe that, for a unit vector e, $\langle f, e \rangle$ is the scalar projection of f on the span of e. Thus $D(e) \equiv \sum_{k=1}^{N} |\langle X_k, e \rangle|^2$ is a measure of how well e points in the direction of the vectors X_k, $k = 1, \ldots, N$, as a group. The best direction $e = e_1$ will maximize this sum. To get the next best direction, look for a vector $e \perp e_1$ for which $D(e)$ is maximal, and call it e_2. Generally, having found $e_1, \ldots, e_{p-1}$, find $e = e_p$ orthogonal to these vectors maximizing $D(e)$. A dual viewpoint is given in the exercises.

To analyze this problem using the standard inner product, note that

$$D(e) = \sum_{k=1}^{N} \langle X_k, e \rangle \overline{\langle X_k, e \rangle} = \sum_{k=1}^{N} \langle X_k, e \rangle \langle e, X_k \rangle = \sum_{k=1}^{N} e^H X_k X_k^H e$$

so that

$$D(e) = e^H XX^H e = e^H Q e \tag{10}$$

where

$$X = [X_1 \ \cdots \ X_N] \quad \text{and} \quad Q = XX^H. \tag{11}$$

In statistics, $(1/N)Q$ is called the *sample covariance matrix*. Thus the tasks just outlined are

Maximize $e^H Q e$ subject to $\|e\| = 1$.
Given $e_1, \ldots, e_{p-1}$:
    Maximize $e^H Q e$ subject to $\|e\| = 1$, $e \perp e_1, \ldots, e_{p-1}$.

This is nothing more than the extended maximization problem for the Rayleigh quotient of Q, discussed in Section 5.4. The maximizing vectors are eigenvectors $e_1, \ldots, e_m$ of Q, and the sequence of constrained maxima are $D(e_1) = \lambda_1, D(e_2) = \lambda_2, \ldots, D(e_m) = \lambda_m$, with $\lambda_1 \geq \lambda_2 \geq \cdots \geq \lambda_m$. If m is small and λ_m is not too small, then the eigenvectors and eigenvalues can be computed from Q by standard methods. Otherwise, observe that the (nonzero) eigenvalues of Q are precisely $\lambda_i = \sigma_i^2$ and the eigenvectors are $e_i = U_i$, where $X = USV^H$ is the singular value decomposition of X.

The basis $\{e_k\}_{k=1}^m$ of $\mathbf{C}^m$ found in this process is orthonormal. Its members are sometimes called *features* of the set $\{X_k\}_1^N$. Each $x \in \mathbf{C}^m$ can be expanded as

$$x = \sum_{i=1}^m \xi_i e_i, \quad \xi_i = \langle x, e_i \rangle. \tag{12}$$

The coordinates ξ_i of x in this basis are called *principal components* of x. Equation (12) is a finite-dimensional example of the *Karhunen–Loeve* expansion. The basis $\{e_k\}_{k=1}^m$ is called a *principal component basis*.

In data compression and pattern recognition, this question arises: How many of the ξ_i are actually needed to represent x efficiently? Let $s_p = \sum_{i=1}^p \xi_i e_i$ be the projection on $\text{span}\{e_k\}_{k=1}^p$. Then $\|x - s_p\|^2 = \sum_{i=p+1}^m |\xi_i|^2$. If this is negligible, we can accept s_p as a sufficiently good approximation of x. Let us determine how many terms are needed if x is an arbitrary one of the X_k. To this end, note that $\|x - s_p\|^2 = \sum_{i=p+1}^m e_i^H x x^H e_i$. Then putting $x = X_k$ and summing over all k yields

$$\sum_{k=1}^N \|X_k - s_p(X_k)\|^2 = \sum_{k=1}^N \sum_{i=p+1}^m e_i^H X_k X_k^H e_i = \sum_{i=p+1}^m e_i^H \left(\sum_{k=1}^N X_k X_k^H \right) e_i$$

$$= \sum_{i=p+1}^m e_i^H Q e_i = \sum_{i=p+1}^m \lambda_i.$$

Thus to represent x in the set $\{X_k\}_1^N$, it is sufficient to take p so that $\sum_{i=p+1}^m \lambda_i / \sum_{i=1}^m \lambda_i$ is small. Once the desired p has been determined, either by analysis or experiment, then the projection of a vector x can be computed by the truncated principal components expansion

$$s_p = \sum_{i=1}^p \langle x, e_i \rangle e_i = s_p = \sum_{i=1}^p e_i e_i^H x = [e_1 \cdots e_p][e_1 \cdots e_p]^H x. \tag{13}$$

The fact that p is adequate means that the subspace $\text{span}\{e_k\}_{k=1}^p$ contains most of the information.

EXAMPLE 1. IMAGE COMPRESSION, FEATURE CAPTURING. In this example, we take a simple small bitmap picture, find the principal component basis, and then use (12) with $p = 1, 2, \ldots$ to recreate the picture. This shows how the subspaces $\text{span}\{e_k\}_{k=1}^p$ capture the main features of the original object.

A bitmap can be captured in an *incidence matrix* P, wherein $p_{ij} = 1$ if the pixel is lit and 0 otherwise. In this example the matrix is $N \times N$. The matrix P can be converted to a sequence of data vectors in various ways, e.g., by dividing it into block submatrices and arranging the entries of the blocks into vectors or by using the columns (or rows) as data vectors. In this example, each column P_k will be used as a data vector: $X_k = P_k$, so that $X = P$. Then the SVD of P is computed: $P = USV^H$, so that the principal component basis is given by $e_k = U_k$. The projection of each data vector P_k on span$\{e_k\}_{k=1}^p$ is obtained from (12), which becomes

$$\tilde{P}_k = [U_1 \ U_2 \ \cdots \ U_p][U_1 \ U_2 \ \cdots \ U_p]^H P_k. \tag{14}$$

The reassembled bitmap image is then

$$\tilde{P} = [\tilde{P}_1 \ \tilde{P}_2 \ \cdots \ \tilde{P}_N] = [U_1 \ U_2 \ \cdots \ U_p][U_1 \ U_2 \ \cdots \ U_p]^H P. \tag{15}$$

However, $\tilde{P}$ is not filled with 0s and 1s. A threshold of $\frac{1}{2}$ will be used in the decision to plot: $\tilde{P}_{ij}$ is plotted iff $\tilde{P}_{ij} > \frac{1}{2}$.

A very simple example is used with $N = 50$. This represents just a small portion of a monitor, for example. The original picture is given together with the captured image in P. The circle in the captured picture looks rough because of the crude resolution.

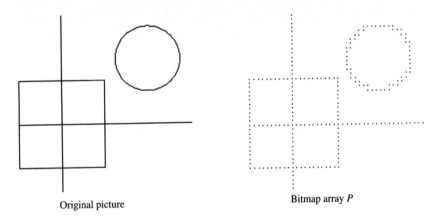

Original picture

Bitmap array P

Computation of the singular values gives

k	σ_k
1	1.0232E+01
2	6.6332
3	4.8347
4	4.3850
5	3.3018
6	2.6354
7	2.3761
8	1.8549
9	1.1921
10	1.0658
11	1.0072
12	0.88016
13	0.52250
14	0.34746
15	0.19780
16	1.1910×10^{-15}

The rest are even smaller, i.e., at or below roundoff. The next sequence of pictures shows the plotting of the projected images for $p = 1, 2, \ldots, 8$. At $p = 8$ the original bitmap is recovered, so that, visually, all of the information is captured in span$\{U_k\}_{k=1}^8$. Mathematically, it is clear from the σ_i that span$\{U_k\}_{k=1}^{15}$ must contain virtually all of the information. Notice how the essential features of P are successively captured. There are a few spurious points at first.

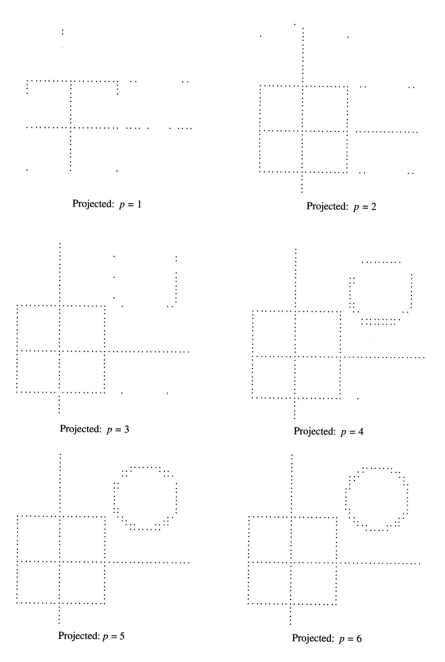

Projected: $p = 1$ Projected: $p = 2$

Projected: $p = 3$ Projected: $p = 4$

Projected: $p = 5$ Projected: $p = 6$

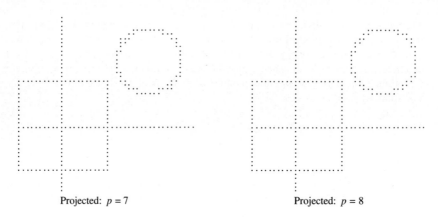

Projected: $p = 7$ Projected: $p = 8$

EXAMPLE 2. DESCRIPTION AND IDENTIFICATION. A simple example will be described without a numerical illustration. This method is widely used in many situations.

Suppose the objective is to classify and describe human profiles. For example, a surgeon may wish to do this for facial reconstruction, a security organization may wish to do this as a means of identification, or an anthropologist may wish to do this to study differences and structure in cultural, age, or genetic groups. A reasonable collection of profiles would be gathered and digitized into vectors X_k, perhaps many. A singular value decomposition on the array $X = [X_1 \cdots X_N]$ is then performed: $X = USV^H$. The singular values σ_i are examined to determine the significant ones; call these $\sigma_1, \ldots, \sigma_p$. The corresponding elements of the principal component basis $e_k = U_k$ are then stored. Each of $U_1, \ldots, U_p$ describes some fundamental structure in human profiles. If a new profile is measured, it can then be expanded in terms of these: $x = \sum_{i=1}^{p} \xi_i U_i, \xi_i = \langle x, U_i \rangle$. (Of course, this is really the projection of x on span$\{U_1, \ldots, U_p\}$.) The principal components $\xi_1, \ldots, \xi_p$ then describe the profile in terms of the basic building blocks of human profiles $U_1, \ldots, U_p$. Perhaps these ξ_i need to be determined only to a few digits to provide a sufficiently accurate description.

Exercises 8.2

1. Show that if A is nonsingular, then $A^{-1} = A^+$.

2. Show that a compact form for A^+ is $A^+ = [V_1 \ V_2 \ \cdots \ V_r]\sigma^{-1}[U_1 \ U_2 \ \cdots \ U_r]^H$.

3. Let A^+ be the pseudo-inverse of A. Show that
   (a) $R(A) = R(AA^+)$.
   (b) $R(A^+) = R(A^H)$.
   (c) AA^+ is the matrix of the projection operator from $\mathbf{F}^m$ onto $R(A)$.
   (d) A^+A is the matrix of the projection operator from $\mathbf{F}^n$ onto $R(A^H)$.
   (e) If $r(A) = n$, then A^+ is a left inverse of A.
   (f) If $r(A) = m$, then A^+ is a right inverse of A.
   (g) If $r(A) = n$, then $A^+ = (A^H A)^{-1} A^H$.
   (h) If $r(A) = m$, then $A^+ = A^H (AA^H)^{-1}$.

4. A matrix $\tilde{A}$ is a *Moore–Penrose generalized inverse* of A iff (i) $A\tilde{A}$ and $\tilde{A}A$ are Hermitian, (ii) $A\tilde{A}A = A$, (iii) $\tilde{A}A\tilde{A} = \tilde{A}$. Prove the following:
(a) If A is nonsingular, then $\tilde{A} = A^{-1}$.
(b) $\tilde{A}$ is unique: If B also satisfies (i), (ii), (iii), then $B = \tilde{A}$.
(c) $A^+ = \tilde{A}$, i.e., A^+ satisfies (i), (ii), (iii).

5. Suppose $\langle x, y \rangle = \sum_{i=1}^{m} x_i \bar{y}_i w_i$ is an inner product on $\mathbf{C}^m$, A is a given $m \times n$ matrix, and $y \in \mathbf{C}^m$. Explain how to use the SVD decomposition to minimize $\|Ax - y\|^2$ in the norm generated by the given inner product. What minimum properties does the x^+ obtained by this process have?

6. Let $\langle x, y \rangle = y^H G x$ be an inner product on $\mathbf{C}^m$. Repeat Exercise 5 for this situation.

7. Given a data set $\{X_1, \ldots, X_N\}$, it is desired to find the best subspaces of approximation. This approach is dual to that of finding the best directions describing the data set discussed in the text. Consider the following sequence of questions.
(a) If f is a unit vector in $\mathbf{C}^m$, let $M = \text{span}\{f\}^\perp$. Show that $|\langle x, f \rangle|$ is the distance from x to M.
(b) Interpret $D(f) \equiv \sum_{i=1}^{N} |\langle X_k, f \rangle|^2$ and explain why it should be minimized.
(c) Let f_1 be a vector minimizing $D(f)$ subject to $\|f\| = 1$. The next step is to minimize $D(f)$ subject to $\|f\| = 1$, $f \perp f_1$. Explain what this means geometrically.
(d) The general step: Suppose $f_1, \ldots, f_{p-1}$ have been found. Minimize $D(f)$ subject to $\|f\| = 1$, $f \perp f_1, \ldots, f_{p-1}$, and denote a minimizer by f_p. Write the sequence of steps as a sequence of constrained minimization problems involving $f^T Q f$.
(e) Suppose the process is complete and $f_1, \ldots, f_m$ have been found. Relate these f_i to the e_j found in the "best directions" process described in the text. Identify the $D(f_j)$ and relate these numbers to the $D(e_k)$ found in the text.

8. Use MATLAB to create incidence matrices of pictures and experiment with their feature extraction/compression as in Example 1.

9. Look for other intriguing feature extraction/data compression problems and try the method of Example 1 on them.

8.3
SCHUR'S THEOREM

Schur's theorem can be regarded as a generalization of the spectral theorem, which asserts that a normal operator is diagonalized in a suitable orthonormal basis. Schur's theorem asserts that an arbitrary operator $T : V \to V$ has an upper triangular matrix in a suitable orthonormal basis. This theorem has many applications, of which only a few will be mentioned here.

SCHUR'S THEOREM. Let $L : V \to V$ be linear, $\dim(V) \neq 0, \infty$, and $\langle , \rangle$ be an inner product on V. Then there exists an orthonormal basis $\{e_k\}_{k=1}^{n}$ for V such that

$$L(e_j) = \sum_{i=1}^{j} t_{ij} e_i, \quad j = 1, 2, \ldots, n, \tag{1}$$

and t_{ii} is an eigenvalue of L for each $i = 1, 2, \ldots, n$.

If $V = \mathbf{C}^n$ and $L(x) = Ax$, then there exists a unitary matrix U and an upper triangular matrix T such that

$$A = UTU^H. \tag{2}$$

Furthermore, the diagonal entries of T are the eigenvalues of A.

Proof. L has an eigenvector e_1 of unit length: $L(e_1) = \lambda_1 e_1$, $\|e_1\| = 1$ by Theorem 5 of Section 4.2. Let $V_1 = (\text{span}\{e_1\})^\perp$, and define S_1 by $S_1(x) = L(x) - \langle L(x), e_1 \rangle e_1$; that is, S_1 is L with the projection along e_1 subtracted out. Thus $S_1(x) \perp e_1$ for all x, by an easy calculation or by Lemma 2 of Section 3.2, so that $S_1 : V_1 \rightarrow V_1$. The same argument may be applied to S_1 acting on V_1 to obtain a unit eigenvector e_2 of S_1: $S_1(e_2) = \lambda_2 e_2$, $e_2 \in V_1$. Observe that $e_2 \perp e_1$ and $L(e_2) = S_1(e_2) + \langle L(e_2), e_1 \rangle e_1 = \lambda_2 e_2 + \langle L(e_2), e_1 \rangle e_1 \in \text{span}\{e_1, e_2\}$. The process continues this way: Having found orthonormal vectors $e_1, \ldots, e_k$, with $L(e_j) \in \text{span}\{e_1, \ldots, e_j\}$ for $j = 1, \ldots, k$, define S_k by $S_k(x) \equiv L(x) - \sum_{i=1}^{k} \langle L(x), e_i \rangle e_i$. Then S_k obeys $S_k : V_k \rightarrow V_k$, where $V_k = (\text{span}\{e_1, \ldots, e_k\})^\perp$, and so has a unit eigenvector e_{k+1}, with $S_k(e_{k+1}) = \lambda_{k+1} e_{k+1}$. Then

$$L(e_{k+1}) = S_{k+1}(e_{k+1}) + \sum_{i=1}^{k} \langle Le_k, e_i \rangle e_i = \lambda_{k+1} e_{k+1} + \sum_{i=1}^{k} \langle Le_k, e_i \rangle e_i, \tag{$*$}$$

so that $L(e_{k+1}) \in \text{span}\{e_1, \ldots, e_{k+1}\}$. The process terminates in an orthonormal basis $\{e_k\}_{k=1}^{n}$. Since $L(e_j) \in \text{span}\{e_1, \ldots, e_j\}$ for $j = 1, \ldots, n$, the matrix T of L in the basis $\{e_k\}_{k=1}^{n}$ is upper triangular. The characteristic polynomial of L is independent of basis, so it is $p(\lambda) = |T - \lambda I| = \prod(t_{ii} - \lambda)$. Thus the t_{ii} are the eigenvalues of L.

To derive the statement for the matrix operator, the change of basis gives $[L]_\beta = T = U^{-1}AU = U^H A U$, since $U = [e_1 \ e_2 \ \cdots \ e_n]$ is unitary.

There is little that can be said for the uniqueness of the Schur factorization. However, the preceding proof shows that there is some flexibility in the selection of the basis $\{e_k\}_{k=1}^{n}$ and therefore in the form of T. This flexibility will be useful in some problems.

Remarks. Let T be the upper triangular matrix of L in Schur's theorem.

1. The order of the eigenvalues on the diagonal of T can be specified, using algebraic multiplicities. This is clear from the construction. At each step, the eigenvalue λ_{k+1} in $(*)$ is on the diagonal of T and therefore must be an eigenvalue of T as well as of S_k. Thus, at any step, one of the remaining eigenvalues of T, counted with its algebraic multiplicity, may be chosen to be the next one on the diagonal of T.

2. There is some flexibility of the e_{k+1} constructed at each step. For example, if the eigenspace $M(\lambda_{k+1})$ contains an eigenvector orthogonal to $e_1, \ldots, e_k$, it could be chosen as e_{k+1}, since for such vectors $L(e_{k+1}) = S_k(e_{k+1})$.

One consequence of Schur's theorem is the interesting fact that, for every square matrix A, there are diagonalizable matrices arbitrarily close to A. In the context of numerical analysis, this means that within (any) machine precision, it is impossible to separate defective and diagonalizable matrices. The idea of the proof is to perturb the diagonal entries of T in Schur's theorem slightly.

THEOREM 1. Let A be an $n \times n$ matrix and $\varepsilon > 0$ be given. Then there is a diagonalizable matrix B with distinct eigenvalues so that $\|B - A\| < \varepsilon$, where $\| \cdot \|$ is the matrix norm $\|M\| = \mathrm{Tr}(M^H M)^{1/2}$.

Proof. Let $A = UTU^H$, where U is unitary and T is upper triangular. Choose numbers d_i so that $\sum_{i=1}^n |d_i|^2 < \varepsilon^2$ and all $t_{ii} + d_i$ are distinct. Put $B = A + UDU^H$. Then $U^H BU = T + D$ so that the eigenvalues of B are $t_{ii} + d_i$. Furthermore, since the trace of an operator is independent of basis, $\|B - A\|^2 = \|U^H DU\|^2 = \mathrm{Tr}(U^H D^H DU) = \mathrm{Tr}(D^H D) = \sum_{i=1}^n |d_i|^2 < \varepsilon^2$.

Application to Matrix Limits

In Chapter 4, it was observed that if $T : V \to V$ is diagonalizable and if all of the eigenvalues of T are smaller than 1 in modulus, then $T^n x \to 0$ for all x. Schur's theorem allows us to extend this to nondiagonalizable operators. Some preparatory material is required. Recall that the norm $\|\cdot\|_1$ on $\mathbf{C}^n$ was defined by $\|x\|_1 = \sum_{i=1}^n |x_i|$. The following definitions give a useful norm on the set of matrices.

DEFINITIONS. Let $A = [A_1 \ A_2 \ \cdots \ A_n]$ be an $n \times n$ matrix, and $x \in \mathbf{C}^n$. Define

$$\|\|A\|\|_1 = \max\{\|A_j\|_1 : j = 1, 2, \ldots, n\} \tag{3}$$

$$\rho(A) = \max\{|\lambda| : \lambda \text{ is an eigenvalue of } A\}. \tag{4}$$

$\rho(A)$ is called the *spectral radius of* A.

It is easy to see that $\|\| \cdot \|\|_1$ satisfies the properties of a norm on the vector space of all $n \times n$ matrices (cf. Sections 3.1 and 4.5). In the discussion following Eq. (2) of Section 4.5, it was shown that

$$\|Ax\|_1 \leq \|\|A\|\|_1 \|x\|_1 \quad \forall x. \tag{5}$$

If one takes $x = \delta_i$, then $Ax = A_i$ and $\|x\|_1 = 1$, so equality holds in (5), when i is chosen such that $\|\|A\|\|_1 = \|A_i\|_1$.

LEMMA 2. For all square matrices A, B,

$$\|\|AB\|\|_1 \leq \|\|A\|\|_1 \|\|B\|\|_1$$

$$\|\|A^n\|\|_1 \leq \|\|A\|\|_1^n \quad n = 1, 2, \ldots.$$

Proof. For each x, $\|ABx\|_1 \leq \|\|A\|\|_1 \|Bx\|_1 \leq \|\|A\|\|_1 \|\|B\|\|_1 \|x\|_1$. Taking $x = \delta_i$ with δ_i so that $\|AB\delta_i\|_1 = \|\|AB\|\|_1$, the first part of the lemma follows. The second part follows using the first part inductively: $\|\|A^n\|\|_1 \leq \|\|A\|\|_1 \|\|A^{n-1}\|\|_1 \leq \cdots \leq \|\|A\|\|_1^n$.

LEMMA 3

$$|a_{ij}| \leq \|\|A\|\|_1, \quad i, j = 1, 2, \ldots, n$$

Proof

$$|a_{ij}| \leq \|A_j\|_1 \leq \|\|A\|\|_1$$

DEFINITION. A sequence of matrices A_n tends to a limiting matrix A if and only if $(A_n)_{ij} \to a_{ij}$. This is denoted by $A_n \to A$ as $n \to \infty$ or by $\lim_{n \to \infty} A_n = A$.

PROPOSITION 4. If $A_n \to A$ as $n \to \infty$, then $BA_n \to BA$, $A_nB \to AB$ as $n \to \infty$ (provided the products are defined).

Proof. This is immediate from the definitions and limit theorems for complex numbers.

PROPOSITION 5

$$A_n \to 0 \Leftrightarrow \|\|A_n\|\|_1 \to 0$$

Proof. If $A_n \to 0$, then clearly each column $(A_n)_j \to 0$; thus each $\|(A_n)_j\|_1$ approaches 0, and the maximum of these, which is $\|\|A_n\|\|_1$, approaches 0 as well. If $\|\|A_n\|\|_1 \to 0$, then by Lemma 3, the entries $(A_n)_{ij} \to 0$ as well.

THEOREM 6

$$A^n \to 0 \text{ as } n \to \infty \Leftrightarrow \rho(A) < 1$$

Proof. First suppose $A^n \to 0$, and let λ be an eigenvalue of A. Therefore $AE = \lambda E$ and $A^nE = \lambda^nE$ for some $E \neq 0$. Since $A^n \to 0$, $A^nE = \lambda^nE \to 0$ by Proposition 4. Hence $|\lambda| < 1$, and so $\rho(A) < 1$. Next suppose that $\rho(A) < 1$ and that A is an $m \times m$ matrix. First factor $A = UTU^H$ by Schur's theorem. Note that $A^n = UTU^HUTU^H\cdots UTU^H = UT^nU^H$. Thus $A^n \to 0 \Leftrightarrow T^n \to 0$ by Proposition 4. Let $D = \text{diag}(\theta, \theta^2, \ldots, \theta^m)$ with $\theta < 1$, and consider $S = D^{-1}TD$. Note that $S^n = D^{-1}TDD^{-1}TD\cdots D^{-1}TD = D^{-1}T^nD$, so that $S^n \to 0 \Leftrightarrow T^n \to 0$. Let S_j be the jth column of S. Then $\|S_j\|_1 = \sum_{i=1}^{j}|s_{ij}| = |t_{jj}| + \sum_{i=1}^{j-1}|t_{ij}|\theta^{-i+j} \leq |t_{jj}| + \theta\sum_{i=1}^{j-1}|t_{ij}| \leq |t_{jj}| + \theta M \leq \rho(A) + \theta M$, where $M = \max_j\{\sum_{i=1}^{j-1}|t_{ij}|\}$. The fact that each t_{jj} is an eigenvalue of A was used. If θ is small enough, then $\rho(A) + \theta M \equiv \mu < 1$. Thus $\|\|S\|\|_1 = \max_j\|S_j\|_1 \leq \mu < 1$. By Lemma 2, $\|\|S^n\|\|_1 \leq \|\|S\|\|_1^n \leq \mu^n \to 0$. Thus $S^n \to 0$ by Proposition 5.

The next theorem specifies when the limit of A^k exists, in general.

THEOREM 7. Suppose A is $n \times n$ and has eigenvalues $\lambda_1, \ldots, \lambda_n$. Then $\lim_{k\to\infty} A^k$ exists iff (a) the eigenvalues satisfy either $|\lambda_i| < 1$ or $\lambda_i = 1$ and (b) if 1 is an eigenvalue, then the algebraic multiplicity of $\lambda = 1$ is the same as its geometric multiplicity.

Proof. The proof is outlined in the exercises together with further information. (See Exercises 7, 8, and 9.)

Application to Iterative Solution of Equations

When the system of equations is large and has some special structure, it is actually more efficient and accurate to solve the system iteratively than by Gauss elimination. Sometimes these methods can be used to solve a system whose matrix is far too large to put into the computer's memory. It is therefore important to know when an iterative method converges. An important theorem concerning this will be given and a very simple example exhibited. For other methods, consult a decent numerical analysis book.

EXAMPLE 1. MATRIX ITERATIVE METHODS. Consider the system $Ax = b$, i.e., $\sum_{j=1}^{n} a_{ij}x_j = b_i$. Assume $a_{ii} \neq 0$ for all i. If the ith equation for x_i is solved, then

$$x_i = a_{ii}^{-1}\left(b_i - \sum_{j=1, j\neq i}^{n} a_{ij}x_j\right) \qquad i = 1, 2, \ldots, n. \qquad (*)$$

To put this in matrix form, set $A = D + \tilde{A}$ where $D = \text{diag}(a_{11}, a_{22}, \ldots, a_{nn})$ and $b = (b_1, b_2, \ldots, b_n)$. Then (*) is equivalent to

$$x = D^{-1}(b - \tilde{A}x). \qquad (**)$$

The iteration used is as follows: Given a current estimate x of the solution, put these x_i into the right side of (*); the computed values (on the left) are the next estimate of the solution. This is called *Jacobi's method*.

In general, to develop an iterative method, one begins with the system $Ax = b$ and rearranges it into the form

$$x = Gx + c, \qquad (6)$$

where G is $n \times n$ and c is a column, in such a way that both systems have the same solution. In Jacobi's method, $G = -D^{-1}\tilde{A}$, $c = D^{-1}b$. G is called the *iteration matrix*. The *Gauss–Seidel* and *successive overrelaxation* methods fall under the purview of this discussion, but G and c for these methods will not be derived. For (6), the iteration is given by the following procedure: Given x_0, compute

$$x_{k+1} = Gx_k + c, \quad k = 0, 1, 2, \ldots. \qquad (7)$$

To discuss the convergence, let $e_k = x_k - x$, where x is the solution of (6), so that e_k is the vector of errors. Subtracting (6) from (7) yields $e_{k+1} = Ge_k$. Hence

$$e_n = Ge_{n-1} = G^2 e_{n-2} = \cdots = G^n e_0. \qquad (8)$$

Thus the iteration (7) converges for all $x_0 \Leftrightarrow e_n \to 0$ for all $x_0 \Leftrightarrow G^n \to 0$. By Theorem 6, this occurs iff $\rho(G) < 1$. Thus all discussions of convergence hinge on estimating the spectral radius of the iteration matrix. For a discussion of the methods just mentioned and a proof of their convergence, see Varga (1962) or Young (1971).

Schur's theorem may also be used to prove the next theorem.

CAYLEY–HAMILTON THEOREM. Let A be an $n \times n$ matrix and $p(\lambda)$ the characteristic polynomial of A. Then $p(A) = 0$.

Proof. The proof is outlined in the exercises.

There are many other applications of Schur's theorem. For a good account of these, see Horn and Johnson (1985).

Exercises 8.3

1. (a) Give an example of a nonzero matrix B and a sequence of matrices S_n such that BS_n has a limit but S_n does not.
   (b) Formulate conditions on B so that, whenever BS_n has a limit, so does S_n.

2. Suppose that $L : V \to V$ is Hermitian in the inner product $\langle,\rangle$ on V. Use Schur's theorem to show that L has an orthogonal eigenbasis.

3. Let $A = UTU^H$ be a factorization of A according to Schur's theorem. Show that each subspace $W_k = \text{span}\{U_1, \ldots, U_k\}$ is an invariant subspace of A, i.e., $A(W_k) \subset W_k$.

4. (a) Suppose $\rho(A) < 1$. Show that $\sum_{i=0}^{\infty} A^i = (I - A)^{-1}$. (*Hint:* Consider $(I - A)(1 + A + \cdots + A^k)$.)

   (b) Suppose some eigenvalue of A satisfies $|\lambda_i| \geq 1$. Show that $\sum_{i=0}^{\infty} A^i$ diverges.

5. Define a sequence of vectors by $x^{(n+1)} = Ax^{(n)} + f$, with $f \neq 0$ and A $m \times m$.

   (a) If $\rho(A) < 1$, show that $x^{(n)}$ converges and find its limit.

   (b) Show that $x^{(n)}$ may have a limit, when some eigenvalues obey $|\lambda_i| \geq 1$ and others obey $|\lambda_j| < 1$. Find a 3×3 example.

6. (a) Suppose

$$A = \begin{bmatrix} I & P \\ 0 & S \end{bmatrix},$$

   where S is a matrix all of whose eigenvalues are less than 1 in absolute value. Find $\lim_{k \to \infty} A^k$.

   (b) Suppose $A = PBP^{-1}$, where

$$B = \begin{bmatrix} I & Q \\ 0 & S \end{bmatrix},$$

   P is nonsingular, and S is a matrix all of whose eigenvalues are less than 1 in absolute value. Find $\lim_{k \to \infty} A^k$.

The next three exercises describe precisely when $\lim_{k \to \infty} A^k$ exists when some of the eigenvalues do not satisfy $|\lambda| < 1$. They constitute a proof of Theorem 7. Shur's theorem and the remarks following it will be useful.

7. Suppose A has eigenvalues $\lambda_1, \ldots, \lambda_n$, $\lambda_i = 1$ for $i = 1, \ldots, s$ and $|\lambda_i| < 1$ for $i = s + 1, \ldots, n$. Suppose that the eigenspace $M(1)$ has dimension s. Show that $B = \lim_{k \to \infty} A^k$ exists, and find B.

8. Suppose A has eigenvalues $\lambda_1, \ldots, \lambda_n$, $\lambda_i = 1$ for $i = 1, \ldots, s$ and $|\lambda_i| < 1$ for $i = s + 1, \ldots, n$. Suppose that the eigenspace $M(1)$ has dimension $r < s$. Show that $\lim_{k \to \infty} A^k$ does not exist.

9. (a) Suppose A has an eigenvalue λ_1 such that $|\lambda_1| > 1$. Show that A^k has no limit as $k \to \infty$.

   (b) Suppose A has an eigenvalue λ_1 such that $\lambda_1 \neq 1$, $|\lambda_1| = 1$. Show that A^k has no limit as $k \to \infty$.

10. **Cayley–Hamilton theorem.** Let A be an $n \times n$ matrix and $p(\lambda)$ the characteristic polynomial of A. Then $p(A) = 0$. (*Hints:* Show that $p(A) = Up(T)U^H$ if $A = UTU^H$ is the factorization of A in Schur's theorem. Write $p(\lambda) = (-1)^n(\lambda - \lambda_1)(\lambda - \lambda_2) \cdots (\lambda - \lambda_n)$. Examine what happens in the product $(T - \lambda_1 I)(T - \lambda_2 I)$. It will become clear what happens in $p(T)$.)

8.4
JORDAN CANONICAL FORM

The Jordan canonical form of an operator or matrix plays a very important role in understanding the structure of matrices and linear operators. An example of how this knowledge can be applied to the study of differential equations is given. Unfortunately, the Jordan canonical form is almost impossible to obtain using numerical methods for most matrices, since even minuscule changes in the entries of a matrix can produce a completely different Jordan form. However, we will indicate a simple way to construct the Jordan form that works at least theoretically.

First the theorem is stated, some consequences deduced, and the method of construction given. Then an application of the theorem to linear systems of differential equations is described.

A *Jordan block* is an $m \times m$ matrix of the form

$$J_1(\lambda) = [\lambda] \qquad \text{if } m = 1$$

$$J_m(\lambda) = \begin{bmatrix} \lambda & 1 & & & \\ & \lambda & 1 & & \\ & & \ddots & & \\ & & & \lambda & 1 \\ & & & & \lambda \end{bmatrix} \qquad \text{if } m > 1,$$

where the unspecified entries are zero.

THEOREM 1. Let $T : V \to V$ be a linear operator on the complex n-dimensional vector space V. Then there is a basis $\beta = \{e_k\}_{k=1}^n$ for V such that

$$[T]_\beta = \begin{bmatrix} J_1 & & & 0 \\ & J_2 & & \\ & & \ddots & \\ 0 & & & J_s \end{bmatrix} \equiv J, \quad \text{where } J_i = J_{m_i}(\lambda_i), i = 1, 2, \ldots, s. \tag{1}$$

When $T(x) = Ax$, there exists a nonsingular matrix P such that

$$P^{-1}AP = J. \tag{1'}$$

Each λ_k is an eigenvalue of T, and $m_1 + m_2 + \cdots + m_s = n$. The λ_k exhibited need not be distinct, and the m_i need not be distinct. s is the number of linearly independent eigenvectors of A. Let λ be a fixed eigenvalue of T. Then the number of Jordan blocks with $\lambda_i = \lambda$ is the geometric multiplicity of λ, and the sum of the sizes of these blocks is the algebraic multiplicity. P may be taken to be real if A and all λ_i are real. The *Jordan form J* is unique up to permutations of the blocks.

For example, J may look like (omitting 0s)

$$\begin{bmatrix} 1 & & & & & & & \\ & 2 & & & & & & \\ & & 1 & 1 & & & & \\ & & & 1 & & & & \\ & & & & 2 & 1 & & \\ & & & & & 2 & 1 & \\ & & & & & & 2 & 1 \\ & & & & & & & 2 \end{bmatrix}$$

Some of the consequences of the theorem will be examined next. The leading columns of the blocks J_k are given by: $j_1 = 1$ and $j_k = m_1 + \cdots + m_{k-1} + 1$. From the definition of $[T]_\beta$, (1) is equivalent to

$$T(e_{j_k}) = \lambda_{j_k} e_{j_k}, \quad T(e_j) = \lambda_{j_k} e_j + e_{j-1} \quad \text{for } j = j_k + 1, \ldots, j_k + m_k - 1. \quad (2)$$

The sequence $\{e_{j_k}, e_{j_k+1}, \ldots, e_{j_k+m_k-1}\}$ is called a *Jordan chain* of length m_k associated with the eigenvalue λ_{j_k}, e_{j_k} is called the *leading vector* of the chain, and $e_{j_k+m_k-1}$ is called the *end vector* of the chain. If $m_k = 1$, the lead and end vectors are the same. If $T(x) = Ax$, then $P = [e_1 \; e_2 \; \cdots \; e_n]$, and (2) can also be seen from equating columns in $AJ = JP$. Thus if (2) holds for the basis $\{e_k\}_{k=1}^n$, then (1) is valid. Such a basis is called a *Jordan basis*.

Remark. Each leading vector e_{j_k} is an eigenvector of T with eigenvalue λ_{j_k}. If λ is an eigenvalue of T, then the eigenspace $M(\lambda)$ is given by $M(\lambda) = \text{span}\{e_{j_k} : \lambda = \lambda_{j_k}\}$. The reasoning behind this is fleshed out in the exercises. (See Exercises 5, 6, and 7.)

COROLLARY 2. If A is diagonalizable with eigenvalues λ_i, then $J = \text{diag}(\lambda_1, \lambda_2, \ldots, \lambda_n)$ and $P = [e_1 \; e_2 \; \cdots \; e_n]$ where $Ae_i = \lambda_i e_i$.

Proof of Theorem 1. The proof proceeds by induction on the dimension of V. The theorem is easily seen to be true when $n = 1$. Suppose it is true for $\dim(V) < n$. Theorem 5 of Section 4.1 asserts that T has at least one complex eigenvalue. Let λ be an eigenvalue of T and define $T_\lambda = T - \lambda I$ so that $N(T_\lambda)$ is an eigenspace of T. Let $W = R(T_\lambda)$ and $r = \dim(W)$. By the rank and nullity theorem, $r < n$ and $\dim(N(T_\lambda)) = n - r$. If $r = 0$, the proof is finished. Observe that W is an invariant subspace of $V : T_\lambda(W) \subset W$. Let S be T_λ restricted to W, so that $S : W \to W$. By the induction hypothesis, S has a Jordan basis $\beta = \{e_1, \ldots, e_r\}$ in W. Observe that each $0 \neq x \in N(S)$ is in W and is an eigenvector of S corresponding to the eigenvalue 0. Let $p = \dim(N(S))$. By the preceding remark, S has p Jordan chains whose leading vectors are eigenvectors of S. Let $\{f_1, \ldots, f_p\} \subset \beta$ be the end vectors of these chains. Since $f_i \in W$, there are vectors g_i so that $T_\lambda(g_i) = f_i$. Let $q = n - s - p$. Select vectors $z_1, \ldots, z_q$ in $N(T_\lambda)$ that are not in W and that are independent of $e_1, \ldots, e_r$ using the method of Lemma 5 of Section 1.4. This yields a linearly independent set $\{e_1, \ldots, e_r, z_1, \ldots, z_q\}$. Now consider the set of n vectors $\gamma = \{e_1, \ldots, e_r, z_1, \ldots, z_q, g_1, \ldots, g_p\}$. We first show that γ is linearly independent. Suppose $\sum_{i=1}^r a_i e_i + \sum_{i=1}^q b_i z_i + \sum_{i=1}^p c_i g_i = 0$. Then $0 = T_\lambda(0) = \sum_{i=1}^r a_i T_\lambda(e_i) + \sum_{i=1}^q b_i T_\lambda(z_i) + \sum_{i=1}^p c_i T_\lambda(g_i) = \sum_{i=1}^r a_i T_\lambda(e_i) + \sum_{i=1}^p c_i f_i$. $T_\lambda(e_i)$ is either 0 or in a Jordan chain, but not the end member of a Jordan chain. Thus the vectors $T_\lambda(e_i)$ do not overlap with the vectors f_j, whence all $c_i = 0$. This means that $\sum_{i=1}^r a_i e_i + \sum_{i=1}^q b_i z_i = 0$. By construction, $\{e_1, \ldots, e_r, z_1, \ldots, z_q\}$ is linearly independent, so that all $a_i = b_i = 0$. To finish the proof, verify that γ is a Jordan basis for V. To do this, observe that if $\gamma = \{h_k\}_{k=1}^n$, then either $T(h_i) = \mu h_i$ or $T(h_i) = \mu h_i + h_{i-1}$ for some μ and each i.

To formulate a method of construction of a Jordan basis, let e_1 be the leading vector of a Jordan chain of length m associated with the eigenvalue λ. From (2), we have

$$(T - \lambda I)e_1 = 0, \quad (T - \lambda I)e_2 = e_1, \ldots, \quad (T - \lambda I)e_m = e_{m-1}. \quad (3)$$

The first equality simply says e_1 is an eigenvector of T corresponding to the eigenvalue λ. Once e_1 is known, solve $(T - \lambda I)e_2 = e_1$ for e_2. Continue the process, finally obtaining e_m. Trying another step will yield no solution. Since $T - \lambda I$ is singular, the chain will not be unique if $m > 1$.

Thus, a simple recipe to find J is to first find as many linearly independent eigenvectors (say k) of T as possible. For each of these, construct its Jordan chain. If $m_1 + \cdots + m_k = n$, the process is finished. Two examples are given next when $T(x) = Ax$ and so $P^{-1}AP = J$, where $P = [e_1\, e_2\, \cdots\, e_n]$. The second shows that this simple idea for constructing P may run into some difficulties. A more definitive approach consists of looking at the null spaces of $(T - \lambda_i I)^k$, $k = 1, 2, \ldots$, but the rule is moderately complex.

EXAMPLE 1. JORDAN BLOCKS WITH DISTINCT λ_i. Find the Jordan form of

$$A = \begin{bmatrix} 1 & 1 & 2 \\ 0 & 1 & 2 \\ 0 & 0 & 3 \end{bmatrix}.$$

Computation yields $|A - \lambda I| = -(\lambda - 1)^2(\lambda - 3)$. For $\lambda = 1$, consider

$$A - I = \begin{bmatrix} 0 & 1 & 2 \\ 0 & 0 & 2 \\ 0 & 0 & 2 \end{bmatrix}.$$

Solving $(A - I)x = 0$ yields (to a scalar multiple) $x = e_1 = (1, 0, 0)^T$. Solving $(A - I)x = e_1$ gives $e_2 = x = (0, 1, 0)^T$. The eigenspace for $\lambda = 3$ must be one-dimensional, and solving $(A - 3I)x = 0$ yields $x = e_3 = (3, 2, 2)^T$. Thus

$$P = \begin{bmatrix} 1 & 0 & 3 \\ 0 & 1 & 2 \\ 0 & 0 & 2 \end{bmatrix} \quad \text{and} \quad J = \begin{bmatrix} 1 & 1 & 0 \\ 0 & 1 & 0 \\ 0 & 0 & 3 \end{bmatrix}.$$

$P^{-1}AP = J$ is verified by computation.

EXAMPLE 2. JORDAN BLOCKS WITH REPEATED λ_i. Find the Jordan form of

$$A = \begin{bmatrix} 2 & 0 & 1 \\ -1 & 1 & -1 \\ -1 & 0 & 0 \end{bmatrix}.$$

Here $|A - \lambda I| = (1 - \lambda)^3$, and

$$A - I = \begin{bmatrix} 1 & 0 & 1 \\ -1 & 0 & -1 \\ -1 & 0 & -1 \end{bmatrix}.$$

Solving $(A - I)x = 0$ yields the two independent eigenvectors $f_1 = (1, 0, -1)^T$ and $f_2 = (0, 1, 0)^T$. Each linear combination of f_1 and f_2 is in the same eigenspace. Lamentably, the Jordan chains starting with f_1 and f_2 are only of length 1 since the equations $(A - I)x = f_i$, $(i = 1, 2)$ are inconsistent. However, $(A - I)x = f_1 + cf_2$ will have a solution if $c = -1$, e.g., $x = (1, a, 0)^T$, where a is arbitrary. Taking $e_1 = f_1 - f_2 = (1, -1, -1)^T$, $e_2 = (1, 0, 0)^T$ gives a length 2 chain. Let $e_3 = f_2$. This yields a Jordan basis with

$$P = \begin{bmatrix} 1 & 1 & 0 \\ -1 & 0 & 1 \\ -1 & 0 & 0 \end{bmatrix}$$

so that J must be

$$\begin{bmatrix} 1 & 1 & 0 \\ 0 & 1 & 0 \\ 0 & 0 & 1 \end{bmatrix}.$$

That $J = P^{-1}AP$ is verified by computation.

Discontinuity of the Jordan form

As an example to show why there are difficulties with numerical determination, consider

$$A = \begin{bmatrix} 1 & \varepsilon \\ 0 & 1 \end{bmatrix}.$$

$\lambda = 1$ is the only eigenvalue, and A is not diagonalizable, so the Jordan canonical form must be (if $\varepsilon \neq 0$)

$$J = \begin{bmatrix} 1 & 1 \\ 0 & 1 \end{bmatrix}.$$

P can be found by setting $P = [P_1 \, P_2]$ and solving $AP = PJ$, i.e. (equating columns), $AP_1 = P_1$, $AP_2 = P_1 + P_2$. One solution is

$$P = \begin{bmatrix} 1 & 1 \\ 0 & \varepsilon^{-1} \end{bmatrix},$$

and $P^{-1}AP = J$ verifies the assertion. If $\varepsilon = 0$, then A is diagonalizable, and the Jordan canonical form is $J = I$. What this means is that the Jordan form is *not* a continuous function of the initial matrix. This 2×2 case is not an accident. For $n \times n$ matrices, Theorem 1 of Section 8.3 asserts that for every matrix A, there are diagonalizable matrices arbitrarily close to A. Since the Jordan form of a diagonalizable matrix has only 1×1 blocks (and so no 1s above the diagonal), the Jordan form, as a function of A, is extremely discontinuous on the set of all $n \times n$ matrices. This is worse than instability in calculations; the problem is seriously ill-posed, and it would be difficult to trust numerical results without considerable numerical evidence that the form attained is correct. In a few cases, this is possible.

e^{At} and Application to Systems of Differential Equations

Consider the system

$$x' = Ax, \quad ' = d/dt \tag{4}$$

with initial conditions

$$x(0) = x_0, \tag{5}$$

where A is an $n \times n$ matrix. In Section 4.4, it was shown that, when A is diagonalizable, the solution is

$$x = e^{At}x_0, \tag{6}$$

where $e^{At} = P \cdot \text{diag}\,(e^{\lambda_i t}) \cdot P^{-1}$, and P is a matrix whose columns are an eigenbasis for A with corresponding eigenvalues λ_i. It will be shown next that Eq. (6) is true for nondiagonalizable A, provided e^{At} is suitably defined.

DEFINITION. For a square matrix A,

$$e^A = \sum_{k=0}^{\infty} \frac{A^k}{k!}, \quad \text{where } A^0 = I. \tag{7}$$

Equation (7) means that $S_n \equiv \sum_{k=0}^{n} A^k/k! \to e^A$ (entrywise). According to Proposition 5 of Section 8.3, this is equivalent to $\|\|S_n - e^A\|\|_1 \to 0$. From the properties of the matrix norm, $\|\|\sum_{k=0}^{n} A^k/k!\|\|_1 \le \sum_{k=0}^{n} \|\|A\|\|_1^k/k! \le e^{\|\|A\|\|_1} < \infty$, and by an argument entirely similar to that used for ordinary power series, this implies that the series in (7) converges for every A.

PROPERTIES OF e^A AND e^{At}

(a) $e^0 = I$
(b) $e^A e^B = e^{A+B}$ provided $AB = BA$
(c) $(e^A)^{-1} = e^{-A}$
(d) $\dfrac{d}{dt} e^{At} = A e^{At} = e^{At} A$

Outline of the proof. Property (a) is clear, and (b) is shown in just the same way as it is proved for complex numbers, i.e., by expanding the power series on the right by the binomial theorem and showing that it is equal to the product of the two power series on the left. Convergence considerations are accounted for by using the matrix norm $\|\| \ \|\|_1$. The whole discussion is virtually a word-for-word copy of the proof for complex numbers, with the exception that not all matrices commute. Property (c) comes from (a) and (b), with $B = -A$. Property (d) is obtained from differentiating the power series (7) termwise:

$$\frac{d}{dt} e^{At} = \sum_{k=1}^{\infty} k t^{k-1} \frac{A^k}{k!} = A \sum_{k=1}^{\infty} \frac{(tA)^{k-1}}{(k-1)!} = A e^{At}.$$

Just as for property (b), the proof that this is possible carries over literally from complex (or real) analysis.

The next theorem follows immediately from the properties of e^{At}.

THEOREM 3. $x = e^{At}x_0$ satisfies $x' = Ax$, $x(0) = x_0$.

Proof

$$\frac{d}{dt} x = \left(\frac{d}{dt} e^{At} \right) x_0 = A e^{At} x_0 = Ax; \quad x(0) = e^{0t} x_0 = x_0.$$

The following propositions show that the series definition agrees with the definition for diagonalizable matrices. The proofs are left as exercises. (See Exercises 9 and 10.)

PROPOSITION 4. If $\Lambda = \mathrm{diag}(\lambda_1, \ldots, \lambda_n)$, then

$$\sum_{k=0}^{\infty} \frac{(\Lambda t)^k}{k!} = \mathrm{diag}(e^{\lambda_1}, \ldots, e^{\lambda_n}) = e^{\Lambda t}.$$

PROPOSITION 5. Suppose $AE_i = \lambda_i E_i$, $P = [E_1 \, E_2 \, \cdots \, E_n]$, and P is nonsingular. Then

$$\sum_{k=0}^{\infty} \frac{(At)^k}{k!} = Pe^{\Lambda t}P^{-1}.$$

Remark. Equation (7) is usually not a good way to compute e^A on a computer. Computing A^k can quickly become expensive if A is large. If A is badly scaled, there are problems with overflow or underflow in A^k as well as with accuracy and the expense of many matrix multiplications. See Golub and Van Loan (1989).

Use of the Jordan Canonical Form in the Study of $x' = Ax$

In Section 4.4, the method of solving $x' = Ax$ was to make a change of basis/coordinates $Py = x$ so that the new matrix was diagonal. Now a change of basis will be made so that the new matrix is the Jordan canonical form. If P is the matrix in Theorem 1 and $x = Py$, then

$$y' = P^{-1}x' = P^{-1}Ax = P^{-1}APy = Jy. \tag{8}$$

The general solution is $y = e^{Jt}c$, for some column c. To compute e^{Jt}, note that

$$J^n = \begin{bmatrix} J_1^n & & & 0 \\ & J_2^n & & \\ & & \ddots & \\ 0 & & & J_s^n \end{bmatrix}.$$

Thus

$$e^{Jt} = \sum_{k=0}^{\infty} \frac{(Jt)^k}{k!} = \begin{bmatrix} \sum (J_1 t)^n/n! & & & 0 \\ & \sum (J_2 t)^n/n! & & \\ & & \ddots & \\ 0 & & & \sum (J_s t)^n/n! \end{bmatrix}$$

$$= \begin{bmatrix} e^{J_1 t} & & & 0 \\ & e^{J_2 t} & & \\ & & \ddots & \\ 0 & & & e^{J_s t} \end{bmatrix}. \tag{9}$$

Next, e^{Jt} will be computed, where $J = J_i = J_{m_i}(\lambda_i)$. Let $m = m_i$. If $m = 1$, then $e^{Jt} = e^{\lambda t}$. Now suppose $m > 1$. Then $J = \lambda I + N$, where $N = J_m(0)$. N has a line of 1s just above the main diagonal. Computing N^2, N^3, etc., one sees that the line of 1s above the main diagonal moves toward the upper right corner of these matrices one step at a time, and finally that $N^m = 0$. Further, λI and N commute, so

$$e^{Jt} = e^{\lambda It + Nt} = e^{\lambda It} e^{Nt} = \text{diag}(e^{\lambda t}, e^{\lambda t}, \ldots, e^{\lambda t}) \sum_{k=0}^{m-1} (Nt)^k / k!$$

$$= e^{\lambda t} \begin{bmatrix} 1 & t & t^2/2 & \cdots & \cdots & t^{m-1}/(m-1)! \\ 0 & 1 & t & \cdots & \cdots & t^{m-2}/(m-2)! \\ & & & \cdots & \cdots & \\ & 0 & & & 1 & t \\ & & & & 0 & 1 \end{bmatrix} \qquad (10)$$

For example, if $J = J_4(\lambda)$, then

$$N = \begin{bmatrix} 0 & 1 & 0 & 0 \\ 0 & 0 & 1 & 0 \\ 0 & 0 & 0 & 1 \\ 0 & 0 & 0 & 0 \end{bmatrix}, \qquad N^2 = \begin{bmatrix} 0 & 0 & 1 & 0 \\ 0 & 0 & 0 & 1 \\ 0 & 0 & 0 & 0 \\ 0 & 0 & 0 & 0 \end{bmatrix},$$

$$N^3 = \begin{bmatrix} 0 & 0 & 0 & 1 \\ 0 & 0 & 0 & 0 \\ 0 & 0 & 0 & 0 \\ 0 & 0 & 0 & 0 \end{bmatrix}, \qquad N^4 = 0,$$

and

$$e^{Jt} = e^{\lambda t}[I + tN + t^2 N^2/2 + t^3 N^3/6] = e^{\lambda t} \begin{bmatrix} 1 & t & t^2/2 & t^3/6 \\ 0 & 1 & t & t^2/2 \\ 0 & 0 & 1 & t \\ 0 & 0 & 0 & 1 \end{bmatrix}.$$

From the preceding computations, we can deduce the following important theorem from differential equations.

THEOREM 6. If the real part of every eigenvalue of A is negative, then every solution of $x' = Ax$ tends to 0 as t tends to ∞.

Proof. In fact, $x = P^{-1}y$, where Eqs. (6), (7), and (8) describe y. Let $\lambda = -\alpha + i\beta$ be an eigenvalue. For each polynomial $p(t)$, $e^{-\alpha t} p(t) \to 0$ as $t \to \infty$, and therefore $e^{\lambda t} p(t) \to 0$. Hence every entry of $e^{Jt} \to 0$. From this, $e^{At} \to 0$, and therefore so does $x = P^{-1}y$.

Exercises 8.4

1. Find the Jordan form J of the following matrices A and a matrix P so that $P^{-1}AP = J$.

(a) $\begin{bmatrix} 0 & 0 \\ 1 & 0 \end{bmatrix}$ (b) $\begin{bmatrix} 1 & 2 \\ 0 & 1 \end{bmatrix}$ (c) $\begin{bmatrix} 3 & -1 & 2 \\ -1 & 2 & -1 \\ -1 & 1 & 0 \end{bmatrix}$ (d) $\begin{bmatrix} 2 & 1 & 0 \\ -1 & 0 & 0 \\ -1 & -1 & 1 \end{bmatrix}$

(e) $\begin{bmatrix} 2 & 0 & -1 \\ 2 & 1 & -1 \\ 1 & 0 & 0 \end{bmatrix}$ (f) $\begin{bmatrix} 0 & 1 & 0 \\ 0 & 0 & 1 \\ 1 & 1 & -1 \end{bmatrix}$

2. Find the Jordan form of the following block matrices $A = \begin{bmatrix} A_1 & 0 \\ 0 & A_2 \end{bmatrix}$.

(a) $A_1 = [3]$, $A_2 = \begin{bmatrix} 2 & 1 \\ -1 & 0 \end{bmatrix}$ (b) $A_1 = \begin{bmatrix} 2 & 1 \\ -4 & -2 \end{bmatrix}$, $A_2 = \begin{bmatrix} 0 & 1 \\ -1 & 0 \end{bmatrix}$

(c) $A_1 = \begin{bmatrix} 5 & 4 \\ -1 & 1 \end{bmatrix}$, $A_2 = \begin{bmatrix} 3 & -1 \\ 1 & 1 \end{bmatrix}$ (d) $A_1 = \begin{bmatrix} 3 & -4 \\ 1 & -1 \end{bmatrix}$, $A_2 = \begin{bmatrix} 2 & -2 & 3 \\ -1 & 1 & -1 \\ -1 & 2 & -2 \end{bmatrix}$

3. Characterize all defective 2×2 matrices $\begin{bmatrix} a & b \\ c & d \end{bmatrix}$.

4. Let $A = uv^H$, where u and v are nonzero vectors in $\mathbf{C}^n$. Find the Jordan form of A.

5. Suppose $e_1, e_2, \ldots, e_m$ is a (nontrivial) Jordan chain corresponding to the eigenvalue λ. Let $W = \text{span}\{e_1, e_2, \ldots, e_m\}$. Show
   (a) W is an invariant subspace: $T(W) \subset W$.
   (b) $\{e_1, e_2, \ldots, e_m\}$ is linearly independent.
   (c) If $T(e) = \mu e$, $e \neq 0$ and $e \in W$, then $\mu = \lambda$ and $e = ce_1$ for some c.
   (d) $T^m(x) = 0$ for all $x \in W$.

In Exercises 6 and 7, let $\{e_1, e_2, \ldots, e_n\}$ be a Jordan basis for T and j_k the first column of the kth Jordan block. m_k is the length of the kth chain and s is the number of Jordan chains (i.e., Jordan blocks).

6. Let W_k be the span of the Jordan chain with leading vector e_{j_k}. Show that
   (a) $V = W_1 \oplus W_2 \oplus \cdots \oplus W_s$, i.e., each $x \in V$ has the form $x = \sum_{i=1}^{s} w_i$, $w_i \in W_i$, and $w_1, \ldots, w_2$ are unique.
   (b) $T(W_i) \subset W_i$ for each i, i.e., each W_i is an invariant subspace of T.

7. Let λ be an eigenvalue of T and e a corresponding eigenvector. Show that e is a linear combination of those $e_{j_1}, e_{j_2}, \ldots, e_{j_s}$ for which $\lambda = \lambda_{j_k}$.

8. Show that $\{e_{j_1}, e_{j_2}, \ldots, e_{j_s}\}$ is a maximal linearly independent set of eigenvectors.

9. Prove Proposition 4.

10. Prove Proposition 5.

11. Find e^{Jt} for each of the following matrices J.

(a) $\begin{bmatrix} 2 & 1 \\ 0 & 2 \end{bmatrix}$ (b) $\begin{bmatrix} 0 & 1 \\ 0 & 0 \end{bmatrix}$ (c) $\begin{bmatrix} 2 & 0 & 0 \\ 0 & -1 & 1 \\ 0 & 0 & -1 \end{bmatrix}$ (d) $\begin{bmatrix} 2 & 1 & 0 & 0 & 0 \\ 0 & 2 & 0 & 0 & 0 \\ 0 & 0 & 3 & 1 & 0 \\ 0 & 0 & 0 & 3 & 1 \\ 0 & 0 & 0 & 0 & 3 \end{bmatrix}$

12. Solve $x' = Ax$ for the following A.

(a) $\begin{bmatrix} 1 & 1 \\ 0 & 1 \end{bmatrix}$ (b) $\begin{bmatrix} 1 & 1 & 0 \\ 0 & 1 & 1 \\ 0 & 0 & 2 \end{bmatrix}$

13. Suppose that $AB = BA$. Show that $e^{A+B} = e^A e^B$.

14. Show that the sequence A^n is bounded, provided that all eigenvalues of A obey the following: $|\lambda_i| \leq 1$, and if some $|\lambda_j| = 1$, then all Jordan blocks associated with λ_j are 1×1. Is the converse true?

15. Let $\tilde{\lambda}$ be an eigenvalue of A. Show that all Jordan blocks associated with $\tilde{\lambda}$ are 1×1 iff the algebraic multiplicity of $\tilde{\lambda}$ equals the geometric multiplicity of $\tilde{\lambda}$.

APPENDIX

Answers
to Selected Problems

CHAPTER 0

Section 0.1

1. (a) $A = \begin{bmatrix} 2 & 1 \\ 1 & -3 \end{bmatrix}$, $b = \begin{bmatrix} 2 \\ 0 \end{bmatrix}$

(c) $A = \begin{bmatrix} 1 & 2 & 0 & 0 \\ 2 & 1 & 3 & 0 \\ 0 & 2 & 1 & 3 \\ 0 & 0 & 2 & 1 \end{bmatrix}$, $b = \begin{bmatrix} 0 \\ 1 \\ 2 \\ -1 \end{bmatrix}$

(d) Each row of A is $[1 \ 4 \ 9 \ 16 \ 25 \ 36]$, $b = (1, 1, 1, 1, 1)^{\mathrm{T}}$

Section 0.2

$A \sim B$ means A row-reduces to B.

1. (a) $A \sim \begin{bmatrix} 1 & 0 & 1 \\ 0 & 1 & -2 \\ 0 & 0 & 0 \end{bmatrix}$, $x_1 = -x_3$, $x_2 = 2x_3 \Rightarrow x = (-x_3, 2x_3, x_3)^{\mathrm{T}}$

(c) $A \sim \begin{bmatrix} 1 & 0 & 0 \\ 0 & 1 & 0 \\ 0 & 0 & 1 \\ 0 & 0 & 0 \end{bmatrix}$, $x = 0 = (0, 0, 0)^{\mathrm{T}}$

(f) $A \sim \begin{bmatrix} 1 & -2 & 0 & 1 \\ 0 & 0 & 1 & 2 \\ 0 & 0 & 0 & 0 \end{bmatrix}$, $x_1 = 2x_2 - x_4$, $x_3 = -2x_4 \Rightarrow x = (2x_2 - x_4, x_2, -2x_4, x_4)^{\mathrm{T}}$

Section 0.3

1. (a) (i) $x = (-1 - x_3, 3 + 2x_3, x_3)^T$; x_3 arbitrary
   (ii) inconsistent
   (c) (i) $x_1 = \frac{8}{3} - \frac{1}{3}x_3 - x_4$, $x_2 = -\frac{1}{3} + \frac{2}{3}x_3$; x_3, x_4 arbitrary
   (ii) inconsistent

3. $[A \mid b]$ row-reduces to a matrix with m nonzero rows for each b. By Theorem 2, $Ax = b$ is therefore consistent for each b.

5. By Theorem 2, there must be free variables; i.e., $r < n$. Thus when the system is consistent, there must be infinitely many solutions.

Section 0.4

1. (a) undefined   (b) $\begin{bmatrix} 12 & 5 & 13 \\ 0 & 1 & 5 \end{bmatrix}$ (c) $\begin{bmatrix} -2 & 2 & 0 \\ 10 & 7 & 7 \end{bmatrix}$ (d) undefined

   (e) undefined   (f) $\begin{bmatrix} -2 & 7 \\ -2 & 1 \\ -8 & -2 \end{bmatrix}$ (g) $\begin{bmatrix} 11 \\ 12 \\ 7 \end{bmatrix}$ (h) $\begin{bmatrix} 4 \\ 3 \\ 2 \end{bmatrix}$

   (i) undefined   (j) [9]   (k) $\begin{bmatrix} 2 & 4 & -2 \\ 4 & 8 & -4 \\ 1 & 2 & -1 \end{bmatrix}$ (l) $[-4 \ -3 \ -11]$

3. *Note:* There are several ways to prove each of these.
   (a) $a_i^T = (A^T)_i = A^T\delta_i$.
   (b) Row i of $AB = \delta_i^T AB = a_iB$, by part (a).
   (c) $(AB)_{ik} = \sum_{j=1}^{n} a_{ij}b_{jk}$, and $\sum_{j=1}^{n}(A_jb_j)_{ik} = \sum_{j=1}^{n}(A_jb_j)_{ik} = \sum_{j=1}^{n} a_{ij}b_{jk}$

7. $AE = [AE_1 \ AE_2 \cdots AE_n]$ and $ED = [d_1E_1 \ d_2E_2 \cdots d_nE_n]$ by (2) and (3). Equate corresponding columns.

9. Take $y = \delta_i$, $x = \delta_j$ to get $a_{ij} = b_{ij}$.

11. $PDQ^T = [d_1P_1 \ d_2P_2 \cdots d_nP_n]Q^T = \sum_{i=1}^{n} d_iP_iQ_i^T$

Section 0.5

1. (a) $A^{-1} = \begin{bmatrix} -1 & 2 \\ 1 & -1 \end{bmatrix}$ (b) $A^{-1} = \begin{bmatrix} 0 & -1 & 1 \\ -1 & 1 & 0 \\ 1 & 0 & 0 \end{bmatrix}$

   (c) $A^{-1} = \frac{1}{2}\begin{bmatrix} 1 & -2 & 3 \\ -1 & 2 & -1 \\ 1 & 0 & -1 \end{bmatrix}$ (d) $A^{-1} = \begin{bmatrix} 1 & 0 & -2 & 3 \\ 0 & 1 & -1 & 1 \\ 0 & 0 & 1 & -3 \\ 0 & 0 & 0 & 1 \end{bmatrix}$

3. (a) $E^2 = I$, $E^{-1} = E = E^T$
   (b) Interchange columns i and j of A.
   (c) (i) $A = BE^{-1} = BE$, and invert this.

   (ii) Otherwise, $A^{-1} = E^{-1}B^{-1} = EB^{-1} = \begin{bmatrix} 1 & 2 & 1 \\ 0 & 2 & 1 \\ 0 & 1 & 1 \end{bmatrix}$.

5. *Hint:* Use Proposition 1 and (3) of Section 0.4.

7. $P^{\mathrm{T}}GP = I \Rightarrow P^{-1} = P^{\mathrm{T}}G \Rightarrow PP^{-1} = PP^{\mathrm{T}}G = I$

$\quad P^{\mathrm{T}}GP = I \Rightarrow (P^{\mathrm{T}})^{-1} = GP \Rightarrow (P^{\mathrm{T}})^{-1}P^{\mathrm{T}} = GPP^{\mathrm{T}} = I$

11. *Hint:* Show that C is nonsingular and then that B and A are nonsingular.

16. *Hint:* Multiply $Ax = y$ on the left with L.

CHAPTER 1

Section 1.1

No solutions are given for this section.

Section 1.2

1. (*a*) yes, subspace of $\mathbf{R}^3$

  (*c*) yes, subspace of $\mathbf{R}^3$

  (*e*) yes, subspace of $\mathbf{C}^n$

  (*g*) no, $0 \notin M$

  (*i*) yes, subspace of $\mathscr{F}(\mathbf{R})$

  (*k*) yes, subspace of $C(\mathbf{R})$

  (*m*) yes, column space of $A = \begin{bmatrix} 2 & 3 \\ 1 & -5 \end{bmatrix}$.

4. (*a*) true

  (*c*) true

  (*d*) false

  (*f*) False: Every linear combination of $\{\sin(kt)\}_{k=1}^n$ is bounded.

7. *Hint:* Show that every t^j is a linear combination of $\{(t - a)^k\}_{k=0}^n$ using the binomial theorem.

Section 1.3

1. (*a*) linearly dependent

  (*b*) linearly independent

  (*c*) linearly dependent

2. (*b*) linearly independent

  (*c*) linearly dependent

  (*d*) linearly independent

  (*g*) linearly independent

7. *Hint:* Use Euler's identities and Proposition 5.

15. *Hint:* Can $Bx = 0$ with $x \neq 0$?

Section 1.4

1. (*a*) $\{1, (t - 1)^2, (t - 1)^3\}$, $\dim(M) = 3$

  (*c*) $\{t, t^3\}$, $\dim(M) = 2$

2. (a) $\{(2, 0, 1, 0)^T, (0, 2, -1, 0)^T, (0, 0, 0, 1)^T\}$, $\dim(M) = 3$

3. (a) $\{(3, 0, 1, 2)^T, (0, 3, -2, -1)^T\}$, $\dim(M) = 2$

(b) $\{(1, 0, 0, \frac{2}{5})^T, (0, 1, 0, \frac{3}{5})^T, (0, 0, 1, \frac{1}{5})^T\}$, $\dim(M) = 3$

4. (a) $\{(1, -2, 1)^T\}$, $\text{Dim}(M) = 1$

(b) $\dim(M) = 0$; M has no basis.

Section 1.5

1. The vectors are linearly indendent. The coordinate column is $(0, 5, -3)^T$.

3. $P = \begin{bmatrix} 1 & 0 & 1 \\ 0 & 1 & -2 \\ 1 & -1 & 1 \end{bmatrix}$, $Q = \frac{1}{2}\begin{bmatrix} 1 & 1 & 1 \\ 2 & 0 & -2 \\ 1 & -1 & -1 \end{bmatrix}$, $1 + t = f_0 + f_1$

5. $\cos^3(x) = \frac{1}{4}[e^{ix} + e^{-ix}] = \frac{1}{8}[e^{-3ix} + 3e^{-ix} + 3e^{ix} + e^{3ix}] = \frac{1}{4}[3\cos(x) + \cos(3x)]$

Hence (a) $[\cos^3(x)] = \frac{1}{4}(0, 3, 0, 0, 0, 1, 0)^T$, (b) $[\cos^3(x)] = \frac{1}{8}(1, 0, 3, 0, 3, 0, 1)^T$

7. (a) Use Proposition 5.

(b) $R = P = \begin{bmatrix} 0 & 0 & 0 & 1 \\ 0 & 0 & 1 & -3 \\ 0 & 1 & -2 & 3 \\ 1 & -1 & 1 & -1 \end{bmatrix}$, $S = P^{-1} = \begin{bmatrix} 1 & 1 & 1 & 1 \\ 3 & 2 & 1 & 0 \\ 3 & 1 & 0 & 0 \\ 1 & 0 & 0 & 0 \end{bmatrix}$

(c) $[1]_\gamma = (1, 3, 3, 1)^T$, $[1 + x + x^2]_\gamma = (3, 6, 4, 1)^T$

CHAPTER 2

Section 2.1

1. Not one to one since $D(1) = 0$. Onto: $g(x) = \int_0^x f(t)\, dt$ solves $D(g) = f$

3. Not one to one and not onto

6. Not one to one and not onto

Section 2.2

2. (i) (a) $\{(3, 2, -1)^T\}$, $n(T) = 1$, $r(T) = 2$

(b) T is neither one to one nor onto.

3. (i) (a) $\{(1, 1, 0, 1)^T, (1, 0, 1, -1)^T\}$, $N(T) = R(T) = 2$

(b) Not one to one, not onto.

(iii) (a) The columns of A are a basis for $R(T)$. $r(T) = 3$, $n(T) = 0$

(b) T is one to one and onto.

5. $a \notin \{1, 2, \ldots, n\}$.

13. $r(T) = r(F)$ for both (a) and (b).

Section 2.3

1. (i) (a) $A = \begin{bmatrix} 1 & 1 & 2 \\ 1 & -1 & 1 \\ 1 & 5 & 4 \end{bmatrix}$ (b) $B = \begin{bmatrix} -1 & 0 & -4 \\ 1 & 4 & 8 \\ 0 & 1 & 1 \end{bmatrix}$

3. (a) $A = \begin{bmatrix} 1 & 1 & 2 \\ 1 & -1 & 2 \\ 0 & 0 & 1 \end{bmatrix}$ (b) $B = \begin{bmatrix} 0 & 2 & 2 \\ 1 & 0 & 1 \\ 0 & 0 & 1 \end{bmatrix}$

(c) $N(T) = \{0\}$, $n(T) = 0$, $r(T) = 3$

(d) T is one to one, onto, and invertible.

(e) $p(x) = -\frac{1}{2}(1 + x - 2x^2)$.

5. (a) $A = \begin{bmatrix} 0 & 1 & 1 & 1 & 1 \\ 0 & 0 & 2 & 3 & 4 \\ 0 & 0 & 0 & 3 & 6 \\ 0 & 0 & 0 & 0 & 4 \end{bmatrix}$ (b) $B = \begin{bmatrix} 0 & 1 & 0 & 0 & 0 \\ 0 & 0 & 2 & 0 & 0 \\ 0 & 0 & 0 & 3 & 0 \\ 0 & 0 & 0 & 0 & 4 \end{bmatrix}$

(c) $N(\Delta) = \mathrm{span}\{1\}$, $r(\Delta) = 4$, $R(\Delta) = \mathbf{P}_3$. Δ is onto but not one to one. Δ is not invertible.

CHAPTER 3

Section 3.1

1. $< f, g >$ is an inner product on $\mathbf{P}_2$ but not on $\mathbf{P}_3$. It is not definite on $\mathbf{P}_3$.

6. (a) $35.3°$ (b) $30°$

13. *Hint:* In the proof of the Cauchy–Schwarz inequality, consider the case $| < x, y > | = \|x\|\|y\|$.

Section 3.2

4. (a) $(1, 1, 1)^T$, $(1, 0, -1)^T$, $(1, -2, 1)^T$

5. (a) $(1, -1, 0, 0)^T$, $(1, 1, -2, 0)^T$, $(1, 1, 1, -3)^T$

6. (a) $Q = \dfrac{1}{\sqrt{10}} \begin{bmatrix} 1 & 3 \\ 3 & -1 \end{bmatrix}$, $R = \dfrac{1}{\sqrt{10}} \begin{bmatrix} 10 & -2 \\ 0 & 4 \end{bmatrix}$

14. (a) $t, t^2 - \frac{3}{4}t$

Section 3.3

1. (a) An orthogonal basis for M is $e_1 = (1, 1, 0, 1)^T$, $e_2 = (1, 0, 1, -1)^T$,

$e_3 = (-1, 1, 1, 0)^T$. The Gram matrix is $G = \begin{bmatrix} 3 & 3 & 3 \\ 3 & 15 & 9 \\ 3 & 9 & 9 \end{bmatrix}$. The projections are

(i) $s = (0, 2, 1, 1,)^T$ (ii) $s = (4, 1, 0, 1)^T$ (iii) $s = 0$

(b) $[\Pi] = \dfrac{1}{3} \begin{bmatrix} 3 & 0 & 0 & 0 \\ 0 & 2 & 1 & 1 \\ 0 & 1 & 2 & -1 \\ 0 & 1 & -1 & 2 \end{bmatrix}$

4. (a) (i) $s = 0$ (iii) $s = \frac{1}{4}(3, 3, -1, -5)^{\mathrm{T}}$

(b) $[\prod] = \dfrac{1}{4}\begin{bmatrix} 3 & -1 & -1 & -1 \\ -1 & 3 & -1 & -1 \\ -1 & -1 & 3 & -1 \\ -1 & -1 & -1 & 3 \end{bmatrix}$

10. (a) An orthogonal basis is $e_1 = t$, $e_2 = t^2$, $e_3 = t^3 - \frac{3}{5}t$. The normal equations, using $\{t, t^2, t^3\}$, are

$$\begin{bmatrix} 2/3 & 0 & 2/5 \\ 0 & 2/5 & 0 \\ 2/5 & 0 & 2/7 \end{bmatrix} c = \begin{bmatrix} 0 \\ 2/3 \\ 0 \end{bmatrix}.$$

The minimizer is $p = \frac{5}{3}t^2$, and $\|h - p\| = \frac{2\sqrt{2}}{3}$.

13. (a) $q = 0.9957 + 1.148x + 0.5486x^3$

(b) $p = 0.9963 + 1.104x + 0.5367x^3$

Section 3.4a

1. (b) $x = (\frac{5}{6} - 2c, \frac{1}{2} + c, c)^{\mathrm{T}}$, c arbitrary, $[\prod] = \dfrac{1}{6}\begin{bmatrix} 5 & 2 & -1 \\ \cdot 2 & 2 & 2 \\ -1 & 2 & 5 \end{bmatrix}$

(d) $x = \frac{1}{4}(-1, 6, -2)^{\mathrm{T}}$, $[\prod] = \dfrac{1}{4}\begin{bmatrix} 3 & 1 & 1 & -1 \\ 1 & 3 & -1 & 1 \\ 1 & -1 & 3 & 1 \\ -1 & 1 & 1 & 3 \end{bmatrix}.$

3. (a) $[\prod] = \dfrac{1}{3}\begin{bmatrix} 3 & 0 & 0 & 0 \\ 0 & 2 & 1 & 1 \\ 0 & 1 & 2 & -1 \\ 0 & 1 & -1 & 2 \end{bmatrix}.$

10. (a) Let $W = \mathrm{diag}(w_1, w_2, \ldots, w_n)$. Then $[\prod] = A(A^H W A)^{-1} A^H W$.

(b) Let $N = \mathrm{diag}(\|E_1\|^2, \ldots, \|E_r\|^2)$ and $E = [E_1\, E_2 \cdots E_r]$. Then $[\prod] = E N^{-1} E^H W$. Note that $N = E^H W E$. Alternatively, $[\prod] = \left(\sum_{i=1}^{r} E_i E_i^h / \|E_i\|^2\right) W$, which is derived from Fourier sums.

Section 3.4b

2. $m = 997.2$ kg, $\mu = 0.0501$

For $v = 10$ m/s, $E = 128.4$ kJ; for $v = 20$ m/s, $E = 278.0$ kJ.

Section 3.4c

1.   For $\mathbf{P}_3$,   For $\mathbf{P}_4$,

(a) $p = \frac{3}{32} + \frac{1}{2}x + \frac{15}{32}x^2$ $p = \frac{1}{4}P_0 + \frac{1}{2}P_1 + \frac{5}{16}P_2 - \frac{3}{32}P_4$

(b) $p = 1/3\pi + \frac{1}{2}x + (4/3\pi)x^2$ $p = (1/\pi)T_0 + \frac{1}{2}T_1 + (2/3\pi)T_2 - (2/15\pi)T_4$

(c) $p = \frac{2}{21} + \frac{1}{2}x + \frac{3}{7}x^2$ $p = 0.04329 + \frac{1}{2}x + 0.8636x^2 - 0.4091x^4$

Here, $P_0 = 1$, $P_1 = x$, $P_2 = \frac{1}{2}(3x^2 - 1)$, $P_3 = \frac{1}{2}(5x^3 - 3x)$, $P_4 = \frac{1}{8}(35x^4 - 30x^2 + 3)$

$T_0 = 1$, $T_1 = x$, $T_2 = 2x^2 - 1$, $T_3 = 4x^3 - 3x$, $T_4 = 8x^4 - 8x^2 + 1$

6. Using $x = (t - 1930)/10$, where t is in years, some approximations are as follows:

n	Polynomial	Relative error	Year 2000 prediction
2	$26.77 + 6.335x + 1.120x^2$	0.0313	126.0
3	$27.58 + 4.303x + 2.001x^2 - 0.09790x^3$	0.0303	122.2

Section 3.5

1. (a) $s = (1, -1, -1, 5)^T$, $r = (3, 0, -2, -1)^T$

3. (a) $\{t^2 - \frac{3}{5}, t^3 - \frac{3}{5}t\}$ is an orthogonal basis for $M^\perp$.

(b) Projection on $M : s = \frac{5}{3}t^2$; projection on $M^\perp : r = 1 - \frac{5}{3}t^2$

6. (a) $\{(-2, 1, 1)^T\}$

(b) $\{(1, 0, 2)^T, (0, 1, -1)^T\}$

(c) $b_1 = (-2, 1, 1)^T$

(d) Same answer as part (b)

Section 3.6

1. (a)   (i) $G = \begin{bmatrix} 2 & 0 & 2/3 \\ 0 & 2/3 & 0 \\ 2/3 & 0 & 2/5 \end{bmatrix}$ (ii) $G = \begin{bmatrix} 1 & 1/2 & 1/3 \\ 1/2 & 1/3 & 1/4 \\ 1/3 & 1/4 & 1/5 \end{bmatrix}$

(iii) $G = \begin{bmatrix} 5 & 0 & 5/2 \\ 0 & 5/2 & 0 \\ 5/2 & 0 & 17/8 \end{bmatrix}$ (iv) $G = \begin{bmatrix} 5 & 5/2 & 15/8 \\ 5/2 & 15/8 & 25/16 \\ 15/8 & 25/16 & 177/128 \end{bmatrix}$

CHAPTER 4

Section 4.1

Note: One-dimensional eigenspaces are not designated by "span."

3. (a) $\lambda_1 = -1$, $e_1 = \begin{bmatrix} 1 \\ -1 \end{bmatrix}$; $\lambda_2 = 1$, $e_2 = \begin{bmatrix} 1 \\ 1 \end{bmatrix}$.

(e) $\lambda = 1, (1, 1, 0)^T, \lambda = 3, M(3) = \text{span}\{(3, 1, 0)^T, (0, 1, 3)^T\}$

(h) $\lambda_1 = -1$, $e_1 = (1, 2, -1, 0)^T$; $\lambda_2 = 1$, $e_2 = (1, 0, 1, 0)^T$; $\lambda_3 = 2$,
$e_3 = (1, -1, -1, 0)^T$; $\lambda_4 = 3$, $e_4 = (0, 0, 0, 1)^T$

5. (a) $\lambda_1 = -1$, $e_1 = \begin{bmatrix} 0 & 1 \\ -1 & 0 \end{bmatrix}$, $\lambda_2 = 1$, $M(1) = \text{span}\left\{\begin{bmatrix} 1 & 0 \\ 0 & 0 \end{bmatrix}, \begin{bmatrix} 0 & 1 \\ 1 & 0 \end{bmatrix}, \begin{bmatrix} 0 & 0 \\ 0 & 1 \end{bmatrix}\right\}$

(c) $\lambda_1 = 1$, $e_1 = 1 - x$; $\lambda_2 = 2$, $e_2 = 1$

Section 4.2

1. (a) $\lambda_1 = e^{i\theta}, e_1 = (1, -i)^T; \lambda_2 = e^{-i\theta}, e_2 = (1, i)^T$. If $\theta = n\pi$, $\lambda_1 = \lambda_2$ and $M(e^{i\theta}) = \mathbf{C}^2$.

   (b) $R^n x = a \cdot e^{in\theta} e_1 + b \cdot e^{-in\theta}$

   (c) $R^n x = [\cos(n\theta) - 2\sin(n\theta), \sin(n\theta) + 2\cos(n\theta)]^T$

3. (a) $T^{-1}(\delta_1) = \frac{1}{3}(1, 0, -2, 2)^T, T^2(\delta_1) = \delta_1$

   (b) $T^{-1}(\delta_4) = \frac{1}{6}(4, 1, 3, 1)^T, T^2(\delta_4) = (0, -1, 1, 2)^T$

   (c) $[T^2] = \begin{bmatrix} 1 & 0 & 0 & 0 \\ 0 & 2 & -1 & -1 \\ 0 & -1 & 2 & 1 \\ 0 & -1 & 1 & 2 \end{bmatrix}$

4. (b) $\lambda_1 = \lambda_2 = 1, e_1 = (1, 0, 1)^T, e_2 = (0, 1, 1)^T, \lambda_2 = 4, e_3 = (1, -1, 3)^T.$ $\{e_1, e_2\}$ may be replaced by any basis for their span.

5. (a) $P = \begin{bmatrix} 1 & 0 & 2 \\ -1 & 1 & 1 \\ 1 & 1 & -1 \end{bmatrix}, \Lambda = \text{diag}(-1, 2, 2).$ The last two columns of P span $M(2)$ and

   may be replaced by any basis for $M(2)$.

   (b) $P = \begin{bmatrix} 1 & 1 \\ 2 & 1 \end{bmatrix}, \Lambda = \text{diag}(1, 2).$

11. (a) $\lambda = 1, e = \begin{bmatrix} 0 & 1 \\ -1 & 0 \end{bmatrix}, \lambda = 3, M(3) = \text{span}\left\{\begin{bmatrix} 1 & 0 \\ 0 & 0 \end{bmatrix}, \begin{bmatrix} 0 & 1 \\ 1 & 0 \end{bmatrix}, \begin{bmatrix} 0 & 0 \\ 0 & 1 \end{bmatrix}\right\}.$ T is diago-

    nalizable, $r(T) = 4, N(T) = \{0\}$.

13. (a) (i) Show that, if $x \perp b = (1, -2, 1)^T$, then $Ax \perp b$.

    (ii) $\lambda_1 = -1, e_1 = (1, 0, -1)^T; \lambda_2 = 1, e_2 = (3, 2, 1)^T.$ T is diagonalizable.

Section 4.3

1. (a) $A^{-1} = \frac{3}{2}I - \frac{1}{2}A$ (b) $2^A = 2A$

   (c) $e^{At} = (2e^t - e^{2t})I + (e^{2t} - e^t)A = \begin{bmatrix} 2e^{2t} - e^t & e^t - e^{2t} \\ 2(e^{2t} - e^t) & 2e^t - e^{2t} \end{bmatrix}$

   (d) $f(A) = \begin{bmatrix} 3 & -2 \\ 4 & -3 \end{bmatrix}$ (e) $\sqrt{3A - 2I} = A$

3. (a) (i) $A^{-100} = I$ (ii) $A^{101} = A$ (iii) $2^A = \frac{1}{4}\begin{bmatrix} 2 & -6 & -6 \\ 3 & 11 & 3 \\ -3 & -3 & 5 \end{bmatrix}$

   (b) $\lambda^2 - 1$

   (c) (i) $A^{-1} = A$ (ii) $2^A = \frac{5}{4}I + \frac{3}{4}A$ (iii) $(A + I)^4 = 8I + 8A$

7. (a) $\lambda_1 = -2, e_1 = 2x^2 - 1; \lambda_2 = 1, e_2 = x; \lambda_3 = 2, e_3 = 1$

   (b) $2^{T/2}(x^2) = \frac{1}{2}x^2 + \frac{3}{4}$

Section 4.4

1. (a) $x = c_1 \begin{bmatrix} 1 \\ 1 \\ 2 \end{bmatrix} + \left\{ c_2 \begin{bmatrix} 1 \\ 1 \\ 0 \end{bmatrix} + c_3 \begin{bmatrix} 0 \\ 1 \\ 1 \end{bmatrix} \right\} e^{2t}$. The second pair of columns may be replaced by

any pair of columns that are a basis for $M(2)$.

(b) $x = c_1 \begin{bmatrix} 0 \\ 1 \\ -1 \end{bmatrix} e^{-2t} + c_2 \begin{bmatrix} 2 \\ -3 \\ 1 \end{bmatrix} + c_3 \begin{bmatrix} 2 \\ -1 \\ 1 \end{bmatrix} e^{2t}$

3. (a) $x = (e^t - 2) \begin{bmatrix} 1 \\ 1 \end{bmatrix} + \frac{1}{2}(e^{2t} + 1) \begin{bmatrix} 1 \\ 2 \end{bmatrix}$

Section 4.5

1. (a) By rows: Each λ_k is in the union of $|\lambda - 7| \le 3, |\lambda - 10| \le 2, |\lambda - 6| \le 2$ (not disjoint).
By columns: Each λ_k is in the union of $|\lambda - 7| \le 2, |\lambda - 10| \le 3, |\lambda - 6| \le 2$ (not disjoint).
The following figures illustrate the Gershgorin disks.

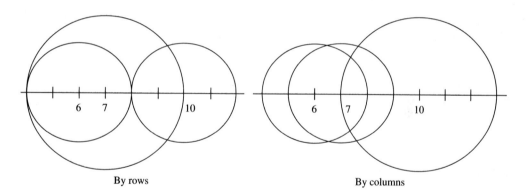

By rows   By columns

(b) Two eigenvalues satisfy $|\lambda - 6| \le 3$, one satisfies $|\lambda| \le 2$.
2. (a) One eigenvalue lies in each of $|\lambda| \le 0.1, |\lambda + 3| \le 0.4$, and $|\lambda - 5| \le 0.3$ (disjoint).
3. (a) The disks are disjoint. To improve $|\lambda - 1| \le 0.3$, take, for example, $d_1 = 60, d_2 = d_3 = 1$ to get $|\lambda - 1| \le 5 \times 10^{-4}$, using rows.

Section 4.6

4. $A = \frac{1}{3} \begin{bmatrix} 2 & 1 & 0 \\ 1 & 1 & 1 \\ 0 & 1 & 2 \end{bmatrix}$. At the nth step, $x_n = \frac{1000}{3}(1, 1, 1)^T + 500(\frac{2}{3})^n(1, 0, -1)^T$. Thus

(rounded to integers), $x_5 = (399, 333, 268)^T$, $x_{10} = (342, 333, 325)^T$. The limiting population is $\frac{1000}{3}(1, 1, 1)^T \approx (333, 333, 333)^T$.

9. (a) $a_n = c_1 + c_2(-1)^n$, a_n oscillates
   (b) $a_n = c_1(\frac{1}{2})^n + c_2(\frac{1}{4})^n$, $a_n \to 0$

CHAPTER 5

Section 5.1

2. T is Hermitian and unitary.
3. T is skew-Hermitian.
8. $T^*(ae_0 + be_1 + ce_2) = -(7.5a + 2.5c)e_0 + (22.5a + 7.5c)e_2$.
25. The relevant identity is $\int_\Omega v\Delta u\, dA = \int_\Omega u\Delta v\, dA + \int_{\partial\Omega}((\partial u/\partial n)v - (\partial v/\partial n)u)\, ds$, where $\partial\Omega$ denotes the boundary of Ω. Replace v by $\bar{v}$ and use the boundary contitions.

Section 5.2

1. (a) Unitary, Hermitian, and diagonalizable. $\lambda_1 = \lambda_2 = -1$, $e_1 = x$, $e_2 = 1 - x^2$; $\lambda_3 = 1$, $e_3 = 1 - 5x^2$
   (d) Diagonalizable, but not normal.
3. Diagonalizable, but not normal. For normal operators eigenvectors in different eigenspaces must be orthogonal, but here $< v_2, v_4 > \neq 0$.
7. For T, $\lambda_1 = 0$, $e_1 = 1 - 15x^2$; $\lambda_2 = 1$, $e_2 = x$; $\lambda_3 = 3$, $e_3 = 2 - 3x^2$. T is Hermitian by Example 3 of Section 5.1 or because the λ_k are real and the e_k are orthogonal. Each $f(T)$ has the same eigenbasis $\{e_1, e_2, e_3\}$. The corresponding eigenvalues are listed:
   (a) $\lambda_k = 1, 2, 8$. Hermitian.
   (b) $\lambda_k = 1, 1, -1$. Hermitian and unitary.
   (c) $\lambda_k = 1, -1, -1$. Hermitian and unitary.
   (d) $\lambda_k = -\frac{1}{2}, -1, 1$. Hermitian.
15. (b) For $\lambda = 0$, take $e_1 = v$. For $\lambda = 1$, take arbitrary orthogonal $e_2, e_3, \ldots, e_n \perp v$.

Section 5.3

1. (a) Skew-Hermitian and unitary. $P = \dfrac{1}{\sqrt{2}}\begin{bmatrix} 1 & 1 \\ -i & i \end{bmatrix}$, $\Lambda = \mathrm{diag}(-i, i)$.

   (b) Unitary. $P = \dfrac{1}{\sqrt{2}}\begin{bmatrix} 1 & 1 \\ 1 & -1 \end{bmatrix}$, $\Lambda = \mathrm{diag}(1, i)$.

   (g) Hermitian. $P = \begin{bmatrix} 1/\sqrt{3} & 1/\sqrt{2} & -1/\sqrt{6} \\ -1/\sqrt{3} & 1/\sqrt{2} & 1/\sqrt{6} \\ 1/\sqrt{3} & 0 & 2/\sqrt{6} \end{bmatrix}$, $\Lambda = \mathrm{diag}(-1, 2, 2)$.

4. (a) A is Hermitian. $U = \dfrac{1}{\sqrt{2}}\begin{bmatrix} -i & i \\ 1 & 1 \end{bmatrix}$, $U^H AU = \mathrm{diag}(1, 3)$.

   (b) In each case, $V = U$.
      B is Hermitian and unitary; $\Gamma = \mathrm{diag}(-1, 1)$.
      C is skew-Hermitian and unitary; $\Gamma = \mathrm{diag}(i, -i)$.
      D is Hermitian; $\Gamma = \mathrm{diag}(1, \frac{1}{2})$.

   (c) $D = U\mathrm{diag}(1, \frac{1}{2})U^H = 2(A + I)^{-1} = \dfrac{1}{4}\begin{bmatrix} 3 & -i \\ i & 3 \end{bmatrix}$

8. $x = (0, e^{2t} - 1, e^{2t})^{\mathsf{T}}$

31. (a) $\lambda_1 = 0, E_1 = \frac{1}{3}(2, -1, 1)^{\mathsf{T}}; \lambda_2 = 1, E_2 = \frac{1}{3}(1, 1, -1)^{\mathsf{T}}; \lambda_3 = 2, E_3 = \frac{1}{3}(1, 1, 2)^{\mathsf{T}}$.
$\{E_1, E_2, E_3\}$ is orthonormal in the inner product generated by G, so $P^{\mathsf{T}}GP = I$, $P^{\mathsf{T}}AP = \operatorname{diag}(0, 1, 2)$, where $P = [E_1 \ E_2 \ E_3]$.

Section 5.4

1. (d) $-5 \leq \lambda_1 \leq \lambda_2 \leq \lambda_3 \leq 19$, by Gershgorin's theorems. With $x = (1, 1, 1)^{\mathsf{T}}$, $R(x) = 15\frac{2}{3} \leq \lambda_3$. With $x = (1, -4, 1)^{\mathsf{T}}, 0.4 > R(x) \geq \lambda_1$.

5. $\operatorname{Min}\{\|Ax\|^2/\|x\|^2 : x \neq 0\} = \lambda_1^2$ and is attained at $x \in M(\lambda_1), x \neq 0$.
$\operatorname{Max}\{\|Ax\|^2/\|x\|^2 : x \neq 0\} = \lambda_n^2$ and is attained at $x \in M(\lambda_n), x \neq 0$.

Section 5.5

Because starting vectors are unspecified, only final answers are given. The solutions for the eigenvectors have length 1, unless otherwise specified.

1. $\lambda_1 = 11.82, e_1 = (0.3420, 0.6186, 0.07074)^{\mathsf{T}}$
$\lambda_3 = 0.07689, e_3 = (0.5883, 0.4460, -0.6745)^{\mathsf{T}}$

7. (i) $\lambda_1 = 19.51, e_1 = (0.0021, 0.6871, 0.7266)^{\mathsf{T}}$
$\lambda_2 = 2.677, e_2 = (0.6644, -0.5565, 0.5244)^{\mathsf{T}}$
$\lambda_3 = -0.1914, e_3 = (0.7647, 0.4671, -0.4440)^{\mathsf{T}}$

8. These eigenvectors are normalized so that $e^{\mathsf{T}}Ge = 1$.
(i) $\lambda_1 = 13.65, e_1 = (-0.0331, 0.4185, 0.3602)^{\mathsf{T}}$
$\lambda_2 = 1.325, e_2 = (0.8590, -0.2794, 0.1456)^{\mathsf{T}}$
$\lambda_3 = 0.02765, e_3 = (0.5109, 0.4968, -0.2215)^{\mathsf{T}}$

Section 5.6

1. (a) Integrate by parts twice and use the boundary conditions.
(b) With $f_i = 1 - x^{i+1}, i = 1, 2$,

$$A = \begin{bmatrix} 4/3 & 3/2 \\ 3/2 & 9/5 \end{bmatrix}, G = \begin{bmatrix} 8/15 & 7/12 \\ 7/12 & 9/14 \end{bmatrix}. \text{ This gives}$$

$\Lambda_1 = 2.486, \tilde{e}_1 = 1.394 f_1 - 0.394 f_2 \ (\lambda_1 = 2.467)$
$\Lambda_2 = 23.56, \tilde{e}_2 = 12.11 f_1 - 11.11 f_2 \ (\lambda_2 = 22.21)$
(c) $n = 4, \Lambda_k = 2.467, 22.22, 63.74, 148.38$
Actual: $\lambda_k = 2.467, 22.21, 61.69, 120.9$

3. (c) (i) $A = \begin{bmatrix} \frac{1}{2} & -\frac{13}{60} \\ -\frac{13}{60} & \frac{1}{6} \end{bmatrix}, G = \begin{bmatrix} 4\log(2) - \frac{11}{4} & \frac{83}{15} - 8\log(2) \\ \frac{83}{15} - 8\log(2) & 16\log(2) - \frac{133}{12} \end{bmatrix}$

$\lambda_1 = 20.54 \quad \Lambda_1 = 20.61 \quad \tilde{e}_1 = f_1 - 1.252 f_2$
$\lambda_2 = 82.17 \quad \Lambda_2 = 96.34 \quad \tilde{e}_2 = f_1 - 1.813 f_2$

(ii) $A = \begin{bmatrix} \frac{1}{3}\log^3(2) & \frac{1}{6}\log^4(2) \\ \frac{1}{6}\log^4(2) & \frac{2}{15}\log^5(2) \end{bmatrix}, G = \begin{bmatrix} \frac{1}{30}\log^5(2) & \frac{1}{60}\log^6(2) \\ \frac{1}{60}\log^6(2) & \frac{1}{105}\log^7(2) \end{bmatrix}$

$\lambda_1 = 20.54 \quad \Lambda_1 = 20.81 \quad \tilde{e}_1 = f_1$
$\lambda_2 = 82.17 \quad \Lambda_2 = 87.42 \quad \tilde{e}_2 = f_1 - 2.885 f_2$

CHAPTER 6

Section 6.1

1. (a) $y^H A x = \bar{y}_1 x_1 + 3i\bar{y}_1 x_2 - 3i\bar{y}_2 x_1 + 2\bar{y}_2 x_2$
   $x^H A x = |x_1|^2 + 6\,\mathrm{Re}(i\bar{x}_1 x_2) + |x_2|^2$
   (b) $y^T A x = y_1 x_1 + 4y_1 x_2 + 2y_2 x_2$
   $x^T A x = x_1^2 + 4x_1 x_2 + x_2^2$

   The symmetric matrix is $\begin{bmatrix} 1 & 2 \\ 2 & 1 \end{bmatrix}$.

2. (b) $A = \begin{bmatrix} 1 & 3 \\ 3 & -1 \end{bmatrix}$

   (c) $A = \begin{bmatrix} 1 & 0 \\ 6 & -1 \end{bmatrix}$

3. (a) $b(x, y) = 3\int_0^1 x'y'\,dt + x(0)y(0) + \frac{1}{2}[x(1)y'(1) + y(1)x'(1)]$

   (b) Let $\phi(x, y) = 3\int_0^1 x'y'\,dt + x(0)y(0) + x(1)y'(1)$. $[\phi] = B = \begin{bmatrix} 1 & 0 & 0 \\ 1 & 4 & 4 \\ 2 & 5 & 6 \end{bmatrix}$

   (c) $A = \dfrac{1}{2}\begin{bmatrix} 2 & 1 & 2 \\ 1 & 8 & 9 \\ 2 & 9 & 12 \end{bmatrix}$

Section 6.2

1. (a) Indefinite: $q((1, 1)^T) > 0 > q((1, -1)^T)$
   (c) Positive definite: $q(x) = (x_1 + x_2)^2 + x_2^2 > 0$, unless $x = 0$
   (f) Positive, not definite: $q(x) = 0$, for $x = t(t - 1)(t - 2)$

20. (a) (i) $u = 1 - 0.07407x + 1.444x^2 - 1.481x^3$
    (ii) $u = 1 + 0.07407 f_1 + 0.1481 f_2 - 0.1111 f_3$

    (b) Exact: $u = 1 + \frac{2}{9}x - 3(x - \frac{2}{3})_+^2$, where $z_+ = \begin{cases} z & z \geq 0 \\ 0 & z < 0 \end{cases}$

    At the points $0, \frac{1}{3}, \frac{2}{3}, 1$, the errors are less than 0.01.

Section 6.3

1. (a) (i) $\alpha_k = 1, 2, 5; e_k = \frac{1}{\sqrt{2}}(0, 1, 1)^T, \frac{1}{\sqrt{3}}(1, -1, 1)^T, \frac{1}{\sqrt{6}}(2, 1, -1)^T$, respectively.
   (ii) $M = [e_1\ e_2\ e_3]^{-1}$
   (iii) Positive definite
   (iv) $\max\{Q(x) : \|x\| = 1\} = 5$ at $\pm e_3$, $\min\{Q(x) : \|x\| = 1\} = 1$ at $\pm e_1$
   (v) Ellipsoid with semiaxes $1, 1/\sqrt{2}, 1/\sqrt{5}$

(f)   (i) $\alpha_k = -1, 2, 2; e_k = \frac{1}{\sqrt{3}}(1, -1, 1)^T, \frac{1}{\sqrt{2}}(1, 1, 0)^T, \frac{1}{\sqrt{6}}(-1, 1, 2)^T$
      (ii) $M = [e_1\ e_2\ e_3]^{-1}$
      (iii) Indefinite
      (iv) $\max\{Q(x) : \|x\| = 1\} = 2$ at $x = ae_2 + be_3$, with $a^2 + b^2 = 1$;
           $\min\{Q(x) : \|x\| = 1\} = -1$ at $\pm e_1$
      (v) Hyperboloid of one sheet, rotated about the e_1 axis

5. (a) q minimizes at $x_0 = (-1, 5, 3)^T$ and min $q = -43$. The eigenvalues and eigenvec-
        tors of $[q]$ are $\lambda_k = 1, 1, 4, e_k = \frac{1}{\sqrt{2}}(0, 1, 1)^T, \frac{1}{\sqrt{6}}(2, 1, -1)^T, \frac{1}{\sqrt{3}}(1, -1, 1)^T. q = 21$
        is a prolate ellipsoid with center x_0 rotated about the e_3 axis.

Section 6.4

1. (a) (i) $A = \begin{bmatrix} 1 & -1 & 1 & 2 \\ -1 & 3 & 3 & -4 \\ 1 & 3 & 10 & 0 \\ 2 & -4 & 0 & 10 \end{bmatrix}$

        (ii) $P = M^{-1}, M = \begin{bmatrix} 1 & -1 & 1 & 2 \\ 0 & 1 & 2 & -1 \\ 0 & 0 & 1 & 2 \\ 0 & 0 & 0 & 1 \end{bmatrix}$, and $w_k = 1, 2, 1, 0$

        (iii) $P = [E_1\ E_2\ E_3\ E_4] = \begin{bmatrix} 1 & 1 & -3 & 5 \\ 0 & 1 & -2 & 5 \\ 0 & 0 & 1 & -2 \\ 0 & 0 & 0 & 1 \end{bmatrix}$, $w_k = 1, 2, 1, 0$

        (iv) Positive, not definite
        (v) Three λ_k are positive, one is zero.

3. (a) (i) $A = \frac{1}{2}\begin{bmatrix} 2 & 1 & 2 \\ 1 & 4 & 9 \\ 2 & 9 & 12 \end{bmatrix}$

        (ii) Complete the squares. $P = \begin{bmatrix} 1 & -1/2 & -7/15 \\ 0 & 1 & -16/15 \\ 0 & 0 & 1 \end{bmatrix}$ with $w_k = 1, 15/4, 11/15$.

        This gives $e_0 = 1, e_2 = t - \frac{1}{2}, e_3 = t^2 - \frac{16}{15}t - \frac{7}{15}$.
        (iii) Positive definite

5. Jacobi's method applied to Exercise 1(a) fails since A is singular. Applied to 1(d) it yields

   $P^T AP = \text{diag}(1, 1, \frac{1}{9})$, with $P = \begin{bmatrix} 1 & -1 & 1/9 \\ 0 & 1 & -1/9 \\ 0 & 0 & 1/9 \end{bmatrix}$.

   Jacobi's method applied to Exercise 3(a) yields $f_1 = 1, f_2 = \frac{1}{15}(-2 + 4t)$,
   $f_3 = \frac{1}{11}(-7 - 16t + 15t^2); w_k = 1, 4/15, 15/11$.

Section 6.5

1. (a) (i) g is positive definite.
   (ii) $\lambda_k = 0, 1, 2; E_k = \frac{1}{3}(1, -1, 1)^T, (0, 0, 1)^T, \frac{1}{3}(1, 2, -2)^T, k = 1, 2, 3$

   (iii) $M = P^{-1} = P^T[g] = \begin{bmatrix} 2 & -1 & 0 \\ 0 & 1 & 1 \\ 1 & 1 & 0 \end{bmatrix}$

   Note: $P^T[g]P = I, P^T[f]P = \text{diag}(0, 1, 2)$, where $P = [E_1\ E_2\ E_3]$
   (iv) $\min\{q(x) : n(x) = 1\} = 0$ at $x = \pm E_1$
   $\max\{q(x) : n(x) = 1\} = 2$ at $x = \pm E_3$
   (b) (i) f is positive definite.
   (ii) $\lambda_k = -1, 1, 1; E_k = \frac{1}{3}(1, -2, 1)^T, \frac{1}{2}(0, 0, 1)^T, \frac{1}{6}(2, 2, 1)^T, k = 1, 2, 3$

   (iii) $M = P^{-1} = P^T[f] = \begin{bmatrix} 1 & -1 & 0 \\ 0 & -1 & 2 \\ 2 & 1 & 0 \end{bmatrix}$

   Note: $P^T[f]P = I, P^T[g]P = \text{diag}(-1, 1, 1)$, where $P = [E_1\ E_2\ E_3]$
   (iv) $\min\{q(x) : n(x) = 1\} = -1$ at $x = \pm E_1$
   $\max\{q(x) : n(x) = 1\} = 1$ at $x = aE_2 + bE_3, a^2 + b^2 = 1$

CHAPTER 7

Section 7.1

2. (i) (a) $V = mgL[1 - \cos(\theta)] \approx \frac{1}{2}mgL\theta^2, T = \frac{1}{2}mL^2\dot{\theta}^2$
   (b) $L\ddot{\theta} + g\theta = 0$
   (ii) (a) $V = \frac{1}{2}gL_1(m_1 + m_2)\theta_1^2 + \frac{1}{2}m_2gL_2\theta_2^2,$
   $T = \frac{1}{2}(m_1 + m_2)L_1^2\dot{\theta}_1^2 + m_2L_1L_2\dot{\theta}_1\dot{\theta}_2 + \frac{1}{2}m_2L_2^2\dot{\theta}_2^2$. Both are approximate.

   (b) $\begin{bmatrix} L_1^2(m_1 + m_2) & L_1L_2m_2 \\ L_1L_2m_2 & m_2L_2^2 \end{bmatrix}\ddot{\theta} + \begin{bmatrix} gL_1(m_1 + m_2) & 0 \\ 0 & gL_2m_2 \end{bmatrix}\theta = 0, \theta = \begin{bmatrix} \theta_1 \\ \theta_2 \end{bmatrix}$

4. The differential equations of motion are $M\ddot{c} + Kc = 0$ in each case.
   (a) $m_{ij} = \int_0^L \rho f_i f_j\, dx + mf_i(L)f_j(L), \quad f_i(0) = 0,$
   $k_{ij} = \tau\int_0^L f_i' f_j'\, dx$
   (b) $m_{ij} = \int_0^L \rho f_i f_j\, dx, \quad f_i(0) = f_i'(0) = 0,$
   $k_{ij} = \int_0^L EI f_i'' f_j''\, dx + kf_i(L)f_j(L)$

Section 7.2

1. (a) $\omega_k = 1, 2, \sqrt{10}; E_k = (1, 1, 1)^T, (1, 0, -1)^T, (1, -2, 1)^T, k = 1, 2, 3$
   (b) $\theta(t) = \sum_{k=1}^3 [a_k \cos(\omega_k t) + b_k \sin(\omega_k t)]E_k$
2. (a) Clearly $V \geq 0. V = 0 \Rightarrow x_1 = x_1 - x_2 = x_2 - x_3 = 0 \Rightarrow$ all $x_i = 0$.
   (b) $\omega_k = 1, \sqrt{2}, 2; e_k = (2, 2, 1)^T, (2, 0, -1)^T, (2, -4, 1)^T, k = 1, 2, 3$
   $x(t) = \sum_{k=1}^3 [a_k \cos(\omega_k t) + b_k \sin(\omega_k t)]e_k$
   (c) Mode 1: $\omega = 1$ (all particles), relative amplitudes 2:2:1
   Mode 2: $\omega = \sqrt{2}$, relative amplitudes 2:0:1, m_1 and m_3 out of phase, m_2 stationary
   Mode 3: $\omega = 2$, relative amplitudes 2:4:1, m_2 out of phase with m_1 and m_3
   (d) $x(t) = [4 - 3\cos(\omega_1 t)]e_1 + [3 - 4\cos(\omega_2 t)]e_2 + \frac{1}{2}[1 - \cos(\omega_3 t)]e_3$

6. (a) $x = (a_1 + b_1 t) \begin{bmatrix} 1 \\ -1 \\ 1 \end{bmatrix} + [a_2 \cos(t) + b_2 \sin(t)] \begin{bmatrix} 0 \\ 0 \\ 1 \end{bmatrix} + [a_3 \cos(\sqrt{2}t) + b_3 \sin(\sqrt{2}t)] \begin{bmatrix} 1 \\ 2 \\ -2 \end{bmatrix}.$

7. (a) $\omega_1 = 5.199$, $e_1 = (-0.08323, 1, -0.07990)^{\mathrm{T}}$
$\omega_3 = 0.5240$, $e_3 = (0.6331, 0.8627, 1)^{\mathrm{T}}$

9. (b) (i) $\tilde{\omega}_1 = 1.577$, $\tilde{e}_1 = 1 - x + 0.8271x(1-x)$
$\tilde{\omega}_2 = 5.673$, $\tilde{e}_2 = 1 - x - 0.3627x(1-x)$
(ii) $\tilde{\omega}_1 = 1.571$, $\tilde{e}_1 = 1 - 1.3939x^2 + 0.3939x^3$
$\tilde{\omega}_2 = 4.854$, $\tilde{e}_2 = 1 - 12.11x^2 + 11.11x^3$
Exact: $\omega_1 = 1.5708$, $\omega_2 = 4.7124$

13. (a) (i) No solution, $\omega^2 = 4 = \lambda_3$
(ii) $x = (2, 0, -1)^{\mathrm{T}} \cos(\sqrt{2}t)$
(iii) $x = -(3, 25, 15)^{\mathrm{T}} e^{i\sqrt{2}t}$

Section 7.3

1. (a) $C = 2M + \frac{1}{2}K$. $x = e^{-2t}(c_1 + c_2 t)\begin{bmatrix} 1 \\ 1 \end{bmatrix} + e^{-4t}[d_1 \cosh(2t) + d_2 \sinh(2t)]\begin{bmatrix} 1 \\ -1 \end{bmatrix}$

(e) $KM^{-1}C = \begin{bmatrix} 1 & -1 & 0 \\ -1 & 3 & 2 \\ 0 & 2 & 2 \end{bmatrix}$ is Hermitian.

$x = (1 + t)\begin{bmatrix} 1 \\ 1 \\ -1 \end{bmatrix} + e^{-t/2}[\cos(\tfrac{\sqrt{3}}{2}t) - \tfrac{1}{\sqrt{3}}\sin(\tfrac{\sqrt{3}}{2}t)]\begin{bmatrix} 2 \\ -1 \\ 1 \end{bmatrix} + \tfrac{2}{\sqrt{7}}e^{-t/2}\sin(\tfrac{\sqrt{7}}{2}t)\begin{bmatrix} 1 \\ 1 \\ 2 \end{bmatrix}$

2. (a) $C = M + 2K$

$x = (a_1 + b_1 e^{-t} + t)\begin{bmatrix} 1 \\ 1 \end{bmatrix} + \{e^{-2t}[a_2 \cosh(\tfrac{\sqrt{10}}{2}t) + b_2 \sinh(\tfrac{\sqrt{10}}{2}t)] - \tfrac{2}{3}\}\begin{bmatrix} 1 \\ -2 \end{bmatrix}$

3. (a) (i) $x = -\cos(2t)\begin{bmatrix} 1 \\ 1 \end{bmatrix}$ (ii) $x = (1 - 2i)e^{2it}\begin{bmatrix} 1 \\ -1 \end{bmatrix}$

(d) (i) $x = -5e^{2it}\begin{bmatrix} 1 \\ 1 \\ -1 \end{bmatrix}$ (ii) $x = 2\sin(t)\begin{bmatrix} 2 \\ -1 \\ 1 \end{bmatrix}$ (iii) $x = 2(1 + i)e^{-2it}\begin{bmatrix} 1 \\ 1 \\ 2 \end{bmatrix}$

Section 7.4

1. For a general set of basis functions $\{f_k\}_{k=1}^n$, the differential equations are $M\dot{c} + Kc = 0$, where $m_{ij} = \int_0^1 f_i f_j \, dx$, $k_{ij} = \int_0^1 f_i' f_j' \, dx$.
(a) Hat functions. See Example 2 for M, K. For the initial conditions, use interpolation: $c_k(0) = u_0(k/n)$.
(b) $f_k = x^k(1-x)$. Here $m_{ij} = 2/(p+1)(p+2)(p+3)$, $k_{ij} = 2ij/(p-1)p(p+1)$, $p = i + j$. For the initial conditions, use $Mc(0) = a$, where $a_k = \int_0^1 u_0 f_k \, dx$.

2. See Exercise 1(b) for M and K. Use (8) for initial conditions.

CHAPTER 8

Section 8.1

1. *Note:* As has been noted in the text, singular value decompositions may have a substantial amount of nonuniqueness.

(a) $U = \dfrac{1}{5}\begin{bmatrix} 3 & 4 \\ 4 & -3 \end{bmatrix}, S = \begin{bmatrix} 20 & 0 & 0 & 0 \\ 0 & 10 & 0 & 0 \end{bmatrix}, V = \dfrac{1}{2}\begin{bmatrix} 1 & 1 & 1 & 1 \\ 1 & -1 & -1 & 1 \\ 1 & 1 & -1 & -1 \\ 1 & -1 & 1 & -1 \end{bmatrix}$

(b) $U = \begin{bmatrix} 1/\sqrt{6} & 1/\sqrt{3} & 1/\sqrt{2} \\ 1/\sqrt{6} & 1/\sqrt{3} & -1/\sqrt{2} \\ -2/\sqrt{6} & 1/\sqrt{3} & 0 \end{bmatrix}, S = \begin{bmatrix} \sqrt{6} & 0 \\ 0 & \sqrt{3} \\ 0 & 0 \end{bmatrix}, V = \begin{bmatrix} 1 & 0 \\ 0 & 1 \end{bmatrix}$

(e) $U = \begin{bmatrix} 1/\sqrt{3} & 0 & -2/\sqrt{6} \\ -1/\sqrt{3} & -1/\sqrt{2} & -1/\sqrt{6} \\ 1/\sqrt{3} & -1/\sqrt{2} & 1/\sqrt{6} \end{bmatrix}, S = \begin{bmatrix} 3 & 0 & 0 \\ 0 & 2 & 0 \\ 0 & 0 & 0 \end{bmatrix},$

$V = \begin{bmatrix} 1/\sqrt{3} & 0 & -2/\sqrt{6} \\ 1/\sqrt{3} & -1/\sqrt{2} & 1/\sqrt{6} \\ 1/\sqrt{3} & 1/\sqrt{2} & 1/\sqrt{6} \end{bmatrix}$

12. (a) $\sigma_k = |\lambda_k|$, when $\lambda_k \neq 0$

Let $\gamma_k = \text{sgn}(\lambda_k)$, where $\text{sgn}(z) = z/|z|$ if $z \neq 0$ and $\text{sgn}(0) = 0$. Put $\Gamma = \text{diag}(\gamma_1, \ldots, \gamma_n)$.
(b) $U = P$: Take $V = P\Gamma, S = |\Lambda|$
(c) $V = P$: Take $U = P\Gamma, S = |\Lambda|$

Section 8.2

1. Note that $S^+ = S^{-1}$. Verify that $AA^+ = I$, or compute $A^{-1} = (USV^H)^{-1}$.
5. Put $W = \text{diag}(\sqrt{w_1}, \sqrt{w_2}, \ldots, \sqrt{w_m})$, $z = Wy, B = WA$. Then $\|y - Ax\|^2 = \|z - Bx\|_s^2$, where $\| \ \|_s$ is the norm generated by the standard inner product. This reduces the minimization problem to one using the standard inner product. Use the SVD for B to find $x = x^+$ for the vector z. $\|x^+\|_s$ is minimal.

Section 8.3

1. (b) B should be nonsingular.
6. (a) $A^k \to \begin{bmatrix} I & P(I - S)^{-1} \\ 0 & 0 \end{bmatrix}$ (b) $A^k \to P\begin{bmatrix} I & Q(I - S)^{-1} \\ 0 & 0 \end{bmatrix}P^{-1}$

Section 8.4

1. Note that there is a great deal of flexibility in the construction of a Jordan chain, so that there is a substantial degree of nonuniqueness.

(a) $P = \begin{bmatrix} 0 & 1 \\ 1 & 0 \end{bmatrix}, J = \begin{bmatrix} 0 & 1 \\ 0 & 0 \end{bmatrix}$

(c) $P = \begin{bmatrix} 1 & -1 & -1 \\ 0 & 1 & 0 \\ -1 & 1 & 0 \end{bmatrix}, J = \begin{bmatrix} 1 & 0 & 0 \\ 0 & 2 & 1 \\ 0 & 0 & 2 \end{bmatrix}$

(f) $P = \begin{bmatrix} 1 & 1 & 1 \\ 1 & -1 & 0 \\ 1 & 1 & -1 \end{bmatrix}, J = \begin{bmatrix} 1 & 0 & 0 \\ 0 & -1 & 1 \\ 0 & 0 & -1 \end{bmatrix}$

2. (b) $P = \begin{bmatrix} 1 & 0 & 0 & 0 \\ -2 & 1 & 0 & 0 \\ 0 & 0 & 1 & 1 \\ 0 & 0 & i & -i \end{bmatrix}, J = \begin{bmatrix} 0 & 1 & 0 & 0 \\ 0 & 0 & 0 & 0 \\ 0 & 0 & i & 0 \\ 0 & 0 & 0 & -i \end{bmatrix}$

11. (b) $e^{Jt} = \begin{bmatrix} 1 & t \\ 0 & 1 \end{bmatrix}$

(c) $e^{Jt} = \begin{bmatrix} e^{2t} & 0 & 0 \\ 0 & e^{-t} & te^{-t} \\ 0 & 0 & e^{-t} \end{bmatrix}$

12. (a) $x = e^t \begin{bmatrix} 1 & t \\ 0 & 1 \end{bmatrix} x_0$

BIBLIOGRAPHY

Abramowitz, M., and I. A. Stegun. *Handbook of Mathematical Functions.* New York: Dover, 1972.

Achiezer, N. I., and I. M. Glazman. *Theory of Linear Operators in Hilbert Space.* New York: Ungar, 1961.

Anderson, E., et al. *Lapack Users Guide.* Philadelphia: SIAM, 1992.

Birkhoff, G. and G. Rota. *Ordinary Differential Equations.* 2nd ed. Lexington, Mass.: Xerox, 1969.

Ciarlet, P. G. *The Finite Element Method for Elliptic Problems.* Amsterdam: North-Holland, 1978.

Courant, R., and D. Hilbert. *Methods of Mathematical Physics, Vol. 1.* New York: Interscience, 1953.

Davis, P. J. *Interpolation and Approximation.* New York: Blaisdell, 1963.

Dongarra, J. J., C. B. Moler, J. R. Bunch, and G. W. Stewart. *Linpack Users Guide.* Philadelphia: SIAM, 1979.

Dunford, N., and J. T. Schwartz. *Linear Operators, Part I: General Theory.* New York: Interscience, 1958.

Friedman, B. *Principles and Techniques of Applied Mathematics.* New York: Dover, 1990.

Gantmacher, F. R. *The Theory of Matrices.* 2 vols. New York: Chelsea, 1959.

Gantmacher, F. R. *Applications of the Theory of Matrices.* New York: Interscience, 1979.

Golub, G., and C. Van Loan. *Matrix Computations.* 2nd ed. Baltimore: Johns Hopkins University Press, 1989.

Halmos, P. R. *Finite Dimensional Vector Spaces.* 2nd ed. New York: Van Nostrand, 1958.

Hoffman, K., and R. Kunze. *Linear Algebra.* 2nd ed. Englewood Cliffs, N.J.: Prentice Hall, 1971.

Horn, R. A., and C. A. Johnson. *Matrix Analysis.* New York: Cambridge University Press, 1985.

Householder, A. S. *The Theory of Matrices in Numerical Analysis.* New York: Blaisdell, 1964.

Kemeny, J. G. *Finite Markov Chains.* New York: Van Nostrand, 1960.

Lancaster, P., and M. Tismenetsky. *The Theory of Matrices with Applications.* 2nd ed. New York: Academic Press, 1985.

Lebedev, N. N. *Special Functions and Their Applications.* New York: Dover, 1972.

Liusternik, L. A., and V. J. Sobolev. *Elements of Functional Analysis.* New York: Ungar, 1961.

Luenberger, D. G. *Dynamical Systems: Theory, Models, and Applications.* New York: Wiley, 1979.

Oja, E. *Subspace Methods of Pattern Recognition.* New York: Wiley, 1983.

Ortega, J. M. *Matrix Theory: A Second Course.* New York: Plenum Press, 1987.

Parlette, B. *The Symmetric Eigenvalue Problem.* Englewood Cliffs, N.J.: Prentice Hall, 1980.

Roberts, F. S. *Discrete Mathematical Models with Applications to Social, Biological, and Environmental Problems.* Englewood Cliffs, N.J.: Prentice Hall, 1976.

Rutishauser, H. "Computational Aspects of F. L. Bauer's Simultaneous Iteration Method," *Num. Math.* 13 (1969): 4–13.

Sokolnikoff, I. S. *Tensor Analysis.* 2nd ed. New York: Wiley, 1964.

Stewart, G. W. *Introduction to Matrix Computations.* New York: Academic Press, 1973.

Strang, G., and G. J. Fix. *An Analysis of the Finite Element Method.* Englewood Cliffs, N.J.: Prentice Hall, 1973.

Taylor, A. E., and D. C. Lay. *Introduction to Functional Analysis.* 2nd ed. New York: Wiley, 1980.

Varga, R. S. *Matrix Iterative Analysis.* Englewood Cliffs, N.J.: Prentice Hall, 1962.

Weinberger, H. F. *Variational Methods for Eigenvalue Approximation.* Philadelphia: SIAM, 1974.

Wilkinson, J. H. *The Algebraic Eigenvalue Problem.* Oxford, England: Clarendon Press, 1965.

Young, D. *Iterative Solution of Large Linear Systems.* New York: Academic Press, 1971.

Index